A. Mete Isıkara and Andreas Vogel (Eds.)

Multidisciplinary Approach to Earthquake Prediction

Progress in Earthquake Prediction Research

Edited by Andreas Vogel

Volume 1
Andreas Vogel (Ed.),
Terrestrial and Space Techniques
in Earthquake Prediction Research

Volume 2
A. Mete Işıkara and Andreas Vogel (Eds.),
Multidisciplinary Approach
to Earthquake Prediction

A. Mete Işıkara and Andreas Vogel (Eds.)

Multidisciplinary Approach to Earthquake Prediction

Proceedings of the International Symposium on Earthquake Prediction in the North Anatolian Fault Zone held in Istanbul, March 31 – April 5, 1980

With 246 figures

Springer Fachmedien Wiesbaden GmbH

CIP-Kurztitelaufnahme der Deutschen Bibliothek

Multidisciplinary Approach to Earthquake Prediction:
proceedings of the Internat. Symposium on Earthquake Prediction in the North Anatolian Fault Zone, held in Istanbul, March 31–April 5, 1980 / A. Mete Işıkara and Andreas Vogel (eds.).

(Progress in earthquake prediction research; Vol. 2)
ISBN 978-3-528-08482-0 ISBN 978-3-663-14015-3 (eBook)
DOI 10.1007/978-3-663-14015-3

NE: Işıkara, Ahmed Mete [Hrsg.]: International Symposium on Earthquake Prediction in the North Anatolian Fault Zone ‹1980, Istanbul›; GT

1982

Ursprünglich erschienen bei Friedr. Vieweg & Sohn Verlagsgesellschaft mbH, Braunschweig 1982
Softcover reprint of the hardcover 1st edition 1982

Produced by W. Langelüddecke, Braunschweig

ISBN 978-3-528-08482-0

Contents

Editorial IX
Editorial Advisory Board X
Addresses of the Opening Session XI

1. Geological history of the North-Anatolian Fault Zone

A.M.C. Şengör, K. Burke, J.F. Dewey
Tectonics of the North Anatolian Transform Fault 3

H. Bergougnan, C. Fourquin
Paleo-, Tardi- and Neotectonic Mechanisms of the Present North Anatolian Fault Zone in the Light of the Structural History of the Eurasian Margin in the Pontic Ranges 23

P.L. Hancock, A.A. Barka
Structural Evidence for Left-Lateral Displacement on the North Anatolian Fault Zone During the Plio-Pleistocene 43

Y. Yilmaz, A.M. Gözübol, O. Tüysüz
Geology of an Area in and around the Northern Anatolian Transform Fault Zone between Bolu and Akyazi 45

C.R. Allen
Comparisons Between the North Anatolian Fault of Turkey and the San Andreas Fault of California 67

2. Seismotectonics: seismicity statistics according to historical and instrumental records, focal mechanism, relations between seismic activity and neotectonics

S.B. Üçer, S. Crampin, A. Miller
Identification of Swarm Activity using the MARNET Telemetered Seismometer-Network 89

R. Ateş
Earthquake Activity on the North Anatolian Fault Zone 95

M. Erdik, S. Öner
A Rational Approach for the Probabilistic Assessment of the Seismic Risk Associated with the North Anatolian Fault 115

A.A Barka, P.L. Hancock
Seismotectonic Aspects of the North Anatolian Fault Zone Between Bolu and Havza 129

A.A. Barka, P.L. Hancock
Relationship Between Fault Geometry and Some Earthquake Epicentres within the North Anatolian Zone 137

P.W. Burton, R.W. McGonigle, K.C. Makropoulos, S.B. Üçer
Preliminary Studies of Seismic Risk in Turkey, and the Occurrence of Upper Bounded and Other Large Earthquake Magnitudes 143

M. Tokay
Faults and Recently Active Breaks Along the North Anatolian Fault Zone Between Gerede and Ilgaz 173

T. Ahjos, H. Korhonen, J. Saari
Some Aspects of the Seismicity in the North Anatolian Fault Zone 185

B. Sikošek
A Comparative Analysis of the Seismotectonic Characteristics of the North Anatolian Fracture Zone and the Fracture System of Inner Dinarides 193

U Kuran
Fatigue Crack Propagation Along the Anatolian Faults and Levant Coast and Earthquake Prediction 207

H. Soysal, D Kolçak, S. Sipahioğlu
Some Aspects of the North Anatolian Fault Zone Derived from the Comparison of its Instrumental Data with Historical Information 223

A. Ercan
A Statistical Analysis of the Major and Microearthquakes along the East-Anatolian Fault 239

3. Pre-earthquake strain and triggering effects from continuous records

U. Yaramanci
Possible Use of Tilt and Tidal Measurements for Earthquake Prediction 261

E. Druta
Normal Earthquake Gravitational Precursors from Earth-Tide Data 269

A Aytun
Creep Measurements in the Ismetpaşa Region of the North Anatolian Fault Zone 279

4. Pre-earthquake strain and deformation from repeated high-precision geodetic surveys

A. Aksoy
Possible Use of the Turkish National Triangulation Network for the Study of Crustal Movements in the North Anatolian Fault Zone 295

N.K. Demirel, Z. Karahan, E. Öztürk, F. Pirselimoglu, E. Ugur, K. Uysal
Testing for Survey of Recent Crustal Deformation in Akyazi/Adapazari 301

S.M. Nakiboglu, E. Clerici, A.M. Isikara
Detection of Crustal Motion by Repeated Geometric Observations 307

D.G. Vlachos
Land Deformation Control Network in the Epicentral Region of the 1978 Earthquakes in Northern Greece 315

R. Baumer
Methods of Variance Analysis for the Evaluation of Geodetic Control Nets in Relation to Crustal Movements 325

R. Kelm
A Geometric-Gravimetric Estimation Procedure for Most Accurate Determination of Relative Point Displacements in Local Areas Proposal and First Study of Sensitivity 335

B. Richter
Suggestions for High Precision Gravity Measurements in Geodynamic and Earthquake Prediction Research 351

5. Physical state and processes of changing physical rock properties in the earthquake source region

T. Rikitake
Physical Parameters in the Earthquake Source Region and Their Temporal Changes 361

S. Crampin
Polarization Anomalies as Diagnostics of Dilatancy 405

R. Evans, M. Doyle, S. Balamir Üçer, A. Miller, S. Crampin
An Experiment to Investigate Polarization Anomalies in North Anatolia 409

M. Pantović
Possible Information on local Features of the Focal Region Based on Geomagnetic Field Data Recorded After the Main Shock 413

K. Aric, G. Duma, H. Friedmann, R. Gutdeutsch
Investigations of Geophysical Parameters in the Area of Carinthia and Friuli in Relation to Seismicity 427

A. Mete Işikara, N. Orbay, M. Pantović
Note on the Application of the PCA Method – A Possible Approach to the Problem of Elimination of Non-Local Geomagnetic Field Changes 435

6. Space techniques in geodynamics and earthquake prediction research

A. Vogel
Application of Space Technology in Earthquake Prediction Research – A Short Review 441

P. Wilson
The Background to High Mobility Space Geodetic Techniques for Geodynamic and Earthquake Prediction Research 447

C. Gerstenecker
The Installation of a Spaceborne Ranging System for the Detection of Crustal Movements in the North Anatolian Fault Zone 449

A. Robson, H. Houet
The Meteosat Data Collection System and its Potential as a Data Relay for Geophysical Measurements 457

7. Theory of earthquake generation and prediction

D.L. Turcotte
Stress Accumulated Mechanisms on Strike-Slip Faults 471

B.K. Atkinson
Fracture Mechanics Modelling of Earthquake Generating Processes 487

S.M. Nakiboglu, A.M. Isikara
Analysis of Stress and Strain Around a Transform Fault 523

K.E. Kasapoglu
A Multiple Mode of Faulting Mechanism Proposed for the North Anatolian Fault and Prediction of Related Earthquakes 539

C. Froidevaux, L. Fleitout
A Thermal Window in the Lithosphere Underneath North Anatolia and California 549

8. Interdisciplinary approaches to earthquake prediction

A. Vogel
The Concept of a Multidisciplinary Approach to Earthquake Prediction 555

R. Meissner
Deterministic Earthquake Prediction – Present Stage and Future Developments 563

W. Thatcher
Earthquake Research on the San Andreas Fault Zone, California 577

Editorial

Studies of the sources of earthquakes, the only way towards a successful prediction, requires a broad interdisciplinary approach and cooperation between scientists in various fields of earth sciences. Seismotectonic studies, observation of strain accumulation and investigation of the physical state and temporal changes of rock properties in the earthquake source region are mainly the lines to be followed. Modelling of earthquake generating processes constrained by the complexity of precursory phenomena is a necessary basis of prediction research.

The North-Anatolian fault zone is an area of extremely high earthquake risk. Large and destructive earthquakes have occurred during the last decades. A clearly defined migration of earthquake foci indicates that destructive earthquakes are most likely to occur in a most populated and industrialized part of Turkey. The chances "to capture an earthquake" by proper selection of observation sites and networks and to find criteria for earthquake prediction by observation of earthquake generating processes seem rather promising in the case of the North-Anatolian fault zone.

An international and interdisciplinary conference was held in Istanbul. Experts in the fields of geophysics, geology and geodesy from many countries in Europe and overseas concerned with earthquake research and hazard assessment took part. They met to present results of earlier research, to discuss current research activities relevant to earthquake prediction in the North-Anatolian fault zone and to work out and organize a program for earthquake prediction research in the most riskful part of the fault zone.

The conference held at the Department of Geophysics of the University of Istanbul was organized in cooperation with the Department of Geophysics, Free University of Berlin, sponsored by the Faculty of Earth Sciences of the University of Istanbul and the Turkish National Union of Geodesy and Geophysics, cosponsored by the UNESCO, the Parliamentary Assembly of the Council of Europe, the IASPEI Commission on Earthquake Prediction, the European Seismological Commission and under the auspices of the Scientific and Technical Research Council of Turkey, the Earthquake Research Institute, the General Directorate of Mapping, and the Turkish National Committee on Earthquake Engineering.

The editors are thankful for the financial support which was given by the Free University of Berlin to carry out the editorial work for the present proceedings. They also thank Mrs. Elena Börner for carefully correcting and retyping a number of manuscripts.

A. Mete Işıkara
Istanbul

Andreas Vogel
Berlin

Editorial Advisory Board

Addresses of the Opening Session

Address of Prof. Dr. I. Özdogan, Head of the Geophysics Department, Faculty of Earth Sciences, University of Istanbul

As a member of the Organization Committee I would like to say welcome to our most distinguished colleagues who are assembled here to discuss the recent developments of earthquake prediction.

As long as science and technology are yet unable to stop the action of destructive natural forces, we are obliged to concentrate our efforts on protection and on finding means to minimize their destructive effects. Prediction, that is forecasting of an event, is one of the most effective and efficient means that helps us to fulfil the above mentioned requirement.

The most destructive of all natural agents in my country are the earthquakes. Detailed determination of the seismotectonical conditions of an earthquake zone, that means intensive studies of its seismicity, its geological structure, geophysical parameters and geodetical features, are the means of prediction studies. The results of recent studies carried out in the field have indicated that we have reasons to be optimistic. Let us hope that the future research will lead us to more reliable inferences.

Since Turkey is a country of high earthquake activity the protection of the investmen should always be considered, while attempts are being made for the development of the country. We cannot afford the constructions which have costed billions of Turkish Liras, to be destructed by an earthquake in a few seconds. In a country of extreme seismic activity like Turkey, where the rate of population increase is very high, where new nuclear power plants are in prospection, where new major dams are being constructed, the task of the scientists is to elucidate other civil servants and politicians on the dimensions of earthquake destruction and to convince the general public on means of protection.

We would like to acknowledge our thanks to UNESCO, to the Parliamentary Assembly of the Council of Europe, to the Earthquake Prediction Commission of IASPEI, to the European Seismological Commission for their substantial support to organize this conference. We owe our due thanks to the Scientific and Technical Research Council of Turkey, to the General Directorate of Mapping, to the Earthquake Research Institute, to the Turkish National Committee on Earthquake Engineering and to the National Committee of IUGG for their help in organizing this conference.

As the Department of Geophysics, we are happy to have the occasion to organize such a conference where, we hope, different appeals on earthquake prediction will be discussed. My greatest thanks are due to all members of the department who laboriously carried out the task of organization in cooperation with the Department of Geophysics, Free University Berlin.

The abundance and the diversified nature of the communications, as well as the eminency of the participants are promising success of the meeting.

I would like to remind that the Organization Committee is willing to help all the participants in all instances. I anticipate that all of you will be bearing pleasant memories at the end of the meeting on the time when you will leave Istanbul.

Address of Prof. Dr. Ö. Öztunali, Dean of the Faculty of Earth Sciences, University of Istanbul

I have the honour to wish you a cordial welcome in the name of the Faculty of Earth Sciences of the University of Istanbul to this international Conference on Earthquake Prediction.

Predictions, the knowledge of future events, have attracted the interest of mankind since early times. Because of this thirst of knowledge predictions of natural disasters once were the main concern of classical astrology.

Gradually, however, because of the progress in sciences and technology we became independent of clairvoyants in foretelling various occurrences on our earth. Thus the weather forecast is no longer a matter of magic power.

Earthquakes, these tremendous and terrific natural phenomena, have devastated several civilizations here in our country. With the development of human society the extent of disastrous consequences of earthquakes is increasing. To mention an example I remind you of the dangers of a nuclear power station when affected by an earthquake.

We are not able to prevent earthquakes. However, like all natural phenomena earthquakes depend on physical parameters which earth scientists may be able to discover, measure and calculate. When we continue to exchange our experiences and to accumulate our knowledge we certainly will find criteria to predict earthquakes. And this will be a great benefit to mankind.

What has been done and what should be done in the future will be discussed by experts in various earth sciences who came to attend and contribute to this conference.

I should like to express my sincere gratitude to all persons and institutions who have contributed to the organization of this conference. To emphasize the importance of this conference I like to end my talk with a sentence of ELBURNI, who in my opinion should be considered as the founder of modern earth sciences. Already in the twelfth century, nine hundred years ago, he expressed an important view:
Of the most important tasks of man the most essential one is to develop his knowledge of the earth, because he has no other place to live.

Address of Prof. Dr. E. Düren, Deputy Rector of the University of Istanbul

Our University of Istanbul is extremely happy to be able to welcome you here on the occasion of the international Conference on Earthquake Prediction Research in the North Anatolian fault zone.

This is not only a scientific achievement for our young Faculty of Earth Sciences but also a meeting of vital importance for our country. Very severe and destructive earthquakes have occurred in Anatolia during the centuries, and they have become a main reason for the perishment of many civilizations one after another.

Since we have suffered under severe earthquakes also during the last decades, it is a great hope for my country to see the rising of a possibility in predicting earthquakes. If the observation and study of the North-Anatolian fault zone can contribute to an interdisciplinary approach to earthquake prediction, we shall be most happy with it.

I would like to congratulate the staff of our Faculty of Earth Sciences for the work they have done to organize this international Conference here and wish a successful meeting for all participants and an enjoyable stay in Istanbul for our guests.

Address of Prof. Dr. Eberhard Lämmert, Freie Universität Berlin

It is a great pleasure for me to be present here in this venerable hall of the University of Istanbul at the opening ceremony of this international conference on earthquake research, especially as the Free University of Berlin has participated extensively in the preparation and organization of this conference. All of you who are assembled here today, the representatives of various disciplines of natural science, have chosen to work in a field in which new findings may be employed directly to the benefit of mankind and even to the saving of human lives. For this reason alone the hopes and expectations of many countries are based on the success of your work.

Since the beginning of civilization man has accepted the challenges of nature in order to preserve and improve the possibilities of human life. Furthermore man has also yearned and striven to foresee and foretell his destiny in order to try to avert impending misfortune and to take advantage of forthcoming opportunities.

This symposium has set itself bold scientific targets: You plan to localize future earth movements and predict their possibly destructive force. Seldom does a scientist experience, as you do, the satisfaction of knowing that the precise use of data and the careful evaluation of empirical observations are of a direct benefit to mankind. Since it first became possible to systematize the motions of the continents, of the blocks and plates of the earth's crust, the possibilities of predicting tectonic forces and volcanic eruptions have improved extraordinarily. In order to attain still greater accuracy this symposium is dedicated to exact theoretical research without, however, ignoring empirical aspects. For this reason you have chosen to hold your conference in a country in which the effects of former earth tremors and of current tectonic activity are manifest.

A further purpose of my visit to Turkey is the signing of a cooperative agreement between our universities. This agreement will establish close ties between field work carried out here and laboratory research conducted in our country. Such cooperation will be to our mutual advantage and, above all, in the interest of students and young scientists in both our countries.

The worldwide success of scientific efforts to the benefit of mankind depends to a considerable extent on such contacts and on productive cooperation as invisioned here. It is cooperation on the institutional level as exemplified by the formal partnership agreement between our universities and the convening of a conference such as this one to which you have come from many different countries.

May I thank the University of Istanbul who is the host of this meeting and all Turkish authorities, who are sponsoring this conference for their generous hospitality.

I am glad that our Free University of Berlin, represented by Prof. Vogel, is contributing towards the success of this symposium, and I wish all of you present here that your work will contribute to the advancement of earthquake research.

Address of Turhan Sökmen, Director of the Turkish General Directorate of Mapping

This is indeed a pleasure to be here on this first symposium on Earthquake Prediction Research in the North Anatolian Fault Zone. The symposium is a timely one and the topic is of great scientific and social interest.

The General Directorate supports the studies dedicated to reducing earthquake hazards in our country as well as in other countries subjected to earthquake risk. We are also aware of the international studies in this field.

The General Directory of Mapping of the Ministery of Defence is now initiating an extensive program of geodetic measurements. We will start the pilot measurements this year in a site selected by the Turkish National Executive Committee on Earthquake Prediction.

We will do everything that is required in supporting these studies. We are hopeful that the results will help both Turkey and other countries.

Address of Prof. Dr. A. Aksoy, President of the Turkish National Union of Geodesy and Geophysics

On behalf of the Turkish National Union of Geodesy and Geophysics I wish to welcome you to this meeting.

As the earthquake problem is a very actual one for Turkey we notice with great satisfaction that this conference on "Earthquake Prediction Research in the North Anatolian Fault Zone" has been given great attention by so many international organizations.

I hope that this conference may create the basis for international cooperation between various earth sciences in order to promote earthquake prediction research in the North Anatolian Fault Zone. The Turkish National Union of Geodesy and Geophysics will always be ready to support and to sponsor this cooperation.

I should also like to express my thanks on behalf of the Turkish National Union of Geodesy and Geophysics to the Dean and the members of the Faculty of Earth Sciences of the University of Istanbul for their efforts in organizing this conference.

Address of Prof. Dr. T. Rikitake, Tokyo, President of the IASPEI Commission on Earthquake Prediction

It is my great pleasure to say a few words about the importance of earthquake prediction on behalf of UNESCO and the Commission on Earthquake Prediction of IASPEI.

An earthquake prediction programme was started in Japan in 1965. After that, similar programmes were initiated in the USA, the USSR and the People's Republic of China. The Haicheng earthquake of magnitude 7.3, that occurred in Liaoning Province, northeastern part of China, on February 4, 1975 was the very first destructive earthquake that was imminently foretold. Many lives were saved because of the earthquake warning issued several hours prior to the occurrence of the earthquake.

It is my understanding that Turkish people have been suffering from many large earthquakes over a long period in history. I hope, therefore, to initiate an earthquake prediction programme in this country, perhaps with the help of international organizations, if necessary. I think this conference will provide the very starting point of the programme.

Personally speaking, it is my great delight that I could come back to Istanbul again. I still remember the nice time when I worked at the Technical University of Istanbul and the University of Istanbul in 1960 and 1961.

Address of Prof. Dr. R. Yarar, President of the Turkish National Committee for Earthquake Engineering

On behalf of the Turkish National Committee for Earthquake Engineering I have the pleasure of welcoming the distinguished scientists attending this meeting.

I am very happy to express my appreciation to the supporting organizations of this conference, and I would like to thank the chairman of the organizing committee and his colleagues for organizing such a beneficial scientific meeting.

I hope this seminar will help to exchange knowledge among the scientists in the field of engineering seismology, and also increase the understanding and close relations among the participants of our countries.

The Turkish National Committee for Earthquake Engineering is greatly pleased about the efforts to start interdisciplinary earthquake prediction research in Turkey. People in this country are in continuous danger and the economic and social life is extensively affected by earthquakes.

Taking into account the increasing number of earthquakes in Turkey as well as in neighbouring countries in recent years, both governments and scientists are forced to take new measures.

We know that to a certain extent it is impossible to escape the destruction caused by earthquakes. Engineering measures are still insufficient and there are no means of predicting when and where the next earthquake will occur. For this reason we should coordinate our limited possibilities and cooperate with institutes beyond our boundaries in order to reach the final objective of earthquake prediction.

The Turkish National Committee for Earthquake Engineering will be willing to contribute with all its resources to the earthquake prediction research in Turkey.

It is my strong belief that this seminar will contribute significantly to the study and solution of the basic problems of earthquake prediction and to the final goal of disaster prevention.

Address by Prof. Dr. N. Toksöz, Head of the Geophysical Laboratory, MIT

Earthquake Hazards

Earthquakes are a major natural hazard that affect Turkey and many other parts of the world. There are more than 10,000 earthquakes that occur each year, an average of about 30 earthquakes per day shake some part of the world. Fortunately, most of these are small earthquakes and do not cause major damage. Many of the earthquakes occur away from population centers. However, there are many populated areas, both around the Pacific and in the Alpine-Himalayan belt of Europe and Asia, which are subject to earthquake hazard, and Turkey is one of them. In the past ten years, damaging earthquakes hit the populated areas in Japan, China, Central Asia, Iran, Turkey, Greece, Italy, the United States, Mexico, Guatemala, Nicaragua, Colombia, Peru, Indonesia, among other countries. Nearly one million people have lost their lives due to these earthquakes. The economic devastation, human suffering and social consequences of these earthquakes are beyond measure.

What is also tragic is that developing countries suffer most from the effects of earthquakes. In 1976, the Tang-shan earthquake might have killed 1/2 million people in China. Again in 1976, one large earthquake in Guatemala killed or injured more than 1 percent of the population and left homeless more than 10 percent of the total population of the country. In the same year, the Caldiran earthquake in Turkey completely devastated all the villages close to the fault zone. In comparison to these, the large earthquakes in Japan in 1978, and in the USA in 1971 and 1979, caused extensive property damage but few casualties or injuries.

Hazards in Turkey

Let me now briefly discuss the situation in Turkey.
Seismically, Turkey is one of the most active continental regions of the world. Most of the country is subject to great earthquake risk. Every tectonic regime and every type of fault can be found in some part of the country. There are about 1 000 earthquakes a year that are recorded in Turkey. When we look at the seismicity since 1900, we find that there is, on the average, one potentially destructive earthquake per year.

Recent earthquakes in Caldiran, Lice, Bingöl, Varto, and Gediz caused tremendous damage and casualties. Fortunately, most of these earthquakes were away from the population centers. For many places earthquakes are not a one-time occurrence. When we examine the 2000 year history of earthquake activity in Turkey, we find that Erzincan was completely destroyed 5 times before the 1939 earthquake. Niksar, another town in the North Anatolian fault zone, experienced at least 14 destructive earthquakes during its history. Similar figures can be quoted for Antakya in Southern Turkey or many places in Western Turkey.

As grave as the historic data are, the present situation is even more dangerous. Today, 75 % of the population of Turkey lives in first, second or third degree earthquake zones. Only a very small fraction of the Turkish people can ignore the earthquake hazards. Most of the population lives in buildings that are not earthquake resistant. Even moderate earthquakes cause major damage and casualties. 75 % of the Turkish industry is located in first or second degree earthquake zones. The economic and social impacts of a major earthquake in these areas are beyond description. Among all other problems, Turkey must face up to the earthquake problem and find ways to reduce the hazard.

What can be done?

Earthquake hazard reduction is both a social and technical problem. The first and most important step in reducing earthquake danger is the realization by the people, by the government, and by the leadership of a country that

a) earthquake hazard exists, and
b) with proper planning and scientific efforts, the hazard can be reduced.

The countries that have made the most progress are the ones that have taken these steps. The lower casualties in Japan and in California are due to early realization of the earthquake danger and proper planning for it. Many other countries are making progress on the topic. Our being here at this conference is evidence that Turkey is serious about the earthquake hazard problem.

Theoretically, it is possible to reduce the earthquake hazard greatly. However, there are practical limitations. For example, it is not possible to move 3/4 of the people of Turkey away from the active fault zones. These zones, because of water and topography, are prime areas for habitation. Nor is it possible to tear down and rebuild the 90 % of the structures in seismic zones that are not earthquake resistant. Even after earthquakes, the replacing of damaged buildings with earthquake resistant structures have not been successful.

Earthquake prediction

One development in recent years that holds promise for reducing casualties and earthquake damage is earthquake prediction. There is ample evidence that measurable physical precursors may precede earthquakes, and especially large earthquakes. These precursors have been used for successful earthquake predictions in China, the Soviet Union, Japan and the USA. A successful prediction of the Haicheng earthquake in China in 1975 may have saved the lives of 1/4 million people. There were definite precursors of the Caldiran earthquake. Spring flows increased, oil seeped, sounds were heard, lakes bubbled, some animals were restless before the earthquake. These were phenomena similar to those observed in Japan and China. Had there been instruments and scientists in Van or Caldiran region the earthquake might have been predicted and lives saved. The prediction process is a complicated one. Although many scientists today believe that earthquakes can be predicted, they also agree that much more research and testing need to be done before an effective prediction scheme can be implemented. Realizing the importance of prediction several countries have implemented national research programs on earthquake prediction. With its major seismic risk Turkey must undertake this task with full commitment.

Earthquake prediction requires an interdisciplinary approach, comprehensive studies, long-term field monitoring, and a variety of measurements. Large earthquakes occur infrequently and identification of successful precursors takes time. From all we know, different regions or earthquakes may have different precursors. Thus, prediction of earthquakes will require long-term measurements in Turkey. Without these, the problem cannot be solved by technology transfer from other places.

The point I would like to emphasize is that Turkey needs to develop a national program for earthquake prediction. Earthquakes constitute a hazard to the nation. From its geologic and tectonic setting, Turkey is an ideal place for prediction studies. Fault zones are accessible. Different tectonic regimes are present. In many respects it is the world's best laboratory for studying earthquake phenomena. The North Anatolian Fault Zone is unique in Eurasia for prediction studies.

International cooperation is a must in earthquake prediction. Earthquakes are a global phenomenon. Their understanding and prediction require global effort and strong collaboration. Development of an effective prediction capability will depend on testing a multitude of ideas and very sophisticated measurements. Turkey should encourage and facilitate international cooperation. The benefits of such cooperation both to science and to Turkey are indeed great.

I speak to this distinguished audience with mixed experience. I am a Turk, yet I live and work in Boston 10,000 km away. I have studied earthquakes in the United States, in Turkey, on the moon with the Apollo Program and on Mars with the Viking. Today I try to run a MIT-Turkish collaborative earthquake research project in Turkey. To have a full support for such cooperative programs I have to appeal:

To Turkey and to the Turkish Government to realize that earthquakes are the single most dangerous natural hazard in Turkey;

To the Turkish Scientists and Scientific Institutions that an effective national program must be set-up for earthquake prediction and hazard reduction; and

To my international colleagues that they need to be realistic and understanding about the benefits and difficulties of cooperative programs in Turkey.

I am very hopeful that this meeting will signal the true beginning of a broad earthquake prediction program in Turkey.

I would like to express my gratitude to our hosts, to the organizers, to Dr. Isikara and Dr. Vogel, to UNESCO and other sponsors, and most importantly to all of you, the participants.

1 Geological history of the North Anatolian Fault Zone

Tectonics of the North Anatolian Transform Fault

A. M. C. Şengör, Kevin Burke and John F. Dewey

Department of Geological Sciences, State University of New York at Albany, Albany, NY 12222, USA.

Abstract--The North Anatolian Transform Fault is the active right-lateral strike-slip zone that extends from Karliova in the east to the Gulf of Saros in the west and that runs parallel with the Black Sea coast of Turkey. It separates the Anatolian scholle in the south from the Eurasian plate in the north. It originated as a result of the partial closure of the southern branch of Neo-Tethys along the Bitlis/Zagros suture during the medial Miocene, but was mainly nucleated on the suture zone of the northern branch of Neo-Tethys that had closed earlier during the Palaeocene/Eocene. The fault today is characterized by a distinctive rift morphology that extends from Karliova to Mudurnu with only two local interruptions by the right-stepping en échelon offsets of Erzincan and Reşadiye, where two pull-apart basins formed. The seismicity of the fault is characterized by bursts of activity separated by quiet periods of about 150 years. The North Anatolian transform fault appears to have originated sometime between the Burdigalian and the Pliocene and since has accumulated a throw of 85 ∓5 km in the east, which decreases to about 20 km in the west. Large, southwestwards striking splays that take off from the fault segment it along strike and are believed to be responsible for the westerly decrease of its throw along the main strand. The fault is the result of the Anatolian scholle's being driven westward away from the intracontinental zone of convergence of eastern Anatolia.

INTRODUCTION AND HISTORY OF RESEARCH

The North Anatolian transform fault (NATF) is a right-lateral fault zone taking up most of the motion between the Eurasian plate and the Anatolian scholle (Dewey and Şengör, 1979). The fault runs from Karliova in the east to the Gulf of Saros in the west trending subparallel with the Black Sea coast of Anatolia (Allen, 1969; Ketin, 1969; McKenzie, 1972) (Fig. 1). Ketin (1948a) recognized that the NATF is a structure accommodating the westward movement of Anatolia and suggested that another fault would be found bounding Anatolia on the southeast. He further implied that the Kozan earthquake might have been on the second fault. In 1972 Arpat and Saroglu first described this East Anatolian Fault and McKenzie (1972, 1976) accommodated Ketin's suggestion

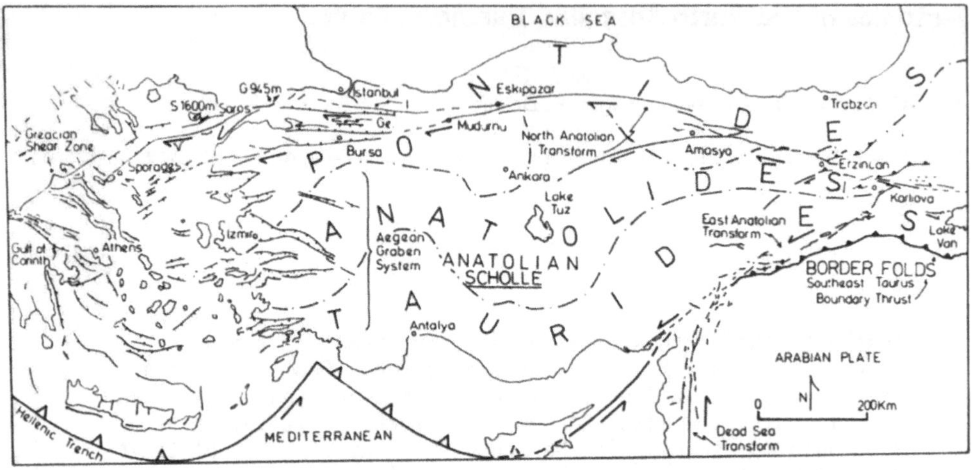

Fig. 1--Simplified map showing active plate and scholle boundaries and Ketin's (1966a) palaeo-tectonic subdivisions of Turkey. Heavy lines with half-arrows are strike-slip faults; lines with black triangles are thrust faults; lines with hachures are normal faults; lines with open triangles are subduction zones; simple solid lines are unspecified faults; stippled regions are depressions; broken lines with dots in between specify the boundaries of palaeo-tectonic subdivisions. i is Izmit/Lake Sapanca Graben, G is Ganosdağ, Ge is Gemlik Graben and S is the island of Samothraki. Figures represent elevations above sea level. After Şengör (1979).

in a plate tectonic framework. Şengör (1979) has summarized early work on the NATF during the nineteen-twenties and thirties which generally recognized it as a major structure associated with Alpine orogenesis (e.g. Nowack, 1928; Salamon-Calvi, 1936; Pamir, 1950), but it was not until a westward progressing series of disastrous earthquakes on the fault began in 1939 that interest in it as an active structure developed (e.g. Pamir and Ketin, 1940, 1941; Paréjas et al., 1942; Pamir and Akyol, 1943; Blumenthal, 1945a, b). Ketin's (1948a) study is a synthesis of the observations stimulated by the disastrous earthquakes and is a landmark not only in recognizing the westward motion of Anatolia but also in its emphasis on the late origin of the NATF (no earlier than Miocene) in contrast to the older Cenozoic dominantly convergent orogenic history of Turkey. This approach is amplified in later summaries (Ketin and Roesli, 1953; Ketin, 1957, 1969, 1976) and in regional (Pavoni, 1961) and plate tectonic (McKenzie, 1972; Dewey and Sengör, 1979; Şengör, 1979, 1980) syntheses.

Emphasis in this review is on the tectonic evolution of Turkey for the last 100 million years as this provides a perspective within which the evolution of the NATF can be viewed. Some properties of the fault are briefly reviewed and its evolution is outlined.

OUTLINE OF THE NEO-TETHYAN CONVERGENT HISTORY OF TURKEY: ESTABLISHMENT OF THE TECTONIC ENVIRONMENT OF THE NORTH ANATOLIAN TRANSFORM FAULT

Figure 1 shows the palaeo-tectonic subdivisions of Turkey as defined by Ketin (1966a) and the present (neo-tectonic) boundaries of the Anatolian scholle. Although the NATF is largely confined to the northernmost of these palaeo-tectonic zones, the Neogene events of southeastern Turkey appear to be directly responsible for its origin; moreover a variety of pre-existing structures such as the eastern termination of the Anatolide/Tauride platform around the Munzur Mountains and the character of the "basement" of eastern Anatolia seem to have exercised a profound influence on the geometry and the evolution of the structure. It is therefore appropriate to review the geological history of the immediate area of the fault zone and its more extensive neighborhood, over the period in which the peculiarities of the present environment have been established.

The palaeo-tectonic subdivisions of Turkey (Ketin, 1966a) correspond to particular tectonic environments generated during the convergence that eliminated the Neo-Tethyan oceans. Although the generation of the Neo-Tethyan oceans in Turkey is of great interest from the viewpoint of the palaeo-tectonic evolution of the country, that part of the geologic history of the area lies beyond the scope of this review. For the entire Tethyan history of Turkey and particularly for the details of the following summary the reader is referred to the recent review by Şengör and Yilmaz (in press), from which the following paragraphs have been largely condensed.

During the late Cretaceous (post-Albian) (Figure 2a) subduction and overall convergence began in Turkey. The intra-Pontide ocean (Şengör and Yilmaz, in press) began contracting along a north-dipping subduction zone beneath the Rhodope-Pontide fragment which resulted in the late Cretaceous island arc volcanism of the northwestern Pontides. South of the Sakarya Continent subduction also began at about the same time along the northern margin of the present Izmir-Ankara Zone (Brinkmann, 1966). Both of these subduction zones joined the eastern Pontide/Lesser Caucasus consuming margin, presumably through a TTT-triple junction somewhere near Ankara. Throughout the late Cretaceous the entire Pontides were the site of calc-alkaline volcanism and, along their southern border, mélange accumulation. The Black Sea began opening behind the Pontide island arc as a back-arc basin during the late Cretaceous (Letouzey, et al., 1977).

The Anatolide/Tauride platform with its Bitlis/Pötürge appendage south of the Inner Tauride ocean was the site of quiet, predominantly carbonate, platform deposition between the late Triassic and the Senonian. During the later Senonian a part of the oceanic crust and upper mantle of the northern branch of the Neo-Tethys (Vardar Ocean in figure 2a) was obducted from the north onto the northern margin of the Anatolide/Tauride platform. Continuous sedimentation from the Jurassic until the Eocene along the Menderes Massif-Kayseri axis indicates that the ophiolitic nappes did not originally go any farther south than this line, in contrast to the inferences of some earlier models (e.g. Ricou et al., 1975) that had interpreted the entire eastern Mediterranean/peri-Arabian Platform ophiolites as parts of the same, northerly obduction event. Later, during the Tertiary, parts of the northern ophiolite nappe (Bozkir Nappe; see Şengör and Yilmaz, in press) were carried farther south, in a "pick-a-back" fashion, during the internal imbrication of the Anatolide/Tauride Platform. We interpret the southern, initial late Cretaceous subduction zone as having had a rather uniform northerly dip (perhaps with the exception of the Alanya Massif area), because it resulted in the late Campanian collision of the Bitlis/Pötürge fragment with the Arabian Platform, causing the obduction of the Baer-Bassit, Hatay (Delaune-Mayere et al., 1977), Koçali (Perinçek, 1979) and Cilo ophiolites onto the northern margin of the Arabian Platform and the Supra-Bitlis/Pötürge ophiolite nappe onto the Bitlis/Pötürge Platform; all of these ophiolite massifs are bounded by northward-dipping thrusts. Westwards, along strike from southeastern Turkey, the

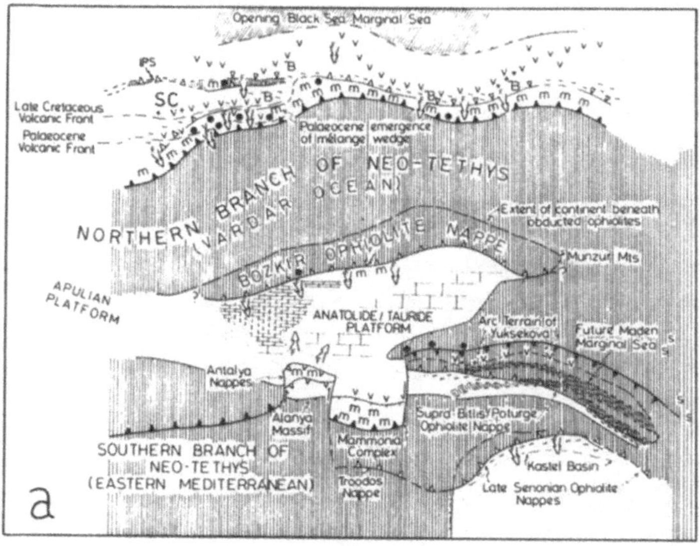

Fig. 2a

Fig. 2--Palaeo-tectonic sketch maps taken from Şengör and Yilmaz (in press) with very slight modification showing the tectonic evolution of Turkey and immediately neighboring regions. White: continental area; closely ruled: oceanic area. Fine brick pattern is pelagic whereas coarse brick pattern is neritic, dominantly carbonate domains, v's are arc volcanics, black dots blueschists, m's mélange and F's flysch. Lines with right-side-up white triangles are thrust fronts, those with upside-down white triangles are rétrocharriage fronts (triangles on the lower plate!) Lines with black triangles are subduction zones (triangles on the upper plate). Lines with half arrows are transform faults, whereas hachured lines are passive or rift margins. Open arrows show generalized sediment dispersal directions. a - Late Cretaceous-Palaeocene tectonics. IPS is the Intra-Pontide suture, whereas SC is the Sakarya Continent. b - Early to Medial Eocene tectonics. Widely spaced, vertical ruling represents the area buried beneath the Taurus allochthons that will become the Anatolides. MM is the Menderes Massif, KM is the Kirsehir Massif. LN, AN, B-HN and HN are the Lycian, Antalya, Beyşehir-Hoyran and the Hadim nappe systems, respectively. AM is the allochthonous Alanya crystalline. c - Late Eocene-Early Miocene tectonics. Figure is age in m.y. d - Medial Miocene-Pliocene tectonics. Upside-down v's are 'Tibetan-type' volcanics. J stands for evaporites, A/CB is Adana-Cilicia basin, HG is the Hatay Graben, DST is the Dead Sea Transform. PM and BM are the Pötürge and the Bitlis allochthonous crystallines. ZS is the Zagros suture.

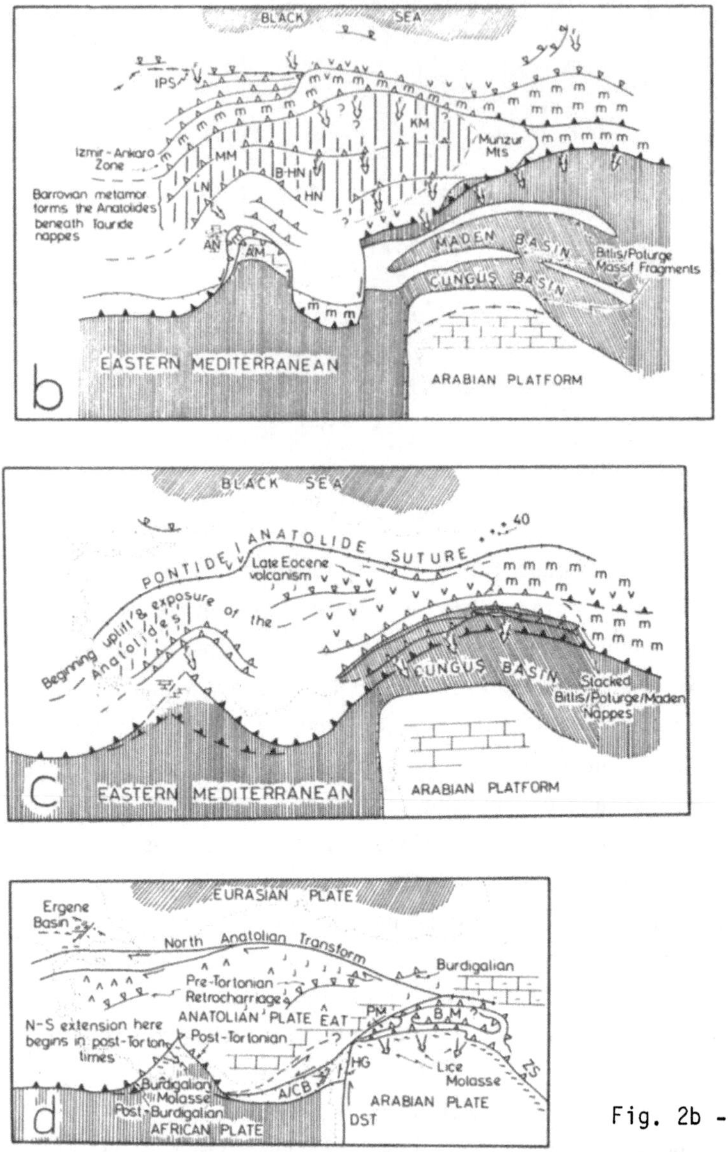

Fig. 2b - d

Troodos ophiolite and the Mamonia Nappes were also emplaced southwards during the late Cretaceous (Lapierre, 1975). Immediately after the Bitlis/Pötürge-Arabia collision in southeastern Turkey, continuing convergence along the southern plate boundary in Turkey began to be accommodated by south-dipping subduction (i.e. by "flipping" of the originally north-dipping subduction zone across which the Supra-Bitlis/Pötürge ophiolite nappe had been emplaced) north of the Bitlis/Pötürge fragment that gave rise to latest Cretaceous-early Palaeocene arc magmatism of the Yüksekova complex. Associated with this subduction zone is the latest Maastrichtian rifting that disrupted the ophiolite-laden Bitlis/Pötürge Complex and thus initiated the opening of the future Maden and Çüngüs marginal basin complexes. Also during the late Cretaceous the Alanya Massif overrode the future Antalya Nappes and terminated sedimentation in the Pamphylian basin (Dumont et al., 1972; Delaune-Mayere et al., 1977).

During latest Palaeocene to medial Eocene time (Figure 2b) the Anatolide/Tauride platform collided with the Pontide active continental margin and almost immediately thereafter large-scale, south-vergent internal imbrication of the former began. A consequence of this deformation was that the previously obducted Bozkir ophiolite nappe was carried farther south with respect to the Anatolide/Tauride platform on the backs of the Lycian, Beyşehir-Hoyran and Hadim nappes which represent slices resulting from the imbrication of the platform. The portions of the platform lying immediately to the south of the imbricate nappes were overriden by and buried beneath them. Following Dürr's (1975) interpretation of the metamorphism of the Menderes Massif, Şengör (1979) argued that the Anatolides represent the metamorphic products of this burial and hence they must be of early to medial Tertiary age as had been pointed out earlier by Ketin (1959, 1966a).

While south-directed thrust-stacking was going on in the Anatolide/Tauride platform, extensive rétrocharriage characterized the Pontides in general during the late Palaeocene-medial Eocene interval. Most of the generally steeply south-dipping early Tertiary and later thrusts of the Pontides are the products of this rétrocharriage phase, which steepened and also often overturned the pre-existing, originally north-dipping structures.

In southeastern and eastern Turkey the Yüksekova subduction zone ceased its activity sometime during the medial(?) Palaeocene and the overall Eurasia/Arabia convergence, as well as the continuing opening of the Maden and Çüngüş basins began to be taken up by a north-dipping subduction zone along the northern margin of the Inner Tauride Ocean. East of the Munzur Mountains the future East Anatolian Accretionary Complex, a giant mélange wedge similar to that of the present Makran (Farhoudi and Karig, 1977), was growing mainly over this subduction zone.

In southern Turkey the Alanya nappe with its tectonic cushion composed of the Antalya Nappes was emplaced during the late Palaeocene-medial Eocene (pre-Lutetian) interval.

During the late Eocene-early Miocene interval (Figure 2c) the overall north-south shortening of the Turkish orogen continued while the metamorphosed Menderes and Kirşehir massifs (the Anatolides) were uplifted and unroofed. In the southeast the Maden basin and the Inner Tauride Ocean finally closed and the main Eurasia/Africa convergence began to be taken up along a uniformly north-dipping sinuous subduction zone.

During Serravallian/Tortonian times (Figure 2d) normal-thickness Arabian continental crust finally collided with Anatolia and its easterly extension, the East Anatolian Accretionary Complex. In the resulting foredeep on the Arabian platform the Lice molasse was deposited and deformed with the autochthonous carbonate rocks of the foreland to give rise to the Border Folds (Ketin, 1966a).

As is seen in figures 2A through D, this collision had a very profound effect on the tectonics of Turkey. In fact, it introduced such drastic changes that Şengör (1980) has used it to separate the neo-tectonic evolution of Turkey from the palaeo-tectonic evolution.

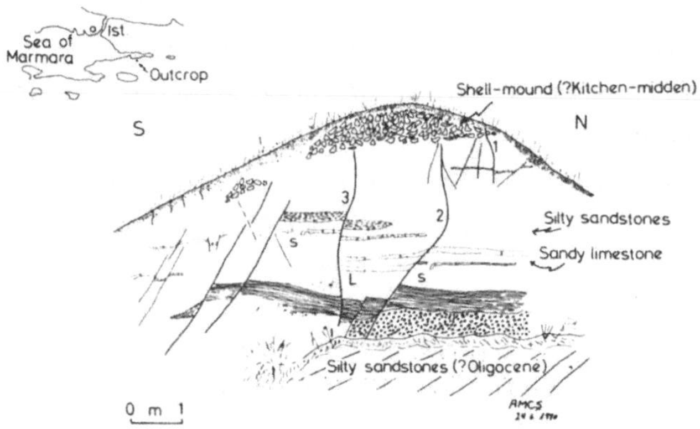

Fig. 3--Sketch of an outcrop showing faulting related to the activity of the North Anatolian transform fault in Tyrrhenian II sediments along the southern shore of the Gulf of Izmit. Faults 2 and 3 have such a curved cross-section and little deformation on either side that they are interpreted as having strike-slip throw (in and out of the sketched face). L indicates laminations of syn-faulting deposits. The outcrop was located in a quarry on the south side of the Yalova-Izmit highway opposite the main entrance to the Karamürsel military base, but has been destroyed by continued quarrying since our visit.

The NATF is the most spectacular neotectonic element imposed on the complex structure of Turkey that was established through the 100 m.y. long history events outlined in this section.

THE FAULT

NATF is marked at the surface over most of its 1200 km length by distinctive topography. Its trace in a broad zone (~5-15 km) is delineated by such features as sub-parallel, anastomosing faults, offset, captured and dammed streams, sag ponds, major valleys with shutter ridges and isolated hills (often exposing shattered, deformed and faulted rock) landslides and lakes caused by such landslides. Where historical earthquakes have broken the ground, as for example around Erzincan, very fresh scarps and scarplets are still preserved. A complex network of young joints and mesoscopic faults characterizes the Neogene-Quaternary fills of many of the basins aligned along the trace of the fault (Irrlitz, 1972; Hancock and Barka, 1980) (Fig. 4). Features of these kinds distinguish the fault as a fairly continuous zone from Karliova to about Mudurnu. NATF joins the East Anatolian transform fault (EATF) at the Karliova "triple junction" (see Şengör, 1979) and does not continue eastwards beyond the junction point (see, for example, Allen, this volume, fig. 20). Although several earthquakes east of Karliova produced right-lateral surface breaks along the strike of NATF (e.g. Varto: Ambraseys and Zatopek, 1968) such breaks do not have the continuity and uniformity of those that occurred along the transform trace. Moreover, many of the east Anatolian earthquakes show varying components of thrusting (e.g. McKenzie, 1972) and are more appropriately considered as elements of the east Anatolian compressional regime (Şengör, 1980).

From Karliova to Erzincan the fault zone is continuous, characterized by Quaternary offsets, although the fault has not produced a known break during historical times in this segment (Allen, this volume). Near Erzincan, the continuity of the fault trace is interrupted by a right-stepping en-échelon offset which localized the Erzincan pull-apart basin (Akkan, 1964; Ketin, 1976) (Fig. 1), a locus of young volcanism (Baş, 1979). Between Erzincan and Reşadiye the trace of the fault is again continuous, characterized by fresh scarps produced by the catastrophic 1939 Erzincan earthquake, sag-ponds, springs with local travertine deposits and deformed stream valleys (Seymen, 1975). To the west of Erzincan Tatar (1975) mapped recent

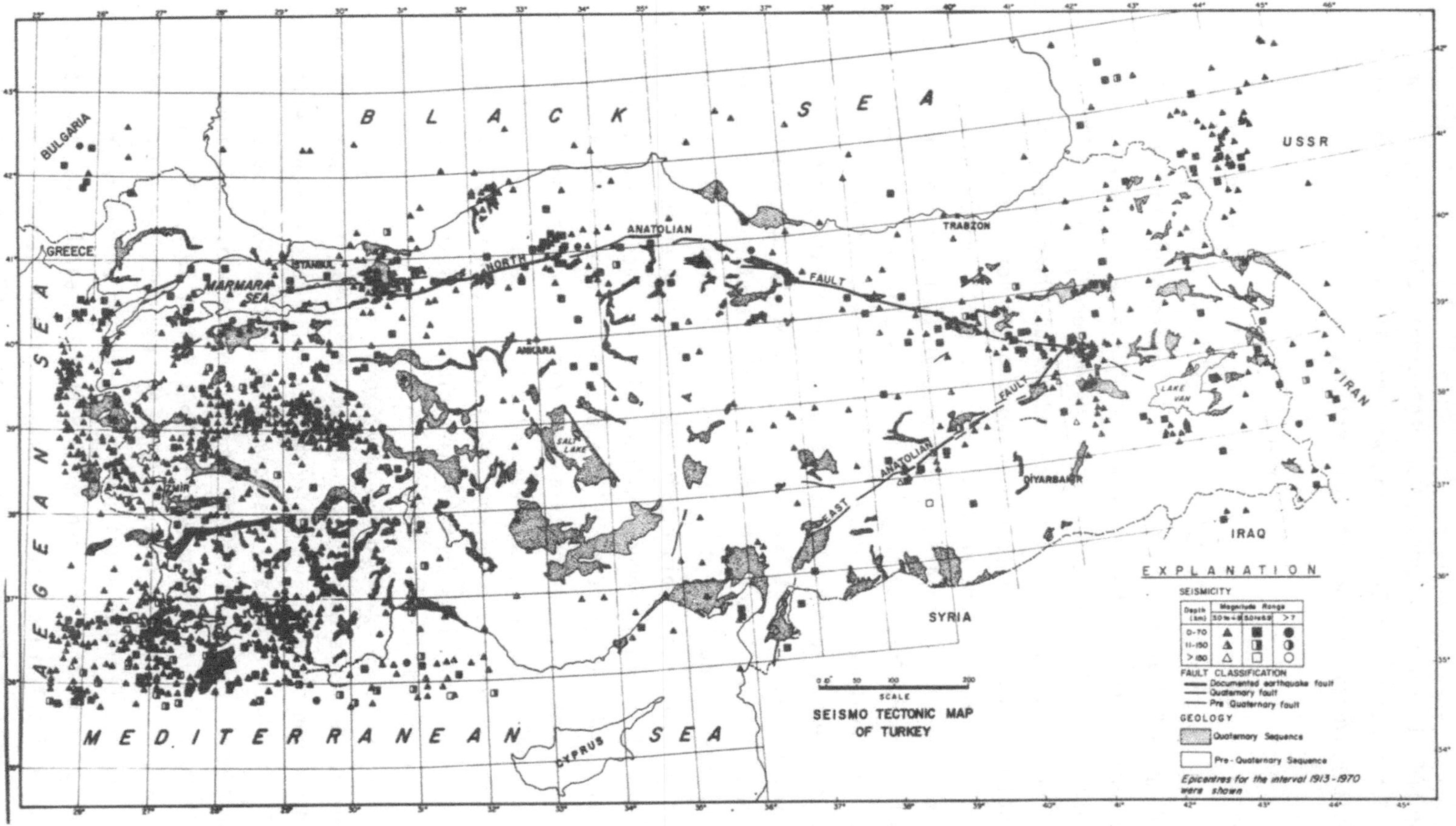

Fig. 4--Seismotectonic map of Turkey showing the distribution of epicenters for the interval 1913-1970 and their relations to some major neotectonic structures in Turkey.

Anticlines deforming Pliocene sediments, which have apparently been produced as a result of right-lateral shear along the fault zone. Between Reşadiye and Erbaa a right-stepping en-échelon offset interrupts the fault trace for the second time and gives rise to another pull-apart (Fig. 1) also characterized by recent basaltic volcanism (Seymen, 1975). An interesting observation regarding the Reşadiye offset is that the trace coming from the east seems not to be simply relayed by the more northerly one coming from the west, but continues, with a general south-westerly, but curved (concave to the southeast), strike all the way to the east of Ankara in central Anatolia (Fig. 1; Mr. Esen Arpat, personal communication, 1979). A similar, but less well-known situation also seems to exist in the case of the Erzincan pull-apart. Such major splays taking off southwestwards from the NATF produce segmentation along strike and progressively reduce the cumulative offset along the fault's main strand westwards (from about 85 km in the east to nearly 20 km in the west: see Şengör and Canitez, in press). In Central Anatolia these major splays may become part of the ova tectonic regime recognized by Şengör (1979, 1980).

Between Amasya and Eskipazar the trace of the fault is uninterrupted with superb rift morphology (Erinç, et al., 1961a) and locally mappable subparallel fault families (Tokay, 1973). Hancock and Barka (1980) studied several of the Neogene basins in this segment (see their figure 1b) with respect to recent mesoscopic structures affecting the young fill of these basins. Based on their mapping of northwest striking conjugate high-angle thrusts, northeast striking conjugate gravity faults and joints and conjugate vertical joints enclosing an acute angle about a northeast trending bisector, they conclude that there was left-lateral motion on the fault zone during part of its neotectonic evolution. However, their observations can also be explained by considering local complications within the shear zone itself or those imposed by the ova regime without resorting to a wholesale reversal of motion on NATF. A full-scale reversal would have had profound effects not only on the local structure of the fault itself but on the neotectonics of the entire eastern Mediterranean region, for which we know of no evidence.

West of Eskipazar the fault shows its first signs of dividing into the subparallel northern and southern strands that have been distinguished farther west around the sea of Marmara. Grabens and closed depressions (e.g. the fault-wedge basin of Çaga: Erinç et al.,

1961b; Sengör, 1979) indicating some measure of extension become common between the fault strands as the Sea of Marmara is approached. Within the Izmit-Lake Sapanca graben, a part of the northern strand, along the southern shore of the Gulf of Izmit, Erinç (1956) described a remarkable Pleistocene section deposited within the graben. Fig. 3 illustrates an outcrop of this section that we visited in June 1980 showing faulting. These Tyrhennian II sediments consist of brown and buff-colored silty sandstones with interlayered dark brown shales and whitish sandy limestones. They overlie deformed ?Oligocene, olive-green, silty sandstones with a basal conglomerate. This outcrop is topped by two shell-mounds. Only a small fault with about a 2 cm throw (Fig. 3, 1) cuts the shell mound. Different faults penetrate to different levels in the outcrop indicating a history of repeated faulting within the Quaternary. Some of the faults (e.g. fig. 3, 2) were growth faults at least during part of their history. Possible blow-out structures (S) may indicate past earthquakes. Such observations may help to contribute to our knowledge of the seismic repeat time along the transform, as Sieh's (1978) remarkable work has shown along the San Andreas fault (see also Allen, this volume).

On the western shore of the Sea of Marmara the northern strand reappears south of Ganosdag. Between Ganosdag and Saros, the northern strand is characterized by a continuous strike slip trace, delineated by an impressive rift topography formed by a zone of faulting and gauge development (Kopp et al., 1969). Here the fault cuts through and is perhaps localized by the westernmost portion of the Intra-Pontide suture (Şengör and Yilmaz, in press), composed of a zone of imbrication that structurally interleaved serpentinized ultramafics and Eocene flysch in a steeply to moderately north-dipping monoclinal structure (Kopp et al., 1969 and our own observations). Just east of where the northern strand comes ashore a left-stepping en-échelon offset just south of Ganosdag gives rise to a restraining bend along the fault which is the cause of the anomalous elevation of Ganosdag (Şengör, 1979).

Except for the short segment between Ganosdag and Saros (which broke during the 1912 Mürefte earthquake: Ambraseys, 1970), the North Anatolian transform zone in and around the Sea of Marmara is characterized by distinctive horst and graben morphology, reflecting the strong influence of the Aegean extensional regime here (Dewey and Şengör, 1979). However, fault-plane solutions (McKenzie, 1978), surface breaks during earthquakes (Ketin, 1966b) and outcrop geometries of faults (see Fig. 3) all suggest that the predominant displacement here is strike-slip.

Overall, NATF forms a broad belt of numerous, sometimes parallel, sometimes anastomosing strike-slip faults. Within the 'rift zone' of the fault local lithologies often appear extensively crushed and mixed; the low resistance of these fault rocks to subaerial erosion is responsible for much of the rift morphology. This rift morphology extends from Karliova to Mudurnu with only two minor interruptions by the right-stepping en-échelon offsets of Erzincan and Reşadiye and eventually merges with the horst and graben regime of the Aegean west of Mudurnu.

The seismicity of NATF (Fig. 4) is among the better known aspects of the structure characterized by episodes of earthquake activity separated by quiet periods of about 150 years (Ambraseys, 1970). The present period of activity began with the 1939 Erzincan quake and progressed generally from east to west in an unparallelled regularity as first pointed out by Ketin (1948a, 1948b). Fault-plane solutions of the major shocks along the NATF have been published mainly by Canitez and Üçer (1967a, 1967b), McKenzie (1972, 1978) and Dewey (1976). Between Karliova and Mudurnu the fault-plane solutions show consistently pure right-lateral strike-slip. East of Karliova, earthquakes within the east Anatolian compressional regime give mainly thrust and strike-slip solutions with one N-S striking normal faulting solution (Dr. D.P. McKenzie, personal communication) in northwestern Iran. At the western end of the transform fault plane solutions give mainly strike-slip (McKenzie, 1978) with some subordinate normal (e.g. Canitez and Toksöz, 1971) and thrust (e.g. Papazachos, 1976) displacements consistent with the complex tectonics of the northern Aegean (Dewey and Şengör, 1979).

Şengör (1979) and Şengör and Canitez (in press) have reviewed evidence concerning the age and the cumulative offset of NATF. Geological and geomorphological data along the fault indicate that the present fault formed sometime between the Burdigalian and the Pliocene. Recently the observations by Hancock and Barka (1980) confirmed and elaborated Irrlitz's (1972) earlier conclusion that the Pontus Formation (Upper Miocene-Lower Pleistocene) has been deposited in basins controlled by the activity of the fault along its course. Because of the unfortunate lack of a very detailed Neogene-Quaternary stratigraphy of the continental successions along the trace of the structure we do not as yet know if the entire fault originated at once or if different segments came into existence at slightly different times.

The cumulative offset of the fault is fairly well established in the eastern sector, where mapping by Bergougnan (1975, 1976), Seymen (1975) and Tatar (1975) revealed it to be around 85 km ∓ 5 km. In the west the offset is more difficult to map, because here the fault very much parallels pre-existing structures associated with the Intra-Pontide suture (Şengör and Yilmaz, in press). Nevertheless, it is generally accepted that the cumulative offset in the western sector is much less than that in the east, perhaps as little as 20 km (Bergougnan et al,, 1978). This westerly decrease of offset is probably the result of the numerous splays that take off from the main trunk and enter central Anatolia to become parts of the ova regime.

NATF originated during the mid to late Miocene interval along with other neotectonic structures of the country. In the next section we briefly summarize the neotectonic evolution of Turkey and surrounding regions, largely after Şengör (1980), in an attempt to clarify the origin and the evolution of NATF.

EVOLUTION OF THE NORTH ANATOLIAN TRANSFORM FAULT

Fig. 5A depicts the tectonic situation in Turkey and its surrounding areas during the early Miocene. It is essentially a simplified version of the picture illustrated in Fig. 2c. Here the closure of the northern branch of Neo-Tethys and the formation of the crystalline axis of Anatolia, the Anatolides, are in their final stages. The southern branch is also about to be closed along the Bitlis/Zagros suture zone. In Fig. 5B Eurasia/Arabia collision has already occurred (∓ Serravallian) by the terminal closure of the Çüngüş Basin along the Bitlis suture and Eurasia/Arabia convergence has begun to be taken up by intracontinental deformation, mainly of the East Anatolian Accretionary complex. At this stage (post-Tortonian times) intra-continental convergence-related excessive thickening in eastern Anatolia (Şengör and Kidd, 1979) resulted in the expulsion of an 'Anatolian *scholle*' along the newly generated, conjugate strike-slip faults, namely NATF and EATF.

The eastern boundaries of this *scholle*, represented by NATF's eastern segment and EATF, very closely follow the original boundaries of the Anatolide/Tauride Platform. So, although the southern boundary of the Pontides and with it the northern ophiolitic suture of Turkey

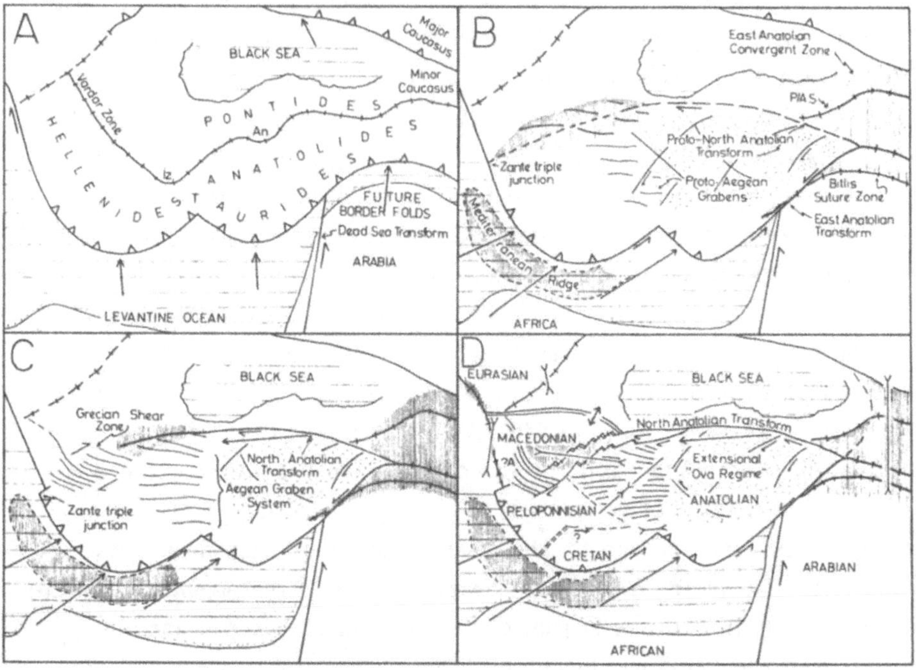

Fig. 5--Tectonic evolution of the North Anatolian Transform Fault. Lines with white triangles are subduction zones (with triangles on the upper plate), lines with half arrows are transform faults, and lines with ladder-pattern are sutures and/or zones of intracontinental convergent high strain. Widely spaced horizontal ruling represents oceanic domains, whereas closely spaced vertical ruling designates intraplate convergent strain. A - early Miocene tectonics, B - medial to late Miocene tectonics, C - Pliocene tectonics, D - present tectonics.

strikes northeast east of Erzincan and is apparently discordant to the NATF's trace, the latter here seems to be still using an old structure, namely the edge of the Anatolide/Tauride Platform. Şengör's (1979) original statement that NATF did not necessarily nucleate on the suture should therefore be modified to apply only to the northern edge of the suture zone (i.e. the former subduction zone) and not to the whole of it. However, to what degree NATF has been localized by which structures and/or lithologies of the pre-existing suture(s) in northern Turkey is a problem yet to be addressed. On the other hand, the sharp south-westerly bend in the course of the fault between Saros and Zante, where it is represented by the broad Grecian shear zone (Şengör, 1979) is definitely an entirely discordant structure where it cuts the Hellenide structures almost at right angles. It created there an obstruction against the westerly flight of the Anatolian scholle that gave rise to a roughly east-west compression in the northern and central Aegean area. The relief of this east-west compression by north-south extension formed the extensional regime of the Aegean area (fig. 5C). The present neotectonic picture is illustrated by Fig. 5D. The extensional zones north and west of the Macedonian scholle may be the result of Anatolia's attempting to rip this piece away from Eurasia and a still smaller Albanian scholle (A in Fig. D) from the former. The neotectonic evolution of the Ergene Basin in Thrace has also been influenced by this regime (see Kopp et al., 1969).

The extensional regime of the Aegean continues eastwards into central Anatolia where it creates the ova regime. The amount of north-south extension in the ova region diminishes eastwards and attains a zero value at the Karliova junction.

In conclusion, it appears that since the Tortonian the dominant factor that has governed the neotectonic evolution of the entire eastern Mediterranean domain has been the collision and continued convergence between Eurasia and Arabia along the Bitlis/Zagros suture zone. This collision created all of the neotectonic structures in Turkey, of which NATF is only an element that helps the westerly migration of the Anatolian scholle. If we are to understand its evolution and significance, we should consider it not in isolation as, unfortunately, has been often the case until recently, but in its natural habitat in conjunction with other elements of the neotectonic system of the entire eastern Mediterranean. Because, unless the kinematic whole of the system is known, there is little chance that parts thereof can be completely understood.

REFERENCES

Akkan, E., 1964: Erzincan ovasi ve çevresinin jeomorfolojisi. Ankara Üniv. Dil Tarih-Cogr. Fak. Yay., 153, 104.

Allen, C.R., 1969: Active faulting in northern Turkey. Contr. No. 1577, Div. Geol. Sci., Calif. Inst. Techn., 32

Ambraseys, N.N., 1970: Some characteristic features of the North Anatolian Fault Zone. Tectonophysics, 9, 143-165.

Ambraseys, N.N. and Zatopek, A., 1968: The Varto-Üstükran (Anatolia) earthquake of 1966 August 19: a summary of a field report. Bull. Seism. Soc. America, 58, 47-102.

Arpat, E. and Şaroglu, F., 1972: The east Anatolian fault system: thoughts on its development. Bull. Mineral Res. Explor. Inst. Turkey, 78, 33-39.

Baş, H., 1979: Petrologische und geochemische Untersuchungen an subrezenten Vulkaniten der nordanatdischen Störungszone (Abschnitt: Erzincan-Niksar), Türkei. Dissertation, Univ. Hamburg, 116.

Bergougnan, H., 1975: Relations entre les edifices pontique et taurique dans le Nord-Est de l'Anatolie: Bull. Soc. Géol. France, 17, 1045-1057.

Bergougnan, H., 1976: Structure de la Chaîne pontique dans le Haut-Kelkit (Nord-est de l'Anatolie). Bull. Soc. Géol. France, 18, 675-686.

Bergougnan, H., Fourquin, C. and Ricou, L.-E., 1978: Les deux tronçons et le double jeu de la faille nord-anatolienne dans la tectonique recente du Moyen-Orient. C.R. Acad. Sci. Paris, 287, 1183-1186.

Blumenthal, M.M., 1945a: Ladik deprem hatti. MTA Mecmuasi, 1/33, 153-174.

Blumenthal, 1945b: Die Kelkit-Dislokation und ihre tektonische Rolle. MTA Mecmuasi, 2/34, 372-386.

Brinkmann, R., 1966: Geotektonische Gliederung von Westanatolien: N. Jb. Geol. Paläont. Mh., 10, 603-618.

Canitez, N. and Toksöz, N., 1971. Focal mechanism and source depth of earthquakes from body- and surface-wave data. Bull. Seism. Soc. America, 61, 1369-1379.

Canitez, N. and Üçer, B., 1967a: Computer determinations for the fault-plane solutions in and near Anatolia. Tectonophysics, 4, 235-244.

Canitez, N. and Üçer, B., 1967b: A catalogue of focal mechanism diagrams for Turkey and adjoining areas. ITÜ Maden Fak. Arz Fizigi Enst. Yay. 25, 111

Delaune-Mayere, M., Marcoux, J., Parrot, J.F., and Poisson, A., 1977: Modèle d'evolution Mésozoique de la paléo-marge téthysienne au niveau des nappes radiolaritiques et ophiolitiques du Taurus Lycien, d'Antalya et du Baër-Bassit in Biju-Duval, B. and Montadert, L., eds. Structural History of the Mediterranean Basins, Editions Technip, Paris, 79-94.

Dewey, J.F. and Şengör, A.M.C., 1979: Aegean and surrounding regions: complex multi-plate and continuum tectonics in a convergent zone. Geol. Soc. America Bull., 90 (I), 89-92.

Dewey, J.W., 1976: Seismicity of northern Anatolia. Bull. Seism. Soc. America, 66, 843-868.

Dumont, J.F.; Gutnic, M.; Marcoux, J.; Monod, O. and Poisson, A., 1972: Le Trias des Taurides occidentales (Turquie) Définition du basin pamphylien: un nouvean domaine à ophiolites à la marge externe de la chaine taurique. Zeitschr. Deutsch Geol. Gesell., 123, 385-410.

Dürr, S., 1975: Über Alter und geotektonische Stellung des Menderes-Kristallins/SW-Anatolien und seine Aequivalente in der mittleren Aegaeis. Marburg/Lahn, 107.

Erinç, S., 1956: Yalova civarinda bahri Pleistosen depolari ve taracalari. Revue de Geogr. Turque, 12, 188-192.

Erinç, S.; Bilgin, T. and Bener, M., 1961a: Gerede civarinda akarsu şebekesi. Istanbul Üniv. Cogr. Enst. Dergisi, 7, 90-99.

Erinç, S.; Bilgin, T. and Bener, M., 1961b: Çaga Depresyonu ve Bogazi. Istanbul Üniv. Cogr. Enst. Dergisi, 7, 170-173.

Farhoudi, G. and Karig, D.E., 1977: Makran of Iran and Pakistan as an active arc system. Geology, 5, 664-668.

Hancock, P.L. and Barka, A.A., 1980: Plio-Pleistocene reversal of displacement on the North Anatolian fault zone. Nature, 286, 591-594.

Irrlitz, W., 1972: Lithostratigraphie und tektonische Entwicklung des Neogens in Nordostanatolian. Beih. Geol. Jb., 120, 111

Ketin, I., 1948a: Über die tektonisch-mechanischen Folgerungen aus den grossen anatolische Erdbeben des letzten Dezenniums. Geol. Rundsch., 36, 77-83.

Ketin, I., 1948b: Die grossen anatolischen Erdbeben in den letzten zehn Jahren. Urania, 11, 230-236.

Ketin, I., 1957: Kuzey Anadolu Deprem Fayi. ITÜ Dergisi, 15, 49-52.

Ketin, I., 1959: Türkiye'nin orojenik gelişmesi: M.T.A. Dergisi, 53, 78-86.

Ketin, I., 1966a: Tectonic units of Anatolia. Bull. Mineral Res. Explor. Inst. Turkey 66, 23-34.

Ketin, I., 1966b: 6 Ekim 1964 Manyas depremi esnasinda zeminde meydana gelen tansiyon çatlaklari. Türkiye Jeoloji Kurumu Bülteni, 10, 1-2.

Ketin, I., 1969: Über die nordanatolische Horizontalverschiebung. Bull. Mineral Res. Explor. Inst. Turkey, 72, 1-28.

Ketin, I., 1976: San Andreas ve Kuzey Anadolu Faylari arasinda bir karsilastirma. Türkiye Jeoloji Kurumu Bülteni, 19, 149-154.

Ketin, I. and Roesli, F., 1953: Makroseismische Untersuchungen über das nordwest-anatolische Beben vom 18. März 1953. Eclog. geol. Helvet., 46, 187-208.

Kopp, K.-O., Pavoni, N. and Schindler, C., 1969: Das Ergene-Becken. Beih. Geol. Jb., 76, 136.

Lapierre, H., 1975: Les formations sédimentaires et éruptives des Nappes de Mamonia et leurs relations avec le Massif du Troodos (Chypre occidentale) Mém. Soc. Geol. France, nouv. sér. 54, 131.

Letouzey, J.; Biju-Duval, B.; Dorkel, A.; Gonnard, R.; Kristchev, K.; Montadert, L. and Sungurlu, O., 1977: The Black Sea: a marginal basin. Geophysical and geological data in Biju-Duval, B. and Montadert, L., edts., Structural History of the Mediterranean Basins, Editions Technip, Paris, 363-376.

McKenzie, D.P., 1972: Active tectonics of the Mediterranean region. Geophys. Jour. R. Astron. Soc., 30, 109-185.

McKenzie, D.P., 1976: The East Anatolian Fault: a major structure in eastern Turkey. Earth Planet. Sci. Lett., 29, 189-193.

McKenzie, D.P., 1978: Active tectonics of the Alpine-Himalayan belt: the Aegean Sea and surrounding regions. Geophys. Jour. R. Astron. Soc., 55, 217-254.

Nowack, E., 1928: Die wichtigsten Ergebnisse meiner anatolischen Reisen. Zeitschr. Deutsch. Geol. Gesell., 80, B304-B312.

Pamir, H.N., 1950: Les séismes en Asie Mineure entre 1939 et 1944: La Cicatrice Nord-Anatolienne. Proc. Int. Geol. Congr. London, XIII, 214-218.

Pamir, H.N. and Akyol, I.H., 1943: Çorum ve Erbaa depremleri. Türk Cografya Dergisi, 2, 1-7.

Pamir, H.N. and Ketin, I., 1940: Das Erdbeben in der Türkei vom 27 und 28 Dezember 1939. Geol. Rundsch., 31, 77-78.

Pamir, H.N. and Ketin, I., 1941: Das anatolische Erdbeben Ende 1939. Geol. Rundsch., 32, 279-287.

Papazachos, B.C., 1976: Seismotectonics of the northern Aegean area. Tectonophysics, 33, 199-209.

Paréjas, E.; Akyol, I.H. and Altinli, E., 1942: Le tremblement de terre d'Erzincan du 27 Decembre 1939. Revue Fac. Sci. Univ. Istanbul, BV1, 177-222.

Pavoni, N., 1961: Die Nordanatolische Horizontalverschiebung. Geol. Rundsch., 51, 122-139.

Perinçek, D., 1979: Guidebook for Excursion "B", Interrelations of the Arab and Anatolian Plates: First Geological Congress, Middle East, Ankara, Turkey, 34.

Ricou, L.-E.; Argyriadis, I. and Marcoux, J., 1975: L'Axe calcaire du Taurus, un alignement de fenetres arabo-africaines sous des nappes radiolaritiques, ophiolitiques et metamorphiques. Bull. Soc. Geol. France, 17, 1024-1044.

Salomon-Calvi, W., 1936: Die Fortsetzung der Tonalelinie in Kleinasien. Yüksek Ziraat Enst. Çalişmalarindan, 9, 11-13.

Şengör, A.M.C., 1979: The North Anatolian transform fault: its age, offset and tectonic significance. Geol. Soc. London Jour., 136, 269-282.

Şengör, A.M.C., 1980: Türkiyenin Neotektoniginin Esaslari. Türkiye Jeoloji Kurumu Konferanslar Serisi, 40.

Şengör, A.M.C. and Canitez, N., in press: The North Anatolian Fault. ICG Working Group 3 Report, Ch. 5.

Şengör, A.M.C. and Kidd, W.S.F., 1979: Post-collisional tectonics of the Turkish-Iranian Plateau and a comparison with Tibet. Tectonophysics, 55, 361-376.

Şengör, A.M.C. and Yilmaz, Y., in press: Tethyan evolution of Turkey: a plate tectonic approach. Tectonophysics (75, 181-241 (1981)).

Seymen, I., 1975: Kelkit vadisi kesiminde Kuzey Anadolu Fay Zonunun tektonik özelligi I.T.Ü. Maden Fak. Yay., 192.

Sieh, K.E., 1978: Prehistoric large earthquakes produced by slip on the San Andreas fault at Pallett Creek, California. Jour. Geophys. Research, 83, 3907-3939.

Tatar, Y., 1975: Tectonic structures along the North Anatolian Fault Zone, northeast of Refahiye (Erzincan) Tectonophysics, 29, 401-410.

Tokay, M., 1973: Kuzey Anadolu Fay Zonunun Gerede ile Ilgaz arasindaki kisminda jeolojik gozlemler. Kuzey Anadolu Fayi ve Deprem Kuşagi Simpozyumu, MTA Yay., Ankara, 12-29.

Paleo-, Tardi- and Neotectonic Mechanisms of the Present North Anatolian Fault Zone in the Light of the Structural History of the Eurasian Margin in the Pontic Ranges.

H. Bergougnan and C. Fourquin

Groupe d'Etudes Géologiques de l'Université de Reims (G.E.G.U.R) Faculté des Sciences, 51 100 Reims, France
C.N.R.S., ERA N° 9 „Asie Alpine occidentale".

Summary

The present seismically active North Anatolian Fault Zone (NAFZ) is a recent accident which cannot be regarded as the major Alpine suture. Yet, this neotectonic accident whose present trace is shaped by the stress system established since the Tortonian (after the Red Sea opening) is not an original structural feature; it is indeed composite, consisting of two parts running along sections of two great lineaments which recorded the relative motions of Eurasia in relation to Africa and Arabia during all Alpine history: a paleotectonic stage from Tethyan opening to the late Cretaceous collision of the two blocks; a tarditectonic stage from this collision to the Read Sea opening; and finally a neotectonic stage up to the present period. Therefore, these two lineaments played different rôles during different periods, acting either as a left or right strike-slip or transform fault, or as a longitudinal structural feature playing an important structural rôle in Hercynian or Alpine orogeny, respectively.

I - INTRODUCTION

The NAFZ has been regarded, sometimes as an essential structural feature of Alpine orogeny in Anatolia, sometimes, on the contrary, as a recent accident cutting the major structures of this building. We shall prove that neither of those two reverse outlooks can account for what actually took place historically in this fracture zone.

First we shall compare the two Alpine realms (the Pontide and the Tauride) and the NAFZ. This comparison will prove that the latter is superimposed on two older lineaments which played an active rôle as early as the Hercynian period but particularly when the Alpine range was built up.

We shall successively examine, for either of the two lineaments, this complex rôle, i.e. as a left or right strike-slip fault, or as a longitudinal fault, or at times as a transform fault.

The uniform course of the NAFZ clearly appears in the latest evolutionary stage, not yet completed. This evolution can be subdivided into three major stages: the paleotectonic stage, extending from the Hercynian period (or earlier?) to the Alpine collision of the two Tethyan sides in late Cretaceous time; the tarditectonic stage, extending from this collision to the aftermath of the Red Sea opening which established a new stress pattern, dominant in these regions after the Tortonian, and finally the neotectonic stage, from the Tortonian to the present.

Disparate and successive tectonic mechanisms produced the two lineaments whose traces were later followed by the present NAFZ. It is unlikely that the NAFZ, all along its trace, could have played the rôle of an essential boundary as has been proposed. However, at specific times, and along restricted segments, it unquestionably was an important structural element.

II - ROLE OF THE NAFZ IN THE TURKISH GEOLOGICAL FABRIC

The Anatolian geological framework is seen as the juxtaposition of European and African blocks including, on their respective margins, the two belts of the Alpine orogeny, namely the Pontic realm in the North (European margin) and the Tauric one in the South (Arab-African margin). In between, the essential ophiolitic suture zone runs parallel to those two structures, like a scar left by the Tethyan ocean. To grasp the different geometrical and historical connections between the main features of this and the NAFZ, it is essential to show their respective boundaries and geographical traces clearly.

As far as the NAFZ is concerned, we have long known its trace which, marked by earthquakes is clearly visible as a topographic feature. However, it proves to be more difficult to define the extension of the European framework southward and, to locate precisely the suture or contact with the Tauric block, because in many places the ophiolites are overthrust. Because of this difficulty writers until recently, were divided into those who saw the NAFZ as an essential boundary between the two Alpine structures, a boundary playing an essential rôle in the paleogeography and the structural development of the Alpine range, and those who saw it as a mere neotectonic accident cutting across Anatolia.

In the first group are E. NOWACK (1928), followed by W. SALOMON-CALVI (1940) who regarded the NAFZ as "a geosynclinal scar ... a contact between the Alpidic and Dinaric realms", a zone of collision between the European and African continents. ARNI (1939), PAVONI (1961) and OZKOCAK (1969) placed here the rise of the north Anatolian ophiolites which gave birth to Bailey and Mc. Callien's "coloured" or "ANKARA" melange. In the Erzincan area, where the NAFZ meets the ophiolitic zone, G. ATAMAN, E. BUCKET, Ü.Z. CAPAN (1975) recognized it as the sign of a subduction plane diving under the Pontide active until Miocene, and they saw the present strike-slip fault as the horizontal new movement of this Benioff paleoplane. Other writers such as EGERAN (1945), KRAUS (1958), TOKEL (1973) explained the neotectonic movement as the logical result of the paleogeographical rôle of this fault zone which, since the Mesozoic, marked the boundary of Europe, and since the collision with the Arab-African continent, formed a suture zone. Another group of geologists did not regard the NAFZ as an essential structural element in the Alpine stage of Anatolia, and saw it only as a weak zone playing no active rôle until a later period (BLUMENTHAL, 1945; PAMIR, 1950; TOKAY, 1973). According to N. PINAR and E. LAHN, (1950) it has nothing to do with the actual "East Anatolian scar ... a median line in the East Anatolian orogenic framework followed by the "green" zone". I. KETIN (1966, 1968, 1969) was the first to interpret it as a young seismically active fault zone which first appeared after the Alpine orogeny. This is the definition A.M.C. ŞENGÖR (1979) refers to. The surveys we conducted in the Pontides first supported this point of view (BERGOUGNAN, FOURQUIN, 1976). Northern Anatolia makes up indeed a coherent whole (FOURQUIN, 1975; BERGOUGNAN, 1975, 1976; BERGOUGNAN, FOURQUIN, 1980), whose main features are Mesozoic history, frequently interrupted by tectonic events and disrupted by ceaseless magmatism since Liassic time. In addition, there are some paleogeographical features and faunas (fig. 1) belonging to the north Tethyan paleobiogeographical provinces (BASSOULLET, BERGOUGNAN, ENAY, 1975; ENAY, 1972; ENAY, 1976). Those features characterize the Irano-Balkanic region (BERGOUGNAN, FOURQUIN, 1979) which is the Eurasian active margin of the Alpine system in the East Mediterranean, and facing the Dinaro-Zagro-Tauric passive margin which is part of the Arab-African platform. Therefore, we can fix the southern boundary of the European craton rather precisely. It clearly appears (fig. 2) that:

1) In the western Pontides, this boundary is far more south than the NAFZ along the IZNIK-HAVZA section.

2) Between Havza and Erzincan, the fault cuts the Pontic structure crosswise, separating the ancient metamorphic Tokat massif in the West, from the volcanic granitic range that runs along the Black Sea in the East (BERGOUGNAN, 1975; FOURQUIN, 1975) which both together belong to the Pontic belt. Therefore we can rule out the possibility that the NAFZ may have played - <u>all along its trace</u> - the rôle of a plate boundary during Mesozoic time. Hence, the present authors originally concluded that it was an exclusively tardi- to neotectonic offset.

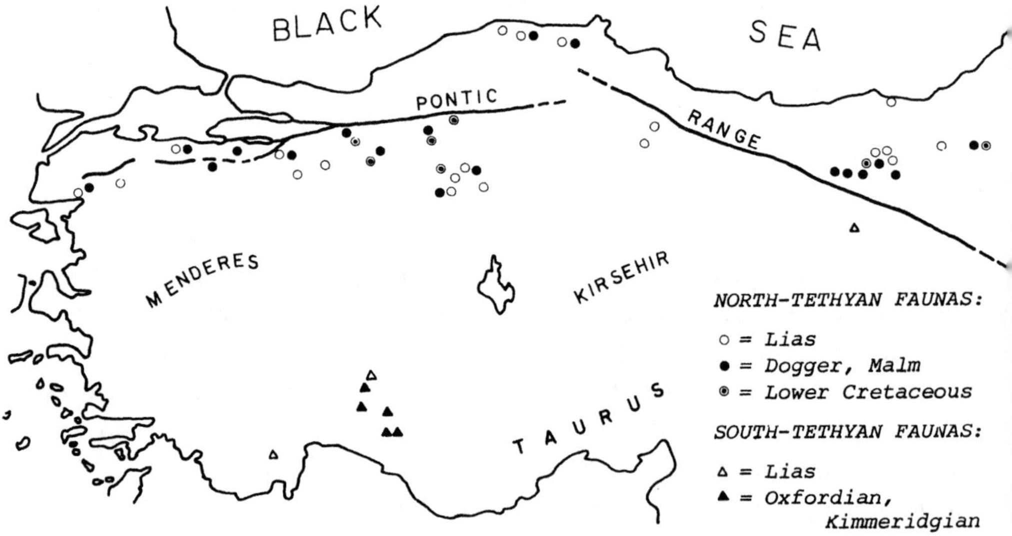

Fig.: 1. North-Anatolian Fault and distribution of the north and south-Tethyan mesozoic faunas

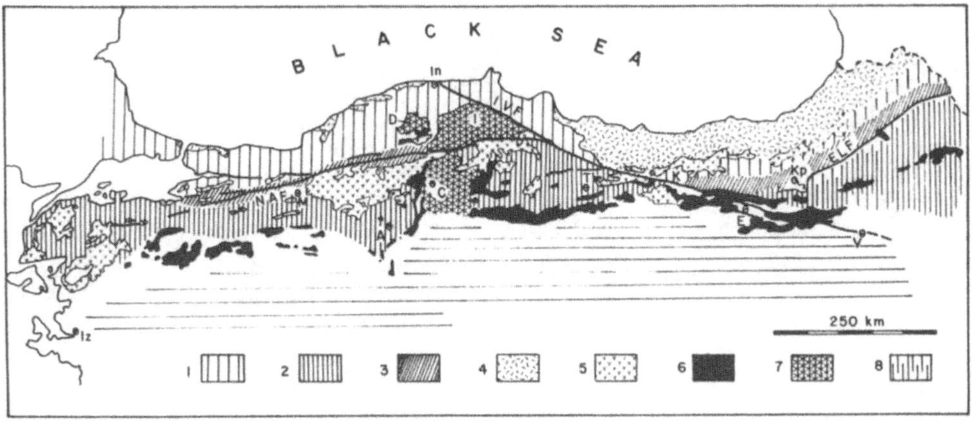

1)Outer Pontides; 2) Inner Pontides; 3) Pelagic furrows during the Malm-Neocomian; 4) Upper Cretaceous calc-alcaline volcanism of the north-Pontic belt; 5) Late Cretaceous and Tertiary calc-alcaline and acid volcanism of the south-Pontic belt; 6) Medio-Tethyan ophiolitic suture; 7) Ophiolitic and metamorphic nappes of Çankiri and Ilgaz-Daday massifs; 8) North-Tethyan Alpine area with a Gondwanian basement.

IVF: Inebolu Varto fault (eastern branch of the North Anatolian Fault)
NAF: Western branch of the North-Anatolian Fault
ELF: Erzincan-Lyssogozok faults

Fig.: 2. North-Anatolian Fault zone and geological framework of the northern Anatolia (from BERGOUGNAN and FOURQUIN 1980)

Since then, our outlook has evolved as we took into account either new surveys, or facts that had been established but overlooked.

1) In the West FOURQUIN (1975) presented proof of the paleotectonic movement in the Hercynian ages and during the Mesozoic followed by tarditectonic movement from Maastrichtian to Miocene along the IZNIK-HAVZA section of the NAFZ. The author believes therefore that there existed an older lineament that is connected with the Balkan fractures (the Maritsa fault), i.e. the "A" lineament or West Pontic lineament.

2) In the Central portion after proving the existence of the gread Dinaric thrusts over the west Pontides (allochtonous ILGAZ and DADAY massifs) (Fig. 2 and 4), FOURQUIN (1977) remarked that they remain within the perpendicular line of the

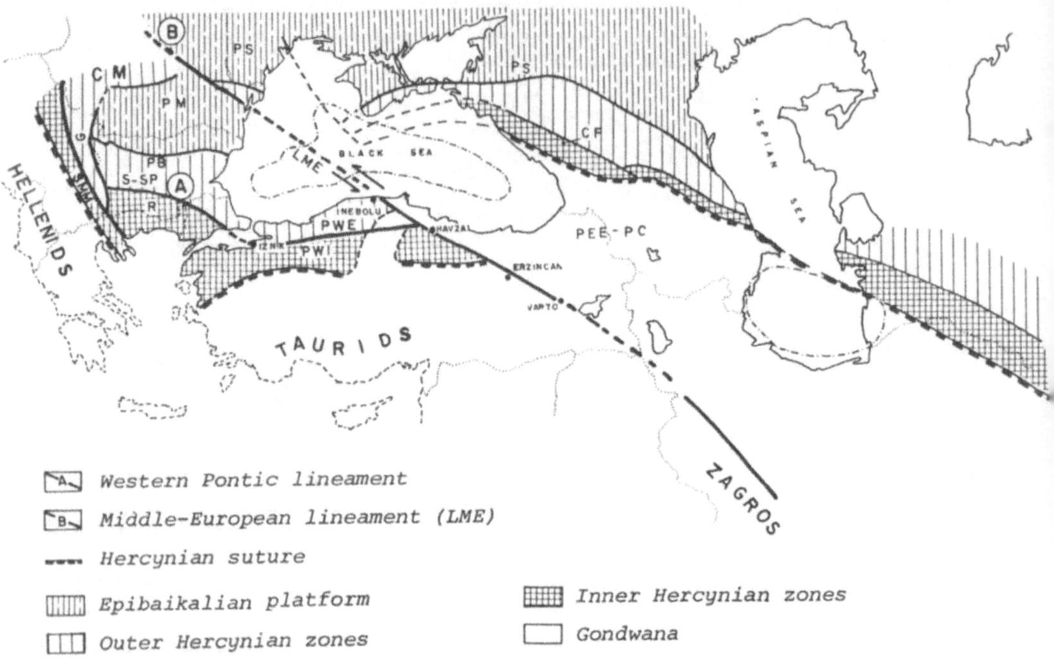

Fig.: 3. Hercynian organisation and paleotectonic rôle of the western-Pontic (A) and middle-European (B) lineaments (for abbreviations, s. last page)

INEBOLU-HAVZA axis, straight in the prolongation of the present active HAVZA-VARTO section of the NAFZ. This alignment recognized in the Black Sea (Moiseyev Range) by Russian writers (BALAVADZE, 1966; CHEKUNOV, 1973), extends into Dobrudja (Peceneaga fault), then along the Szamos line which is marked along its entire margin with positive anomalies, and finally along the Elbe line (STILLE, 1953). This alignment constitutes an actual Meso-European lineament, the "B" lineament.

3) In the East and South-East (Fig. 4), the position of Erzincan ultrabasic massifs indicates a sort of "ophiolitic crossroads", one branch of which forks off to South Transcaucasia (Sevan and Vedi ophiolites, SHIKHALIBEYLI, 1964-1967; KNIPPER, 1971; AZIZBEKOV and RUSTAMOV, 1973). The other branch forks off to Zagros. The offset of the former in relation to the North Anatolian ophiolitic zone (which represents a suture

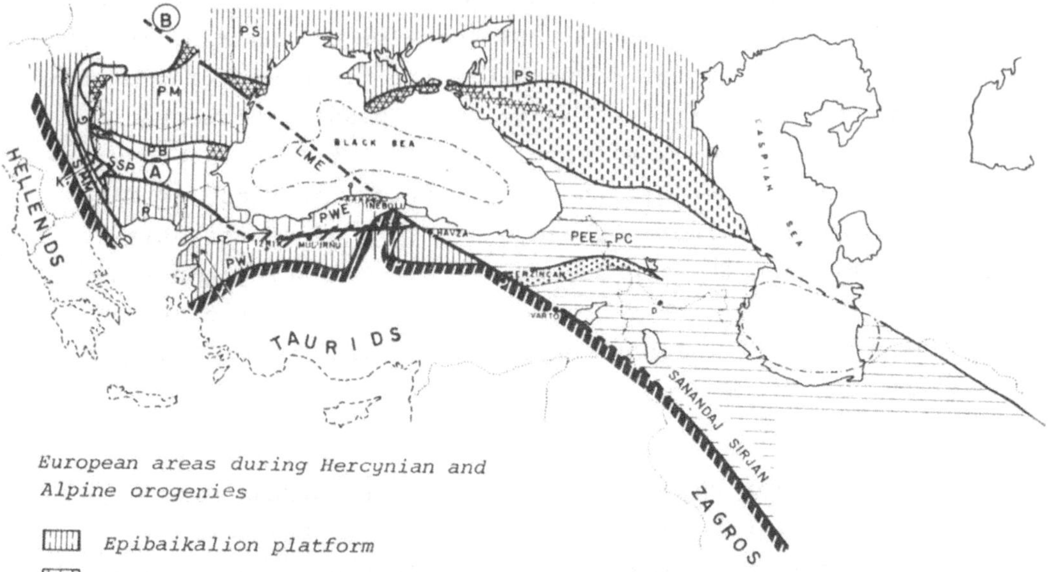

European areas during Hercynian and Alpine orogenies

- *Epibaikalion platform*
- *Western outer Pontides*
- *Western inner Pontides*
- *Caucasian range*

Areas being Gondwanian during the Hercynian time and European during Alpine time

- *Gondwano-European*
- *Arabo-African areas during the Hercynian and Alpine orogenies*

- *Medio-Tethyan ophiolites*
- *Erzincan-Sevan ophiolites*
- *Peri-scythian and perimoesian furrows (Trias-Upper Jurassic?)*
- *Kralstide and Mudurnu pelagic furrows*
- *Ophiolitic thrusts over the European margin*

Fig.: 4. Mesozoic paleogeography and paleotectonic rôle of the western-Pontic (A) and middle-European (B) lineaments (the latter moving as a left strike-slip fault or a transform fault) (for abbreviations, s. last page) between the European and African blocks) looks to be at least 100 kms long over the HAVZA-VARTO section of the NAFZ (BERGOUGNAN, 1975). This amount is probably the minimum offset of this section if the first branch (Erzincan-Sevan) represented a major suture between the two blocks. If not, the major suture runs along the other branch, thus meeting the trace of NAFZ which, along this section, plays the rôle of an essential boundary between the two Alpine belts; evaluating the importance of the offset would thus become quite impossible: considerations of the tectonics of South Transcaucasia will enable us to decide the question.

III THE "A" LINEAMENT OR WEST PONTIC LINEAMENT according to FOURQUIN, (1977); FOURQUIN in BERGOUGNAN and FOURQUIN, (1980) and in ADAMAYA et al., (1980)

a) Hercynian paleotectonic rôle (Fig. 3)

The Paleozoic series that are not or only slightly metamorphosed and are overlain by Carboniferous molasses above the Sudetian unconformity, constitute the "North Variscan belt" of West Pontides. This belt extends westwards into the Balkan Srednagora and Stara Planina, then into South Carpathian Mountains, and through the Supergetic, Getic and Danubian units. At present the active IZNIK-HAVZA section of the NAFZ is the southern boundary of this Hercynian belt; as a matter of fact, along this axis it gives way to the "South Variscan belt" which was metamorphosed at different periods and intruded by granites. It contains ophiolites, which gives it "Inner-Hercynian" features that can be compared with those of Serbo-Macedonian and Rhodopian massifs. Therefore the present active IZNIK-HAVZA section closely follows a much more ancient lineament separating the inner and outer Variscan zones.

b) Alpine paleotectonic rôle

During the Mesozoic, this lineament marked the boundaries between Pontic terranes with different paleogeographical features (Fig. 4).

In the South there are thin, finely detrital, lacunar neritic sediments with no volcanism at all (Western inner Pontides).

In the North there are sediments that are in places pelagic and detrital and including thick volcanic intercalations (Western outer Pontides) of late Cretaceous age. All along the lineament itself, a narrow furrow existed that remained active from the Jurassic to the Cretaceous, creating a subsiding pelagic sequence with Kimmeridgian submarine volcanic intercalations and without any unconformity. According to Y. YILMAZ (1980) the occurrence of some ophiolites in this area should be considered as the sign of a minor branch of the Tethyan ocean.

c) Tarditectonic rôle (Miocene)

The IZNIK-HAVZA section probably played no significant rôle as a strike-slip fault in the Miocene. However, the paleogeographic survey of the Galatian andesitic massif shows that this volcanic arc was active in latest Cretaceous and in the early Tertiary namely in continental environment in its inner part, and an off-shore environment to the north. The facies change takes place in the area of those areas of crustal weakness which are also often used as a pathway by volcanic lavas (FOURQUIN, 1975).

The IZNIK-HAVZA section in this area is therefore in a back-arc position, and more precisely in the position of a boundary-flexure between continental and marine volcanism.

d) Conclusion

The IZNIK-HAVZA section of the NAFZ closely follows a more ancient longitudinal lineament which played an essential rôle in the building of Hercynian and Alpine ranges in those areas. Its near-by western extension is the Maritsa which sets apart the Rhodope and Srednagora massifs and seems to play a similar rôle in the building of the Balkan range. Farther West, it can at present not be determined whether the IZNIK-HAVZA Fault is similar to the lineament (the Kraistides linear zone) of Bulgarian writers which separates the Serbo-Macedonian Massif from the Getic Massif (South Carpathian mountains). This lineament has the same paleotectonic significance, because the longitudinal translations of the Serbo-Macedonian Massif along this lineament have disrupted their primary relationships.

IV THE "B" LINEAMENT OR MESO-EUROPEAN LINEAMENT

From middle and East Anatolia, the NAFZ suddenly changes its course and deviates by 30° South-East, after the Havza meridian, to Erzincan and Varto. It meets and closely follows another lineament that can be seen from Dobrudja to Zagros. Therefore, the NAFZ consists of <u>two distinct sections.</u>

Let us examine first the North-West part of this other lineament, from the Dobrudja to the North Anatolian ophiolitic suture, then its South-East part along the suture itself.

1) The North-West part
   according to FOURQUIN, (1977); FOURQUIN in BERGOUGNAN and FOURQUIN (1980) and in ADAMAYA et al. (1980)

a) The offset of the Hercynian range

The Hercynian framework reveals a <u>left motion</u> along the "B" lineament since the same Pontic zonation of outer and inner zones exists further north in the Dobrudja, Crimea (outer zone) and Greater Caucasus (outer and inner zones). If we take into account the Black Sea opening and the later right slip we can estimate the offset to be 450 kms long (actual offset = apparent offset + right offset - Black Sea opening).

That works for the Mesozoic paleogeographical zones directly connected with Hercynian structures. The Perimoesian furrow (Triassic to early Cretaceous) in the South Carpathian Mountains (Danubian terrains) and in the Balkans (Kotel area) sees its counterpart slipped northwards in the Dobrudjan, Crimean, Caucasian Periscythian furrow.

The offset of the Hercynian and Mesozoic zones can be accounted for through a late working of the "B" lineament either as a strike-slip fault at eo-Cretaceous or as a tardi-Hercynian strike-slip fault, then as a transform fault until the eo-Cretaceous. In the first case, this motion would be the result of the left lateral drift of Africa during the Jurassic and of the blockage of this drift in early Cretaceous because of the collision between Arabia and Europe in the East. In this case, there would be no connection at all between the East and West Pontides until the eo-Cretaceous, and Gresten

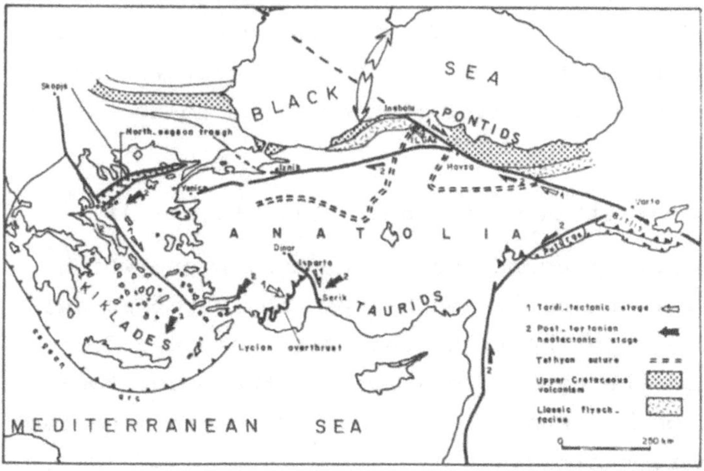

Fig.: 5. Tardi-tectonic and neotectonic stages (Tertiary and Quaternary) of the two branches of the North-Anatolian Fault in the general stress-pattern of the Middle-East (after BERGOUGNAN, FOURQUIN and RICOU, 1978 modified)

Liassic in Inebolu in particular would be in no way similar to the Gümüşhane Liassic terrane (East Pontides). In the second case, the offset would be the result of the East-West stresses which produced the Ural range in Permian times.

Another argument supporting the activity of the "B" lineament at that same period is the magmatic activity of the Oslo graben connected with a faulting pattern having the same direction. During the Mesozoic, this lineament might act as a transform fault. Lastly, it cannot be ruled out that both mechanisms worked.

b) East boundary of Tauric thrusts over the Pontic west margin (Fig. 4)

During the "Laramian stage", some Mediotethyan material (including ultrabasites, submarine lavas with Halobia-bearing limestones, radiolarites and Cretaceous formations) was thrust over the European Pontic margin (Bailey and Mc Callien's "Ankara" mélange or "coloured" mélange). This material is found along with metamorphic units of Tauric affinities (metamorphic Daday and Ilgaz massifs). This significant overthrust is clearly limited in the East by the "B" lineament which very likely channelled this translation movement to the North/North-West.

c) Tarditectonic right strike-slip fault (Fig. 5, st. 1)

This slip in the opposite course of the Hercynian one, which must have begun after the collision between the European and African blocks, offsets on the right the thick volcanic detrital formations of the upper Cretaceous in the outer Pontides as well as probably the thick flyschoid volcanic series of Liassic in the Gresten-facies.

2) The South-East part. The Mesoeuropean lineament: a channel for the Tethyan opening in the Triassic according to BERGOUGNAN in BERGOUGNAN and FOURQUIN (1980) and in ADAMYA et al. (1980) and Fig. 4

The fact that the Mesoeuropean lineament should "rendez-vous" the Erzincan ophiolites is not a mere chance and it accounts for some authors' errors, especially the one of mistaking the NAFZ motion with that of the Tethyan break.

We have underlined the confusion that makes some writers think the NAFZ is a boundary of the two Alpine belts. If, in the first part of this report, we have shown that this assertion is not true for the section located West of Erzincan (where the NAFZ cuts through and into the European framework), it nevertheless remains valid for the section located East to the meridian of this town, where the NAFZ follows the major Tethyan break to Zagros (cf. further down).

Another complication comes from the ophiolitic trail (different from the previous one) which forks off the major suture precisely at Erzincan, then runs along the South border of the East Pontic and Minor Caucasian frameworks (LOMIZE, 1970), until it ends in the Lyssogorsk pass area (Erzincan-Sevan ophiolites). In no case could those ophiolites play the rôle of a major suture between the European and African blocks: whether they come from the major suture itself and have been thrust to the North-East over the European block, or represent a different ophiolitic corridor, coming from a minor break or a minor branch of the main opening. As a matter of fact, the Mesozoic sedimentary sequences that are close to them (like the one of Djoulfa which forms the sedimentary cover of the Daralagoez massif in South Transcaucasia) have here again European affinities, as much because of their facies as of the faunas they contain (BONNET, 1922, 1947). This is so despite the fact that they lie on a Paleozoic basement of the Gondwano-European type namely the so-called "Gondwano-European" appendix (Fig. 4). To find out where the two blocks collided, it is necessary to follow the Zagros lineament (the "B" lineament) where the South-West margin of Central Iran (Sanandaj - Sirjan area) represents the European margin facing the Zagros Arab - African range.

For this reason, evaluating the offset of the NAFZ based on the right offset of the South boundary of Europe in the Erzincan area is quite impossible. If the southern border of the Tokat Massif in the West represents the southern fringe of Europe as seems likely, the same does not apply to the edge of the Pontic range in the East which runs along the Erzincan - Sevan intra-European ophiolitic trail. Therefore the possibility of a 100 km long offset suggested by one of the present authors and others (BERGOUGNAN 1975, BERGOUGNAN in BERGOUGNAN and FOURQUIN, 1976; FOURQUIN, 1977; BERGOUGNAN et al., 1978; ŞENGÖR, 1979) is low.

3) Conclusion

Along its HAVZA-VARTO section, the NAFZ follows the Mesoeuropean "B" lineament whose paleo- and later, tarditectonic rôle can be measured as follows:

a) Paleotectonic rôle

In the Triassic time, it channels the Erzincan Tethyan opening as far as the Zagros range.

Before the collision of the European and African blocks (therefore before the "Laramian" or Maestrichtian stage), it offsets on the left the Hercynian structures for over 450 kms, if we take into account the later Black Sea opening. Just before and during the collision, it limited or even channelled the Tauric overthrusts to the North/North-West.

b) Tarditectonic rôle
   Between Maastrichtian and late Miocene, it offsets on the right the Pontic paleogeographical zones

V INDIVIDUALISATION AND UNITY OF THE NAFZ

The neotectonic slip, i.e. the presently operative tectonic movement is shown in Fig. 5 (st. 2). The NAFZ individualisation took place recently when the fault abandoned the HAVZA-INEBOLU section of the Mesoeuropean "B" lineament, and followed the West Pontic "A" lineament along its HAVZA-IZNIK section as a strike-slip fault, channelling the movement to the West of Anatolia. At the same time, the HAVZA-VARTO section of the "B" lineament remains active and increases its offset. This pattern clearly points to two-fold slip of the NAFZ (FOURQUIN, 1977).

Dating the beginning of the slip along both sections is not easy. From late Miocene time, the creation of the Aegean trough running parallel to the expulsion course of Anatolia to the West is the expected result of the new stress pattern. The same holds true for the westward overthrust along the Dinar-Isparta-Serik lineament (post-Tortonian period, see AKBULUT and POISSON 1977), after it had acted as a right strike slip or transform fault, channelling the Lycian overthrusts (age between Langhian and Tortonian, POISSON, 1977) in the earlier tarditectonic stage (BERGOUGNAN, FOURQUIN, RICOU 1978).

According to ŞENGÖR (1979), other structures developed by these movements can be recognized in the basins resulting from an overstretching where the three neoplates (Eurasia, Anatolia, Arabia, cut out by the East Anatolian fault and the "B" lineament) meet: this overstretching regime produces morphologic depressions filled with Neogene and more recent sediments, whose age and thickness, if we knew them, could provide us with very helpful data on the relative movement of these plates; that is what ŞENGÖR calls the "ova regime".

Last, the northward Arabian push that has been stronger since the Red Sea opening and the consequent offset of the Levant Faulting Zone and the East Anatolian Fault (ARPAT and SAROGLU, 1972, 1975) seem to be the driving factors of the neotectonic regime (BRUNN, 1976; BERGOUGNAN, FOURQUIN, RICOU, 1978). They bring about a crush of the edge of the Arabic Plate on its North West front along the East Anatolian Fault and on its North East front along the "B" lineament. The Arabic underthrust under Anatolia is part of this system and is respectively covered by the Pütürge-Bitlis Massifs.

VI CONCLUSION

1) The NAFZ unity only lies in the westward lateral drifting of the Anatolian Plate, a drifting that appeared in late Neogene, probably after the Tortonian. It cannot be defined by one original structural feature, i.e. "a neotectonic accident" appearing at that epoch because this fault follows zones of weakness already existing, namely the traces of the two ancient lineaments that had already deeply broken up Europe. The new stresses re-used the former mosaic in the best possible way.

2) Under these conditions, the offset notion has a meaning only if it is related to a clearly defined period of time setting apart phenomena affecting at least the whole domain concerned. Strictly speaking, the NAFZ's own offset in its uniform course is the neotectonic offset which extends from the Tortonian to the present time. On the contrary, the offset - a more important one - affecting the Mesozoic paleogeographic zones of the volcanogenic upper Cretaceous or of the Gresten Lias, extends from the Maestrichtian collision down to present times. It is 300 km long; tarditectonic offset and neotectonic offset together.

3) "The paleotectonic, tarditectonic, neotectonic differences give an overriding importance to the tectonic stage" (AUBOUIN, 1973) which is represented here by the late Cretaceous collision of the European and African blocks. The different stages are merely successive stages in this major phenomenon: the Alpine orogeny around the Tethyan ocean. Therefore, the pre-

sent NAFZ offset is only a late episode in the history of the relative drifts of Europe in relation to Africa and Arabia, a drift the fault zone has kept recording since the Tethyan opening, since the dawn of the Alpine orogeny.

ACKNOWLEDGMENTS. - We are deeply thankful to the Turkish Departments of Geology and Mining in which we conducted our research successfully over several years, and particularly to the former head of this Institute Dr. SADRETTIN ALPAN.
Thanks to a kind invitation from the Academy of Sciences of the USSR, one of us (Henry BERGOUGNAN) was allowed to visit the Caucasian areas connected with the Anatolian surveys; he is deeply grateful to the geological departments of the Sciences Academies of the Republics of Armenia, Azerbaidjan and Georgia for their kind welcome and for the fruitful contacts he established with them. He is particularly thankful to Dr. A.L. KNIPPER of the Moscow Academy of Sciences who accompanied him in the field and let him benefit from his expert knowledge of the geology of these regions.
Our work was successful because of the generous hospitality extended by the departments headed by Prof. G. ATAMAN at Hacettepe University in Ankara.
Our research could not have been successful without the helpful and indispensable support of "West Alpine Asia" ERS N° 9 in the C.N.R.S.
Last, but not least we warmly thank Prof. BERNARD, the President of Reims University for sponsorship of our team of geological studies (Groupe d'Etudes Géologiques de l'Université de Reims) and for financial support.

Abbreviations used in the Figures 3 and 4

C.M.: South Carpathian
G: Getic zone
C.F.: front range of the Caucasus
P.M. Moesian platform
P.S.: Scythian platform
P.B.: Fore-Balkan
S.SP.: Stara-Planina and Srednagora zones
R.: Rhodop massif
S.M.M.: Serbo-Macedonian massif
P.W.E.: West outer Pontides
P.W.T.: West inner Pontides
P.E.E. - PC: East outer Pontides and Minor Caucasus

REFERENCES

ADAMYA S., LORDKIPANIDZE M.B., ZAKARIADZE G.S. (1977).- Evolution of an active continental margin as exemplified by the alpine history of the Caucasus. Tectonophysics, 40, 183-199.

ADAMYA S., BERGOUGNAN H., FOURQUIN Cl., HAGHIPOUR A., LORDKIPANIDZE M., ÖZGÜL N., RICOU L.E. and ZAKARIADZE C. (1980).- The alpine middle east between the Aegean and the Oman traverses. Colloque C5 "Chaînes alpines issues de la Téthys" XXVIème Congrès Géologique International, Paris.

AKBULUT A. (1977).- Etude géologique au sud d'Eğridir. Thèse 3ème Cycle, Orsay.

ARNI P. (1939).- Relations entre la structure régionale et les gisements minéraux et pétrolifères de l'Anatolie. M.T.A. Mecm. 4/2.

ARPAT E. and SAROĞLU F. (1972).- The East Anatolian Fault system : thoughts on its development. Bull. Mineral. Res. Explor. Irst., Ankara, 78, 33-39.

ARPAT E. and SAROĞLU F. (1975).- Türkiye'deki basi önemli genç tektonik olaylar. Türkiye Jeoloji Kurumu Bülteni, 18, 91-101.

ATAMAN G., BUKET E. and ÇAPAN Ü.Z. (1975).- Could North Anatolian Fault Zone be a paleo-Benioff Zone ? Bull. Mineral Res. Explor. Inst., Ankara, 84, 97-102.

AUBOUIN J. (1973).- Des tectoniques superposées et de leur signification par rapport aux modèles géophysiques : l'exemple des Dinarides ; paléotectonique, tectonique, tarditectonique, néotectonique. Bull. Soc. Géol., Fr., (7), XV, p. 426-460.

AZIZBEKOV S.A. and RUSTAMOV M.I. (1973).- Some characteristic features of ophiolites of the Caucasus ; the Near and Middle East. Intern. Symposium "Ophiolites in the Earth's Crust", Moscou.

BAILEY E.B. and Mc CALLIEN W.J. (1953).- Serpentine lavas, the Ankara Melange and the Anatolian thrust. Trans. Roy. Soc. Edim., 62, 2, n° 11.

BALAVADZE B.K. et *al.* (1966).- Tectonics of the Black Sea and Sea of Azov area. Geotectonics, n° 4, p. 238-246.

BASSOULLET J.P., BERGOUGNAN H., ENAY R. (1975).- Répartition des faunes et faciès liasiques dans l'Est de la Turquie (région du Haut-Euphrate). C.R. Acad. Sci., t. 280, sér. D , p. 583-586.

BERGOUGNAN H. (1975).- Relations entre les édifices pontique et taurique dans le Nord-Est de l'Anatolie. Bull. Soc. Géol. Fr., 7ème sér., XVII, p. 1045-1057.

BERGOUGNAN H. (1976).- Structure de la chaîne pontique dans le Haut-Kelkit (Nord-Est de l'Anatolie. Bull. Soc. Géol. Fr.,18:675-686.

BERGOUGNAN H., FOURQUIN Cl. (1976).- Faille nord-anatolienne et cassure téthysienne. IVème Réunion Annuelle des Sciences de la Terre, Paris, p. 49.

BERGOUGNAN H., FOURQUIN Cl., RICOU L.E. (1978).- Les deux tronçons et le double jeu de la faille nord-anatolienne dans la tectonique récente du Moyen-Orient. C.R. Acad. Sci. Paris, t. 287, p. 1183-1186.

BERGOUGNAN H., FOURQUIN Cl. (1980).- Un ensemble d'éléments communs à une marge active alpine des Carpathes méridionales à l'Iran central : le domaine irano-balkanique. Bull. Soc. Géol. France (7), t. XXII, n° 1.

BLUMENTHAL M.M. (1945).- Die Kelkit-Dislokation südlich Niksar und ihre tektonische Rolle. M.T.A. Mecm. 10/2, p. 372-386.

BLUMENTHAL M.M. (1948).- Un aperçu de la géologie des chaînes nordanatoliennes entre l'Ova de Bolu et le Kizilirmak inférieur. M.T.A. publ. série B, N° 13.

BONCEV E. (1974).- Generals of the Kraïstide - Vardaride lineament. In Mahel M. (ed.), Tectonics of the Carpathian Balkan Regions, 330-332, Dionyz Stur, Bratislava.

BONNET P. (1922).- Sur le Jurassique de la Transcaucasis méridionale. C.R. Somm. Soc. Géol. Fr., N° 17, p. 207-209.

BONNET P. (1947).- Description géologique de la Transcaucasie méridionale (chaînes de l'Araxe moyen). Mém. Soc. Géol. Fr., N° 53.

BRUNN J.H. (1976).- L'arc concave zagro-taurique et les arcs convexes taurique et égéen : collision et arcs induits. Bull. Soc. Géol. Fr., (7), t. XVIII, N° 2, p. 553-567.

CHEKUNOV A.V. (1973).- Main stages in geotectonic development of Northern Black Sea region. Internat. Geology Rev., vol. 16, N° 5, p. 495-508.

EGERAN N. (1945).- Contribution apportée aux connaissances sur la tectonique alpine par les études géologiques et tectoniques effectuées récemment en Turquie. M.T.A. Mecm 10/2.

ENAY R. (1972).- Paléobiogéographie des Ammonites du Jurassique terminal (Tithonique/Volgien/Portlandien s.l.) et mobilité continentale. Geobios, N° 5, fasc. 4, p. 355-407, 13 fig., Lyon.

ENAY R. (1976).- Faunes anatoliennes (Ammonitina, Jurassique) et domaines biogéographiques nord et sud téthysiens. Bull. Soc. Géol. Fr. (7), t. XVIII, N° 2, p. 533-541, et coll. intern. C.N.R.S., Paris, N° 244, p. 337-345.

FOURQUIN Cl. (1975).- L'Anatolie du NW marge sud du continent européen, histoire paléogéographique, tectonique et magmatique. Bull. Soc. Géol. Fr., (7), t. XVII, p. 1058-1070.

FOURQUIN Cl. (1977).- Sur un dispositif en deux tronçons de la faille nord-anatolienne ; implications sur les grandes structures pontiques autochtones et allochtones (Nappe de Cankiri), relation avec le linéament méso-européen. 5ème R.A.S.T., Rennes, p. 235.

KETIN I. (1966).- Tectonic units of Anatolia. Bull. Mineral. Res. Explor. Inst. Ankara, 66, 23-34.

KETIN I. (1968).- Relations between general tectonic features and the main earthquake regions of Turkey. Bull. of Min. Res. Explor. Inst. Turkey, 71, p. 63-67.

KETIN I. (1969).- Über die nordanatolische Horizontalverschiebung. Bull. Mineral Res. Explor. Inst., Ankara, 72, 1-28.

KNIPPER A.L. (1971a).- Constitution and age of serpentinite melange in the Lesser Caucasus. Geotectonics N° 5, p. 275-282.

KNIPPER A.L. (1971b).- Development of Serpentinite Melange in the Lesser Caucasus. Geotectonics N° 6, p. 384-396.

KRAUSS E.C. (1958).- Die Orogene Ostanatoliens und ihre Schubweiten. Bull. Min. Res. Explor. Inst. Turkey, N° 51.

LOMIZE M. (1970).- La place des ophiolites dans la structure tectonique de l'Anatolie orientale et la Transcaucasie. Izvesti ja Vyssikh Ucebnykh Zavedenij Geologija i Razvedka N° 11, p. 32-40.

NOWACK E. (1928).- Die wichtigsten Ergebnisse meiner anatolischen Reisen. Z dt. geol. Ges. 80, 304-312.

OZKOCAK O. (1969).- Etude géologique du massif ultrabasique d'Orhaneli et sa proche bordure. Thèse Doct. Univ. Paris.

PAMIR H.N. (1950).- Les seismes en Asie Mineure entre 1939 et 1944. La cicatrice Nord-Anatolienne. Proc. Int. Geol. Congr. Lond. XIII, 214-218.

PAVONI N. (1961).- Die Nordanatolische Horizontalverschiebung. Geol. Rdsch. 51, 122-139.

PINAR N. and LAHN E. (1955).- Nouvelles considérations sur la tectonique de l'Anatolie (Turquie, Asie Mineure). Bull. Soc. Géol. Fr., (6), V, p. 11-34.

POISSON A. (1977).- Recherches géologiques dans les Taurides occidentales. Thèse Orsay, N° 1902.

RICOU L.E. (1978).- Induction d'arcs et de chaînes rectilignes par décrochements : le Zagros et les Dinaro-Hellénides. C.R. Acad. Sci. Paris, t. 286, série D, p. 1763-1765.

SALOMON-CALVI W. (1940).- Les tremblements de terre d'Erzincan du 21 novembre et du 27 décembre 1939. M.T.A. Mecm. 1/18, p. 27-30.

ŞENGÖR A.M.C. (1979).- The North Anatolian transform fault : its age, offset and tectonic significance. Jl. Geol. Soc. Lond. Vol. 136, p. 269-282.

SEYMEN I. (1975).- Kelkit vadisi kesiminde Kuzey Anadolu Fay Zonunun tektonik özelliği. ITÜMaden Fak Yay., Istambul, 192 p.

SHIKHALIBEYLI E.SH. (1964).- Geologicheskoye stroyeniye i istoriya tektonicheskogo razvitiya vostochnoy chasti Malogo Kavkaza (Geology and History of tectonic development in the east of the Lesser Caucasus), vol. 1, Izd. Akad. nauk AzerbSSR.

SHIKHALIBEYLI E.SH. (1967).- Geologicheskoye stroyeniye i istoriya tektonicheskogo razvitiya vostochnoy chasti Malogo Kavkara (Geology History of tectonic development in the east of the Lesser Caucasus), vol. 3, Izd. Akad. nauk AzerbSSR.

STILLE H. (1953).- Der geotektonische Werdegang der Karpaten. Beih. Geol. Jb. 8, 239 p.

TATAR Y. (1975).- Tectonic structures along the North Anatolian Fault Zone, north-east of Refahiye (Erzincan). Tectonophysics, 29, 401-410.

TOKAY M. (1973).- Kuzey Anadolu Fay Zonunun Gerede ile Ilgaz arasindaki kisminda jeolojik gözlemler. Kuzey Anadolu Fayi ve Deprem Kusagi Simpozyumu. MTA Yayinl., Ankara, 12-19.

TOKEL S. (1973).- Mesozoic and Tertiairy development of the Eastern Pontids and its relation to the North Anatolian Seismic Zone. Congr. Earth. Sc. 50 th. Ann. Turk. Rep. Ankara.

YILMAZ Y., GÖZÜBAL A.M., TÜYSÜZ D. (1980).- Geological evolution of the region around the North Anatolian Fault Zone between Iznik and Bolu. Interdisciplinary conference on Earthquake prediction research in the North Anatolian Fault Zone. Istambul.

Structural Evidence for Left-Lateral Displacement on the North Anatolian Fault Zone During the Plio-Pleistocene

P. L. Hancock and A. A. Barka

Department of Geology, University of Bristol, Queen's Building, University Walk, Bristol BS8 1TR, UK

The approximately 1200 km arc of the North Anatolian fault zone is generally interpreted as an intracontinental transform boundary separating the Black Sea and Anatolian plates (McKenzie 1972, Dewey et al. 1973, Sengör 1979). The offset of geomorphic and man-made features (Allen 1969, Ambraseys 1970) together with fault plane solutions (McKenzie 1972) indicates that present-day displacements on the fault zone are right-lateral. Offset geological features also indicate that there has been a substantial cumulative displacement in a right-lateral sense since the Miocene (Sengör 1979). However, the possibility that there may have been one or more reversals of motion cannot be discounted, because as Mercier (1977) has shown, such reversals are known from the Aegean region which is in continuity with the Anatolian plate.

During a study of all mesoscopic-scale fracture systems cutting Neogene-Quaternary continental clastic sediments in intermontane basins within the North Anatolian fault zone between Çerkeş and Erbaa, we have discovered small faults and joints whose orientations are consistent with their development as accommodation structures in an approximately E-W trending left-lateral shear zone. The synsedimentary and post-sedimentary structures which were generated by secondary directions of compression or extension oblique to the fault zone include: conjugate reverse faults striking NW; conjugate normal faults and shear joints striking NE; and conjugate vertical joints enclosing an acute angle about a NE trending bisector. A more complete account of these meso-fractures, many of which are cut by those related to right-lateral shear, is given elsewhere (Hancock & Barka in press).

Mesofractures related to left-lateral shear are restricted to the Pontus Formation (Irrlitz 1972) and its lateral equivalents which are dated as Late Miocene - Early Pleistocene. Because they have not been observed in Upper Pleistocene-Holocene sediments it is concluded that there has not been effective left-lateral shear since the Early Pleistocene. Synsedimentary and post-sedimentary mesofractures interpreted as being related to right-lateral shear occur throughout the Neogene-Holocene successions. Mesofractures from which left-lateral shear on the fault zone has been inferred are more abundant in the Lower Pontus Formation (U. Miocene - L. Pliocene) and in the western basins of Cerkes-Ilgaz, Tosya and Kargi than they are in the Upper Pontus Formation (U. Pliocene - L. Pleistocene), or the eastern basins of Havza and Tasova-Erbaa. Thus left-lateral shear may have been more effective before the Late Pliocene and along the more western parts of the fault zone which are closer to the Aegean.

Because both synsedimentary and post-sedimentary mesofaults are developed we conclude that displacements on the North Anatolian fault zone occurred intermittently during the Pliocene and early Pleistocene. Their wide distribution suggests that many of them originated during regional reversals of dominantly right-lateral shear, but some of them may be related to local factors such as anti-Riedel overlaps of strike-slip faults within the main fault zone, or directions of stress release following the initiation of other structures.

REFERENCES

Allen,C.R. 1969. Active faulting in northern Turkey. Contr. No.1577, Div. Geol. Sci., Calif. Inst. Tech. 1-32.

Ambraseys,N.N. 1970. Some characteristic features of the Anatolian fault zone. Tectonophysics 9, 143-165.

Dewey,J.F., Pitman,W.C., Ryan,W.B.F. & Bonnin,J. 1973. Plate tectonics and the evolution of the Alpine system. Bull. geol. Soc. Am. 84, 3137-3180.

Hancock,P.L. & Barka,A.A. 1981. Evidence for a Plio-Pleistocene reversal of displacement on the North Anatolian fault zone. Nature, Lond., 286, 591-594.

Irrlitz,W. 1972. Lithostratigraphie und tektonische entwicklung des Neogene in Nordostanatolien. Beih. geol. Jahrb. 120, 1-111.

McKenzie,D. 1972. Active tectonics of the Mediterranean region. Geophys. J.R. astr. Soc. 30, 109-185.

Mercier,J. 1977. L'arc égeén, une bordure déformée de la plaque eurasiatique; réflexions sur un exemple d'étude néotectonique. Bull. Soc. géol. Fr., 7, Ser.19, 663-672.

Sengör,A.M.C. 1979. The North Anatolian transform fault: its age, offset and tectonic significance. J. geol. Soc. Lond. 136, 269-282.

Geology of an Area in and Around the Northern Anatolian Transform Fault Zone Between Bolu and Akyazi

Y. Yilmaz, A. M. Gözübol, O. Tüysüz

Faculty of Earth Sciences, University of Istanbul, Department of Geology, Istanbul, Turkey

Summary

With respect to the northern Anatolian Transform Fault zone there are two groups of rocks of contrasting environment; the Istanbul-Zonguldak tectonic unit in the north, and the Sakarya Continent in the south.
Palinspastic reconstructions of the northern sequence of the Sakarya Continent indicate a Mesozoic continental-margin and an associated ocean-basin regime which was established during the Mesozoic and Early Tertiary time.
The sequence in the north resulted from tectonic amalgamation of an ocean basin and a continent. The latter rests on top of the former on a wide zone of intense cataclasis.
These two assemblages, the northern and the southern, were brought together prior to the late Eocene Epoch. The geological events which brought these tectonic units together do not bear any genetic relationship to the formation of the Northern Anatolian Fault which was initiated in Middle Miocene.

INTRODUCTION

This paper deals with the regional geological characteristics of an area in and around the Northern Anatolian Transform Fault Zone between the towns of Bolu in the east and Akyazi in the west (Fig. 1). Implications of the lithological sequences and regional tectonic development will be discussed. In the light of the demonstrated stratigraphic, structural and regional geological relationship a possible reconstruction of the tectonic units will be given and a plate tectonic model to explain the geological evolution will be proposed.

An area of approximately 14.000 sq km in the region of Bolu, Mudurnu and Akyazi in the Northwestern Anatolia has been studied. The area to the north of the Northern Anatolian Fault Zone forms the southern zone of the Western Pontide tectonic province, and the portion in the south corresponds to the northern part of the Sakarya continent. The history of depo-

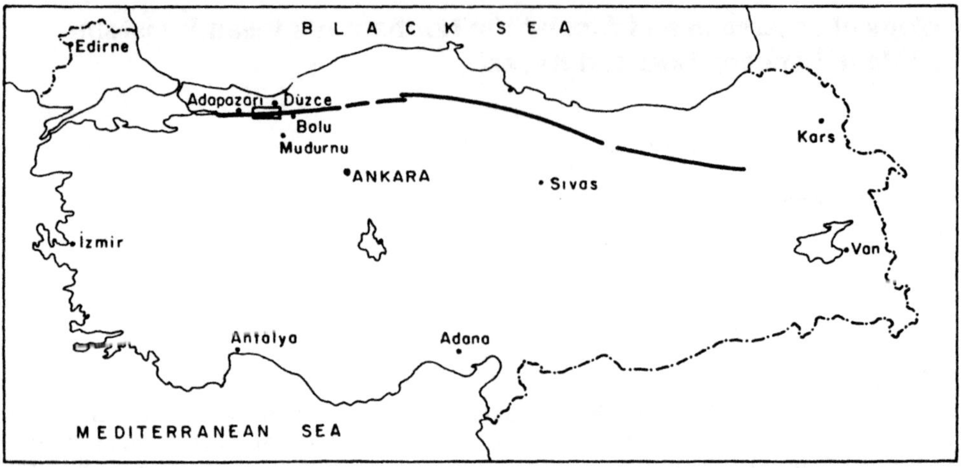

Fig.1 Location map

sition, deformation and metamorphism in the area is presented here, based on the results of geological mapping on a scale of 1:25.000 and on petrographic analyses.

The North Anatolian Transform Fault Zone within the area studied extends in an east-westerly direction. Two groups of rocks of contrasting genesis exist in the area of this fault zone. Throughout this paper the rocks will be referred respectively as the northern assemblage and the southern assemblage.

In the north there is a fairly complete ophiolite association at the base of the sequence suggesting the former existence of an ocean basin. The age of emplacement of the ophiolite has long been regarded as pre-Devonian because of the presence of Paleozoic rocks lying on top of the ophiolite. However, our field investigations have shown that the contact, whereever it is exposed, is tectonic in nature. The contact between the two is not a single plane but a wide zone of intense cataclasis or a zone of tectonic transition.

The slighty metamorphosed Paleozoic unit on top of the ophiolite is unconformably overlain by a melange association in which blocks of pelagic sedimentary rocks indicate a Late Cretaceous age. These deep-sea sedimentary rocks at the base are similar to the uppermost layer of the ophiolite.

REGIONAL SETTING

The area under consideration covers the Almacik mountains, the Mudurnu river valley and the town of Mudurnu and its surroundings. The Northern Anatolian Transform Fault Zone follows the Mudurnu River valley in a roughly east-westerly direction. The Almacik mountains are north of the valley and the town of Mudurnu lies to the south. The southern region is referred to as the Sakarya Continent (ŞENGÖR and YILMAZ, 1980), and the northern region as the Istanbul-Zonguldak tectonic unit or the Western Pontides.

There are some distinct lithological facies changes in the sequences in the north and in the south of the fault zone (Fig. 2). In contrast to the relatively less disturbed platform sediments in the south, there are ophiolitic rocks, a melange association and metasedimentary sequence in the north. This suggests that the Sakarya Continent behaved as an essentially stable tectonic unit while intense tectonism occurred in the north during Late Mesozoic and Early Tertiary.

The southern association with its continuous sedimentary sequence indicates a progressive opening of an ocean basin from a Liassic rupture to a north-facing Atlantic type of continental margin during the Late Cretaceous.
The destruction of the ocean ultimately resulted in the emplacement of remnants of an oceanic lithosphere on its southern margin as slabs or nappes of ophiolite.

The destruction of such an ocean basin was accomplished largely by subduction towards the north beneath the northern plate (Istanbul Zonguldak Paleozoic tectonic unit as described by SENGÖR and YILMAZ 1980). It seems likely that tectonic emplacement of the ophiolite association is also related to this process. Based on our field investigation we propose as a possible model for the emplacement of a chaotic assemblage the retrocharriage of slices of accretionary prisms during and after the subduction

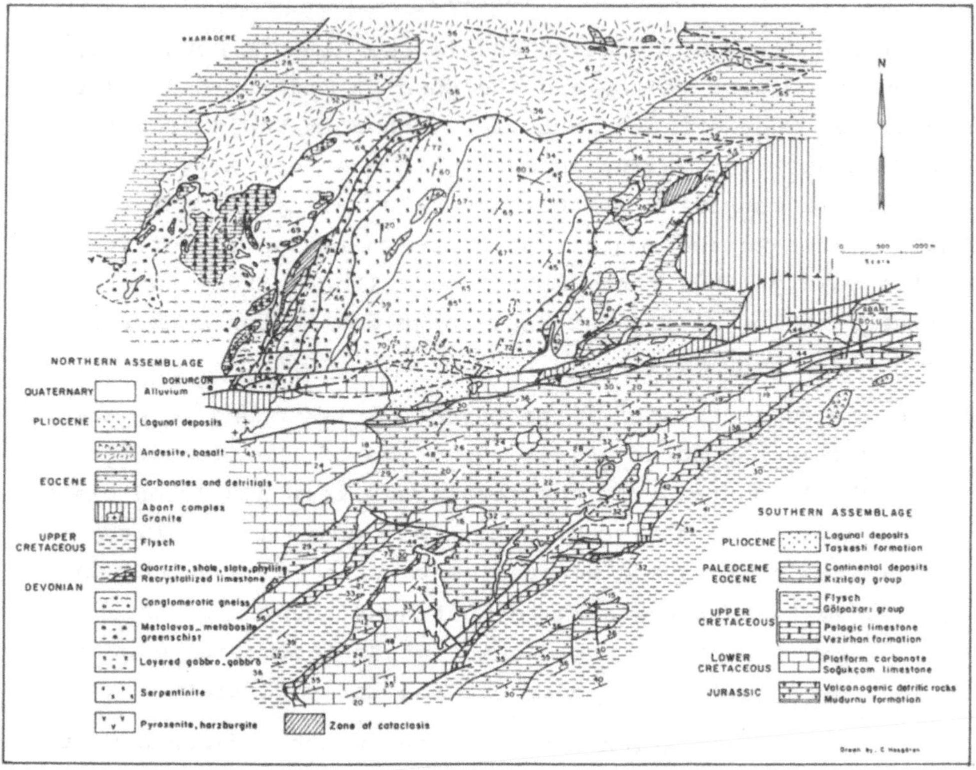

Fig.2 Geological map of the area in and around the Northern Anatolian fault zone between the lake Abant (Bolu town) and Dokurcun (Adapazarı)

THE SOUTHERN ASSEMBLAGE

The southern region is mainly characterized by a thick succession of shallow-water marine sediments of Jurassic age and a deep-water series of Cretaceous age (Fig. 3).

The sequence is generally underlain by a granitic basement. It forms a composite granite varying from deep-seated migmatitic varieties to a high-level, intrusive, post-tectonic granite showing typical igneous textures (YILMAZ, 1977). The age of the granite has been radiometrically determined by COGULU et al. (1965) as 300 m.y. which is further supported by the fact that it is unconformably overlain by Upper Permian arkosic conglomerates in the south of the Iznik-Mekece ridge.

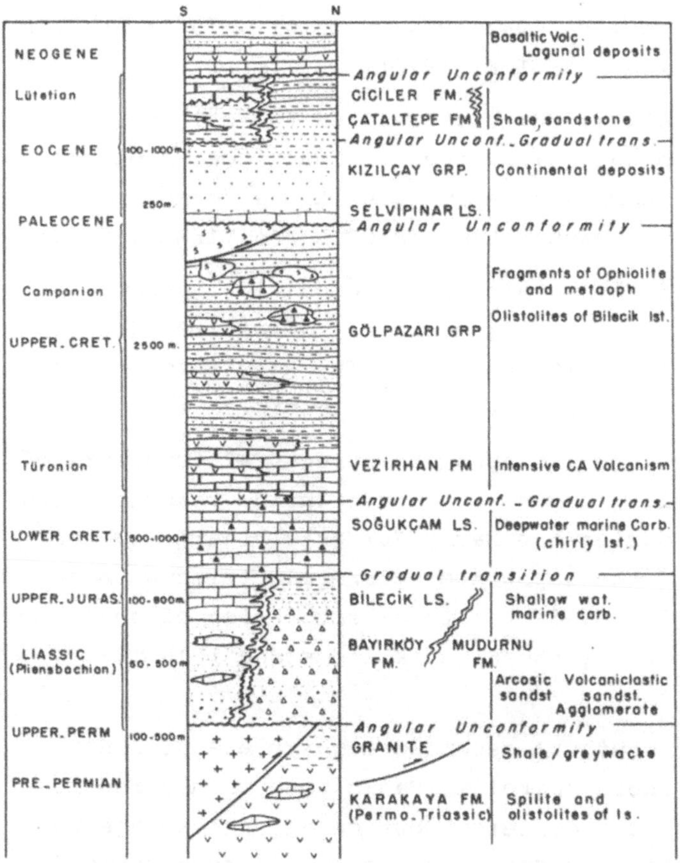

Fig.3 Generalized stratigraphic section of the southern assemblage

Locally the granite is in tectonic contact with a group of spilitic and associated rocks (YILMAZ, 1972) which is known by Turkish geologists as the Karakaya Formation (BINGÖL, 1976, BINGÖL et al. 1973). Within this unit there are some clastic horizons and widespread occurrences of slightly metamorphosed volcanic associations consisting of mafic lavas, tuffs and volcaniclastic rocks of broadly latest Permian-Triassic age.

The spilite encloses a large number of marine carbonate blocks and recrystallized limestones which indicate a Permian age. Some flysch-like sedimentary rocks consisting of fine to coarse-grained sandstones and shales exist within the Karakaya

Formation south of Gürsu town (Bursa), and Ankara, and indicate a Triassic age (ERK 1942; BINGÖL et al., 1973; AKYÜREK, personal communication; LEMOINE, 1978).

Recently ŞENGÖR et al. (1980) have related the origin of these rock groups to an ancient ocean, namely the Paleotethys. We believe that the Karakaya tectonic unit was formed in a backarc basin during Triassic time (ŞENGÖR and YILMAZ, 1980).

Prior to the Lias (Early Jurassic) the granitic basement was thrusted over the Karakaya in a northerly direction. The plane is well exposed in the Bilecik region. The thrust zone is overlain by unconformably Lower Jurassic sediments.

The geological events that shaped the structural framework of the southern assemblage began with a regionwide transgression over this basement during the Early Jurassic. Initially this transgression filled the topographic irregularities with an arkosic sandstone, followed by a thick succession of shallow-water marine carbonates ("the Bilecik limestone") (ALTINLI, 1973; SANER, 1977).

To the north of the southern assemblage, near the present location of the Northern Anatolian Fault Zone in Mudurnu town and neighbouring areas a different but coeval assemblage was deposited. The Jurassic sequence in Mudurnu is represented mainly by thick, fine to coarse elastic sediments and intercalated lava flows and pyroclastic rocks. This sequence is cut by altered diabase dykes (GÖZÜBOL, 1978). The volcanic rocks show alkaline and tholeiitic affinities. These shallow water marine units were deposited in a tectonically active environment with much the same condition as the lower Jurassic of the Eastern Pontides.

The Jurassic sequence of the southern assemblage passes transitionally into a thick sequence of interbedded siliceous limestones. At places, siliciclastic intervals and thinly-bedded clayey limestone indicate that the region gradually subsided. Eventually, a deep water marine environment was established during the Early Cretaceous (SANER, 1977; GÖZÜBOL, 1978).

A red, pelagic, thinly bedded fossiliferous limestone and clastic deposition followed in Early Cretaceous time. It is overlain by flysch-type sediments with shale-siltstone alternations. This unit yields a microfauna indicative of a Turonian Capanian age. Around Mekece exotic, shallow-water carbonate olistoliths are included within the flysch (SANER, 1977). Later, blueschist blocks, ophiolite and meta-ophiolite fragments were transported to the environment of deposition and lie in a matrix of highly contorted shale and sandstone. These blocks, forming a wildflysch, were possibly emplaced by sedimentary rather than tectonic processes. This relationship indicates that while the first period of active thrusting was going on in the northern region (the western Pontides), the emplacement of the scattered slices of ophiolites and associated turbidites on the previous succession occurred in the north. Following this event, major changes occurred in the regime of sedimentation with the flysch sequence being gradually and continously replaced by shallow-water marine carbonates during the Maastrichtian, and then by continental deposits during the Paleocene (Fig. 3). Relatively fast transition from a deep-sea to a continental environment without any apparent break indicates an uninterrupted deposition of regressive nature.

The continental deposits of the Paleocene series are overlain by shallow-water marine Eocene carbonates and detrital sediments. The Eocene sequence coveres some large ophiolites represented mainly by serpentinite and deep-sea sedimentary rocks such as radiolarian cherts and associated volcanic rocks of Late Cretaceous age. The ophiolite associations of the southern assemblage were emplaced in "Late Cretaceous" to Eocene time.

THE NORTHERN ASSEMBLAGE

The sequence in the north is different in many ways from the south. The north boundary of the fault zone is marked by thrusts and associated imbricate and "crush" zones (Fig. 2). In the region 4 tectonic units, each displaying distinct lithological associations and facies have been distinguished. The first period of active thrusting possibly occurred in Late Cretaceous time, and involved also the development of a melange association.

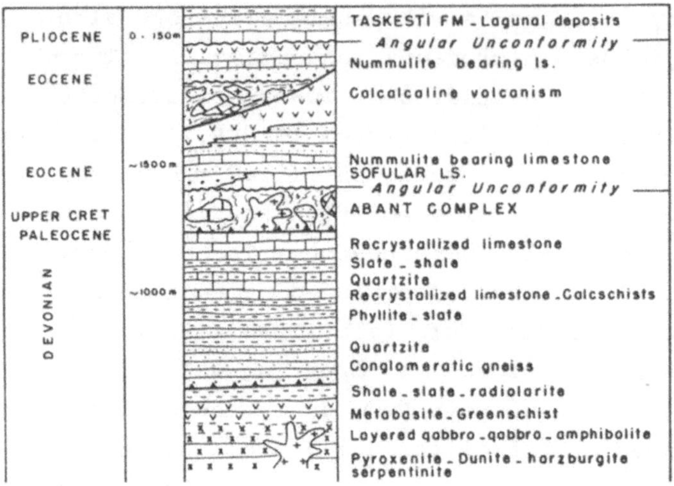

Fig.4 Generalized stratigraphic section of the northern assemblage

The northern assemblage as a whole is structurally complex. It is characterized by reverse-fault mosaics. The thrust region is strongly affected by Alpine deformation. Fig. 4 shows a generalized stratigraphic section of the northern assemblage. At the bottom of the northern assemblage is an ophiolite association which crops out primarily along the deeply eroded valley of the Mudurnu River (Fig. 2). In the ophiolite, deformation increases upward and its top is marked by a zone of intense shearing representing a major tectonic contact. Above this zone there is the lowest slice of the Paleozoic metasedimentary rocks.

A number of superposed ophiolite slices was generated during the thrusting (Fig. 2 and Fig. 5). Field mapping demonstrated that the relationships between the slices (ophiolite) are complicated. Along the Mudurnu valley, an anticline within the ophiolite is recognized. The wavelength of the anticline is about 20-30 km. The complete igneous stratigraphy, characteristic of an ophiolite complex, is recognizable in each slice. Internal deformation is locally intense.

At the core of the anticline there are ultramafic rocks such as dunite, harzburgite and locally developed chromite pods.

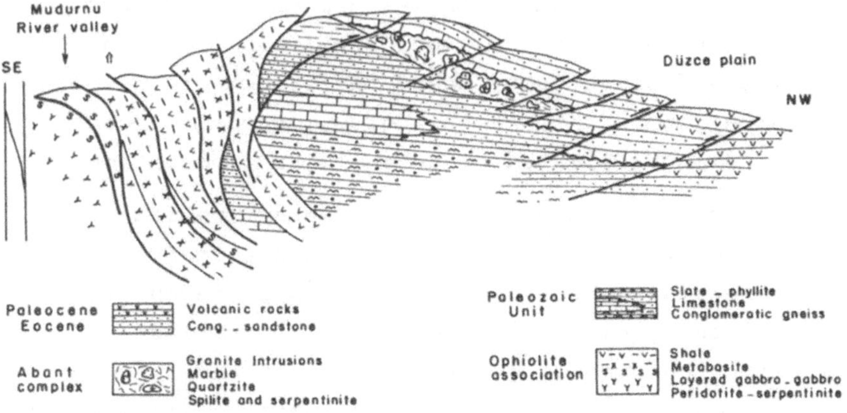

Fig. 5 Schematic cross section across the Almacik mountain showing the large-scale structure and interrelationships of the oceanic and continental assemblages

Locally, basal ultramafic lithologies are highly sheared. Chemical alteration affected the ultramafic rocks in different ways. To the north, along Akcaalan and Kursunlu streams gradual serpentinization of the ultramafic and associated gabbros can be traced for a few hundred meters. Serpentinized ultramafics have produced magnesite (e.g. to the north of Ilica village). Thus a talc-magnesite or magnesite-quartz paragenesis was formed.

The ultramafic rocks grade to layered gabbros through a zone of transition where leucocratic mineral bands appear for the first time, forming anorthosite layers. The layered gabbros are followed by leucogabbros and massive gabbros, with dolerite dykes and microgabbro occurrences. Widespread outcrops of gabbro pegmatites, in which megacrysts of hornblende and plagioclase form an intergrowth texture, are exposed along Ilica village and Bektemurlar road. The gabbros display a variety of pre- and post-emplacement deformational textures.
The gabbros gradually pass into their metamorphosed equivalents along the flanks of the anticline. The result is an amphibole-rich rock with locally developed penetrative deformation. Hornblende and albite are the major constituents of this rock.

This characteristic intrusive complex grades upward into a volcanic member consisting of largely spilitic lavas and intercalated shales and siltstones. Where the effect of the penetrative deformation is intense this group of rocks becomes greenschist and interfoliated slate-phyllite association.

The ophiolite is overlain by a slightly metamorphosed sedimentary sequence of Paleozoic age. The contact of the Paleozoic sequence against the ophiolite is with different lithologies from place to place. However, in a generalized stratigraphic section the exposed lowermost unit of the Paleozoic sequence is a conglomeratic gneiss whose pebbles are typical derivatives of a sialic basement. The dominant pebbles are quartzite. Beside these, there are rare pebbles of slate-phyllite which resemble closely the overlying slate horizons. Therefore, the lowermost unit exposed is not in fact the bottom of the sequence, because neither the pebbles of the quartzite nor that of the slate could have been derived from the underlying ophiolite. Furthermore, no ophiolite fragments or pebbles were found within the Paleozoic sedimentary sequence.

Pebbles of various sizes of the conglomeratic gneiss have been flattened along foliation planes. The rock is mainly derived from quartzitic conglomerate which grades into quartzite. The quartzite is fairly thin and is followed by another thin slate horizon. Locally where the slate level is missing, the sequence is followed by carbonate deposition. The carbonate within this metamorphosed sequence is represented in places by recrystallized limestone, calcschist and semi-marbles. Fossils found within the recrystallized limestone have provided an age of Middle Devonian (Fig. 4). Towards the top, the carbonates laterally interfinger with a slate-phyllite association with a fossil assemblage which further supports the Devonian age assignment. The slate-phyllite unit becomes dominant towards the top. There are rare garnetiferous schists and biotite muscovite-quartz schists within the phyllites. As the metamorphism decreases upward, it is clearly seen that the slate-phyllite series represents a metamorphosed flysch-like sequence in which shale, siltstone and fine-grained sandstone (graywacke) alternate. However, where the effect of metamorphism dominates, slaty cleavage and tight folding obscure most of the primary features.

These slightly metamorphosed sedimentary sequences resemble closely the Paleozoic sequence in the Istanbul and Zonguldak regions. Although the Paleozoic sequence in "the Istanbul-Paleozoic tectonic unit" is unmetamorphosed, its lithological characteristics and its fossil assemblages indicate that the weakly-metamorphosed Mudurnu Paleozoic sequence is in fact the metamorphosed equivalent of the Istanbul-Zonguldak Paleozoic unit.

The Paleozoic sequence in Istanbul represents the development of a south-facing (?) continental margin which was initiated during the Silurian and fully-developed during the Carboniferous time (KIDD et al., in prep). Because of the presence of Paleozoic rocks lying on top of the ophiolites, the Mudurnu Paleozoic sequence has been regarded as pre-Devonian (ABDÜSSELAMOGLU, 1959; BRINKMANN, 1976). But as mentioned above, the contact with the ophiolites wherever it is exposed, is seen to be tectonic in nature. It is a wide zone of intense cataclasis or an imbricate zone. Therefore, based solely on the overlying Paleozoic unit, the stratigraphic position of the ophiolite cannot be established. Besides, along the Mudurnu valley some deep-sea sedimentary rocks have been mapped, e.g. shale and siltstone displaying a number of structural features similar to that of the Paleozoic slate-phyllite association. This association, which is also tectonically overlain by the Paleozoic unit, indicates a Late Cretaceous age.

The following overlying unit is a chaotic assemblage. It is mainly composed of slices, blocks and/or fragments of the upper layers of an ophiolite such as lavas and associated deep-sea sediments namely radiolarian cherts, pelagic limestones, shales etc. as well as quartzite, marble, recrystallized limestone and slate-phyllite. It looks as though this chaotic assemblage is a tectonic mixture of two underlying units, namely the ophiolite and the Paleozoic metamorphic rocks. The size of the blocks varies from a few meters to a few kilometers.

Where the slabs of the ophiolite reach the size of one km as in the north of Abant Lake near Sinekli Yayla, the internal structure, texture, and ordering of the lithologies appear fairly undisturbed.

Some granitic rocks show intrusive contacts with most of the different rock types in the chaotic assemblage. Its crosscutting nature with quartzite and marble blocks can be seen around Dedekan and Findikçukuru villages.

The granitic rocks vary in composition from coarse-grained quartzmonzonitic-granodioritic to tonalitic. They consist mainly of quartz, plagioclase, K-feldspar, hornblende and biotite. The vein and dyke varieties are aplite or porphyritic microgranite.

In many places the granitic rocks display a transitional contact with the spilite, because the spilite grades into tonalite within a few meters. Some crosscutting granitic dykes and veins occur also within the ophiolite sequence near Ilica village. In spite of widespread intrusive contacts of the granite with the Paleozoic sedimentary rocks hardly any relationship of the granite has been seen. Because the intruding granite occurs in units tectonically above and below, its absence in the Paleozoic rocks in between requires an explanation.

The whole chaotic assemblage may well be compared with a gigantic conglomerate with a small-scale matrix. Thus this tectonic unit may be treated as a melange association (SILVER and BEUTNER, 1980).

Locally serpentinite forms a matrix between the other blocks as is the case immediately west of Abant Lake. Blocks of pelagic red limestone within the melange indicate Late Cretaceous age. This age provides evidence for the age of the melange generation, because the tectonical juxtaposition must have occurred within a time spar between Late Cretaceous and Eocene as the melange association is unconformably overlain by Eocene sandy conglomerates and limestones. Scattered occurrences of ophiolitic melange associations may be related to the ordered ophiolite sequence at the bottom. The difference between the two lies in the fact that the former underwent an intensive melange-type deformation but the latter, although sliced internally, is still relatively complete.

Upper Paleocene to Lower Eocene sedimentary rocks unconformably overlie most of the older rocks in the region, including the melange. This stratigraphic relationship provides an upper limit for the age of the emplacement of the chaotic assemblage. Locally the Eocene sequence begins with a basal conglomerate but generally with a shallow water, nummulitic limestone.

The coarseconglomerates are derived from the ophiolites and the Paleozoic rocks. South of the village, the conglomerates interfinger with some red mudstone beds.

Widespread volcanism accompanied the deposition of the Eocene sediments at the northern flanks of the hills facing the Düzce-Hendek plain (Fig. 1). Volcanic activity began with basaltic lavas, intercalated with pyroclastic rocks, mainly agglomerates. Some volcanogenic sandstones and mudstones were also deposited. Olivine basalts and pyroxene basalts are the main varieties of volcanics. Later, basaltic andesites were also extruded.

The thick pile of volcanic rocks and associated volcanogenic sedimentary rocks trends in an east-westerly direction and seems to represent the lateral extension of the Eastern Pontide volcanic Eocene sequence in the west. It was originated in a Chilean-type island arc. Only the lower portion of the sequence is observed in the area studied. Therefore within the limited number of samples collected it is hard to know how the volcanicity varies with time. Towards the east it shows stronger calc-alcaline affinities.

The Upper Cretaceous volcanic facies does not crop out in the area studied. It is widely exposed in a roughly E-W trending axis near the present Black Sea Coast in the north. This spatial arrangement suggests that the volcanic front migrated to the south between Late Cretaceous and Eocene time.

The Upper Eocene sedimentary cover overlying the older rocks extends to the south and crops out along the Northern Anatolian Fault Zone as well as in the southern areas. It may therefore be concluded that the two assemblages developed different lithological sequences only until the late Eocene. Afterwards they were brought together and show a common lithological sequence.

GEOLOGIC DEVELOPMENT

In the light of the description of the two sequences, an attempt has been made to explain the geological evolution of the region.

In the southern region, rifting was probably initiated during the Lias. This is indicated by two lines of evidence, a) a rapid accumulation of conglomerates-coarse sandstones and agglomerates which were deposited in fault-bounded basins, b) volcanic activity producing alkaline and tholeiitic lavas and basic-dyke injections.

By late Jurassic time, the region became a carbonate platform with shallow water, marine limestones (SANER, 1977; ALTINLI, 1973). The steady subsidence continued during the Early Cretaceous as indicated by the deposition of deep water marine carbonates with occasional chert intervals. Upper Cretaceous pelagic limestone and flysch followed during the Turonian.

The presence of complete ophiolite sequences at the bottom of the northern assemblage and the presence of ophiolite slices on the southern platform units suggest the existence of an ocean basin during much of the Mesozoic (Fig. 9). The north-facing Atlantic type Continental margin remained stable until the beginning of the Campanian when platform subsidence was accompanied by gravity emplacement of ophiolite, meta ophiolite and blueschists blocks. There is no evidence for a subduction beneath this margin. The exotic blocks were probably derived from uplifted slices of a melange prism nearby (Fig. 7). After the transportation of the exotic blocks to the area of deposition, the southern marginal sequence was covered first by shallow water and then by continental deposits (Fig. 3). This indicates a fast transition caused by a rapid uplift possibly due to approaching slabs of ophiolite from the north, which were obducted on the platform in Late Cretaceous pre-Eocene time (Fig. 8).

The metamorphic and magmatic associations indicate that the northern margin of the ocean was a site of active subduction initiated in the Late Cretaceous (Fig. 6), and possibly continued intermittently into the Tertiary. The magmatism is related to Late Cretaceous island arc volcanicity. A continued subduction was accompanied by the development of a magmatic arc which is observed all along the present Black Sea Coast. The continuation of this volcanicity in the study area occurred during the Eocene (Fig. 9).

The penetrative deformation forming slaty cleavage and intense folding in the Paleozoic sedimentary unit which is synchronous with low-grade metamorphism was possibly related to this subduction, and the accompanying slicing at the leading edge of the continent. A similar and coaxial penetrative deformation has also developed in the underlying ophiolite slices.

Therefore the two slate-phyllite associations of different origin and age (Devonian and Upper Cretaceous respectively), wherever in contact look similar.

The destruction of the ocean basin caused different effects on the opposite continental margins. In the south it ultimately resulted in the emplacement of oceanic lithosphere (Fig. 8). In the north, at the leading edge of the northern continent, a succession of ocean floor sediments underwent severe tectonism above or adjacent to the subduction zone (Fig. 7). Similar processes may have effected the oceanic lithosphere. Continuing subduction resulted in melange prisms (Fig. 7-9).

Successive thrusts progressively displaced the "ocean-floor" slices and rotated them upwards (Fig. 7). We think these uplifted imbricate slabs supplied material to the southern continental margin during Campanian. Some blueschists fragments were possibly derived from the slices of a melange. It may therefore be concluded that during this period, either the two continents were fairly close to each other, or the ocean basin had undergone an advanced degree of slicing so that some of the slabs came close to the southern margin to provide the exotic ophiolitic blocks.

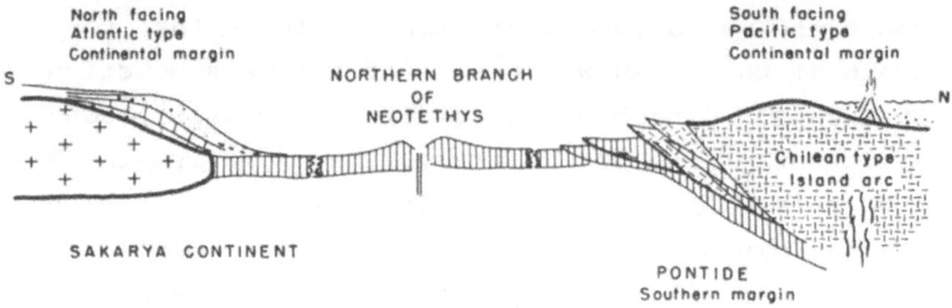

Fig. 6 Early Late Cretaceous

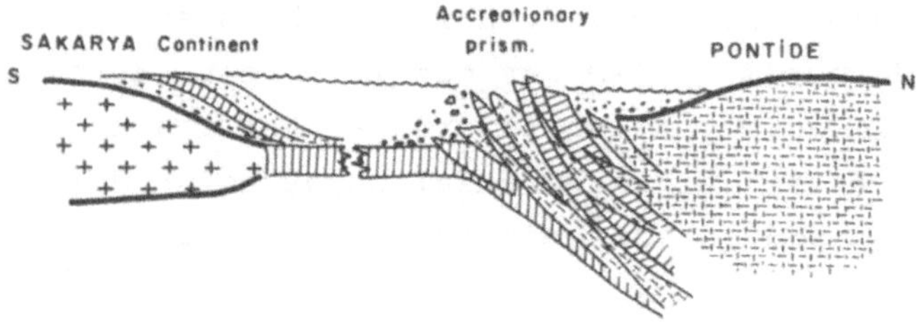

Fig.7 Campanian

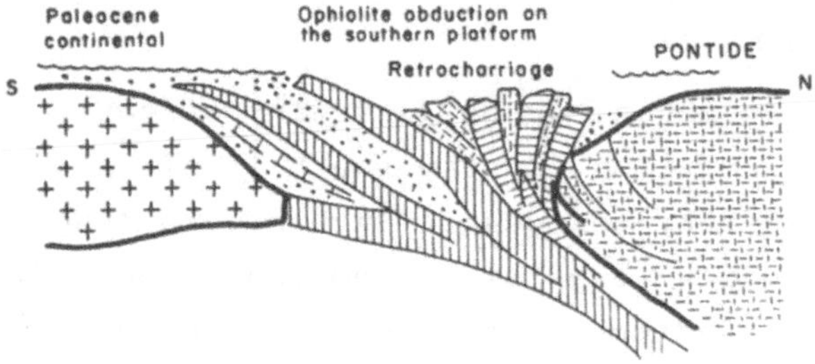

Fig.8 Paleocene

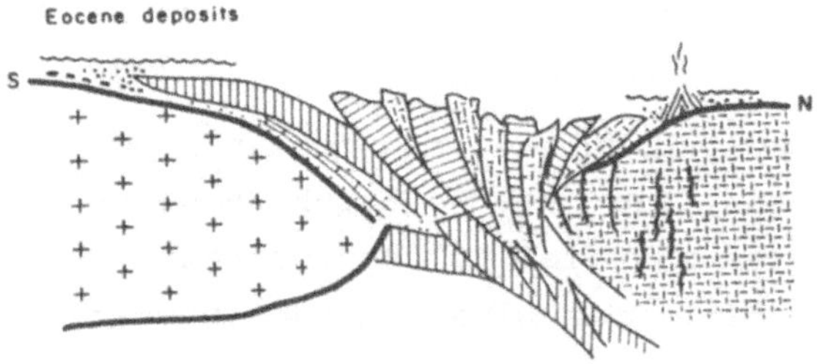

Fig.9 Eocene

Fig. 6 to Fig. 10

Schematic sequential cross-sections showing the successive steps in the evolution of the Almacik mountain and surrounding regions from the Liassic (Early Jurassic) Period to the Holocene

The large-scale thrusting of oceanic lithosphere led to the incorporation of ophiolites in the accreationary prisms, as evidenced at the base of the northern assemblage (Fig. 7-9). This process was possibly accompanied by oceanward migration of the trench (Fig. 9) so that the volcanic front of the Chilean-type island arc in the Western Pontides also moved to the south. Meantime, the northern continent must have suffered crustal compression, slicing and thickening. This compression is indicated by severe thrusting in the northern assemblage (s. accompanying geologic map).

The southern boundary of the northern assemblage with the ophiolites is marked by an imbricated and crushed zone. Therefore the generalized stratigraphic section of the northern assemblage (Fig. 4), based on our field work, may thus be treated as an allochthonous unit (the ophiolite) overlain by some par-authochtonous units (the Paleozoic sedimentary sequence and the cover rocks) (Fig. 5). The overlying sedimentary sequences too are of par-authochthonous nature and together with the ophiolitic rocks underneath have led to an interpretation of tectonic superposition. An intense compressional regime as a result of continuing subduction squeezed the sliced slabs at the edge of the northern continents (Fig. 8), progressively rotating them upwards and afterwards moving them backwards as a retrocharriage (Fig. 9). Thus the slices and slabs were emplaced on the northern continents. These rocks correspond to the chaotic ophiolitic association in the northern assemblage (Fig. 4-5).

The granite intrusion is possibly related to the late phase of subduction. After some crustal thickening in the north, the southern plate, namely the Sakarya Continent, was subducted under the thickened accretionary prism, and under the leading edge of the northern continent. There, partial melting possibly occurred. Therefore, granites were injected mainly into the layers of the ophiolite and their cover of overlying accretionary units. This may explain the distribution of granite injections or their absence in the Paleozoic sedimentary units. At a later phase when some slices or accretionary units were retrocharriaged over the northern unit the Paleozoic sedimen-

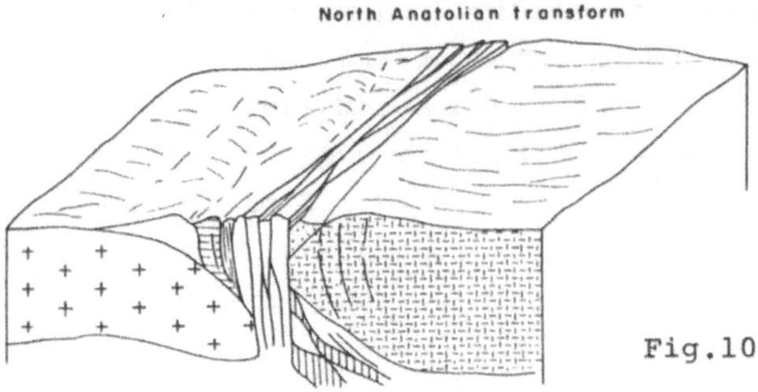

Fig.10 Present

tary rocks were sandwiched between the two "ocean-floor-originated-units" (Fig. 5) and missed the injection of the granite.

The ocean floor subduction was terminated by the complete collision of northern Pontide with the Sakarya Continent. This happened possibly in pre-late Eocene period because the Upper Eocene sedimentary rocks cover the northern and southern assemblages and the suture between them. This conclusion is further supported by the cessation of island arc volcaneity on the northern continent.

The relationship between the northern and southern assemblages can be explained by the presence of an ocean floor between the two continents, namely the Istanbul-Zonguldak tectonic unit in the north, and the Sakarya Continent in the south. These units existed during much of the Mesozoic and the beginning of Tertiary Era. In the north occurred tectonic stacking of a series of contemporaneous sequences, (e.g. different layers of ophiolite and deep-sea sediments) and of sequences derived from originally different environments (e.g. the Paleozoic sedimentary rocks and the ophiolite occurred) (Fig. 5, Fig. 9).

The model that we propose explains satisfactorily the presence of both tectonically shaped slices of ophiolite.

An inferred crustal structural position along the Northern Anatolian Fault Zone between Bolu and Akyazi is shown in Fig. 10. As is seen in this block diagram, the fault follows more or less the suture zone between the two continents. Because of this structural position ATAMAN et al. (1975) assumed that the fault was a paleo-Benioff zone. However, it is widely recognized that the fault originated in Middle Miocene time (KETIN, 1948; ŞENGÖR, 1979). At that time the two continents had already collided with each other, and the ocean floor consumption has long been terminated. Therefore there is not any apparent or direct evidence to relate the formation of the Northern Anatolian fault to the paleo-Benioff zone.

ŞENGÖR (1979) views the generation of the fault as a consequence of collision and following squeezing-in of Eastern Anatolia between the Arabian plate and the Russian plate. Because of intense squeezing in the east, Central and Western Anatolia was forced to move westward. To accomodate this movement the Northern and Eastern Anatolian Faults were formed. The fault in the north followed preferably the ancient suture zone, a zone of weakness. As pointed out by BURCHFIEL (1980) many of the active transform faults follow structural lines formed by older collisional events within Asia. HUANG (1980) has given some good examples. With its branches the North Anatolian Fault Zone affects an approximately 50 km wide zone in the region under consideration, covering areas which are far away from the suture.

ACKNOWLEDGMENTS

A. Okay read the manuscript, A.M.C. Şengör offered some comments on various parts of this study. Their help is greatly acknowledged.

Mrs. E. Yiğitbaş, M. Yildirim, K. Sulu, A. Yalçin, S. Mençer, E. Keskin, F. Kirağasi, and Y. Dilek were involved in some parts of mapping.

REFERENCES

ABDÜSSELAMOĞLU,M.Ş. 1959: Almacıkdağı ile Mudurnu ve Göynük civarının jeolojisi. İstanbul Üniv.Fen.Fak.Monogr. No: 14.94.

ALTINLI,İ.E. 1973: Bilecik Jurasiği: Cumhuriyetin 50. Yılı Yerbilimleri Kongresi Tebliğler. M.T.A. Ankara, p.103-111.

ATAMAN,G. - BUKET,E., ÇAPAN,U.Z. 1975: Kuzey Anadolu Fay zonu bir Paleo-Benioff zonu olabilir mi? M.T.A. mecm. 84, p.112-119.

BINGÖL,E., AKYÜREK,B. and KORKMAZER,B. 1973: Biga Yarımadasının Jeolojisi ve Karakaya formasyonunun bazı özellikleri: Cumhuriyetin 50. Yılı Yerbilimleri Kongresi, Tebliğler, M.T.A. Ankara, p.70-77.

BİNGÖL,E. 1976: Evolution geotectonique de l'Anatolie de l'Quest. Bull.Soc.Geol.France, ser. 7, v.18, p.235-254.

BRİNKMANN,R. 1976: Geology of Turkey: Ferdinand Enke, Stuttgard, pp.158.

BURCHFIEL,B.C. 1980: Plate tectonics and the continents; A review. p.15-25, in "Studies in geophysics" Continental tectonics. National Academy of Sciences. Washington D.C.

ÇOĞULU,H.E., DELALOYE,M. and CHESSEX,R. 1965: Sur l'age de quelques roches plutoniques acides dans la region d'Eskişehir -Turquie: Arch.Sci.Geneve, v.18, p.692-699.

ERK,S. 1942: Bursa ve Gemlik arasındaki mıntıkanın jeolojik etüdü. M.T.A. Yay., Ser. B, no: 9, pp.295

GÖZÜBOL,A.M. 1978: Mudurnu-Dokurcan-Abant (Bolu ili) alanının jeolojisi incelemesi ve Kuzey Anadolu Yarılımının Yapısal Özellikleri (Unpublished doctor's thesis-Univ. Istanbul).

HUANG,T.K. 1980: An outline of the tectonic characteristics of China, p.184-197, in "Studies in geophysics", Continental tectonics. National Academy of Sciences Washington D.C.

KETİN,İ. 1948: Über die tektonisch-mechanischen Folgerungen aus den grosen anatolischen Erdbeben des letzten Dezenniums. Geol.Rdsch. no: 36, p.77-83.

LEMOINE,M. (editor) 1978: Geological Atlas of Alpine Europe and adjoining Alpine areas. Elsevier Publ. co. pp.584.

SILVER,E.A. and BEUTNER,E.C. 1980: Melanges. Geology, 8/1, p.32-35.

SANER,S. 1977: Geyve-Osmaneli-Gölpazarı-Taraklı alanının jeolojisi, eski çökelme ortamları ve çökelmenin evrimi: (Unpublished Doctor's Thesis. Univ. Istanbul, pp.312).

ŞENGÖR,A.M.C. 1979: The North Anatolian transform fault: its age, offset and tectonic signifiance. Bull.Geol.Soc.Lond. v.136, p.269-282.

ŞENGÖR,A.M.C. and YILMAZ,Y. 1980: Tethyan Evolution of Turkey: A Plate tectonic approach. Tectonophysics (in Press).

ŞENGÖR,A.M.C. and YILMAZ,Y. and KETİN,İ. 1980: Remnants of a pre-late Jurassic ocean in northern Turkey: Fragments of Permo-Triassic Paleo-Tethys? Bull.Geol.Soc.Am (in Press).

YILMAZ,Y. 1972: Petrology and structure of the Gümüşhane granite and surrounding rocks, North. Eastern Anatolia: Ph.D. thesis, Univ. London, London, 260 p.

YILMAZ,Y. 1977: Bilecik-Sögüt dolayındaki "eski temel karmaşığı"nın petojenetik evrimi. (Unpublished dissertation Univ. Istanbul), p.169.

Comparisons Between the North Anatolian Fault of Turkey and the San Andreas Fault of California

Clarence R. Allen

Seismological Laboratory, California Institute of Technology, Pasadena, CA 91125, USA

Abstract.--The North Anatolian fault of Turkey is remarkably similar to the San Andreas fault of California in its style of displacement, high seismicity, physiographic expression, neotectonic history, presence of creep, and problems of seismic-hazard evaluation. On the other hand, significant contrasts between the two faults exist in their space-time patterns of seismicity and in their plate-tectonic relationships. Workers in the two areas have much to learn from one another.

Just as the San Andreas fault was first recognized as a throughgoing fault of regional extent as a result of studies that followed the great San Francisco earthquake of 1906 (Lawson et al., 1908), the North Anatolian fault was recognized as a similarly important feature following the remarkable series of earthquakes that began with the great Erzincan earthquake of 1939 (Paréjas et al., 1941; Ketin and Roesli, 1953). It has become clear in the subsequent years that the two fault systems are in fact much more similar than was at first appreciated, and it is the purpose of this paper to summarize and illustrate these similarities, as well as to point out a few significant differences.

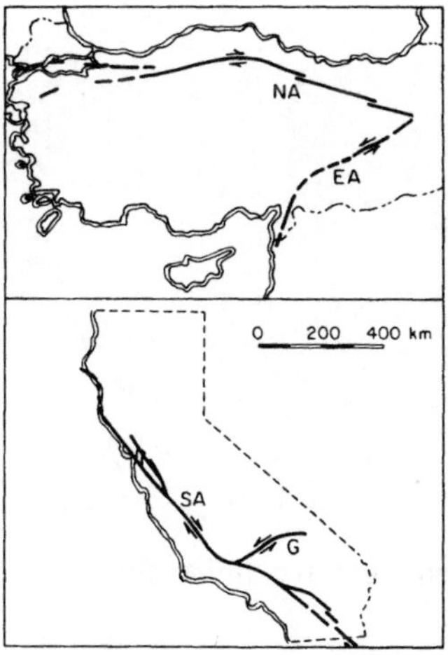

Fig. 1.--Maps of San Andreas (SA) and North Anatolian (NA) faults at same scale, showing relationships to conjugate Garlock (G) and East Anatolian (EA) faults.

Among the specific similarities between the two faults are those that follow, some of which have been earlier discussed by Ketin (1976):

(1) The two faults are transform faults of regional tectonic importance and significant length -- both about 1100 km long (Fig. 1). Both are plate boundaries.

(2) Both faults are dominantly strike slip, at least in their central segments, and both are right lateral in their sense of displacement.

(3) Both faults are associated with major conjugate faults that are left lateral (Fig. 1): the Garlock fault of California and the East Anatolian fault of southeastern Turkey and adjacent areas (Arpat and Saroğlu, 1972, 1975). Furthermore, the angles between the two pairs of faults are almost identical -- both about 50°.

(4) Both faults have generated major historic earthquakes: the 1857 (M = 8+) and 1906 (M = 8.3) earthquakes in California (Fig. 2), and the 1939 (M = 8.0), 1942 (M = 7.3), 1943 (M = 7.6), and 1944 (M = 7.6) events in Turkey, among others (Fig. 3). Numerous large earthquakes are known to have occurred still earlier along the fault in Turkey, back to the 3rd century (Ambraseys, 1970), but virtually no systematic historic records exist for most of California prior to the time of the 1857 event.

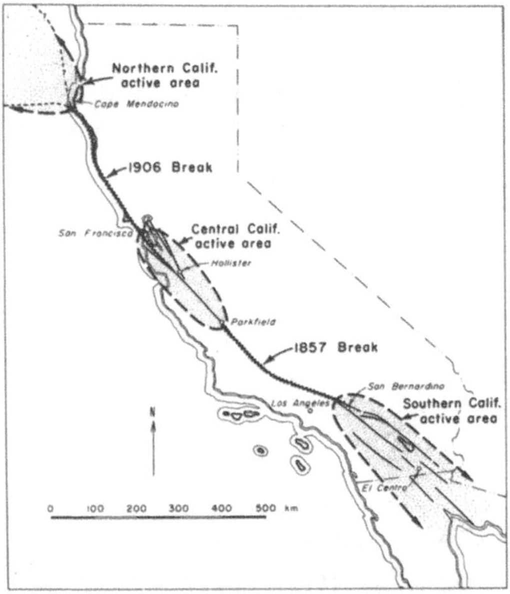

Fig. 2.--Areas of major earthquakes and contrasting seismic behavior along San Andreas fault system. From Allen (1968).

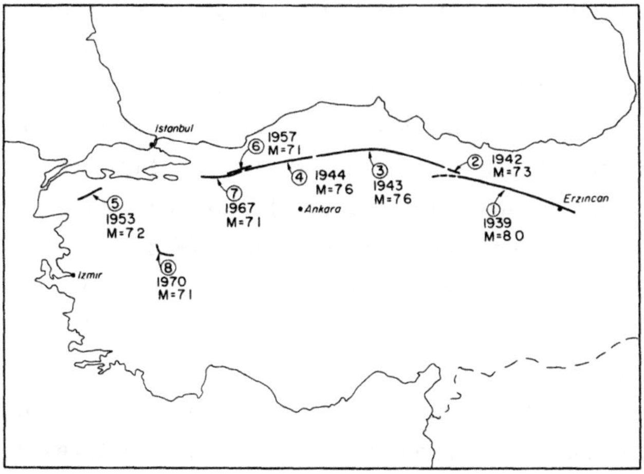

Fig. 3.--Faulting (heavy lines) associated with earthquakes of M = 7.0 and above in Turkey since 1939. From Allen (1975).

(5) Along the southern part of the San Andreas fault, Sieh (1978) has demonstrated a recurrence interval between great earthquakes of about 160 years. Similarly, historic records of earthquakes along the North Anatolian fault suggests bursts of activity about once in every 150 years (Ambraseys, 1970).

(6) Despite the recent large earthquakes on both faults, both still have major unbroken segments that could be interpreted as seismic gaps--particularly likely places for future earthquakes. One of these, although debatable, is the southernmost segment of the San Andreas fault from Cajon Pass to the Salton Sea (Sieh, 1978; Allen, 1980). Similarly, the easternmost segment of the North Anatolian fault, from Erzincan to Karlıova, Bingöl Province, probably has not broken during the 1939 and subsequent series of events.

(7) Both faults have generated large, infrequent earthquakes along those segments where the fault trace is simple and relatively continuous, as contrasted to the more frequent but smaller earthquakes where the fault trace branches and becomes more complicated. For example, three currently active areas of the San Andreas fault are in regions of numerous branching faults, whereas the great 1906 and 1857 earthquakes occurred along simple, relatively continuous segments (Fig. 2)(Allen, 1968). Likewise, the most persistent continuing activity along the North Anatolian zone is at its western end, where it frays-out into a number of branches, whereas the

largest earthquakes have occurred along its simpler central segmen (Fig. 3).

(8) Both faults are characterized by creeping segments and are, in fac the only two fault systems in the world where such behavior has been doc mented. Creep occurs on several branches of the San Andreas fault in t central California active area (Fig. 4)(Nason, 1973), and on the Nort Anatolian fault at İsmetpaşa station, Çankırı Province (Fig. 5)(Aytun 1973). The maximum creep rate on the San Andreas fault is about 3 cm/y (Thatcher, 1979), whereas that at İsmetpaşa station is about 1 cm/y (A. Aytun, personal communication).

(9) Both faults are characterized by conspicuous en-echelon offsets o the active faults traces, so that continuous single surface fractures d not extend along the entire lengths of the fault zones. On the San Andrea fault, such offsets are well documented in the Salton Sea region (Lomnit et al., 1970; Elders et al., 1972), as well as near Parkfield in central California (Dickinson, 1966). Along the North Anatolian fault, major en-echelon offsets occur near Niksar, Tokat Province, and near Erzincan (Fig. 1)--both associated with major valleys resulting from the pull-apart nature of the resulting structures. On both faults the principal en-echelon

Fig. 4.--Creep on Calaveras fault in Hollister, California. Photo in 1967 by C. R. Allen.

Fig. 5.--Creep on North Anatolian fault at İsmetpaşa, Çankırı Province. Photo in 1969 by C. R. Allen.

offsets are right-stepping, a phenomenon which appears to represent a fundamental mechanical property of right-lateral faults (Allen, 1980). On a smaller scale, numerous individual en-echelon surface fractures make up the fault zones. Wallace (1973) has pointed out that individual fractures along the San Andreas fault seldom exceed 8 km in length, and the detailed maps of Arpat and Şaroğlu (1975) suggest a similar situation along the North Anatolian fault.

(10) In areas of rugged topography, both faults are associated with massive landslides that tend to conceal the detailed fault traces. Good examples are found along the San Andreas fault in the Santa Cruz Mountains, south of San Francisco, and along the North Anatolian fault in the mountains north of Ilgaz, Çankırı Province, and Tosya, Kastamonu Province.

(11) Both faults are characterized by impressive zones of crushed and altered rocks, with widths measured in hundreds or thousands of meters. However, these zones are seldom well exposed because of the tendency of rivers to erode preferentially along them to form alluviated valleys. Perhaps the best exposures along the San Andreas fault are those in the freeway cuts near Gorman, 100 km north of Los Angeles. Among the best

exposures along the North Anatolian fault are those in the Kelkit River Canyon about 25 km west of Reşadiye, Tokat Province.

(12) Detailed geomorphic features resulting from late Quaternary surficial displacements on the two faults are so remarkably similar that one could easily mistake localities between the two countries. Most of these are physiographic features typical of active strike-slip faults the world over (e.g., Cotton, 1950; Sharp, 1954; Allen, 1965), but certainly the San Andreas and North Anatolian are among the most spectacular of such faults. This is partly due to their high rates of displacement and partly due to the absence of heavy vegetative cover. Open, grassy terrain, such as characterizes much of both California and Turkey, seems to make such features of recent fault displacement more obvious than in either a more arid or a more humid environment. Among these features are the following:

(a) Fault scarps.--Linear fault scarps associated with late Quaternary earthquakes are abundant along both fault systems and include many features associated with historic earthquakes that are still well preserved (Figs. 6, 7). Evidence is impressive along both faults that historic earthquakes have been typical of earlier events, with faithful repetition of fault movements precisely along earlier fault traces

Fig. 6.--Recent fault scarps (oversteepened slopes in shadows beyond car) along San Andreas fault in Carrizo Plain, California. Photo by C. R. Allen.

Fig. 7.--Recent scarps (beyond jeep) along North Anatolian fault near Abbas Çiftlik, 15 km east of Suşehri, Sivas Province. Photo by C. R. Allen.

Fig. 8.--Scarp of 1958 Borrego Mountain earthquake, on branch of San Andreas fault in southeastern California. Note that 1968 fracture occurred at base of larger Holocene scarp, faithfully reproducing details of earlier fault displacements. Photo by C. R. Allen.

Fig. 9.--1944 scarp (arrow) 5 km east of Gerede, Bolu Province, following trace of earlier displacement. Right-lateral displacement here in 1944 was about 2 m. Photo by C. R. Allen.

(Figs. 8, 9). Additional localities along the North Anatolian fault are illustrated by Allen (1969).

(b) Closed depressions.--Recent fault movements have created numerous closed depressions along both faults (Figs. 10, 11), resulting either from the blocking of normal lines of drainage or from the development of pull-apart depressions between the overlapping ends of en-echelon fault segments. The largest such depression along the San Andreas fault, although probably of multiple origin, is the Salton Sea, whereas Abant Gõl (Lake), Bolu Province, represents the largest closed depression along the main segment of the North Anatolian Fault. Along both faults, physiographic features of recent displacement are better preserved in the passes between river drainage systems than in the alluviated, fault-controlled valleys that mark most of the fault trace. Thus sag ponds are particularly well preserved at drainage divides, where they are rela-

Fig. 10--Sag pond along recent trace of San Andreas fault near Priest Valley, Monterey County. Photo by C. R. Allen.

Fig. 11.--Sag pond along recent trace of North Anatolian fault near Küçükgüzel, 12 km east of Suşehri, Sivas Province. Photo by C. R. Allen.

tively protected from erosion and deposition. In addition to Abant Göl, other such localities along the North Anatolian fault include Umman Göl, on the high divide 19 km northwest of Ilgaz, Çankırı Province; Gölbel, at the major pass 18 km northeast of Osmancık, Çorum Province; and at the main divide east of the Erzincan Valley, about 65 km east of Erzincan. On the East Anatolian fault, 20-km-long Hazar Gölü is a particularly obvious fault-caused depression (Arpat and Şaroğlu, 1975; Allen, 1975).

(c) Rift topography,--Primarily because of erosion of the shattered and altered rocks within the fault zones, aligned rift-like linear valleys tend to delineate strike-slip faults and are particularly obvious from the air. Although first described along the San Andreas fault in California, (e.g., Fig. 12), nowhere is rift topography better displayed than along the North Anatolian fault, such as near Gerede, Bolu Province (Figs. 13, 14)(Erinç et al., 1961). Also diagrammatic is the area adjacent to the Ankara-Samsun highway near Havza, Samsun Province (Figs. 15, 16).

Fig. 12.--Rift valley of San Andreas fault (lower right to upper left) at Cuddy Valley, 110 km northwest of Los Angeles. Note side-hill ridge (arrow) resulting in part from 1857 earthquake. Photo by C. R. Allen.

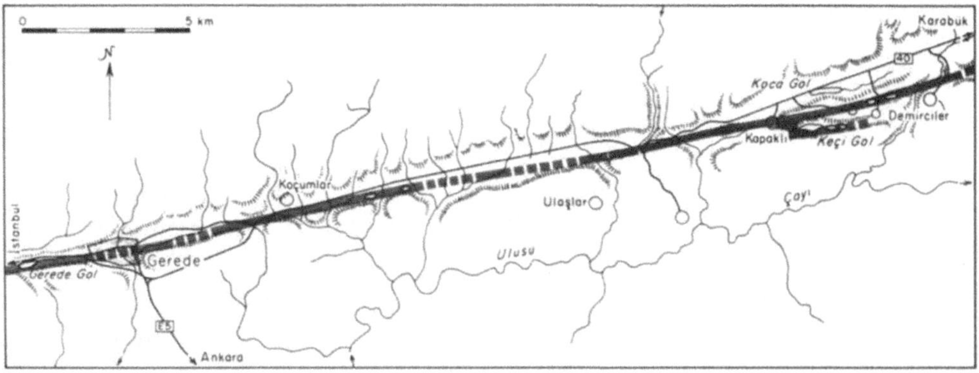

Fig. 13.--North Anatolian fault zone near Gerede, Bolu Province. Shaded zones on this and subsequent maps show locations of principal late Quaternary fault traces.

Fig. 14.--View west along rift valley of North Anatolian fault from a point about 1 km west of Gerede (Fig. 13). Note sag pond in distance. Photo by C. R. Allen.

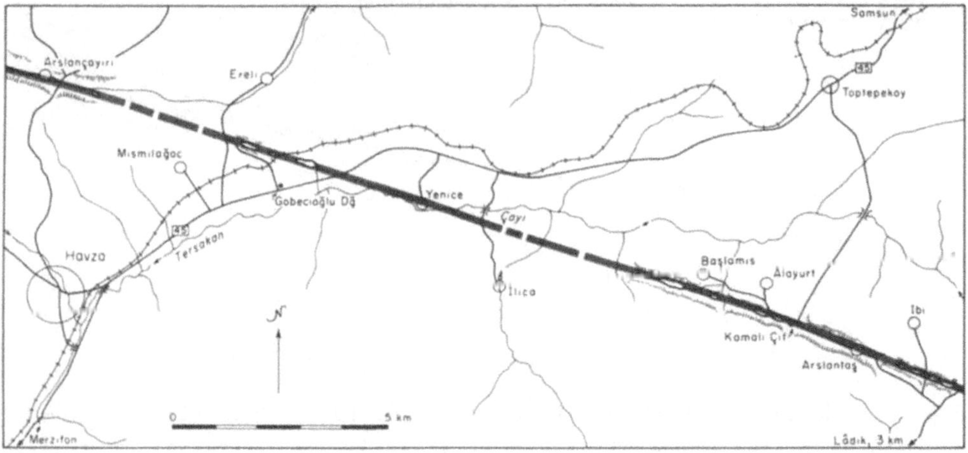

Fig. 15.--Map of North Anatolian fault zone near Havza, Samsun Province.

Fig. 16.--Rift topography near Kamalı Çiftliği (Fig. 15). Note numerous closed depressions. Right-lateral offset here during 1943 earthquake was about 1 m. Photo by C. R. Allen.

(d) Offset streams.--Stream channels that have been tectonically offset by horizontal fault movements have long been recognized along the San Andreas fault in California (e.g., Fig. 17)(Wallace, 1968) and are also present in more subdued form along the North Anatolian fault

Fig. 17.--View northwest along recent trace of San Andreas fault in Carrizo Plain, showing scarps and offset stream channels. Incised channel in foregound is offset 16 m; brush-filled channel beyond is offset 32 m. Right-lateral offset here at time of great 1857 earthquake was about 8 m. Photo by C. R. Allen.

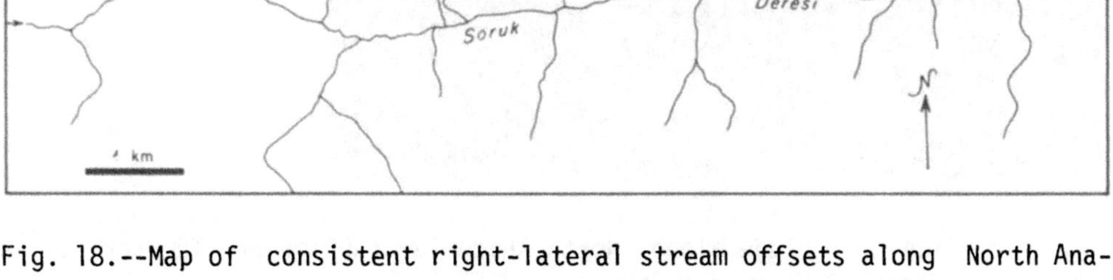

Fig. 18.--Map of consistent right-lateral stream offsets along North Anatolian fault about 20 km northeast of Osmancık, Çorum Province. Shaded zone indicates line of scarps, saddles, and closed depressions.

(Fig. 18). Their apparent lesser degree of development along the North Anatolian fault may be related to its slower rate of displacement. It should also be pointed out that stream offsets are difficult to recognize without aerial photographs, which have not been generally available in Turkey.

(e) Side-hill ridges.--Where active faults cut across steep slopes, such as in passing from one drainage system into another, conspicuous benches or side-hill ridges are often formed. Such features are obvious along both the San Andreas fault (Fig. 12) and the North Anatolian fault (Fig. 19).

(13) Because of the abundance of features of late Quaternary surficial displacements along both the San Andreas and North Anatolian faults, seismic-hazard evaluation is more effectively done by reference to these features than by dependence on epicentral locations of recent earthquakes, as has been argued by Allen (1975). Parts of the East Anatolian fault, for example, are not known to have broken during major historic earthquakes, but the clear features of late Quaternary displacement along this fault --

Fig. 19.--Side-hill ridge along North Anatolian fault near Tümeker, Sivas Province, resulting in part from 1939 earthquake displacement. Photo by C. R. Allen.

comparable to those along the North Anatolian fault (Arpat and Şaroğlu, 1975) -- suggest a similar degree of hazard. Opportunities in Turkey for determining recurrence intervals between major earthquakes by trenching and radiometric dating techniques, such as those pioneered by Sieh (1978), would appear to be just as good as in California.

Several differences between the two faults are also significant. These include the following:

(1) The two ends of the San Andreas fault can be understood in terms of relatively straightforward plate-tectonics processes, but the situation is much less clear with the North Anatolian fault. At its western end, it frays-out into a number of branches whose relationship to plate-tectonics processes still farther west in the Aegean region is a matter of great complexity (Dewey and Şengör, 1979; McKenzie, 1972). And at its eastern end, the North Anatolian fault as a surficial feature ends very abruptly, at the point near Karlıova where it is intersected by the East Anatolian fault (Fig. 20)(Allen, 1969; Arpat and Şaroğlu, 1972, 1975; Seymen and Aydın, 1972). Although the Varto earthquake of 1966 occurred still farther east and appears to represent continuity of regional strain, no continuous surficial feature can be followed east of the Karlıova area. In effect, the North Anatolian and East Anatolian faults appear to cancel-out one another, although clear geomorphic features of late Quaternary displacement are present along both faults within 5 km of their point of juncture (Fig. 20). This remarkable relationship is in clear contrast to the junction of the San Andreas and Garlock faults in California, where the San Andreas fault obviously maintains continuity past the point of intersection.

(2) The history of the San Andreas fault can be visualized in terms of migrating triple junctions (Atwater, 1970), whereas the North Anatolian fault is better visualized as the result of wedging action between two converging continental plates (Şengör, 1979). The total Neogene displacement on the San Andreas fault is thought to be about 300 km (Huffman, 1972), as compared to about 85 km for the North Anatolian fault (Şengör, 1979).

(3) The series of large earthquakes that began in eastern Turkey in 1939 was characterized by a remarkable sequential relationship in space and time, with fault rupture proceeding systematically westward during succeeding earthquakes (Fig. 3), as was first pointed out by Ketin and Roesli (1954). This has not been true in California nor, indeed, anywhere else in the world, at least on such a grand scale.

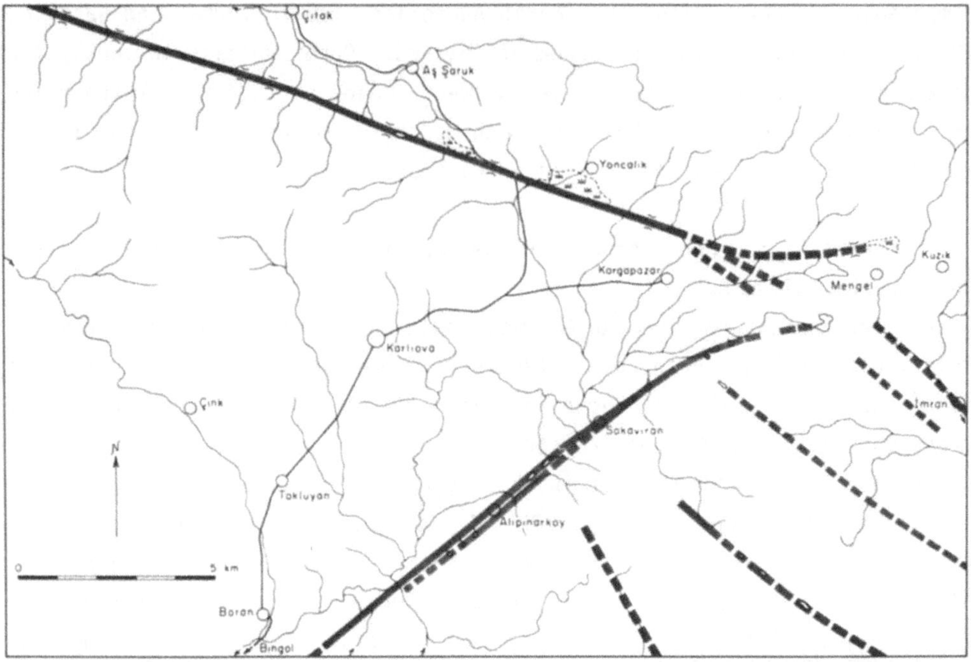

Fig. 20.--Map of juncture of North and East Anatolian faults near Karlıova, Bingöl Province. Locality can be seen on Fig. 1.

(4) The relative plate motion is apparently somewhat higher in California than in Turkey. Relative motion across the fault zone in central California is about 3 cm/yr (Savage and Burford, 1973; Thatcher, 1979), and although there is no direct geodetic evidence from Turkey, other lines of evidence suggest a relative motion along the North Anatolian fault of 1-2 cm/yr (Canıtez, 1973; Arpat and Şaroğlu, 1975). Creep measurement at İsmetpaşa station are consistent with this rate, as mentioned previously.

(5) The creeping zones of the San Andreas fault seem to be limited to those segments with distinctive surficial rock types, particularly an abundance of serpentine (Allen, 1968; Irwin and Barnes, 1975). No such petrologic correlation is known along the North Anatolian fault, although it should be pointed out that there has not yet been a systematic search for creep along the North Anatolian fault except near İsmetpaşa station and at a few localities farther west.

Acknowledgments.--The author's work in Turkey was made possible by the cooperation of the Maden Tetkik ve Arama Enstitüsü (M.T.A.). Field studies were carried out with the generous assistance of İ. Ketin of İstanbul Technical University, Özcan Aksoy of the M.T.A., and Selçuk Sipahioğlu of İstanbul University. California Institute of Technology, Division of Geological and Planetary Sciences Contribution No. 3496.

REFERENCES

Allen, C. R., 1965: Transcurrent faults in continental areas. Roy. Soc. [London] Phil. Trans. A, 258, 82-89.

Allen, C. R., 1968: The tectonic environments of seismically active and inactive areas along the San Andreas fault system. Stanford Univ. Pub. Geol. Sci., 11, 70-82.

Allen, C. R., 1969: Active faulting in northern Turkey. California Inst. Technology, Div. Geol. Sci. Contribution No. 1577, 32 p.

Allen, C. R., 1975: Geological criteria for evaluating seismicity. Geol. Soc. America Bull., 86, 1041-1057.

Allen, C. R., 1980: The modern San Andreas fault. Ch. 15 (p. 511-534) in Ernst, W. G., ed., The geotectonic development of California. Prentice-Hall, New York.

Ambraseys, N. N., 1970: Some characteristic features of the North Anatolian fault zone. Tectonophysics, 9, 143-165.

Arpat, E., and F. Şaroğlu, 1972: The East Anatolian fault system; thoughts on its development. Turkey Mineral Research and Explor. Inst. Bull., 78, 33-39.

Arpat E., and F. Şaroğlu, 1975: Türkiye'deki bazı önemli genç tektonic olaylar. Türkiye Jeol. Kurumu Bül., 18, 91-101.

Atwater, T., 1970: Implications of plate tectonics for Cenozoic evolution of western North America. Geol. Soc. America Bull., 81, 3513-3536.

Aytun, A., 1973: İsmetpaşa istasyonu civarında krip ölçümleri. Kuzey Anadolu Fayı ve Deprem Kuşağı Simpozyumu/MTA Yayinl., Ankara, 114-121.

Canıtez, N., 1973: Yeni kabuk hareketlerine ilişkin çalışmalar ve Kuzey Anadolu Fay problemi. Kuzey Anadolu Fayı ve Deprem Kuşağı Simpozyumu/MTA Yayinl., Ankara, 35-58.

Cotton, C. A., 1950: Tectonic scarps and fault valleys. Geol. Soc. America Bull., 61, 717-758.

Dewey, J. F., and A. M. C. Şengör, 1979: Aegean and surrounding regions: Complex multiplate and continuum tectonics in a convergent zone. Geol. Soc. America, 90, 84-92.

Dickinson, W. R., 1966: Structural relationships of the San Andreas fault system, Cholame Valley and Castle Mountain Range, California. Geol. Soc. America Bull., 77, 707-726.

Elders, W. A., R. W. Rex, T. Meidav, P. T. Robinson, and S. Biehler, 1972: Crustal spreading in southern California. Science, 178, 15-24.

Erinç, S., T. Bilgin, and M. Bener, 1961: Gerede civarında akarsu Şebekesi. İstanbul Üniv. Coğrafya Enstitüsü Dergisi, 7, 90-99.

Huffman, O. F., 1972: Lateral displacement of upper Miocene rocks and the Neogene history of offset along the San Andreas fault in central California. Geol. Soc. America Bull., 83, 2913-2946.

Irwin, W. P., and I. Barnes, 1975: Effect of geologic structure and metamorphic fluids on seismic behavior of the San Andreas fault system in central and northern California. Geology, 3, 713-716.

Ketin, İ., 1976: San Andreas ve Kuzey Anadolu Fayları arasında bir karşılaştırma. Türkiye Jeoloji Kurumu Bül., 19, 149-154.

Ketin, I., and F. Roesli, 1953: Makroseismische Untersuchungen über das nordwestanatolische Beben vom 18. Marz 1953. Eclogae Geol. Helvetiae, 46, 187-208.

Lawson, A. C., and others, 1908: The California earthquake of April 18, 1906; report of the State Earthquake Investigation Commission. Carnegie Inst. Washington Pub. 87.

Lomnitz, C., F. Mooser, C. R. Allen, J. N. Brune, and W. Thatcher, 1970: Seismicity and tectonics of the northern Gulf of California region, Mexico; preliminary results. Geofisica Internac. [Mexico], 10, 37-48.

McKenzie, D., 1972: Active tectonics of the Mediterranean region. Royal Astron. Soc. Geophys. Jour., 30, 109-185.

Nason, R. D., 1973: Fault creep and earthquakes on the San Andreas fault. Stanford Univ. Pub. Geol. Sci., 13, 275-285.

Parĕjas, E., İ. H. Akyol, and E. Altınlı, 1941: 27 Birinci Kânun 1939 Erzincan yerdepremi (Batı kismı). İstanbul Üniv. Fen Fakültesi Mecmuası, B, 6, 187-222.

Savage, J. C., and R. O. Burford, 1973: Geodetic determination of relative plate motion in central California. Jour. Geophys. Research, 78, 832-845.

Şengör, A. M. C., 1979: The North Anatolian transform fault: its age, offset and tectonic significance. Geol. Soc. London Jour., 136, 269-282.

Seymen, İ., and A. Aydın, 1972: The Bingöl earthquake fault and its relation to the North Anatolian fault zone. Turkey Mineral Research and Explor. Bull., 79, 1-8.

Sharp, R. P., 1954: Physiographic features of faulting in southern California. California Div. Mines and Geol. Bull. 170, Ch. 5, 21-28.

Sieh, K. E., 1978: Prehistoric large earthquakes produced by slip on the San Andreas fault at Pallett Creek, California. Jour. Geophys. Research 3907-3939.

Thatcher, W., 1979: Systematic inversion of geodetic data in central California. Jour. Geophys. Research, 84, 2283-2295.

Wallace, R. E., 1968: Notes on stream channels offset by the San Andreas fault, southern Coast Ranges, California. Stanford Univ. Pub. Geol. Sci., 11, 6-21.

Wallace, R. E., 1973: Surface fracture patterns along the San Andreas fault. Stanford Univ. Pub. Geol. Sci., 13, 248-250.

2 Seismotectonics: Seismicity statistics according to historical and instrumental records, focal mechanism, relations between seismic activity and neotectonics

Identification of Swarm Activity Using the MARNET Telemetered Seismometer-Network

S. Balamir Üçer

Kandilli Observatory, Çengelköy, Istanbul, Turkey

Stuart Crampin

Institute of Geological Sciences, Murchison House, West Mains Road, Edinburgh EH 9 3 LA, UK

Alistair Miller

Overseas Development Ministry, London, UK

Abstract

A radiolinked seismic network spanning the Marmara Sea, in combination with ISK conventional station net, allows more accurate epicentre locations within the Marmara region. Several centres of swarm activity have been identified across the Sea. Swarm activity appears to be characteristic of Western Anatolia (Crampin & Üçer, 1975) and sometimes has clear separation from adjacent seismicity.

Kandilli Observatory, Istanbul, operates 13 conventional seismic stations in Western Anatolia writing paper records. Recently, Kandilli has collaborated with the Institute of Geological Sciences, Edinburgh to extend the network by establishing a telemetered network, MARNET, spanning the Marmara Sea (Üçer, Crampin & Miller 1980b). MARNET has five vertical seismometers, with an overall aperture of 200 km, radio-linked in to a three-component instrument at Kandilli, and recorded on analogue magnetic-tape. Playback is by variable-gain variable-speed jet-pen recorder, which permits better phase identification and timing of arrivals than is possible with the fixed-gain fixed-speed chart recorders.

The telemetered network increases the number and accuracy of earthquake locations within the whole of Western Anatolia, but particularly within the Marmara Sea area. However, the number of stations is still far too few to allow accurate determinations of focal depths, even within the Marmara Sea, particularly as the velocity structure of the crust is largely unknown.

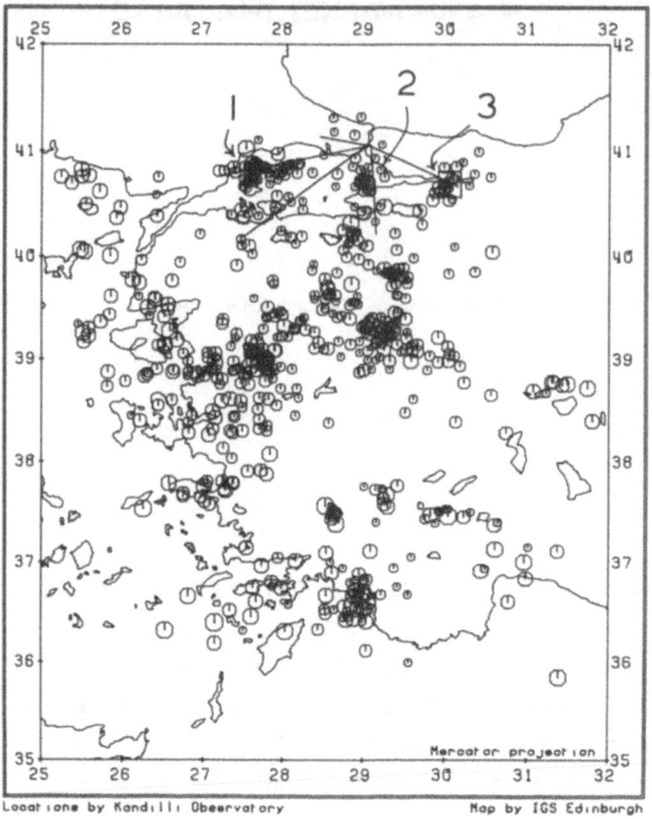

Fig. 1: Seismicity in Western Anatolia during 1978 (after Başarir <u>et al</u>. 1980). Earthquake magnitudes range from 1.7 M_S to 4.4 M_S. The radial lines indicate the MARNET radio-links, and the rectangle around swarm-centre number 3 is the area of the map in Evans et al. (1980) in this volume.

MARNET began operating from September 1978. Fig. 1 is taken from the report on seismic activity in Western Anatolia for the year 1978 by the staff of Kandilli Observatory (Başarir <u>et al</u>. 1980). The map shows the more reliable epicentres determined during routine processing at Kandilli using a computer program written by Üçer (Üçer <u>et al</u>. 1977).

The map shows several centres of swarm activity in the centre of Western Anatolia, as well as three swarms on the East-West line through the Marmara Sea, continuing the line of the most Northerly branch of the Northern Anatolian Fault. Such swarm activity is characteristic of Turkish seismicity, and the activity frequently migrates with time. Kandilli reports for 1977 (Alsan et al. 1980), 1978, and 1979 (Üçer et al. 1980a), show such migration for the swarms in Central Anatolia, but the swarms across the Marmara Sea, numbered 1, 2, and 3 in Fig. 1, have moved very little during this time (the swarm near Izmit, No 3, was only located in late 1978, when MARNET was recording).

Two of the three swarms across the Marmara Sea are very close to the epicentres of large earthquakes in the comparatively recent past. The Mürefte swarm (No 1) is adjacent to the felt epicentre of the Mürefte earthquake (M_S=7.3) of 1912, and the Çinarcik swarm (No 2) is adjacent to the Çinarcik earthquake (M_S=6.4) of 1963 (epicentres listed in Table 1). However, there has not been a major earthquake near the Izmit swarm-centre as far back as records can be traced.

There is very little activity between these three swarm centres: the relative aseismicity being particularly marked between the Izmit and Çinarcik swarms. Such gaps in the activity of otherwise active seismic lineations have frequently been interpreted as indicating seismic gaps, where large earthquakes may be expected in the future. In fact, the Izmit swarm is near the centre of a seismic gap identified by Toksöz, Shakal & Michael (1979). Toksöz et al. were not aware of the Izmit swarm, and it must modify the identification of the gap. The Izmit activity being on the boundary of a possible seismic gap makes it a particularly appropriate area for earthquake prediction investigations (Kelleher & Savino 1975), and the swarm is the site of the Turkish Dilatancy Project.

Table 1

Earthquakes associated with centres of swarm activity in 1978

Mürefte	9 Aug.1912	012900.0	40.6 N 27.2 E	M_S=7.3	(Karnik 1969)
Çinarcik	18 Sep.1963	165809.5	40.80N 29.23E	M_S=6.4	(Crampin & Üçer 1975)

Recent theoretical developments reported by Crampin (1980) in this volume, suggest that polarization anomalies in seismic shear-wavetrains are the most diagnostic feature of seismic waves propagating through a zone of dilatancy. The Izmit swarm, identified from the 1978 Kandilli seismicity maps, seemed to be a suitable place to test the hypothesis, as it is sufficiently small and isolated for most of the activity to be monitored by a small seismic network. The possibility of an adjacent seismic gap makes the experiment particularly significant.

The Turkish Dilatancy Project (Evans, Doyle, Üçer, Miller & Crampin 1980, this volume) used the epicentres in Fig. 1 to locate a 15 km aperture network, TDPNET, of three-component seismometers around the Izmit swarm-centre. The branch of the Northern Anatolian Fault near Izmit, and across the Marmara Sea, appears to be a graben-type structure with deep sedimentary cover within the graben, and TDPNET was sited in the mountains over the Southern half of the activity. In eight weeks operation, several hundred small earthquakes were recorded at epicentral distances between 0 and 20 km from the seismometers. The accuracy of the Kandilli routine locations was essential for the successful deployment and recording of the experiment reported by Evans et al. (1980).

Four reports at this conference (Burton et al. 1980; Crampin 1980; Evans et al. 1980; and this present paper) are the result of twelve years collaboration between Kandilli Observatory and the Institute of Geological Sciences. We wish to thank the many people who have made this collaboration possible; particularly Dr N. Öcal, Chief Seismologist at Kandilli when collaboration was first discussed, and Dr M. Dizer, the present Director of Kandilli Observatory.

This work was supported by Kandilli Observatory, Istanbul, and the Natural Environment Research Council and the Overseas Development Ministry, London, and is published with the approval of the Director of Kandilli Observatory and the Director of the Institute of Geological Sciences.

References

Alsan,E., Tezuçan,L., Başarir,E., Ayhan,E., Ulusan,N., & Üçer,S.B., 1980. Bati Türkiyede deprem etkinligi (Ocak - Aralik 1977), Bull.earthq.Res.Inst.,Ankara, in press.

Başarir,E., Alsan,E., Tezuçan,L., Ayhan,E., Ulusan,N. & Üçer,S.B., 1980. Bati Türkiyede deprem etkinligi (Ocak - Aralik 1978), Bull.earthq.Res.Inst.,Ankara, in press.

Burton,P.W., McGonigle,R.W., Makropoulos,K.C. & Üçer,S.B., 1980. Preliminary studies of seismic risk in Turkey and the occurrence of upper bounded and other large magnitude earthquakes,Proc.Interdis.Conf.Earthq.Pred.Res.in N.Anatolian Fault Zone, Friedr,Vieweg & Sohn, Wiesbaden.

Crampin,S., 1980. Polarization anomalies as diagnostics of dilatancy, Proc.Interdis.Conf.Earthq.Pred.Res.in N.Anatolian Fault Zone,Friedr.Vieweg & Sohn,Wiesbaden.

Crampin,S. & Üçer,S.B., 1975. The seismicity of the Marmara Sea region of Turkey, Geophys.J.R.astr.Soc.,40, 269-288.

Evans,R., Doyle,M.,Üçer,S.B., Miller,A. & Crampin,S., 1980. An experiment to investigate polarization anomalies in North Anatolia, Proc.Interdis.Conf.Earthq.Pred.Res. in N.Anatolian Fault Zone, Friedr.Vieweg & Sohn,Wiesbaden

Karnik,V., 1969. Seismicity of the European area, Part I, D.Reidel Pub.Co., Dordrecht.

Kelleher,J. & Savino,J., 1975. Distribution of seismicity before large strike slip and thrust-type earthquakes, J.geophys. Res.,80, 260-271

Toksöz,M.N., Shakal,A.F. & Michael,A.J., 1979. Space-time migration of earthquakes along the North Anatolian Fault zone and seismic gaps. Pure appl.Geophys.,117, 1258-1270.

Üçer,S.B., Alsan,E., Ulusan,N., Başarir,E., Ayhan,E., Tezuçan,L. & Kaptan, C. 1977. Bati Türkiyede deprem etkinligi (Eylül - Aralik 1976), Bull.earthq.Res.Inst.,Ankara,19, 45-114.

Üçer,S.B., Ayhan,E., Sankakli,N., Alsan,E., Tezuçan,L. & Başarir,E., 1980a. Bati Türkiyede deprem etkinligi (Ocak - Aralik 1979), Bull.earthq.Res.Inst.,Ankara, in press.

Üçer,S.B., Crampin,S. & Miller,A., 1980b. The MARNET radio-linked seismometer network spanning the Marmara Sea, and the seismicity of Western Anatolia, Geophys,J.R.astr.Soc., in preparation.

Earthquake Activity on the North Anatolian Fault Zone

Ref'an Ateş

Earthquake Research Institute, Ministry of Reconstruction and Resettlement, Ankara, Istanbul

SUMMARY

The North Anatolian Fault Zone (NAFZ) of Turkey is a region of high seismicity. It is one of the seismically most active dextral fault systems of the world with a seismicity comparable with the San Andreas Fault in California and the Alpine Fault in New Zealand. The fault zone consists of a belt of shear weaknesses, showing characteristic features of Quaternary and more recent faulting.

Historical earthquake studies have shown that major earthquake sequences associated with faulting have occurred in the past. The last sequence of the fault zone began with the 1939 Erzincan Earthquake which was followed by several major earthquakes - a succession of events in time and space that has been reported nowhere else in the world. Statistics show that this sequence of destructive earthquakes caused a total loss of 50 000 human lives in the last forty years.

This paper, in general, comprises the microearthquake studies around the western and eastern termini of the North Anatolian Fault. The past seismic activity for the western and eastern end is also studied to see the relation between the regional seismicity and recent tectonics.

In the west, the site investigation of the 1912 Earthquake in the Şarköy-Mürefte area of West Marmara indicated that an earthquake fault system extending in northeast-southwest direction was rejuvenated during the 1912 Earthquake. Later on, a microearthquake array was operated around the fault break of 1912, however, it revealed only low seismicity compared to the surrounding regions. This low earthquake activity has been evaluated in view of the recent tectonics of the region.

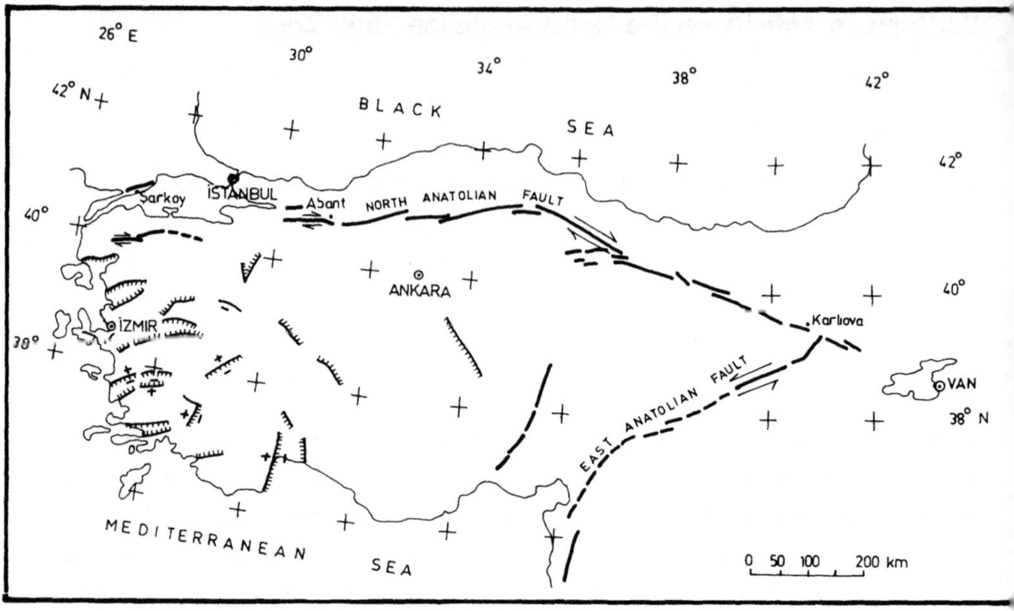

Fig. 1. Major fault systems in Turkey

In the east of the North Anatolian Fault, in order to study the continuation of the North Anatolian Fault into East Anatolia, a microearthquake array is operated in the Lake Van region. Indications of active faulting were studied. The distribution of microearthquakes and structural lineaments were evaluated from the seismotectonic point of view.

INTRODUCTION

The fault zone itself is a unique example of well-defined seismicity. Systematic investigations of the zone therefore may reveal very interesting clues for seismology and earthquake engineering. Fig. 1 shows major fault systems in Turkey. These are the North Anatolian Fault Zone, the East Anatolian Fault Zone and the West Anatolian Graben systems. The North Anatolian Fault is a transcurrent fault, showing right-lateral displacements along an arc of approximately 1000 km in length. It extends from West of Abant to East of Karliova where the tectonic pattern distinctly changes. At the western end of the fault, west of Lake Abant, the system splays out into a series of breaks. The present paper comprises the investigations which were carried out around the western and eastern termini of the fault.

SEISMICITY OF THE WESTERN PART OF THE NORTH ANATOLIAN FAULT

In the western end of the North Anatolian Fault, the Marmara Region, the fault splays out in two main breaks: one passing along the north of Marmara Sea and the other running south of it. The seismic activity of the West Marmara Region during the time 1900-1979 is shown in Fig. 2. The earthquakes for the area between ($39^{o}.5$-$41^{o}.5$)N - ($25^{o}.0$-$28^{o}.0$)E were taken from catalogues and classified according to their magnitude. Major damaging earthquakes and the associated surface faults are also shown in the figure. In our century the southern break of the Marmara Region appears to be associated with major destructive earthquakes. There the Yenice-Gönen Earthquake of March 18, 1953 with its epicentre at $40^{o}.07$N - $27^{o}.39$E is worth to mention from the seismotectonic point of view. The strike-slip fault with right lateral characteristics (Ketin and Roesli, 1953) shows good agreement with the focal mechanism solution reported by McKenzie (1972).
On October 6, 1964 the Manyas Earthquake occurred at $40^{o}.3$N - $28^{o}.2$E with a magnitude of 6.0. The surface displacement features were interpreted as the result of a right handed strike-slip motion along an east-west fault (Ketin 1964). The fault plane solution, however, indicates normal faulting.

Later on, the 1969 earthquake sequence in Western Anatolia (Can 1969) started with the earthquake of March 3rd, having its epicentre at $40^{o}.12$N - $27^{o}.43$E and a magnitude of 5.6. In the town of Gönen, 20 houses collapsed and several people were injured. Astonishing is the fact that the Gönen Earthquake which was closely located to the earthquake of March 18, 1953 showed thrust faulting mechanism.

In addition to these earthquakes, the 1935 Marmara-Erdek (M: 6.7) and 1944 Ayvalik (M:7.1) earthquakes caused considerable damage in the area.

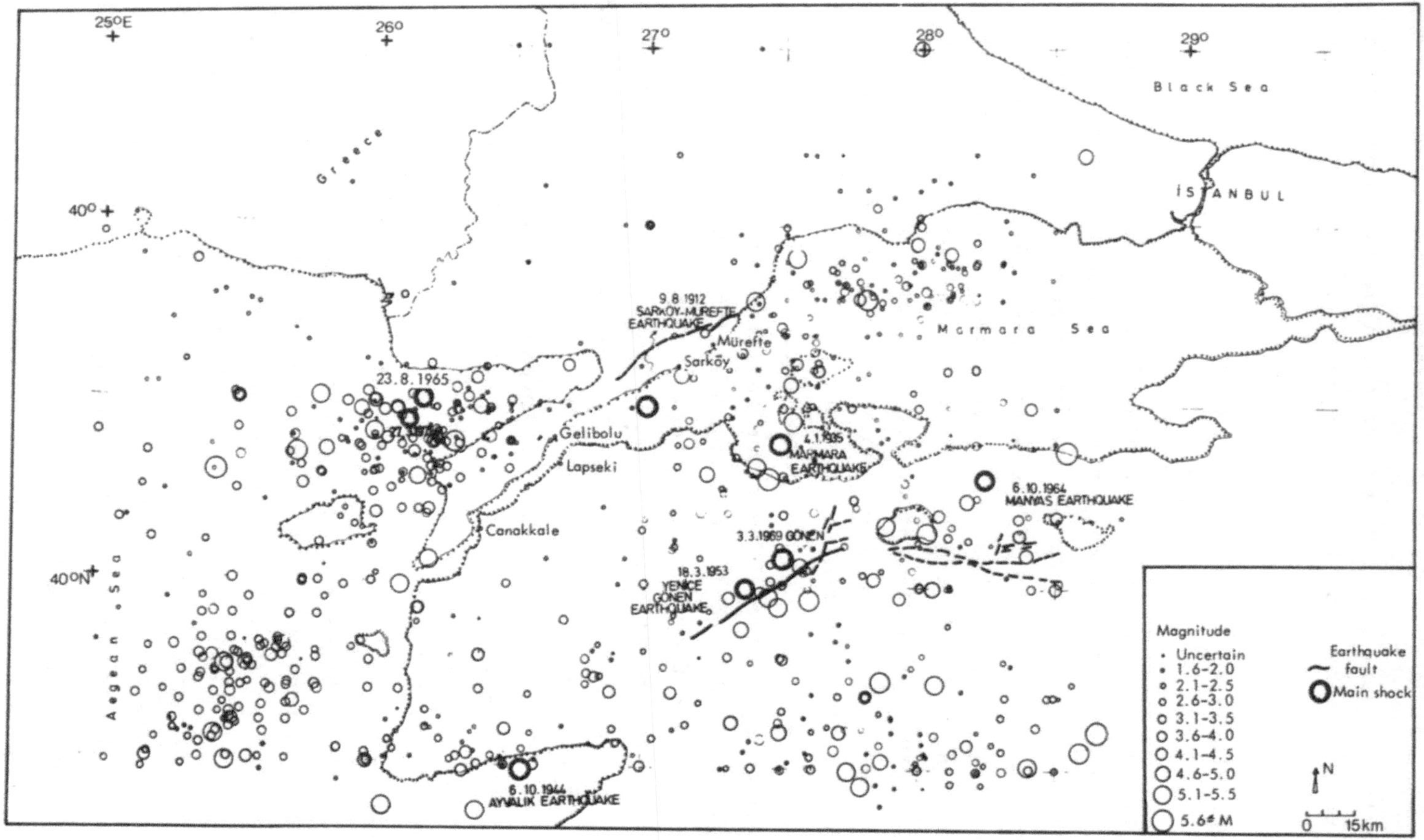

Fig. 2. Seismicity of the western Marmara Region (1900-1979)

On the northern branch of the North Anatolian Fault, which runs north of the Marmara Region, the 1912 Şarköy-Mürefte Earthquake has been the most destructive one of this century. It was associated with heavy damage and large ground ruptures. It appears that the 1912 earthquake with a magnitude of 7.2 causing the casualty of 3000 people has been poorly documented (Can 1975). Ambraseys (1971) drew the attention to the fact that existing earthquake catalogues, in general, have been compiled by copying earlier catalogues without a critical study of the cited accounts. As additional information, from the seismotectonic point of view, it would be very important if the search for original information from the Şarköy Earthquake could be continued.

The earthquake having the epicentre at 38°.65N - 27°.2E had a magnitude of 7.3 and a maximum intensity of X (Karnik 1971). It has been pointed out in the Catalogue of Earthquakes in Turkey and surrounding areas for the years 11 A.D. - 1964 P.D. (Ergin et al. 1967), that the relevant field information for the 1912 Earthquake was taken from the "Descriptive Catalogue for Turkish Earthquakes" (Pinar 1952). In this catalogue it was further referred to an unpublished report of Rothé as follows: "In the epicentral region, Şarköy and Mürefte, 80% of houses became uninhabitable. 183 people were killed and wounded. Cracks of 0.3 - 0.5 m width were observed in the valleys. One of the cracks between Yenice and Kastanbol was 5 m wide 10 m deep and a few kilometers long."

It was not possible to find the original report of Rothé but another report was found (Macovei 1912) in the paper "La Géologie Sismologique", (De Ballore 1918), which contains the same information. Finally, a very valuable field study of Mihailovic (1927) was found and evaluated."

In summer 1976 site investigations were carried out to collect more information about the Şarköy-Mürefte Earthquake of 1912. Evaluation of interviews with local people who lived there by the time of the earthquake as well as our geological investigations in the region led us to conclude that the earthquake fault system extends in the direction from Gaziköy to Kavak Deresi (Fig. 3) along the contact of Eocene-Oligocene

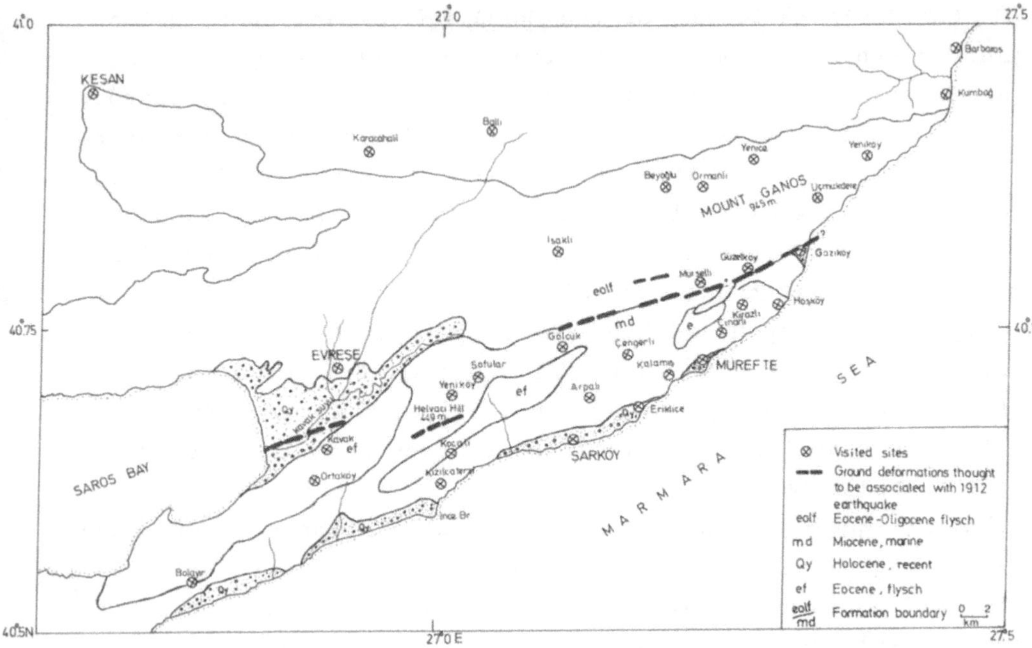

Fig. 3. Size investigation map of the 1912 Earthquake

flysch and Miocene marl which was rejuvenated during the earthquake (Ateş and Tabban 1976). The land deformations which, according to reports, occurred during the 1912 Earthquake are shown in Fig. 3.

The field report of Mihailovic (1927) has been translated to Turkish from Serbian language. It is mentioned in the book that between July 25 to October series of shocks occurred in the region. The main shock occurred on the 9th of August and seven shocks with an intensitiy IX followed the main shock. P arrival times of the main shock and of aftershocks recorded at 68 stations listed in the report may enable researchers to determine focal mechanisms and the aftershock volume. Photographs of ground deformations and damage associated with the earthquake are shown in Fig. 4. The localities where the photographs have been taken are indicated on the map of Fig. 3.

a- Gaziköy,

b- Hoşköy,

c- Güzelköy,

Fig. 4. Field study of the 1912 earthquake

d- Mürselli,

e- Osmanli,

f- Sarköy

Let us look at the past seismic activity of the Şarköy-Mürefte region before considering the microearthquake activity recorded in 1978. Regarding earlier earthquakes,catalogues of earthquakes between 1913 - 1979 were evaluated and the epicentres plotted on the map of Fig. 2. For the years immediately preceding and following the 1912 Earthquake, catalogue data seem rather insufficient. The reason is that during the years from 1912 to 1918, the Balkan Region has been the place of wars.

Until 1970 the number of seismic stations in the western part of the NAFZ were very few. Seismic activity could not have been monitored properly. From 1970 onwards,when the seismic network of West Anatolia became denser, surprisingly not much of an activity has been recorded in the fault region. Especially it appears that seismicity is relatively low before 1975 when an earthquake with a magnitude 5.7 caused considerable damage in Gelibolu-Lapseki area (Fig. 2) southwest of Şarköy. It is noticeable that earthquake activity in the recent years is cumulating at the northeastern and southwestern extension of the Şarköy-Mürefte Fault to the sea, whereas only a few shocks occurred around the fault break of the 1912 Earthquake.

One of the objectives of modern seismology is to monitor seismic gaps which potentially may be the place for the generation of future earthquakes. Therefore it was planned to monitor the earthquakes of Sarköy-Mürefte Fault in more detail.

In the summer of 1978, a small microearthquake recording array of four stations was set up in the fault region extending from Gaziköy in the east to Kavak Deresi in the west (Fig.3).The stations were Karacahalil (KRH), Beyoğlu (BEY), Mürefte (MRT) and Terziköy (TZY) (Fig. 5). It was hoped that data from the fixed seismic network stations of Demirköy (DMK), Edincik (EDC) and Ezine (EZN) would contribute to the determination of earthquakes. The four mobile stations are equipped with a Kinemetrics PS-1A recorder, digital clock, SS-1 seismometer of one second period, radio receiver, radio antenna and accessories. The station can be operated continuously for about two months

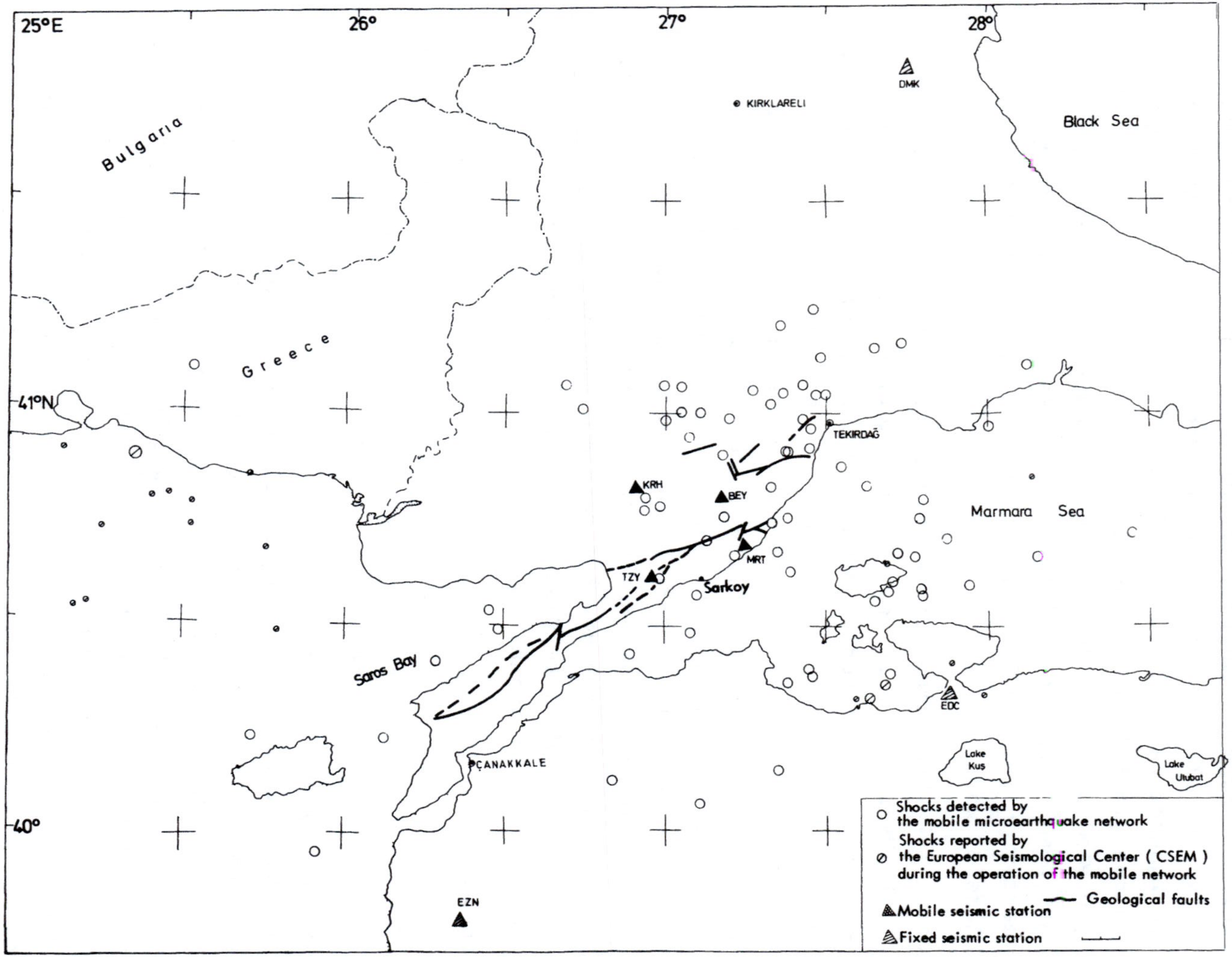

Fig. 5 Preliminary determination of microearthquake activity for the Şarköy-Mürefte Region

in the field. For each earthquake, P and S arrivals as well as durations were read on the records and the "HYPO 71 Revised" computer programme (Lee and Lahr 1975) modified for the IBM 370/145 computer, was used for the determination of focal parameters, employing a variable first layer model. A preliminary determination of earthquakes is shown in Fig. 5. It was not possible to find a detailed neotectonic map of the area studied however, a sketch of faulting is shown in the same figure.

As it can be réalized, the seismotectonic interpretation of any region greatly depends on both accurate neotectonic mapping and reliable determination of earthquake parameters. The latter requires instrument homogeneity for the stations, good quality of the records, errorless timing, correct knowledge of the crustal structure in three dimensions and a good data processing technique. The fixed station readings (DMK, EDC, EZN) caused serious trouble in data processing which may be due to the different type of instruments at mobile and fixed stations and also due to the inconsistency of arrival time readings at the permanent stations, which were not read by ourselves. Furthermore, seismic explosion studies would have been needed to investigate and to take into consideration the local crustal structure.

It can be noticed on Fig. 5 that microearthquakes seem to be concentrated north of the Şarköy-Mürefte Fault. On the preliminary stage of this study, the distribution of microearthquakes may be interpreted in different ways. Either the fault itself does not have a high microearthquake activity at present and the observed activity is associated with the faults south of Tekirdağ, or/and elastic strain is accumulating at a greater depth in the Şarköy-Mürefte region. Additional information on earthquake parameters is however required.

SEISMICITY OF THE EASTERN PART OF THE NORTH ANATOLIAN FAULT

As it has been pointed out earlier, the characteristics of the North Anatolian Fault die out towards the east of Karliova (Fig. 1).

Ketin (1977) suggests a probable extension of the North Anatolian Fault from Varto to Bulanik and furthermore to Erciş west of the Lake Van (Fig. 6). At the eastern part of Lake Van, he distinguishes four parallel zones and suggests the first zone to be the possible continuation of the North Anatolian Fault.

Arpat and Şaroğlu (1977) also suggest an extension of the North Anatolian Fault beyond Varto and a continuation south of Lake Nazik to the Lake Van near Ahlat (Fig. 6). But it should be added that this part of the fault has not been active since the last Quaternary volcanism.

Considering the regional seismicity of the Lake Van area for the years 1913-1978, it appears that most of the damaging earthquakes are recorded teleseismically since there were only a few seismic recording stations in the region. It was thought that it may be useful to analyse the damaging earthquakes around Lake Van in order to evaluate the overall pattern of contemporary seismicity. The earthquakes that took place in the area between (38.0°-40.0°)N-(41.0°-45.0°)E were classified according to their magnitude and plotted on the map of Fig. 6. Major damaging earthquakes and the associated surface faults are also shown in this figure.

In the east of Lake Van, an earthquake swarm occurred from June to December in 1945. Shocks of intensity VI caused damage to the villages in the Van Province.

On May 31, 1946 the Varto Earthquake occurred with an intensity of VII-VIII in the western part of the Lake Van region. An allegedly observed crack in N45°W direction near the completely destroyed village of Karkarot requires verification.

Later, the Varto Earthquake of August 19, 1966 with a magnitude of 6.7 and that of August 20, with a magnitude of 5.3 showed surface features in three parallel zones which are referred to as the southwest, middle and northeast fracture zones. The features trending N65°W display both right-lateral strike slip and extension normal to the fracture (Wallace 1966). Ketin (1966) mentions en-echelon type of fracturing with right-lateral sense of movement in N60°W direction to the south of Leylek Dağ in the Varto area. The earthquake of August 19,

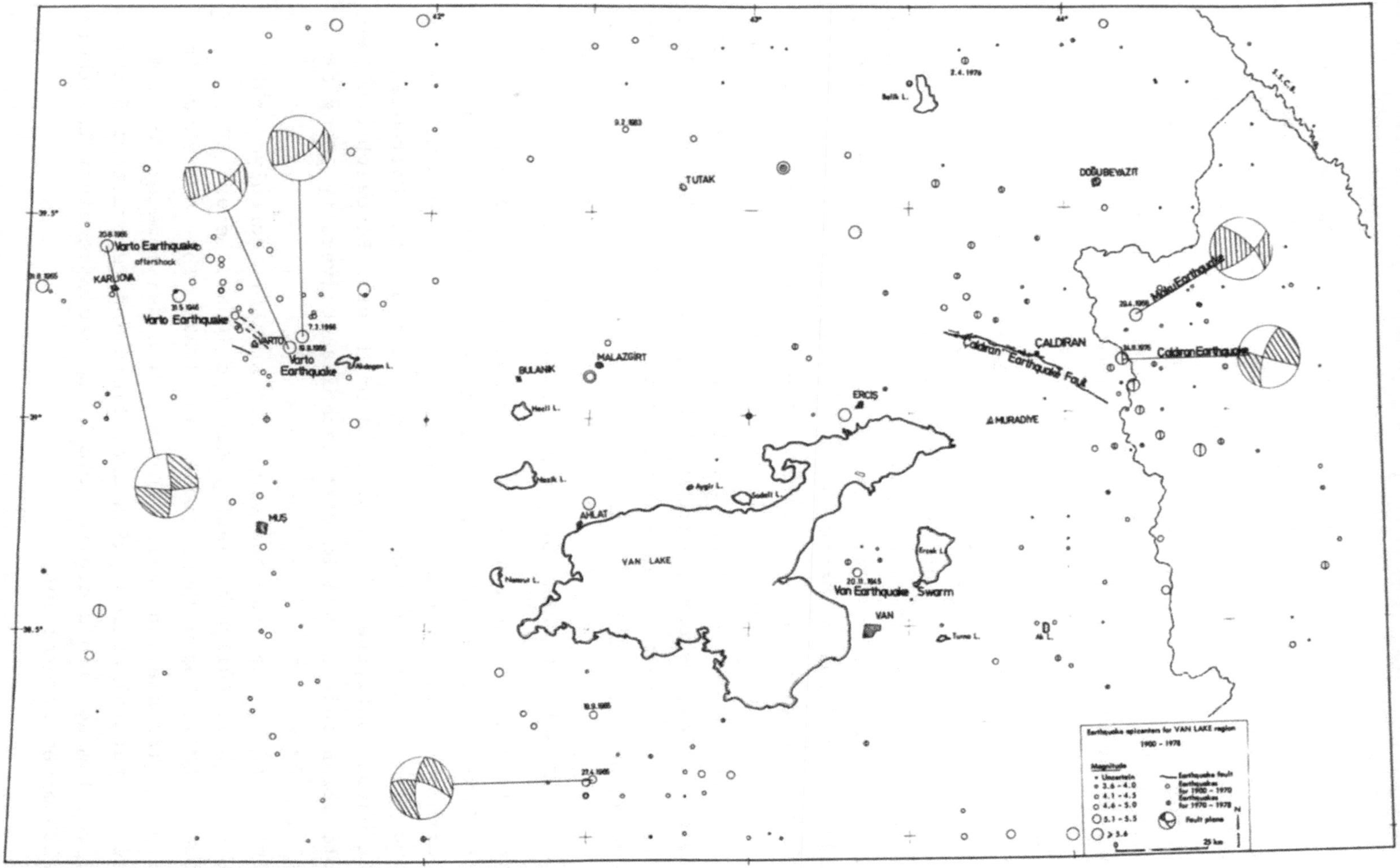

Fig. 6. Seismicity of the Lake Van Region (1900-1978)

1966 with the epicenter at 39°2N-41°6E shows the mechanism of thrust and strike-slip faulting. On the other hand, the major aftershock on August 20, 1966 with the epicenter at 39°4N-40°9E only shows strike-slip fault mechanism (McKenzie 1972).

One of the interesting earthquakes of the region is the Maku Earthquake of Iran on April 29, 1968 with a magnitude of 5.3 which caused damage in Western Azerbeijan. The fault mechanism solution cannot be conclusively evaluated as no surface faulting was observed.

The most recent major earthquake of Eastern Anatolia is the Caldiran Earthquake of November 24, 1976 with a magnitude of 7.3. The earthquake was associated with a 55 km long strike-slip fault of right lateral horizontal displacement as much as 330-350 cm at the western end of the fault (Arpat and Şaroğlu, 1977). The focal mechanism solution which has strike-slip characteristics in right lateral sense with a little thrust component (Toksöz et al., 1978) is in good agreement with the field observations. As can be seen on figure 6, the presently teleseismicly recorded earthquakes and focal mechanism studies are not sufficient enough for a detailed seismotectonic interpretation of East Anatolia.

MICROEARTHQUAKE SURVEY IN THE EASTERN PART OF THE NAFZ

To determine the subsurface configuration of active faults of the Lake Van area and their relation to seismicity, a microearthquake array has been operated for about five months starting from January 1978 by cooperation between the Earthquake Research Institute, the Institute of Mineral Research and Exploration and the Middle East Technical University (Ateş et al., 1979).

The names of the stations are Kalecik (KAL), Muradiye (MUR), Erciş (ERC), Adilcevaz (ADZ), Güntepe (GÜN), Gevaş (GEV), Özalp (ÖZA), Hoşap (HOŞ) and Patnos (PAT). They are shown in Fig. 7. The stations were equipped with seismometers of one second natural period, with amplifiers and recorders and for exact timing with a digital clock and radio receiver together with other accessories.

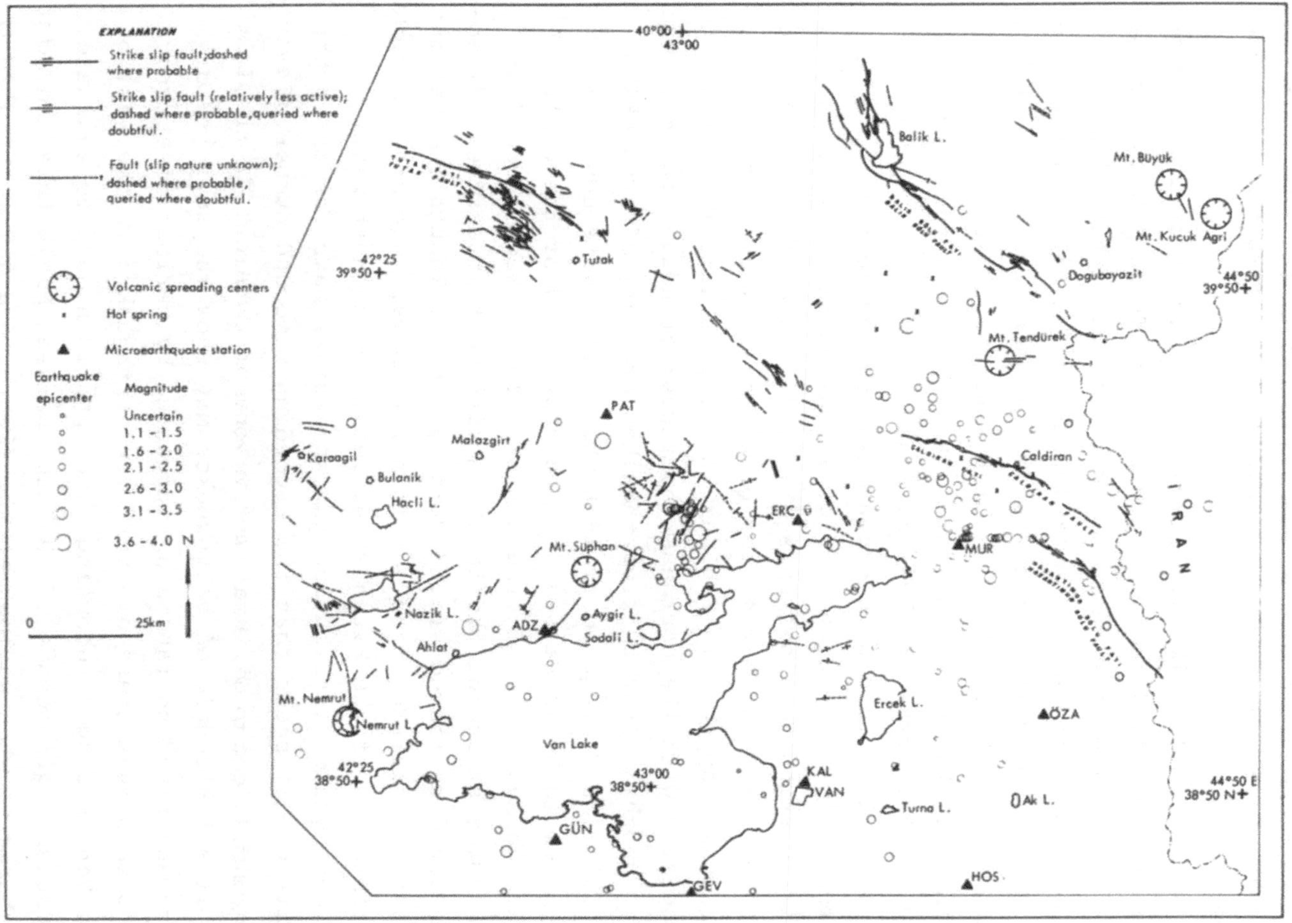

Fig. 7. Microearthquake distribution and active faults of the Lake Van Region

For each shock, P and S arrival times and the duration of shocks were read on the records of the array stations.
The "HYPO 71 Revised" computer program was used for the calculation of focal parameters employing a variable first layer model.
Active faults for the area studied were traced on aerial photographs and verified by ground surveys. Those faults were defined as active which have moved since Quaternary. They were classified according to their length, continuity and the morphological characteristics. Movements which are known, such as along the Balik Gölü Fault, Tutak Fault, Çaldiran Fault and Hasan Timur Fault are indicated by arrows. All these faults show right-lateral strike-slip features.
The spatial pattern of microearthquakes and active tectonics of the region are shown in Fig. 7. Earthquake clusters and their relation to lineaments are evaluated as follows: At the north of Lake Van to the west of Erciş (ERC) a cluster of earthquakes appears in February 1978. This activity can be associated with the northeast-southwest trending faults in the area. Another cluster of earthquakes are late aftershocks of the Çaldiran Earthquake of 1976 which still were continuing. They seem to be located at the northwest end of the surface fault whereas early aftershocks of Çaldiran were located at the southeast end of the fault. A lineament in eastwest direction could be traced around Muradiye (MUR). Though indications of surface faulting were not found, this may suggest a possible continuation of the Hasan Timur Fault to the northwest.
The earthquake distribution in the east of Lake Van is rather scattered. It seems that the recording time and number of events is not sufficient to delineate active zones in the area. In the Lake Van itself a couple of earthquakes may show the activity of east-west trending faults as suggested by various researchers. It appears that there are very few earthquakes in the west of Lake Van. A longer recording time span is needed to evaluate the seismic quiescence of the area between Nazik Lake and Ahlat. Furthermore, when we look at the figures 6 and 7 we find in the north of the region that the recent activity of the Balik Gölü Fault extending in northwest-southeast direction, could be related to the damaging earthquakes of April 12, 1976 and the

Maku Earthquake of April 29, 1968. The second northwest-southeast trend is formed by the Çaldiran Earthquake Fault and its probable continuation to the northwest, the Tutak Fault. The activity of this trend is noticeable.
Besides, the historical earthquakes of the region are being studied in more detail. It is expected that it will bring a wider perspective to this microearthquake evaluation.

CONCLUSION

The North Anatolian Seismic Zone of Turkey is a continental plate boundary located between the Turkish and the Eurasian Plates. The North Anatolian Fault is well defined between Abant and Karliova, then it branches in the regions of Marmara to the west and the Lake Van to the east.
In the Marmara Region, the Şarköy-Mürefte Earthquake of 1912 and its aftershocks caused heavy damage within an area extending from Gaziköy to the northeast, to Kavak to the southwest. This earthquake was associated with a surface faulting of 40 km. The study of 1912 Earthquake shows that it may be feasible to study an earthquake which occurred half a century ago. Many other earthquakes may still be studied using published and unpublished documents and by field inspection.

The Şarköy-Mürefte region appears to be a gap, one of the places of extremely low seismicity on the North Anatolian Seismic Zone. A temporary microearthquake survey lets us suggest that the seismicity pattern of this gap, covering an area of less than 100 km square kilometers should be examined in detail. It would be extremely useful to carry out quantitative measures in order to study the type of earthquakes which occur in this area. Besides, this region is of great socio-economic importance since most industries of the country are located in the Marmara Region.
At present, the seismotectonic interpretation of the East Anatolia is a complex problem. A microearthquake survey and active faulting studies around the Lake Van Region, covering about 400 square kilometers, have emphasized the correlation of micro earthquake activity with the active faults. This correlation of earthquake clusters with lineaments seems very promising for

seismic hazard analysis. A statistical evaluation for the earthquakes in this area is underway. However, it is necessary to establish a dense, permanent seismic network in East Anatolia, supplemented by detailed studies of the local crustal structure and physical properties. This should be of highest priority in future earthquake research.

I would like to thank Professor Vogel for critically reviewing my paper and for his valuable comments.

REFERENCES

Allen, C.R., 1969, Active faulting in Northern Turkey. Div. Geol. Sci. Cont. no. 1577 Cal. Inst. Tech.

Ambraseys, N.N., 1970, Some characteristic features of the Anatolian Fault Zone. Tectonophysics 9, 143-165

Ambraseys, N.N., 1971, Value of historical records of earthquakes. Nature 232, 375-379

Ambraseys, N.N., 1978, A reappraisal of the seismicity-Middle East. Q. Jl. Engng. Geol. v. 11, 19-32

Arpat, E. and Şaroğlu, F., 1977, Some recent tectonic events in Turkey. Bul. Geol. Soc. of Turkey, n. 1 v. 18 (in Turkish)

Ateş, R. and Tabban, A., 1976, Preliminary report for the Şarköy-Mürefte Earthquake of 9th August, 1912. Publ. Earth. Res. Inst.

Ateş, R. and Şaroğlu, F., Türkelli, N., Olguner, A., 1979, Earthquake activity of the Lake Van region. 33. Conf. Geol. Soc. Turkey (in Turkish)

De Ballore Montesus, 1918, La Géologie Sismologique. Paris.
Can, R., 1969, The earthquake sequence of West Anatolia, 1969, a field report. Eng. Seis. Sec. Imp. Coll. publ.

Can, R., 1975, Seismotectonics of the North Anatolian Fault Zone. MPhil. D. thesis, Univ. London, Fac. Eng.

Karnik, V., 1971, Seismicity of the European Area I-II. Praha

Ketin, I., 1964, Géotectonique de Turquie. Comm. Cart. Géol. Monde Moscow, 258-262

Ketin, I. and Roesli, H., 1953, Makroseismische Untersuchungen über das nordwest anatolische Beben vom 18. März 1953. Eclogae geol. Helv., Vol. 46, no. 2 Basel

Ketin, I., 1966, Tectonic units of Anatolia (Asia Minor) Bull. Min. Res. Expl., no. 66, pp. 23-34

Ketin, I., 1977, Brief explanation of geological observations in the Lake Van region and Iranian border. Bull. Geol. Soc. Turkey, no. 20, 79-85 (in Turkish)

Lee, W.H.K. and Lahr, J.C., 1975, HYPO 71 Revised, a computer programme for determining hypocenter, magnitude and first motion pattern of local earthquakes. U.S. Geol. Surv.

Macovei, G., 1912, Sur le tremblement de terre de la mer de Marmara le 9 août 1912. Bull. Sect. Sc. Ac. Roumaine I, no. 1, 9, B

McKenzie, D., 1972, Active tectonics of the Mediterranean region. Geophys. J. R. Astr. Soc. 30, 109-185

McKenzie, D., 1978, Active tectonics of the Alpine-Himalayan Belt: the Aegean Sea and surrounging regions. Geophys. J. R. Astr. Soc. 55, 217-254

Mihailovic, J., 1927, Memoire sur les grandes tremblements de terre de la Mer de Marmara 1912. Publ. Serbian Imperial Académy Beograd 1927

Toksöz, N. and Nabelek, J., Arpat, E., 1978, Source properties of the 1976 Earthquake in east Turkey. Tectonophysics 49, 199-205

EARTHQUAKE CATALOGUES

Pinar, N. and Lahn, E., 1952, Descriptive catalogue of Turkish Earthquakes. Minist. Publ. Works, no. 36

Ergin, K. and Güçlü, U., Uz, Z., 1967, A catalogue of earthquakes for Turkey and surrounding areas (11A.D. to 1964) Publ. Tech. Univ. İstanbul

Alsan, E. and Tezuçan, L., Båth, M., 1975, An earthquake catalogue for Turkey for the interval 1913-1970. Publ. Kand. Obs. and Seism. Inst. Uppsala, no. 7-75

Regional catalogue of earthquakes. Publ. Int. Seism. Center (ISC) Newbury, U.K.

Determination Hypocentrale. Centre Seismologique Europeo-Mediterranéen, Strasbourg-France

A Rational Approach for the Probabilistic Assessment of the Seismic Risk Associated with the North Anatolian Fault

Mustafa Erdik and Semra Öner

Earthquake Engineering Research Institute, Middle East Technical University, Ankara, Turkey

ABSTRACT

The seismic hazard associated with the North Anatolian Fault has been assessed. The results are provided in terms of the expected values of the MSK intensity and the maximum horizontal ground acceleration for different return periods at given spatial locations.

Results are obtained through perusal of (1) artificial complementation of the small magnitude seismic data, (2) locally obtained intensity atteñuation relationships and (3) fault rupture probabilistic model.

This rational approach is also compared with more conventional approaches and differences are indicated.

INTRODUCTION

In assessing the earthquake resistant design parameters of a facility in a region prone to seismic activity, a specification of the maximum ground motion intensity expected at a site over a given duration is needed. These data are usually supplied on the basis of risk analyses. Earthquake risks have to be calculated indirectly, by an extrapolation of existing data. To carry out such an extrapolation, the engineer must be guided by geophysicists on the underlying geophysical theory, which, unfortunately, is still not well understood. Thus the inherent uncertainity of predicting a physical event with only incomplete data is inextricably compounded by an element of uncertainty concerning the status of the theory on which the prediction is based.

In stochastic treatment of earthquake events, the occurrence of earthquakes whose intensity at a site is greater than a given value is generally represented by a homogeneous Poisson process with complete temporal independence. Are earthquake occurrences really random in time or should they occur in regularly predictable cycles? Although there are theories suggesting the latter might be true, particularly for the North Anatolian Fault Zone, the mankind's experience with limited width time windows in observing these events, is the evidence of the random unexpected reccurrence of great earthquakes. Until one can prove the regularity of these events, it is prudent to assume them to be random in time. The representation of earthquake occurrences by a Poisson process will be retained on the grounds of its simplicity, and of the fact that it is probably adequate for the highest intensities, and that seismic design decisions are usually more sensible to the mean number of events than to their distribution in time.

The seismic risk of Turkey in general and the North Anatolian Fault in particular has been subject to many assessments. With the exception of Gürpınar et.al. (1979), all of these studies have (1) assumed the earthquake events as point energy sources in the spatial domain, (2) utilized borrowed attenuation equations with minimal relevance to the regional characteristics, (3) tried to assess the maximum earthquake magnitudes through statistical manipulations with no consideration of the physical limitations of the earthquake process and (4) obtained the reccurrence relationships on the basis of limited and highly deficient data with no attempt to homogenize or artificially complement the deficient reporting periods.

This study aims to analyze the seismic risk associated with the North Anatolian Fault in a rational and consistent manner utilizing (1) the proper fault-rupture models, (2) regional intensity attenuations, (3) fault-length related maximum magnitudes, (4) reccurrence relationships based on complemented data and will comparatively point out the inadequacies inherent in the previous approaches.

NORTH ANATOLIAN FAULT ZONE

North Anatolian Fault Zone is one of the largest tectonic lineaments of the earth's crust exhibiting a very well established correlation with the seismic activity. The recent seismicity of the fault is illustrated by portraying the epicenters of earthquakes of magnitude 5.9 and greater which occurred between 1901 and 1977 on the tectonic map given in Fig 1. McKenzie (1972) traces the North Anatolian Fault between the triple junctions located between the epicenters of 22.7.1967 and 30.7.1967 events on the West and those of 19.8.1966 and 20.8.1966 on the East on the basis of the coherence of fault plane solutions. In his parlance of intra-plate tectonics, the North Anatolian Fault is a dextral strike-slip fault separating the Turkish and Eurasian plates. This definition of the western and eastern termini of the fault is also adapted in this study.

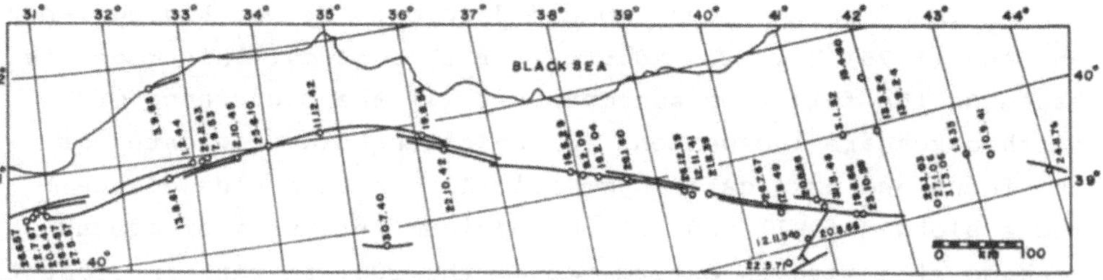

Fig. 1 Macroseismic locations of the large earthquakes occurred in the region considered.

APPLICABLE RISK MODEL

In planning the design of a structure in a region of potential seismic activity, future ground motion expected at a site is needed. The necessary information for the determination of the required site intensity will include: (1) locations, (2) dates of occurrence, (3) magnitudes, (4) intensities, (5) focal depths, (6) strong motion records of all past earthquakes and (7) the attenuation characteristics of the ground motion associated with these events. Since such information is not complete for most zones of seismic activity the required information on the site intensity, therefore, has to be derived on the basis of available information, through the employment of seismic risk models.

Seismic risk analyses aim to assess the probability that a maximum ground motion parameter at a site due to earthquakes from potential seismic sources will exceed a certain value in a given time period. The seismic risk models developed by Cornell (1968), Esteva (1970) and Milne and Davenport (1969) are called "point-source models" since they are based on the assumption that the energy released during an earthquake is radiated from the focus of the earthquakes and the earthquake occurrences are assumed to be spatially and temporally independent. Although the first assumption may be acceptable for small local earthquakes, it would not however, be valid for large events where the total energy released is distributed along a long rupture zone.

The "fault-rupture model" developed by Der Kiureghian and Ang (1975) has been adopted for this study. This model is based on the following assumptions: (1) the occurrence of earthquakes are independent (2) an earthquake originates at the focus and propagates symmetrically on each side of the focus along a fault and (3) the maximum intensity of ground shaking at a site is determined by the rupture that is closest to the site. Note that the last assumption is justified by the elongated shape of isoseismals along the North Anatolian Fault. For this reason the shortest distance of a site from the fault rupture should be the determining factor in attenuation relationships to be utilized along with this fault rupture model. Whereas in the point-source models the attenuation relationships encompassing the hypo-or epicentral distances are to be employed.

For proper implementation of the fault-rupture model the rupture length-magnitude relation, that is the equation that determines the expected total slip length as a function of the magnitude, should be obtained for the fault in question. In this model, relative to the point source models in which the rupture length is neglected or not considered, the length of the rupture zone is very significant since the probabilities associated with long rupture zones propagating close to a site will be a major contribution to the overall risk.

The total length of the rupture zone as related to the magnitude can be given as:

$$\log L = c_1 + c_2 M \qquad (1)$$

where L is the total rupture length in kms associated with the magnitude M and, c_1, c_2 are the regression constants. Perusal of the data on Table 1. provides the following equation for the rupture length-magnitude relationship of the North Anatolian Fault zone:

$$\log L = -3.932 + 0.812\ M \qquad (2)$$

For this relation the correlation coefficient is 0.95 and the standard deviation of rupture length is 0.43.

Table 1. Earthquakes Utilized to Assess the Intensity Attenuation Relationship

Earthquake	Date	Magnitude M	Max. Intensity I_n	Rupture length (km)
Erzincan	26.12.1939	7.9	XI	340
Erbaa	20.12.1942	7.0	IX	40
Ladik	26.11.1943	7.2	X	280
Gerede	01.02.1944	7.2	X	180
Kurşunlu	13.08.1955	6.9	IX	50
Hendek	26.05.1957	7.1	IX	40
Varto	19.08.1966	6.9	IX	25
Adapazarı	22.07.1967	7.2	IX	80
Çaldıran	24.11.1976	7.2	IX	53

RECCURRENCE RELATIONSHIPS

The seismicity of a seismic source is defined through an equation of the type:

$$\log N = a + bM \quad (3)$$

where N is the annual number of earthquakes greater than or equal to the magnitude M and a, b are the regression constants.

For the seismicity of the North Anatolian Fault the earthquake data has been collected from various cotalogues. For the period 1923-1970, Alsan et.al.(1975) catalogue has been found more reliable. This catalogue has been supplemented by the NOAA cataloques for the interval 1970-1976 (Erdik and Özselçuk, 1978). The following frequency-magnitude relationship can be obtained through least-square type regression of the unadjusted, raw reccurrence data of the associated earthquakes:

$$\log N = 2.486 - 0.514\ M \quad (4)$$

Note that this relationship provides a magnitude of about 10.7 corresponding to a reccurrence interval of 1000 years. This unrealistic value stems from the deficiency of data in small magnitudes especially in earlier reporting periods. This deficiency is also evidenced by the small slope of Eq.4. Therefore, to arrive at realistic and rational assessments of the reccurrence relationships one has to combine a short sample of events which is complete in small events with a longer sample of events which is complete in large earthquakes to obtain a complete and homogeneous set of events. For this purpose the data base is supplemented by the historic large events from Ergin, et.al. (1967) and by the findings of the microseismic studies of Üçer, et.al. (1977) and Kaptan and Ata (1978). This extended data set has been further analyzed for completeness and complemented by the method proposed by Stepp (1973). Least-square type regression analysis on this adjusted data set yields the following adjusted reccurrence relationship:

$$\log N = 3.92 - 0.78\ M \quad (5)$$

The sensitivity of the seismic risk assessments to both these reccurrence relationships will be investigated.

ATTENUATION RELATIONSHIPS

In seismic risk analysis, once a causative fault or a tectonic province has been identified and corresponding design earthquake has been associated with it, the problem reduces to the determination of the intensity of site motion resulting from the occurrence of such an earthquake. This involves an understanding of the source mechanism and the attenuation properties of the transmitting medium. The scarcity of the strong-motion acceleration data in Turkey makes it indispensible to assess the attenuation of intensities for a rational seismic risk analysis.

The data base for the intensity attenuation is comprised of isoseismal maps belonging to the large earthquakes associated with the North Anatolian Fault. A list of these earthquakes with pertinent information is provided in Table 1. The iso-seismal maps are compiled from various sources including Ergin et.al. (1967), and Earthquake Research Institute, Ministry of Reconstruction and Resettlement. Intensities in Turkey are given mostly in MSK (Medvedev-Sponheuer-Karnik) and to a lesser degree in MM (Modified Mercalli) scales. For practical reasons and considering the intensity range of the available data both scales will be assumed to be interchangable. Due to the strike slip type mechanism of the North Anatolian Fault Zone the resulting iso-seismals exhibit elliptical character with strong azimuthal dependence. This can be easily observed on the isoseismal maps of Erzincan, Bolu-Gerede and Çaldıran earthquakes given in Fig.2.

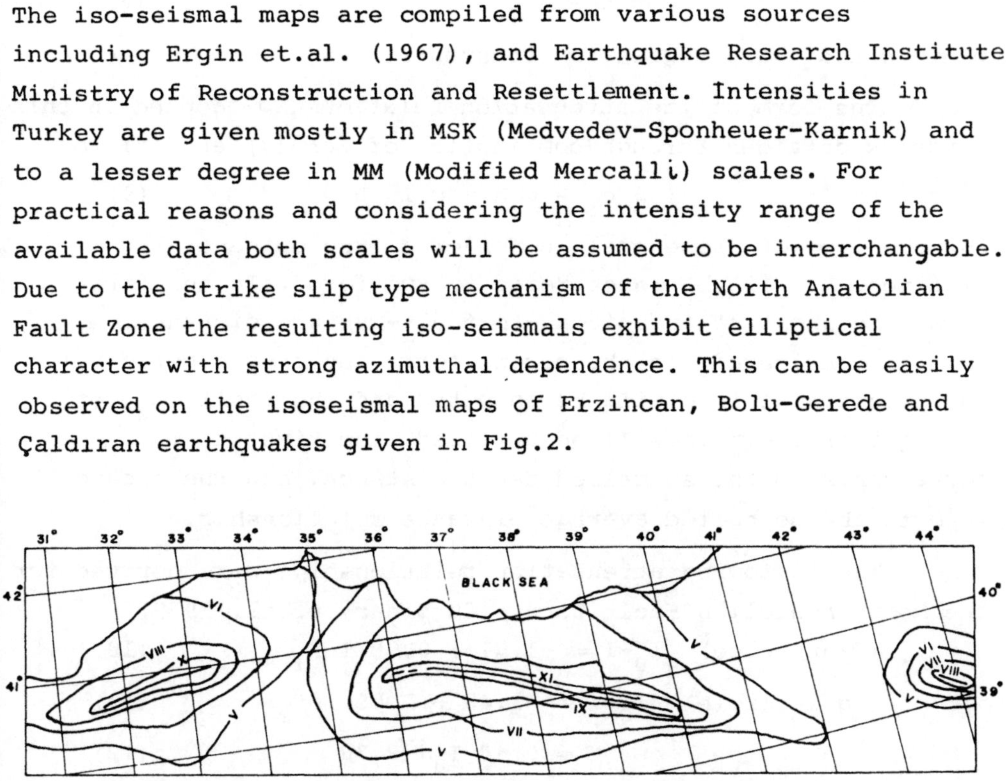

Fig. 2 Isoseismal maps of Erzincan (26.12.39) Bolu-Gerede (1.2.44) and Çaldıran (24.11.76) earthquakes.

For describing the attenuation of intensity with distance many studies (e.g. Howell and Schultz, 1975 and Gupta and Nuttli, 1976 and Chandra, et.al., 1979) have employed relations of the form:

$$I - I_0 = a + bR + c\ln R \qquad (6)$$

Where I is the site intensity at an epicentral distance R, I_0 is the maximum epicentral intensity and a, b, c are regression constants appropriate to the region under consideration. These studies also indicate negligible values for the regression constant b justifying the adoption of the following simpler relationship:

$$I - I_0 = a + c \ln R \qquad (7)$$

The relationship of maximum intensity, I_0, and magnitude, M, has been computed for many areas of the world with the following form:

$$M = d + e\, I_0 \qquad (8)$$

where d and e are regression constants.

The form of the attenuation relationship adopted in this study is obtained through combination of Eqs (7) and (8) to yield:

$$I = c_1 + c_2 M + c_3 \ln R \qquad (9)$$

For the type of isoseismals considered, the epicentral distance, R, will be interpreted in two forms. In the first form, the shortest isoseismal-to-fault-rupture distance, R_1, will be considered. In the second form, the area covered by each isoseismal line will be computed and the radii, R_2, of the equi-area circle will be utilized. The first form corresponds to the so called fault distance, and the second form to the so called average distance relationship.

The following attenuation relationships were derived for the North Anatolian Fault Zone: (Gürpınar, et.al. 1979)

(Transverse) $$I_0 - I = -1.71 + 0.98 \ln R_1 \qquad (10)$$

(Equi-circle) $$I_0 - I = -1.40 + 1.05 \ln R_2 \qquad (11)$$

$$M = 0.48\, I_0 + 2.70 \qquad (12)$$

(Transverse) $$I = -3.92 + 2.08\, M - 0.98 \ln R_1 \qquad (13)$$

(Equi-circle) $$I = -4.22 + 2.08\, M - 1.05 \ln R_2 \qquad (14)$$

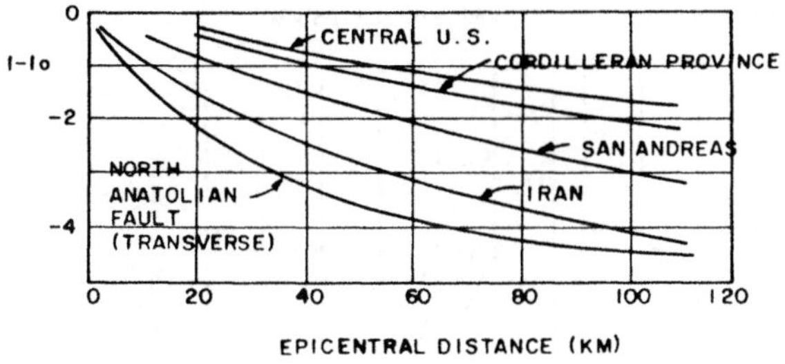

Fig. 3
Comparison of intensity attenuation relationships.

The standard deviation of intensity in Eqs (8) and (9) have been found to be 0.40.

In Fig. 3, Eqs (10) and (11) have been compared to those reported by Chandra, et.al. (1979) for Iran and by Howell and Schultz (1975) for the San Andreas Province, Cordilleran Province and Eastern Province. As it could be seen the attenuation relationships for the North Anatolian Fault Zone indicate faster drop rates compared to the others. This mainly stems from the regional attenuation characteristics and the type of structures from which the intensity assignments are determined.

Although various acceleration attenuations are developed, most of them use the hypocentral distance to provide the site acceleration for a given earthquake magnitude at the source. Attenuation equations given by Donovan (1973) provide horizontal peak site accelerations in terms of the distance to fault and the magnitude, and are indicated below:

$$a_g = 1320\, e^{0.58M}(R + 25)^{-1.52}, \quad \ln a_g = 0.84 \qquad (15)$$

In this equation a_g stands for the peak ground acceleration in gals, R is the distance to fault in kms and M is the magnitude of the earthquake.

RESULTS AND CONCLUSION

For comparative application of the seismic risk methodology adopted herein three test sites located at 25, 75,

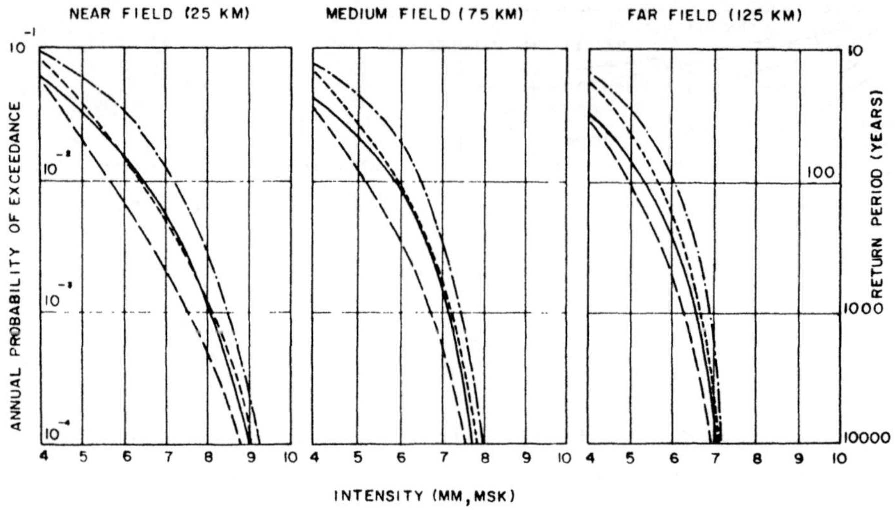

Fig. 4 Risk analysis based on intensity attenuation. (For nomenclature of curves see Fig. 5)

and 125 km perpendicular distances from the mid-point of the North Anatolian Fault are utilized. These sites will respectively represent the so-called near-, medium-, and far-field conditions from the fault.

In Fig.4 the maximum site intensity corresponding to different annual probabilities of exceedance or return periods are plotted for each of the three sites. The intensity attenuation as given by Eq. 13 is employed. For each test case the results are presented for the following combinations of input and risk model.

(A) Adjusted Recurrence Relationship with the Fault Rupture Model

(B) Unadjusted Reccurrence Relationship with the Fault Rupture Model

(C) Adjusted Reccurrence Relationship with the Point Source Model

and (D) Unadjusted Reccurrence Relationship with the Point Source Model

In Fig.5 the same format of presentation is followed for the peak site accelerations corresponding to different annual probabilities of exceedance based on the acceleration attenuation relationship of Donovan, (1973) given by Eq. 15.

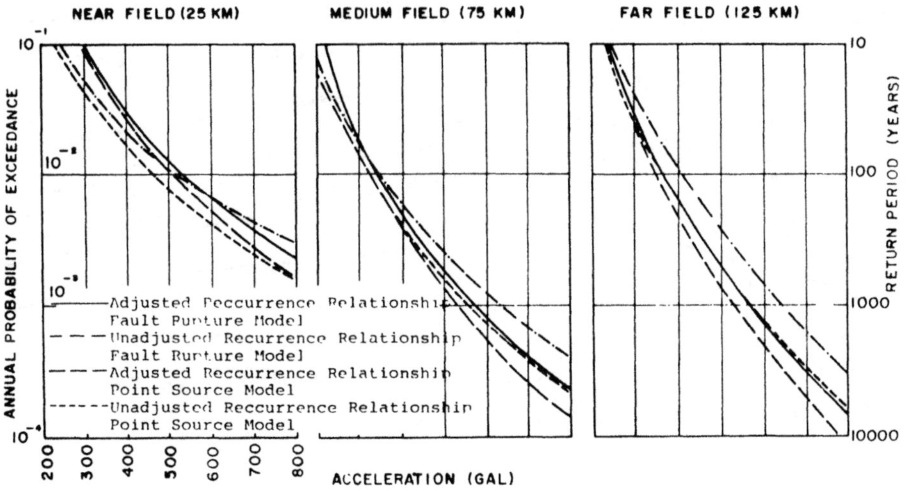

Fig. 5 Risk analysis based on accelaration attenuation.

The correct reccurrence relation and the risk model combination is the one based on the adjusted reccurrence relation and the fault rupture model and the results based on this combination are indicated by solid lines in Figs. 4 and 5.

As it can be seen: irrespective of the risk model employed and the site distance the use of unadjusted reccurrence relationship may seriously overestimate the seismic risk in terms of either intensity or acceleration. In the case of site peak acceleration this finding is more pronounced for higher acceleration or lower annual probability of exceedance levels.

Assuming consistent employment of adjusted reccurrence relationships the use of point-source model without any consideration of the fault rupture and mechanism could seriously underestimate the seismic risk in terms of both the peak acceleration and the maximum intensity. This conclusion is dependent of the site to fault distance.

Note that, although these results of Fig.4 and 5 portray the risk only at some isolated points, the results can be carried to other locations at the same respective distances from the fault line in the middle portions of the

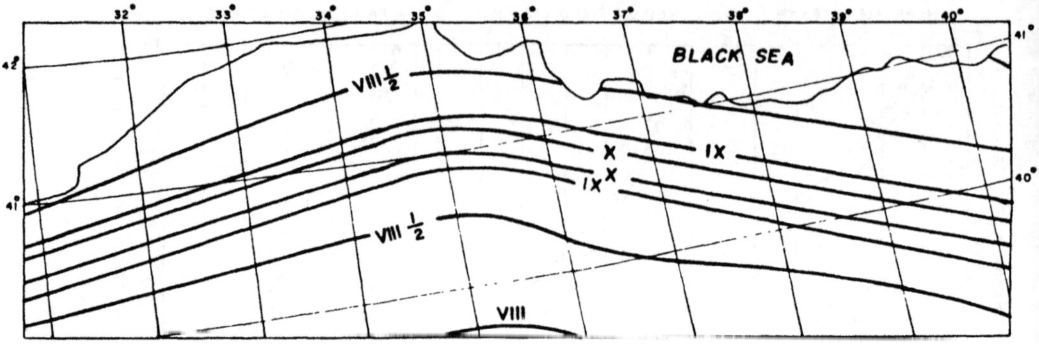

Fig.6 Iso-intensity lines for return period of 475 years. After Gürpınar, et. al., (1979)

fault. This characteristic of the seismic risk results associated with long faults in the presence of no other sources is illustrated in Fig.6 taken after Gürpınar et.al. (1979) portraying the equi-intensity contours of the Northern-Turkey for a return period of 475 year, or for a probability of exceedance of 0.1 in 50 years.

REFERENCES

Alsan, E., L. Tezuçan and M. Bath (1975), An Earthquake Catalogue, for the Interval 1913-1970, Kandilli Observatory, İstanbul.

Chandra, U., J.G. Mc Whorter and AA. Nowroozi (1979), "Attenuation of Intensities in Iran," Bull. Seism. Soc. Am. Vol 69, pp. 237-250.

Cornell, C.A. (1968), Engineering Seismic Risk Analysis: Seism. Soc. Am. Bull., 58, 1503-1606.

Der Kiureghian, A. and A. H-S. Ang (1975), A Line Source Model for Seismic Risk Analysis, University of Illinois Technological Report, 75-2023, Urbana, Ill.

Donovan, N. C. (1973), Earthquake Hazards for Buildings: Building Practices for Disaster Mitigation, Nat'l. Bur. Standards Bldg. Sci. Series, 46, 82-111.

Erdik, M. and A. Özselçuk (1978), Regional Rearrangement of the Earthquake Catalogues for Turkey, Middle East Technical University, EERI, No:76-08.

Ergin, K., U. Güçlü and Z. Uz (1967), Türkiye ve Civarının Deprem Kataloğu, M.S. 11-1964, İstanbul Technical University, İstanbul.

Esteva, L. (1970), Seismic Risk and Seismic Design Decisions: Seismic Design for Nuclear Power Plants, ed. by. R. J. Hansen, MIT, Cambridge, Mass.

Gupta, I.N. and O. W. Nuttli (1976), Spatial Attenuation of Intensities for Central U.S. Earthquakes, Bul. Seism. Sec. Am. Vol. 66. pp. 743-751.

Gürpınar, A., M. Erdik, S. Yücemen and M. Öner (1979), Seismic Risk Analysis of the North Anatolia Based on Intensity Attenuation, Proc. 2nd. Nat. Conf. on Earthquake Engry., Stanford, Cal.

Howell, B.F. and T.R. Schultz (1975), Attenuation of Modified Mercalli Intensity with Distance from the Epicenter, Bull. Seism. Soc. Am. Vol. 65, pp. 651-665.

Kaptan, L. and M. Ata (1979), Altınkaya Baraj Bölgesi ve Çevresinin Depremsel Özelliklerinin Araştırılması. DSİ Report.

Mc Kenzie, D. (1972), Active Tectonics of the Mediterranean Region, Geophy. J., 30, p.90.

Milne, W. G. and Davenport A.G. (1969), Distribution of the Earthquake Risk in Canada, Bull, Seism. Soc. Am, 59, 729-754.

Stepp, J. C. (1973), Analysis of Completeness of the Earthquake Sample in the Puget Sound Area, NOAA Technical Report No. ERL 267-ESL, pp 16-28.

Üçer, S. B. et. al., (1977), Batı Türkiye'de Deprem Etkinliği, Earthquake Research Institute, Bulletin. No. 19, Ankara, Turkey.

Seismotectonic Aspects of the North Anatolian Fault Zone Between Bolu and Havza

A. A. Barka and P. L. Hancock

Department of Geology, University of Bristol, Queen's Building, University Walk, Bristol BS8 1TR, UK

Abstract - The epicentres of 92 shallow earthquakes for the period 1900-75 have been relocated by integrating seismological data with the known distribution of active fault breaks. Four equal energy flux maps prepared from the relocated epicentre map demonstrate that deformation between Bolu and Havza is not restricted to the relatively narrow North Anatolian fault zone. Although seismicity extends to both the north and south of the main fault trace the area of activity to the south is broader. Earthquakes in this southern region may be related to events on fractures whose traces are visible as three sets of lineaments on Landsat images.

INTRODUCTION

The principal aim of this note is to present a relocated epicentre map for the period 1900-75 (Fig.1) and four contoured energy flux maps (Figs.2-5) of the region within and adjacent to the North Anatolian fault zone between Bolu and Havza. The epicentres of 92 shallow focus earthquakes of magnitude greater than 3 were relocated by integrating information available from: (a) earthquake catalogues and other sources of seismological data (Ergin et al. 1967, 1971, Karnik 1969, 1971, Can 1974, Dewey 1976, Alsan et al. 1975, Sipahioglu pers. comm. 1977); (b) the positions of fault breaks determined during surveying or known from records (Ambraseys & Zatopek 1968, Dewey 1976); and (c) field observations of tectonic structures. An account giving details about each epicentre and the procedure used for its relocation is in preparation.

The relocated epicentre map was then used to prepare four equal energy flux maps for the periods 1900-19 (Fig.2), 1920-39 (Fig.3), 1940-59 (Fig.4) and 1960-75 (Fig.5). Energy values were calculated using the formula $\log E = 12.24 + 1.44M$ for earthquakes of magnitude greater than 5, and $\log E = 4.78 + 2.57m$ for those of less than magnitude 5,where E = ergs, M = magnitude and m = 0.56M + 2.9, m being the magnitude based on body waves (Bath 1973, pp. 112 and 116). The contoured maps were derived from the energy data by employing a technique suggested by N.N.Ambraseys (pers. comm. 1978).

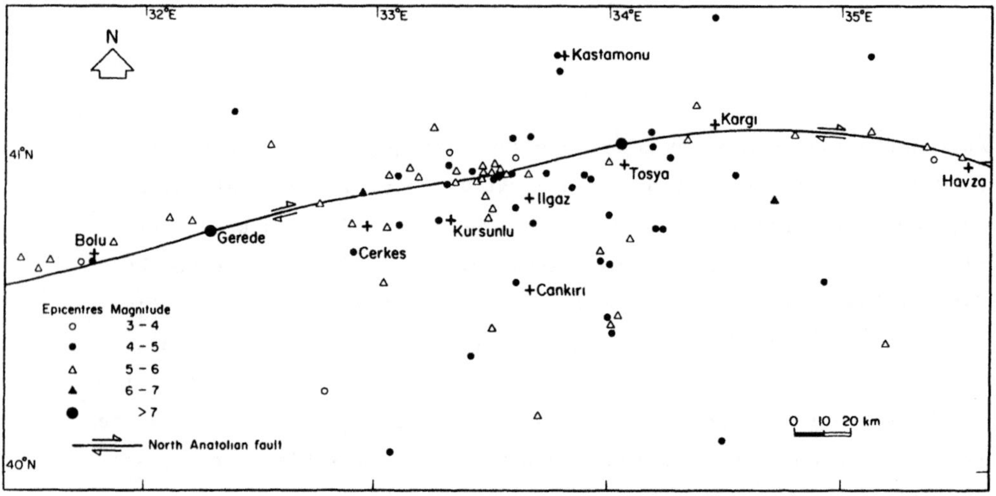

Fig.1. Map of 92 relocated epicentres of shallow focus earthquakes of greater than magnitude 3 which occurred within and adjacent to the North Anatolian fault zone between Bolu and Havza between 1900 and 1975.

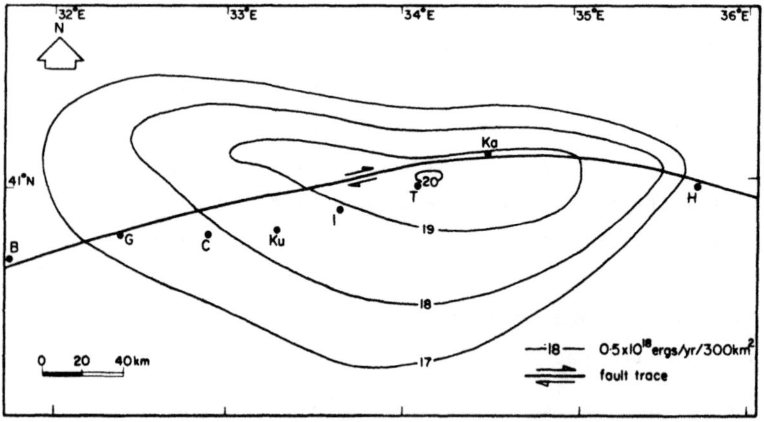

Fig.2. Contoured energy flux map of the region between Bolu and Havza for the period 1900-1919. Towns are indicated by the following letters; B - Bolu, G - Gerede, C - Cerkes, Ku - Kursunlu, I - Ilgaz, T - Tosya, Ka - Kargi, H - Havza.

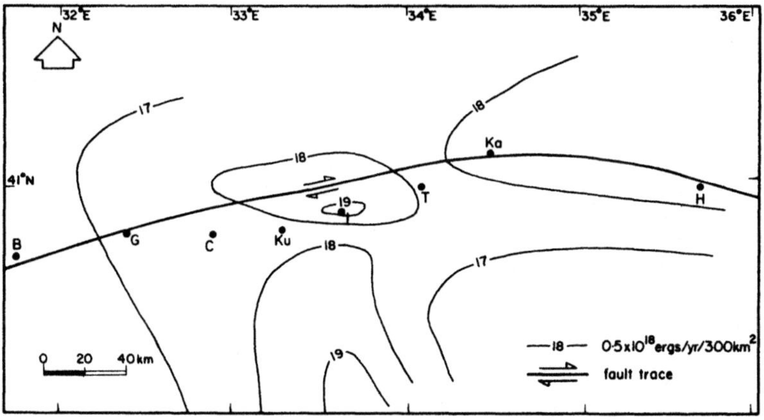

Fig.3. Contoured energy flux map of the region between Bolu and Havza for the period 1920-39. Initials of towns are as in Fig.2.

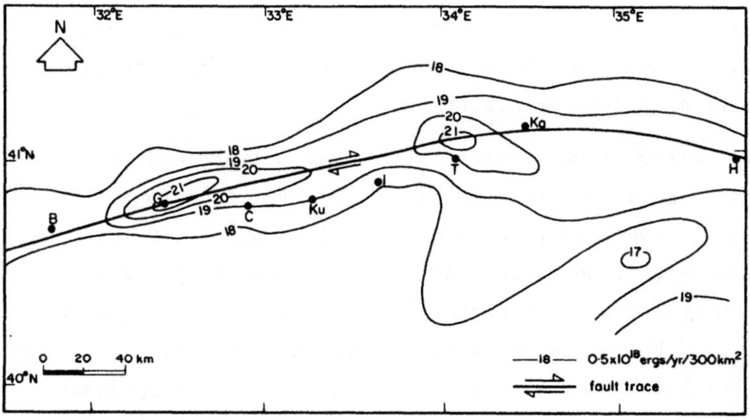

Fig.4. Contoured energy flux map of the region between Bolu and Havza for the period 1940-59. Initials of towns are as in Fig.2.

DESCRIPTION OF THE MAPS

Not surprisingly, the distribution of energy flux contours (Figs. 2-5) indicates that energy release was concentrated mainly along and immediately adjacent to the main trace of the North Anatolian fault. The fault zone between Kursunlu and Ilgaz has experienced numerous small and moderate magnitude earthquakes throughout much of the time

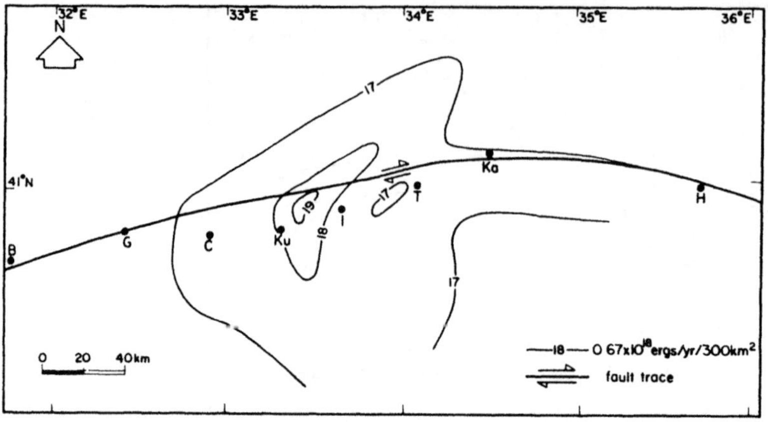

Fig.5. Contoured energy flux map of the region between Bolu and Havza for the period 1960-75. Initials of towns are as in Fig.2.

period under discussion. To the east and west of Ilgaz and Kursunlu respectively seismic activity has been less continuous although there have been a few large magnitude events.

The most direct correlation between the contour pattern and the outcrop of the fault zone is for the period 1940-59 (Fig.4) during which time the 1943 and 1944 events belonging to the sequence of westwards migrating large magnitude earthquakes occurred (Ambraseys 1970, Allen 1975). The high energy release to the southeast of the fault zone during 1940-59 may be related to a tivity on approximately E-W trending faults subparallel to the main fault zone. The correspondence between the energy release pattern and the outcrop of the fault zone is also good for the period 1900-19 (Fig.2), but it has to be recognized that seismological observations during the early part of this century were less reliable than those made in more recent times.

The maps for the periods 1920-39 (Fig.3) and 1960-75 (Fig.5) demonstrate that in addition to activity within and close to the fault zone there was also significant energy release to the north and south. Although activity extended at least 60 km to the north of the main fault trace, the seismic zone to the south is much wider. Most events occurred within an approximately N-S trending belt extending southward from that part of the fault zone lying between Cerkes in the west and Tosya in the east.

DISCUSSION AND TECTONIC IMPLICATIONS

The conclusion that energy release during this century in the Bolu-Havza region has been largely concentrated along the North Anatolian fault zone is not remarkable because, as Allen (1969, 1975), Ambraseys (1970), Dewey (1976), Ketin (1969), McKenzie (1972), Sengor (1979) and Toksoz et al. (1977) have recognized, it is an active intracontinental fault. Ambraseys (1970, 1971) has also shown that the fault zone has been the site of alternating 100-500 year episodes of activity or quiescence during the last 2000 years. Seismicity within the Anatolian plate to the south of the North Anatolian fault has been less commented upon, although numerous epicentres in the study region are shown on the maps of Pamir (1944), Toksoz et al. (1977, fig.1a) and Sengor (1979, fig.2).

The locations and attitudes of the faults on which this intraplate activity occurred have not been positively identified. However, examination of Landsat images at 1:200,000 and 1:500,000 reveals that in the area to the south of Gerede-Tosya there is a pattern of lineaments which it is reasonable to interpret as reflecting fault traces belonging to at least three sets (Fig.6). The greater density of

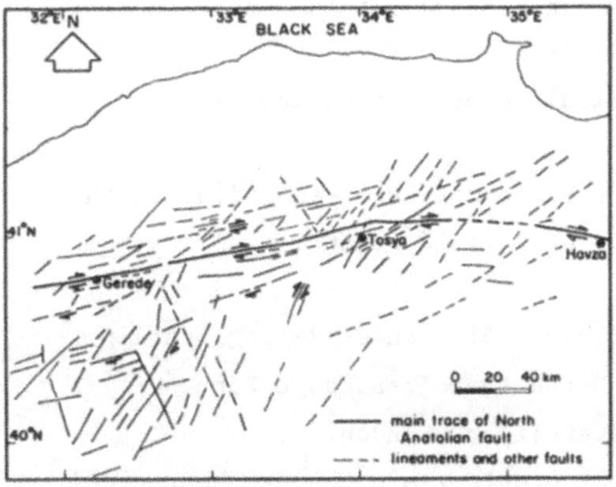

Fig.6. Fracture trace analysis of lineaments and faults visible on 1:200,000 and 1:500,000 Landsat images. The lineaments are interpreted as reflecting at least three sets of active or recently active faults. Faults to the north of the North Anatolian zone which are not directly related to it are not shown.

these traces immediately to the south of Gerede-Tosya may be a consequence of that area being part of the region enclosed by the North Anatolian fault zone where it curves through an arc of 50° between Erbaa and Bolu. Thus the occurrence of the three fault sets within a region of known seismicity suggests that at least this part of the Anatolian plate should not be regarded as rigid. Cummings & Leeds (1977) and Sengor (1979) have put forward similar proposals for the more eastern parts of the Anatolian plate.

Acknowledgements - Professor N.N. Ambraseys suggested the method of preparing the energy flux maps and Alma Gregory draughted the figures. M.T.A. provided a postgraduate award for one of us (A.A.B.) and funded fieldwork for both of us.

REFERENCES

Allen,C.R. 1969. Active faulting in northern Turkey. Contr. No.1577, Div. Geol. Sci., Calif. Inst. Tech. 1-32.

Allen,C.R. 1975. Geological criteria for evaluating seismicity. Bull. geol. Soc. Am. 86, 1041-1057.

Alsan,E., Tazucan,L. & Bath,M. 1975. An earthquake catalogue for Turkey for the interval 1913-1970. Rep. Seismol. Inst. Uppsala 7-75, 1-166.

Ambraseys,N.N. 1970. Some characteristic features of the Anatolian fault zone. Tectonophysics 9, 143-165.

Ambraseys,N.N. 1971. Value of historical records of earthquakes. Nature, Lond. 232, 375-379.

Ambraseys,N.N. & Zatopek,A. 1968. The Varto Ustukran (Anatolia) earthquake of 19 August 1966: summary of a field report. Bull. seism. Soc. Am. 58, 47-102.

Bath,M. 1973. Introduction to Seismology. Birkhauser Verlag, Berlin.

Can,R. 1974. Seismotectonics of the North Anatolian Fault Zone. Unpublished M.Phil. thesis, University of London.

Cummings,D. & Leeds,D.J. 1977. Seismotectonic zoning using theoretical mechanics. Proc. 6th World Conf. Earthquake Eng. 1, 836-842.

Dewey,J.W. 1976. Seismicity of Northern Anatolia. Bull. seism. Soc. Am. 66, 843-868.

Ergin,K., Güçlü,U. & Uz,Z. 1967. A catalogue of earthquakes for Turkey and surrounding area (11 A.D. to 1964 A.D.). Tech. Univ. Istanbul Mining Eng. Fac. Publ. 24, 1-169.

Ergin,K., Güçlü,U. & Aksay,G. 1971. A catalogue of earthquakes of Turkey and surrounding area (1965-1970). Tech. Univ. Istanbul Mining Eng. Fac. Publ. 28, 1-91.

Karnik,V. 1969. Seismicity of the European area: Part 1. Reidel, Dordrecht.

Karnik,V. 1972. Seismicity of the European area: Part 2. Reidel, Dordrecht.

Ketin,I. 1969. Uber die Nordanatolische horizontalverschiebung. Bull. Miner. Res. Explor.Inst., Ankara 72, 1-28.

McKenzie,D. 1972. Active tectonics of the Mediterranean region. Geophys. J. R. astr. Soc. 30, 109-185.

Pamir,H.N. 1944. Une ligne séismogène en Anatolie septentrionale. Rev. Fac. Sci. Univ. Istanbul A9, 143-158.

Sengor,A.M.C. 1979. The North Anatolian transform fault: its age, offset and tectonic significance. J. geol. Soc. Lond. 136, 269-282.

Toksoz,M.N., Arpat,E. & Şaroglŭ,F. 1977. East Anatolian earthquake of 24 November 1976. Nature, Lond. 270, 423-425.

Relationship Between Fault Geometry and Some Earthquake Epicentres within the North Anatolian Zone

A. A. Barka and P. L. Hancock

Department of Geology, University of Bristol, Queen's Building, University Walk, Bristol BS8 1TR, UK

Abstract - The epicentres of the Erzincan (1939), Erbaa (1942) and Tosya (1943) earthquakes are approximately coincident with where rectilinear fault breaks change their direction of trend by up to 20°. Each junction between fault segments may have been a site from which there was propagation of the fault breaks following an episode of locking on one or both segments.

INTRODUCTION

The purpose of this short contribution is to establish a connection between the approximate positions of the epicentres of three 20th century large earthquakes and the geometry of some fault breaks within the North Anatolian fault zone. Figure 1 shows the generalised outcrop of the zone and the locations of the three areas to be discussed. It is customary on small-scale maps, such as Fig.1, to depict the fault zone as a relatively smooth convex-northwards arc with only two major interruptions at Niksar and Erzincan (e.g. Allen 1969, Ambraseys 1970, Sengor 1979). However, as all the above authors recognized and illu-

Fig.1. Location of the North Anatolian fault zone and the three areas discussed in the note.

strated in their more detailed maps, the zone comprises numerous individual offset fault traces, some at an angle to each other. Further, as Ambraseys (1970, fig.11) has shown, it is possible from detailed ground surveying to demonstrate that even these apparently rectilinear breaks are made up of en échelon segments, many shorter than 1000 m.

TOSYA AREA

The fault break of the 7.6 magnitude event of 26 November 1943 passes to the north of the town of Tosya which was severely affected by the shock. According to Can (1974) the epicentre was located at 41°N, 34°E, that is about 5 km west of the town of Tosya. The total length of the fault break was 280 km and the displacement combined 1.1 m of right-lateral slip with a downthrow north of 1.0 m (Ambraseys & Zatopek 1968). Our ground resurvey and photogeological interpretation of the formerly uninvestigated area north of Tosya provided abundant evidence in the form of denuded scarps and offset streams for the position and trend of the 1943 fault break (Fig.2), which occurs within the outcrop of a metamorphic complex. According to Bingöl (1974) the complex comprises rocks of greenschist facies transitional to those of glaucophane facies in addition to ophiolites and bodies of mélange. Landslipping of the steep mountainous terrain has partially obscured

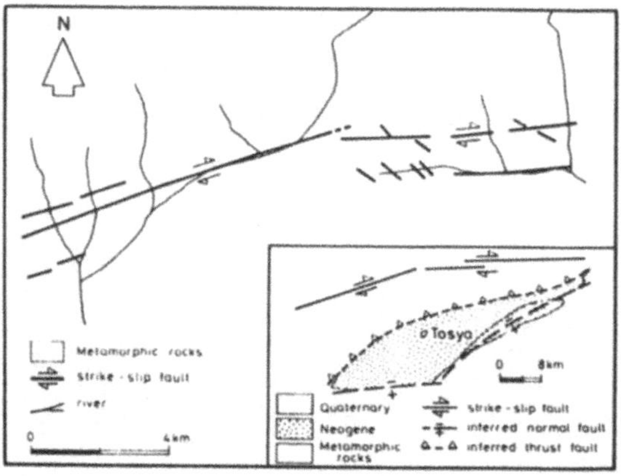

Fig.2. Active fault traces determined from ground surveying and photogeological interpretation in the area to the north of Tosya. The arrow couplets indicate the location of the 1943 fault break. Inset map shows the Neogene basin of Tosya in relation to the 1943 fault breaks.

some of the fault scarps. The rectilinear western fault break trends 073° whereas the eastern fault break trends 087°. The sense of the abrupt 14° change in direction of strike is such that the region on the south side of the fault zone is enclosed by convergent faults.

ERBAA AREA

According to Can (1974) the epicentre of the magnitude 7.3 earthquake of 20 December 1942 was at 40.7°N, 36.6°E between the villages of Zilhor and Tepekisla (Fig.3). The length of the fault break was 50 km and right-lateral displacement in the Erbaa area was about 0.5 m (Ambraseys 1970, fig.4). The location of the epicentre coincides approximately with the junction between rectilinear fault breaks trending 101° and 116°, which were surveyed on the ground and are visible on 1:35,000 aerial photographs. The faults cut an Eocene flysch sequence (Blumenthal 1950) underlying a ridge situated within a basin of Quaternary sediments. Both the Eocene and Quaternary deposits overlie a buried metamorphic complex containing metaspilites, metagreywackes and marbles of greenschist facies (Bingol 1974).

As in the Tosya area the epicentre is close to where the fault breaks abruptly change direction, in this instance by 15°, but in the same sense as at Tosya.

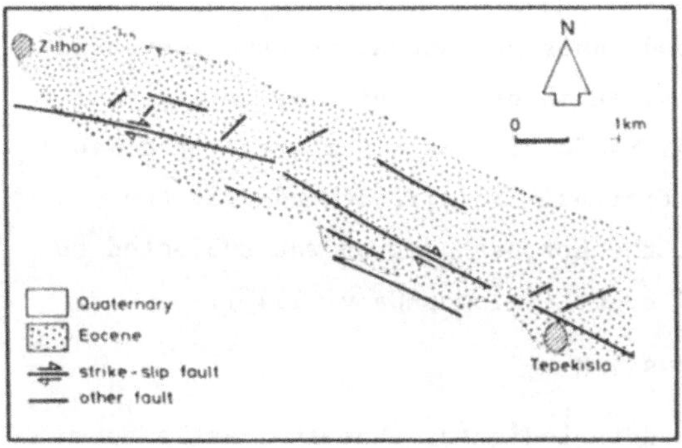

Fig.3. Active fault traces determined from ground surveying and photogeological interpretation of the Zilhor-Tepekisla area north of Erbaa. The arrow couplets indicate the location of the main 1942 fault breaks.

ERZINCAN AREA

The epicentre of the magnitude 8.0 earthquake of 26 December 1939 is located at 39.8°N, 39.38°E (Dewey 1976). Ambraseys (1970, fig.3) reported that the fault break was 350 km long, right-lateral displacement reached 3.7 m, and the northern side of the fault was downthrown by up to 2 m. The epicentre approximately coincides with the junction between straight fault traces trending 105° and 125° (Fig.4). The 20° change in strike occurs adjacent to an outcrop of serpentinite within a region underlain by greenshists and marbles (Tatar1978 , fig.5).

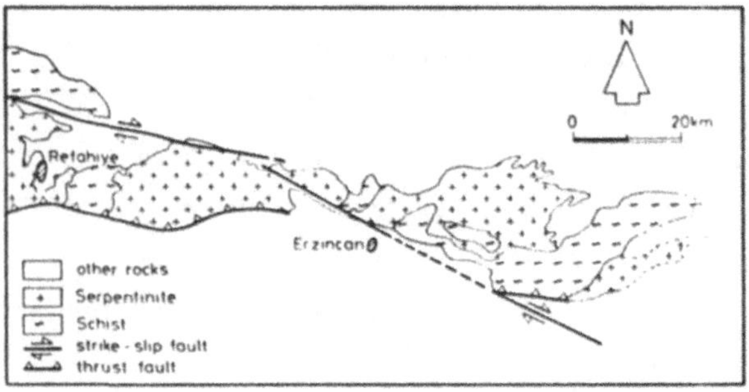

Fig.4. Generalised geology of the Erzincan area (after Tatar 1978, fig.5). The arrow couplets indicate the location of the main 1939 fault breaks.

The presence of the serpentinite body does not appear to have prevented the occurrence of large earthquakes; an inference which is of interest because Scholz (1977) notes that aseismic slip is a feature of the San Andreas fault where it cuts serpentinites. Seismic records for the North Anatolian zone indicate that the Erzincan region was subjected to other large shocks before the 1939 event (Salomon-Calvi 1940).

DISCUSSION

The evidence outlined in this note indicates that at least three of the post 1938 large magnitude shocks on the North Anatolian fault zone appear to have had their epicentres located at, or close to, the junctions of rectilinear fault segments which make angles of up to 20° with each other. In each area at least one fault segment would be oblique to the regional slip vector and hence a component of convergence across that segment may have caused it to lock. Similar arguments have been advanced by Rogers (1973) and Scholz (1977) for those parts of the San Andreas fault in California or Alpine fault in New Zealand across which there is oblique convergence. The Tosya, Erbaa and Erzincan areas could have been sites from which there was propagation of the fault breaks because they are located at the junctions of segments of different trend. Across one segment in each area frictional restraints to sliding would have been lower than in the other.

The reason why the faults in the North Anatolian zone change their strikes in the Tosya, Erbaa and Erzincan areas is unknown. The 10-20 km range of focal depths (Ambraseys & Zatopek 1968) indicates that initial failure occurred in the metamorphic complexes underlying the relatively thin Mesozoic-Cenozoic sedimentary cover. We suspect that fault attitudes are locally controlled by either or both petrological heterogenetics or pre-existing weakness planes introduced by foliations, folds or ductile shear zones. Because the geology of the metamorphic complexes is imperfectly understood it would be worthwhile investigating them in more detail in order to relate their structure to the geometry of the fault breaks.

From the viewpoint of forecasting the location of future large shocks it may be helpful to combine a study of seismic gaps (McCann *et al.* 1979, Toksoz *et al.* 1979) with a closer analysis of the geology of the North Anatolian fault zone, in particular the geometrical relationships between different fault traces.

Acknowledgements - The M.T.A. Institute of Ankara provided a postgraduate award for one of us (A.A.B.) and funded fieldwork for both of us. Jean Bees draughted the figures.

REFERENCES

Allen,C.R. 1969. Active faulting in northern Turkey. *Contr. No.1577, Div. Geol. Sci., Calif. Inst. Tech.* 1-32.

Ambraseys,N.N. 1970. Some characteristic features of the Anatolian fault zone. *Tectonophysics* 9, 143-165.

Ambraseys,N.N. & Zatopek,A. 1968. The Varto Ustukran (Anatolia) earthquake of 19 August 1966: summary of a field report. *Bull. seism. Soc. Am.* 58, 47-102.

Bingol,E. 1974. 1:2,500,000 olcekli Turkiye metamorfisma haritasi ve bazi metamorfik kusaklarin jeotektonik evrimi uzerinde tartismalar. Maden Tetkik Arama Enstit. Dergisi 83, 178-184.

Blumenthal,M. 1950. Orta ve asagi Yesil Irmak bolgelerinin (Tokat, Amasya, Havza, Erbaa, Niksar) jeolojisi hakkinda. Maden Tetkik Arama Enstit. Yayinl. Ser.D, No.4, 1-153.

Can,R. 1974. Seismotectonics of the North Anatolian Fault Zone. Unpublished M.Phil. thesis, University of London.

Dewey,J.W. 1976. Seismicity of Northern Anatolia. Bull. seism. Soc. Am. 66, 843-868.

McCann,W.R., Nishenko,S.P., Sykes,L.R. & Krause,J. 1979. Seismic gaps and plate tectonics: seismic potential for major boundaries. Pageoph 117, 1082-1147.

Rogers,T.H. 1973. Fault trace geometry within the San Andreas and Calaveras fault zones - a clue to the evolution of some transcurrent fault zones. Proc. Conf. Tectonic Problems of the San Andreas Fault System. Stanford Univ. Publs Geol. Sci. 13, 251-258.

Scholz,C.H. 1977. Transform fault systems of California and New Zealand: similarities in their tectonic and seismic styles. J. geol. Soc. Lond. 133, 215-229.

Şengör,A.M.C. 1979. The North Anatolian transform fault: its age, offset and tectonic significance. J. geol. Soc. Lond. 136, 269-282.

Salomon-Calvi,W. 1940. Türkiyedeki zelzelelere muteallik etudler. Maden Tetkik Arama Enstit. Yayinl. Ser.B, No.5, 1-121.

Tatar,Y. 1978. Tectonic investigations on the North Anatolian fault zone between Erzincan and Refahiye. (in Turkish). Publ. Inst. Earth Sci., Hacettepe Univ. 4, 201-236.

Toksöz,M.N., Shakal,A.F. & Michael,A.J. 1979. Space-time migration of earthquakes along the North Anatolian fault zone and seismic gaps. Pageoph 117, 1258-1270.

Preliminary Studies of Seismic Risk in Turkey, and the Occurrence of Upper Bounded and Other Large Earthquake Magnitudes

Paul W. Burton and Robert W. McGonigle

Natural Environment Research Council, Institute of Geological Sciences, Murchison House, West Mains Road, Edinburgh EH 9, UK

K. C. Makropoulos

Seismological Laboratory, University of Athens, Hippokratous 33, Athens, Greece

S. Balamir Üçer

Kandilli Observatory, Çengelköy, Istanbul, Turkey

Summary

Gumbel's third distribution of extreme values is discussed and applied to two files of Turkish earthquakes created from diverse agencies and catalogues. The forecasting ability of both is in agreement and a preliminary map of seismic risk is produced.

Some alternative approaches to evaluating seismic risk, especially through upper bounded and other large magnitude earthquake occurrences, are illustrated using a 4° cell of seismicity centred on 39N, 41E, in Anatolia. Gumbel upper bounds ω to earthquake magnitude occurrence are asymptotic, uncertain, and theoretically correspond to infinite return periods whereas strain energy release diagrams lead to an empirical estimate of the large magnitude M_3 which is equivalent to the total strain energy which may be accumulated in a region. M_3 is less than ω and shows a finite 'waiting time' during which the energy equivalent to M_3 may be accumulated. Combining Gumbel III statistics with an intensity attenuation law defines the 'most perceptible' earthquake magnitude which has the maximum probability of being felt at any intensity level. These 'most perceptible' earthquakes will vary between regional seismicity cells in Turkey, but local values may help as a criterion for choosing engineering design time histories.

1. Introduction and Methodology

Turkey is a country of diverse seismicity, ranging from the apparently low to moderate seismicity in southern and central Turkey to the well known high seismicity of western Turkey and the Anatolian fault system. The problem of assigning return periods to earthquake occurrences is fundamental to all investigations of seismic hazard and alternative statistical models will be applied here to an assessment of the rate of Turkish seismicity. The results presently available are regarded as preliminary; further discussion, comment, and no doubt criticism, will help refine our ensuing assessment of some aspects of seismic risk in Turkey.

The methodology recognises that 'adequate' earthquake histories or catalogues are vital in the assessment of risk and two separate catalogues describing Turkish and surrounding seismicity since 1900 are compared in content and contrasted during this discussion. The first catalogue, referred to as IGS70, is extracted from the global seismicity file described by Burton (1978a); the second catalogue, referred to as TURK70, is more refined than the first and contains many recent relocations of historical earthquakes.

Preferred statistics which will be applied to the catalogues are Gumbel's third distribution of extreme values - Gumbel III - which analyse the part process of seismicity occurrence by considering largest or annual extreme earthquakes only. The procedure adopted (Burton, 1979, in principal comparable to Båth, 1979a,b) divides the regional seismicity into simple 4° square cells, so that the analysis technique may be systematically applied, with the modification (Makropoulos, 1978) that overlapping cells can be cyclically considered in addition to those which are simply adjacent. The three parameters of Gumbel III will be determined for each cell, along with the entire covariance matrix between the three parameters. Largest earthquake magnitudes which may be expected over a 75 year interval will then be predicted with associated uncertainties. Seismic risk maps of both the upper bound to earthquake magnitude occurrence and the contoured largest expected magnitude over 75 years are produced.

Alternative assessments of seismic risk through upper bounded and other large earthquake magnitudes will then be illustrated using the 4° cell of seismicity centred on 39N, 41E, in Anatolia. Finally, these methods will be extended beyond assessing the occurrence probability of particular earthquake magnitudes to assessing the risk of different levels of strong ground motion.

2. Earthquake Histories

2.1 Turkish earthquake history

There is a wealth of historical earthquake data for Turkey (Ambraseys, 1971) and the important catalogue of Pinar & Lahn (1952) gives earthquake locations and damage descriptions for the interval 11AD to 1950, however, seismic parameters of epicentral coordinates, magnitude etc are not included. Epicentral parameters are provided by Ergin, Guclu & Uz, (EGU), (1967) for the interval 11AD to 1964. This catalogue takes the information of Pinar & Lahn (1952) for the historical earthquakes during 11AD to 1903 and assigns epicentral coordinates on the basis of Pinar & Lahn's damage reports. It is held that reliable instrumental recordings commence in 1904 and so the interval 1904 to 1964 contains information based on instrumental data, particularly from five major sources which are also used in other catalogues. Alsan, Tezuçan & Båth (ATB), (1975, 1976) produced a catalogue for Turkey within 35.5°N to 42.5°N and 25.5°E to 45.0°E for the interval 1913 to 1970. Epicentral parameters were recomputed using a systematic scheme, including a consistent redetermination of magnitudes. Data of the British Association for the Advancement of Science was used for 1913 to 1917, but it is held that data are too scanty prior to 1913 to justify recomputation. This catalogue forms the foundation of Erdik & Ozselcuk's (1978) regional rearrangement of earthquake catalogues of Turkey. Several other seismologists, particularly Kárník (VKR), (1968, 1971), and other Seismological Observatories, have contributed to our knowledge of Turkish earthquakes.

With the above information in mind, two separate catalogues of Turkish earthquakes have been formed for comparative studies. Both cover the geographic region 35.5°N to 42.5°N and 25.5°E to 45.0°E and include the interval 1900 to 1970 inclusive.

2.2 IGS70 Catalogue

The IGS70 catalogue of Turkish earthquakes is extracted from a large seismicity file containing epicentral data compiled from major primary reporting agencies e.g. ISC, from individual researchers e.g. Kárnik, and also from local reporting agencies which provide focal data on local earthquakes down to a magnitude which would not be available from teleseismic data alone. In recent years, the major contributor is the ISC, while Kárnik's contribution is considerable for earlier decades. The magnitudes in this file are always on the surface wave magnitude scale.

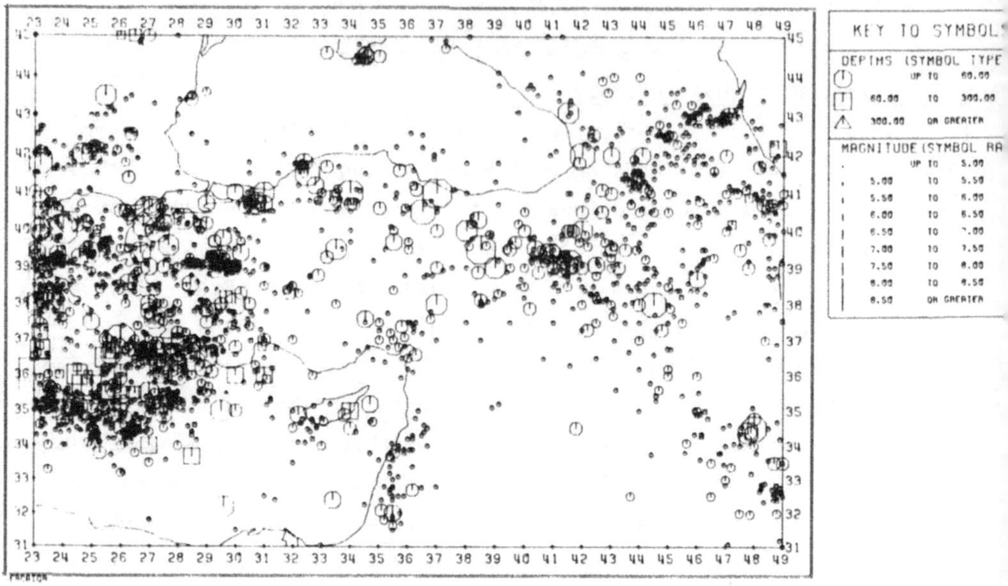

Fig. 1 IGS70: epicentres of all earthquakes for 71 years estimated from the IGS global file of seismicity. All magnitudes are on the surface wave magnitude scale.

The spatial distribution at all magnitude levels in this IGS file is shown in Fig. 1. It should not be expected that all magnitude classes are either completely reported or even equally well reported. Stepp's (1971) method can be used to specify time intervals for which particular magnitude classes are likely to be completely reported. The common assumption that an earthquake sequence in time can be modelled by a Poisson distribution is taken; the mean rate per unit time for a given magnitude range is calculated, ρ, and its variance σ_ρ^2 is ρ/T_0 where T_0 is the time interval here taken in five-year increments throughout the catalogue duration. If the statistical process is stationary then the mean and variance will be time invariant. The standard deviation σ_ρ should behave as $1/\sqrt{T_0}$, and departure from this simple behaviour indicates for a particular magnitude class that reporting is incomplete (or possibly that the time interval on which mean rates are calculated is too short to give a satisfactory estimate of the mean). Fig. 2 shows the results for the IGS70 catalogue.

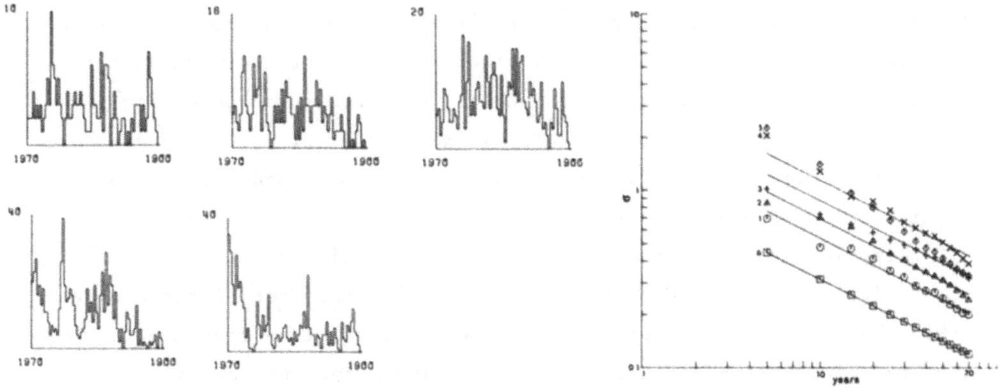

Fig. 2 Completeness analysis of magnitudes reported in IGS70. The histograms from left to right show the annual number of earthquakes in the five magnitude ranges (1) $M_s \geqslant 6.0$, (2) $6.0 < M_s \leqslant 5.5$, (3) $5.5 < M_s \leqslant 5.0$, (4) $5.0 < M_s \leqslant 4.5$ and (5) $4.5 < M_s \leqslant 4.0$. The standard deviation σ of the mean rate of occurrence within each magnitude class is calculated by taking five year increments throughout the catalogue, until the entire catalogue time span is included. Symbols numbered (1) to (5) correspond to the five magnitude ranges. Symbol (6) is the reference line: if the mean rate of occurrence is stable and magnitude reporting is complete within a magnitude range then a line parallel to (6) is obtained; discrepancies can indicate incompleteness or instability in the mean rate of occurrence.

Results of earthquake occurrence ($\rho = N/T_o$) in the five magnitude classes $4.0 \leqslant M < 4.5$, $4.5 \leqslant M < 5.0$, $5.0 \leqslant M < 5.5$, $5.5 \leqslant M < 6.0$, $6.0 \leqslant M$ for five year intervals throughout the 70 years are obtained, and the standard deviation σ_ρ of the mean rate of occurrence when plotted in Fig. 2 indicates that:

Magnitude range	Apparent period of approximate completeness, years
$4.0 \leqslant M < 4.5$	Not apparent
$4.5 \leqslant M < 5.0$	45
$5.0 \leqslant M < 5.5$	50
$5.5 \leqslant M < 6.0$	55
$6.0 \leqslant M$	70

It is also apparent from Fig. 2 that different magnitude ranges require different lengths of time before a stable estimate of the mean rate of occurrence is established; for instance, the magnitude range with maximum up to 5.5 produces a stable estimate in about 15-20 years. The histograms included in Fig. 2 bear out most of these deductions and magnitudes $4.0 \leqslant M < 4.5$ are obviously incomplete, magnitudes $4.5 \leqslant M < 5.0$ being incomplete in the earlier decades of this century. The statistical studies of Gençoglu & Tabban (1973) and Üçer, Ayhan & Alsan (1975) have considered earthquakes with M_s larger than approximately $4\frac{1}{4}$, thus reducing most adverse effects of catalogue incompleteness.

2.3 TURK70 Catalogue

Cooperation between Kandilli Observatory, Istanbul, and the Earthquake Research Institute, Ankara, is producing a further refined earthquake catalogue for Turkey which incorporates the redeterminations of Alsan, Tezuçan & Båth (1975). Earthquakes in the Aegean and western Turkey have been relocated by Makropoulos (MAK), (1978, Makropoulos & Burton, 1980) using joint epicentre determination techniques with control or master events including epicentres substantiated by well documented macroseismic data of the meizoseismal area. These data have all been merged, again for the interval 1900 to 1970.

Magnitude values within this catalogue are taken as surface wave type. Indeed, ATB systematically redetermined magnitudes based mainly on the Wiechert seismograms available from Uppsala, and MAK following the ATB example similarly redetermined magnitudes of earthquakes in Greece, the Aegean and western Turkey based on predominantly Uppsala amplitude readings.

Fig. 3 shows the epicentral distribution of all magnitude earthquakes within TURK70. The Stepp completeness analysis in Fig. 4 again shows a rational pattern of variation in completeness for earthquakes larger than magnitude 4.5. Magnitudes within $5.0 \leqslant M < 5.5$ show apparent completeness for about 5 years more than IGS70. Magnitudes within $4.5 \leqslant M < 5.0$ for TURK70 show an unusual pattern which can be interpreted by roughly fitting <u>different</u> seismicity rates during the past 45 years. The division between seismicity rates can be drawn about 20-25 years ago. Although changes in seismicity rate are particularly well documented in Turkey (Ambraseys, 1971) for historical sequences of hundreds of years, this is probably not the appropriate interpretation here. TURK70 is constructed from various catalogues, as is IGS70, but the largest single contribution to TURK70

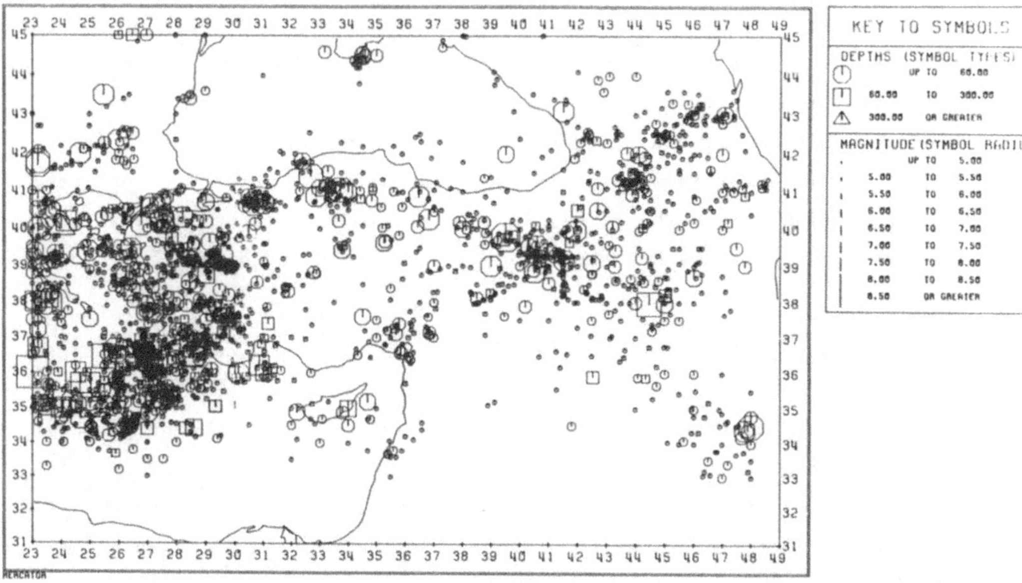

Fig. 3 TURK70: epicentres of all earthquakes for 71 years obtained mainly by combining refined catalogues of Turkish earthquakes eg Alsan, Tezuçan & Båth (1975). All magnitudes are on the surface wave magnitude scale.

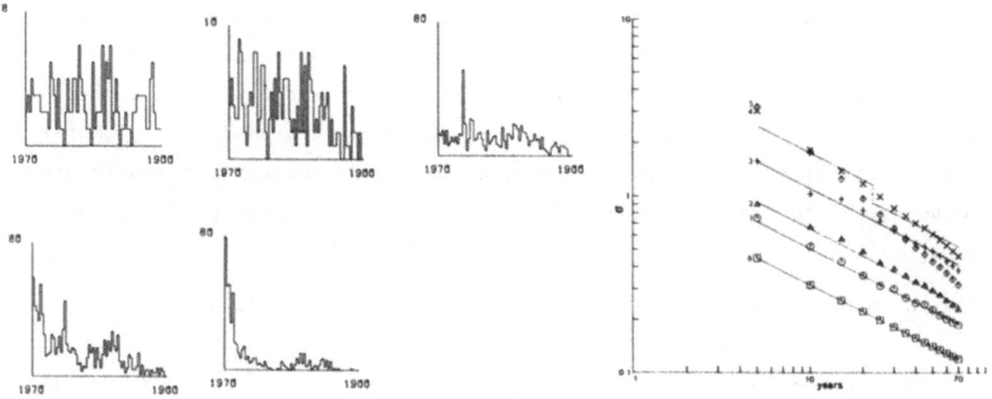

Fig. 4 Completeness analysis of mangitudes reported in TURK70. A description of the sub-illustrations is given in the caption to Fig. 2.

comes from ATB: examination of ATB reveals that well over half of its contents correspond to the twenty years 1950-70, perhaps reflecting real or apparent fluctuations in the true seismicity at lower magnitudes. Inspection of the relevant histograms for $4.5 \leqslant M < 5.0$ confirms this higher seismicity at these magnitudes during 1950-70 (which does not seem to have continued into the 1970s) in contrast to a lower seismicity consistently reported since shortly after the beginning of ATB in 1913.

The two preceding catalogues are included separately for purposes of comparison only, but it is recognised that TURK70 is the catalogue which contains the most refined contributions and awareness of the ideal requirement of homogeneity eg ATB, MAK. Despite these apparent periods of approximate completeness, obtained using Stepp's analysis, it is not expected that such results will apply uniformly throughout Turkey and considerable differences may be expected between localities in east and west Turkey. The ensuing statistical analysis will only use magnitudes $M_s \geqslant 4.5$.

3. Extreme value statistical comparison of Turkish earthquake catalogues for 1900 to 1970 inclusive

In order to assess how the vagaries of catalogue compilation affect the precision of statistical prediction both IGS70 and TURK70 are analysed.

3.1 Extreme value statistics - Gumbel III

Gumbel's (1958) theory of extreme values has the advantage that the entire history of earthquake occurrence need not be analysed for probability of earthquake occurrence, instead the earthquake sequence constructed from the largest annual magnitudes used. Gumbel's third distribution of extreme values - Gumbel III - has been described elsewhere, and is briefly compared with the results obtained from other commonly used techniques later in this paper. Gumbel III is eventually preferred and so it will be used here to compare IGS70 with TURK70.

The third asymptotic distribution of extremes takes the form

$$G^{III}(M) = \exp\left[-\left(\frac{\omega-M}{\omega-u}\right)^k\right] \qquad (1)$$

where the three parameters allow for curvature, k, have an upper bound or magnitude limit, ω, and u is the characteristic value with $G^{III}(u) = 1/e$. $G^{III}(M)$ is the probability that magnitude M is an annual extreme. Methods of estimating Gumbel III parameters have been described by Yegülalp & Kuo (1974), Makropoulos (1978) and Burton (1979); the latter two are folllowed here. The three parameters are obtained using a non-linear least squares curve fitting technique based on Levenburg-Marquadt, and the resulting matrix equation affords the opportunity to obtain the complete covariance matrix between the three parameters, which is of the form:

$$\begin{pmatrix} \sigma^2_{\omega} & \sigma^2_{u\omega} & \sigma^2_{\lambda\omega} \\ \sigma^2_{\omega u} & \sigma^2_{u} & \sigma^2_{\lambda u} \\ \sigma^2_{\omega\lambda} & \sigma^2_{u\lambda} & \sigma^2_{\lambda} \end{pmatrix} \qquad (2)$$

The annual extreme magnitudes M_i (i=1...n) are ranked in ascending size and the 'plotting point' probability at M_i is, following Gringorten (1963),

$$G^{III}(M_i) = (i - 0.44)/(n + 0.12) \qquad (3)$$

The first and last years are then taken as 1900 and 1970 respectively, hence n is here always equal to 71. The technique allows for each individual extreme magnitude to be assigned its own weight. However, because most of the contributions to the catalogue have attempted magnitude homogeneity an uncertainty in magnitude of $\delta M = \pm 0.5$ is assumed throughout. Only annual extremes are considered here, although two or three year extremes could be used. The probability of a particular magnitude <u>not</u> being an annual extreme is $1 - G^{III}(M)$ and the return period in years for an earthquake of magnitude M or larger is $1/(1 - G^{III}(M))$. Rearranging equation (1) gives

$$M = \omega - (\omega - u)\left[-\ell n(G^{III}(M))\right]^{\lambda} \qquad (4)$$

where $\lambda = 1/k$. A largest earthquake with return period 75 years can be obtained from equation (4), this magnitude will be referred to as M_{75}, and confidence bounds on this forecast can be obtained by using either the diagonal elements or the complete covariance matrix.

The catalogues IGS70 and TURK70 are taken to cover the area of $31^o - 45^oN$, $23^o - 49^oE$, although this need not be true in detail if the subsidiary contributing catalogues are considered. This area is divided into 4° cells, the first being $41^o - 45^oN$, $23^o - 27^oE$, centred on 43^oN, 25^oE. Gumbel III parameters and values of the seventy-five-year earthquake M_{75} are thus considered to correspond to this point. The cell is then moved eastwards by 1° etc, thus developing an overlapping grid of point values for Gumbel III parameters and M_{75}. Specific parameter results from a typical cell will be described later, but for comparison of forecasting ability of these two preliminary catalogues composite results only will suffice.

Such an overlapping grid will therefore generate grid point values at 1° intervals throughout 33° - 43^oN, 25^oE - 47^oE. Some grid points will not provide parameter estimates for a variety of reasons, typically through insufficient data. In an effort to facilitate comparison of results deduced from IGS70 and TURK70, both sets of grid point values have been mapped by computer. Computerised attempts at producing contoured maps of these values, and therefore risk related maps, are still not satisfactory (also see Burton, 1978b). Cell size obviously smears out the seismicity to too great an extent where it might be expected to be more localised than indicated by a 4° span, for example, along the Anatolian fault zone. However, comparisons of point values within these maps for characteristic values u, (Figs. 5) upper bounds ω (Figs. 6) and 75 year earthquakes M_{75} (Figs. 7) are informative. Estimates of u typically have standard deviation σ_u around 0.3 to 0.4; estimates of ω have typical σ_ω between 1.0 and 2.0. Given these uncertainties, grid point values show consistency between the overall statistics of the two files, although descriptions of individual earthquakes and file contents can vary considerably.

Rather than only consider a comparison between individual Gumbel III parameters, it is perhaps more important to compare forecasts generated by the two groups: this is so because the same forecast can be generated by a suite of different parameters, and, in principal, it is the forecasting ability which is of first importance. The largest earthquake magnitude expected over a 75-year interval, M_{75}, is shown in Figs. 7.

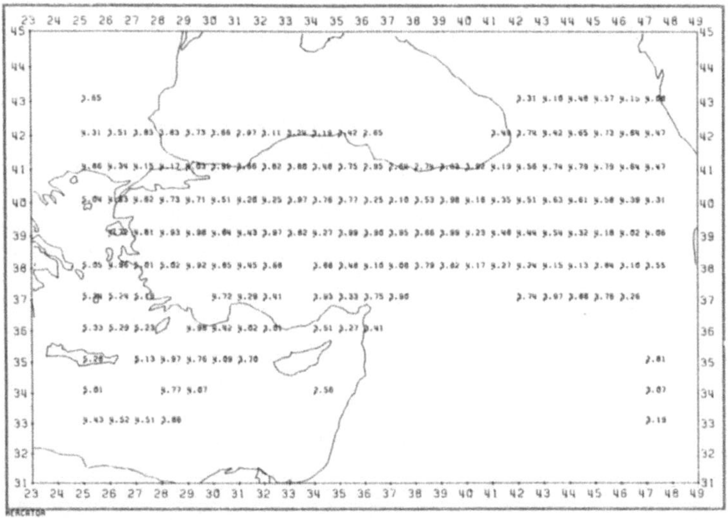

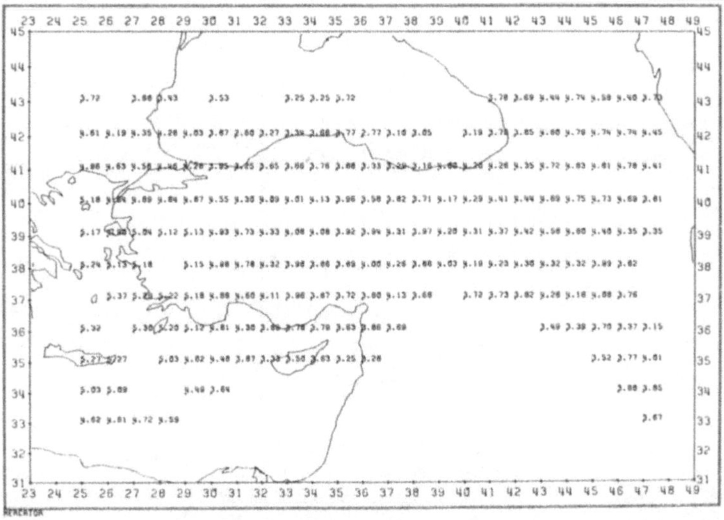

Fig. 5 Modal values of the annual extreme earthquake magnitude: these point values are calculated as the parameter u of Gumbel's third asymptotic distribution of extreme values (Gumbel III) fitted to 4° cells of seismicity extracted from the earthquake files (a) IGS70, and (b) TURK70.

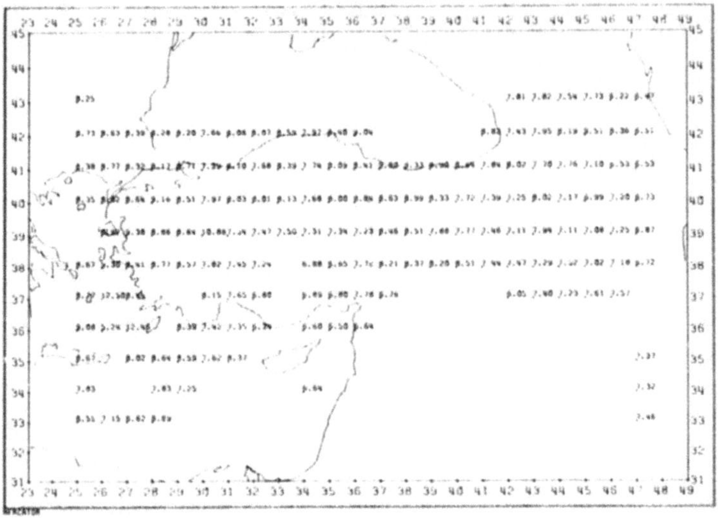

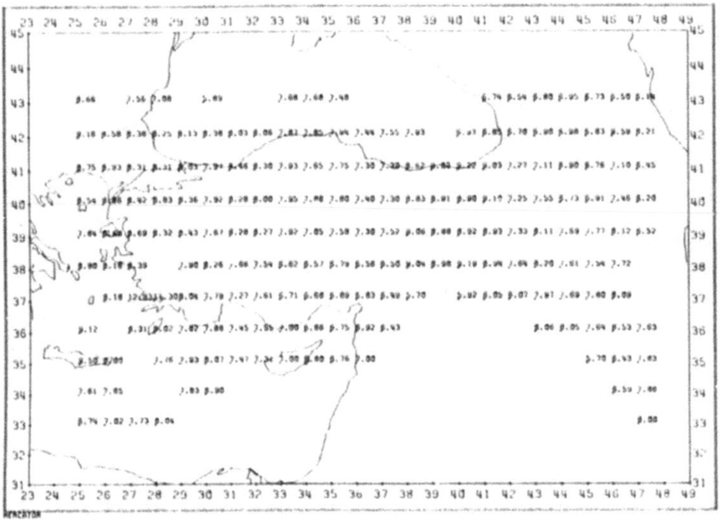

Fig. 6 Asymptotic upper bounds to earthquake magnitude occurrence: these point values are calculated as the uncertain (σ_{ω} typically about one or two magnitudes) parameter ω of Gumbel III fitted to 4° cells of seismicity extracted from the earthquake file (a) IGS70, and (b) TURK70.

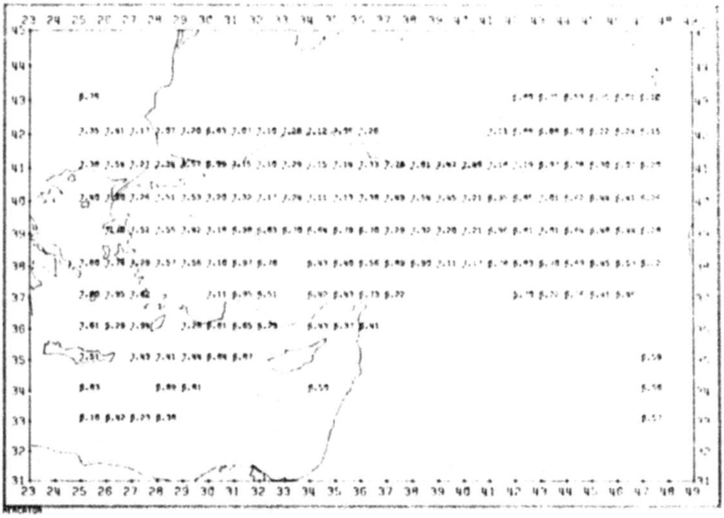

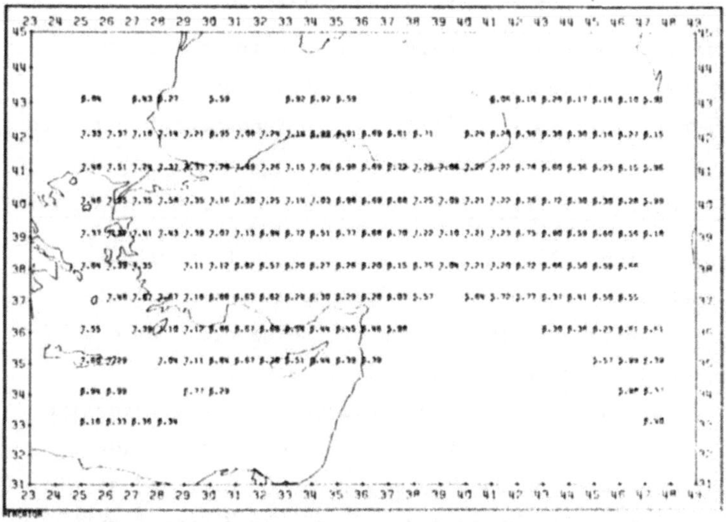

Fig. 7 Values of the largest earthquake magnitude M_{75} expected over an interval of 75 years are forecast using the Gumbel III distribution obtained from the earthquake files (a) IGS70, and (b) TURK70. Scrutiny of individual point values demonstrates that these two sets of 75 year forecasts, although obtained from diverse files, usually give good agreement within their respective forecasting uncertainty.

If those non-overlapping but adjacent 4° cells which provide estimates of M_{75} for both IGS70 and TURK70 are considered alone as examples, we obtain for M_{75} (standard deviations in brackets):

	M_{75}, σ_M	
Cell centre	IGS70	TURK70
43^oN, 25^oE	6.76 (.31)	6.84 (.31)
45^oE	6.25 (.30)	6.16 (.29)
39^oN 29^oE	7.42 (.28)	7.39 (.27)
33^oE	6.70 (.31)	6.72 (.30)
37^oE	7.29 (.30)	6.70 (.29)
41^oE	6.96 (.29)	7.23 (.29)
45^oE	6.48 (.29)	6.60 (.29)
35^oN, 25^oE	7.51 (.28)	7.60 (.28)
29^oE	7.44 (.29)	7.11 (.28)

Forecasts of M_{75} derived from these two catalogues are in accord, and these forecasts, despite being obtained from separately derived catalogues or histories of Turkish seismicity are in fundamental agreement.

4. Alternative large earthquake magnitudes appropriate to risk assessment

The problem of assigning probabilities or return periods to earthquake occurrences is fundamental to risk assessment and usually it is large magnitudes which are the prime concern. Some of the several methods of estimating statistical seismic risk will be briefly described and compared, particularly for larger magnitude occurrences, and alternative concepts of large magnitude risk evaluation will be described including consideration of upper bounds to earthquake magnitude occurrence, strain energy release, and perceptibility of ground motion caused by an earthquake. Of the two earthquake catalogues available here TURK70 is the most refined in its compilation (e.g. ATB attempts to produce a homogeneous catalogue), and this can be extended for eight years by incorporating more recent earthquake reports given by the ISC and the PDE service. In all cases surface wave magnitudes only are incorporated into the extended file used in the following analysis: a full description of the file extension and its application to risk assessment in Turkey will be prepared elsewhere. The 4° cell of seismicity centred on 39N, 41E, in Anatolia will be analysed here using alternative methods of statistical risk assessment to illustrate their differences, but more importantly to demonstrate the way alternative approaches to risk assessment should be complementary.

4.1 Cumulative frequency law assessment

The commonest description of earthquake occurrence is perhaps provided by the Gutenberg-Richter cumulative frequency law where the annual number of earthquakes N_c, with magnitude M or greater is given by:

$$\log_{10} N_c = a - bM \tag{5}$$

Uçer et al (1975) have considered the years 1900 to 1970 for 2° cells within 35.5^o - 42.5^oN, 25.5^oE - 45.5^oE, using equation (5), and obtained values of a and b. Two of their observations should be noted. Uçer et al. carried out this analysis excluding aftershocks, and then including aftershocks; the differences they noted fell within the range of observed standard deviations. They also noted, when N_c was plotted against M, that some cells showed curvature concave towards the origin of the graph, particularly in north east Turkey. This observation foreshadows our later use of Gumbel III.

Seismicity of the illustrative 4° cell in Anatolia, expressed in the form of equation (5), is shown in Fig. 8. The magnitude cut-off around M_s = 4.5 is obviously vital for subsequent analysis, otherwise the sometimes long tail of incomplete low magnitudes will render the results spurious.

Three ways of obtaining the two parameters of equation (5) are commonly considered. If the whole process implied by equation (5) is considered then parameters a and b may be obtained by the usual procedure of least squares, or through the maximum likelihood technique applied by Aki (1965) and Utsu (1966). More circuitously, the part process of annual extreme magnitudes described by Gumbel's first distribution can be obtained - thus excluding aftershock sequencies - and the parameters a and b of equation (5) deduced in the manner described by Epstein & Lomnitz (1966). Gumbel I can be expressed in the linear form

$$-\ell n\left[-\ell n G^{I}(M)\right] = \alpha(M - u) \tag{6}$$

where α and u are constants which relate to a and b through

$$\begin{aligned} a &= \alpha u/\ell n10 \\ b &= \alpha/\ell n10 \end{aligned} \tag{7}$$

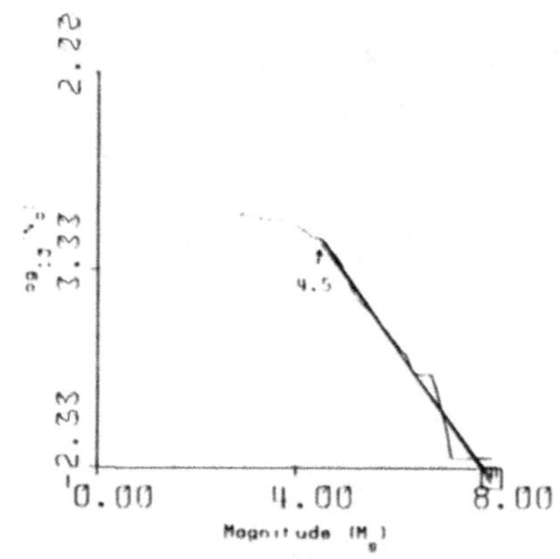

Fig. 8 Whole process statistics. This represents the well known Gutenberg-Richter law $\log N_c = a - bM$, with logarithmic annual frequency $\log N_c$ plotted against surface wave magnitude M_s for the 4° cell of seismicity centred on 39N, 41E, in Anatolia. The cell shows roll-off or incompleteness below about magnitude $M_s = 4.5$. Two lines are drawn: the first line corresponds to those a and b values determined by least squares, the second line corresponds to a and b values determined by maximum likelihood and is indicated by a square symbol at one end. These lines are only fitted to magnitudes $M_s \geq 4.5$.

The two lines on the cell diagram in Fig. 8 are the result of least squares and maximum likelihood fitting to the data. Again, we are comparing forecasting ability rather than merely distribution parameters, so the annual expected modal largest earthquake $M_1 = a/b$ and the 75 year earthquake M_{75} are required. Results for the whole process (all earthquakes) cumulative frequency law expressed by equation (5) are:

Cumulative frequency law

	Least squares processing	Maximum likelihood processing
a	3.57 (0.12)	3.40
b	0.72 (0.02)	0.70
M_1	4.92 (0.50)	4.87
M_{75}	7.51 (0.62)	7.57

where standard deviations have been noted in brackets for the least squares processing. The part process analysis (extreme value earthquakes) expressed through equations (6) and (7) gives least square results:

	Gumbel I law
$1/\alpha$	0.72 (0.07)
M_1 (or u)	4.54 (0.11)
M_{75}	7.66 (0.39)

Inspection of the two lines for the cell of Fig. 8, and the above estimates, reveal little difference between the least squares and maximum likelihood approaches. All three approaches reveal a consistency of forecast within the standard deviations available, however, the cumulative frequency approach would be grossly influenced by the inclusion of incomplete magnitude data below $M_s \approx 4.5$. Inclusion of the 4.5 magnitude cut-off removes most of the curvature seen at low magnitudes in all cell diagrams of Turkish seismicity, including that illustrated in Fig. 8. It should be noted that partial incompleteness for magnitudes above 4.5 may cause curvature at low magnitudes which is not detectable by eye, but corrections can be made for this effect by considering the seismicity rate observed over periods for which such magnitudes are known to be complete. However, curvature for the larger, rarer, earthquake magnitude occurrences, which can be seen in many seismicity cells throughout Turkey, requires analysis by a quadratic or other law of three implicit parameters. A full analysis would include overlapping cells so that more detail of the spatial variation of risk could be mapped, but this is not required here. Båth (1979a) has followed a similar philosophy producing estimates of seismic risk in 2° cells in Turkey deduced from ATB catalogue with its inherent aspiration to homogeneity.

4.2 Gumbel III assessment

Many of the details of this method have already been given and its relevance becomes apparent when the curvature at high and rarer magnitudes is taken into account. The specimen cell diagram is drawn in Fig. 9 where M_s is plotted against $\left[-\ln G^{III}(M)\right]^{\lambda}$ to reduce the data plot to linear form in the ideal case. Corresponding point values of the distribution parameters ω, u and λ are:

$\omega = 8.60\ (1.59)$

$u = 4.36\ (0.15)$

$\lambda = 0.26\ (0.14)$

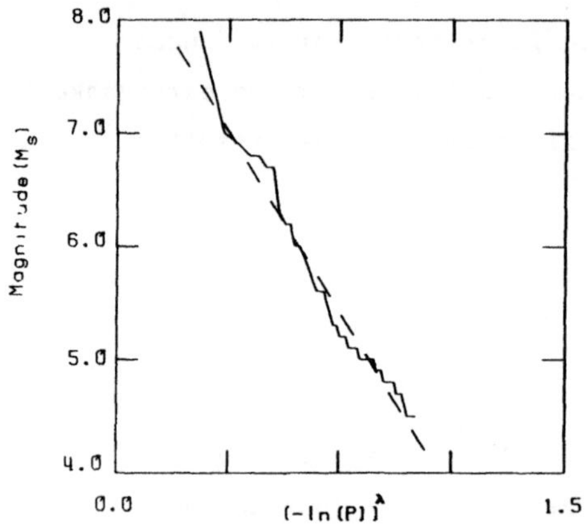

Fig. 9 Part process extreme value statistics. The data-solid line - and the Gumbel III fit to the data-dotted line - are plotted. The ordinate is M_s plotted against $(-\ell n G^{III}(M))^{\lambda}$ as abscissa ($G^{III}(M) = P$). (Data is from the 4° seismicity cell centred on 39N, 41E, in Anatolia).

When point values from cells are sufficiently dense they can be presented in the form of Figs. 5 and 6 or contoured to reveal graduations in these distribution parameters, ω being the upper bound to earthquake magnitude occurrence, albeit with large uncertainty and a theoretically infinite return period.

75 year earthquakes and parameter correlation

The curvature parameter λ is not plotted in this fashion because it does not have a similar direct bearing on seismic risk, but it is closely correlated with the upper bound ω and this in turn has a significant influence on the uncertainties in risk estimates forecast using the 75 year earthquake M_{75}. Matrix (2) gives the form of the complete covariance matrix between parameters, the leading diagonal of which gives the variances on individual parameters ω, u and λ from which parameter standard deviations for a cell can be calculated as above. This covariance matrix for the cell data corresponding to Fig. 9 is:

$$\begin{pmatrix} 2.53 & 0.13 & -0.23 \\ 0.13 & 0.02 & -0.02 \\ -0.23 & -0.02 & 0.02 \end{pmatrix}$$

The important point to note is the correlation between ω and λ consistently revealed in many seismicity cells as a large and negative covariance; here it is $\sigma^2_{\omega\lambda} = -0.23$. Seventy-five year earthquakes M_{75} can be calculated directly from ω, u and λ. Uncertainties σ_M on M_{75} can be calculated from

$$\alpha^2_M \approx \sigma^2_\omega\left(\frac{\partial M}{\partial \omega}\right)^2 + \sigma^2_u\left(\frac{\partial M}{\partial u}\right)^2 + \sigma^2_\lambda\left(\frac{\partial M}{\partial \lambda}\right)^2 + 2\sigma^2_{\omega u}\left(\frac{\partial M}{\partial \omega}\right)\left(\frac{\partial M}{\partial u}\right)\dots \qquad (8)$$

and obtained using just the diagonal elements of matrix (2) or the entire matrix, thereby demonstrating the improved assessment of σ_M resulting largely through the additional adjustment caused by incorporating large and negative values of $\sigma^2_{\omega\lambda}$. The cell data of Fig. 9 leads to:

	σ_M	
M_{75}	matrix diagonal	all matrix
7.23	1.39	0.27

Values of σ_M may be typically halved or quartered by use of the entire covariance matrix.

Preliminary grid point values of M_{75} obtained from overlapping 4° cells can be shown contoured as in Fig. 10. Although preliminary, these results show some improvement in local detail over Map 1 of Burton (1979), for example in the Aegean Sea where finer contour graduations of seismicity now correlate better with the usually accepted seismotectonics. A covariance matrix and uncertainty σ_M is associated with each contoured grid point value of M_{75}.

4.3 Strain energy release

There are two important problems still to be considered which involve, first, the rate of strain energy release and, secondly, motion of the ground caused by earthquake occurrence. A difficulty with the cumulative frequency law of earthquake occurrence is that it extends to infinite magnitudes, unless truncated; on the other hand, although Gumbel III truncates magnitude occurrence, with some uncertainty at the upper bound ω, this truncation is asymptotic and corresponds to theoretically infinite return periods. Neither description seems intuitively true when strain energy input mechanisms are considered, which presumably may give rise to a cycle of strain energy input and release with large or upper bounded earthquakes occurring with finite return periods. To investigate this it is necessary to look at actual rates of strain energy

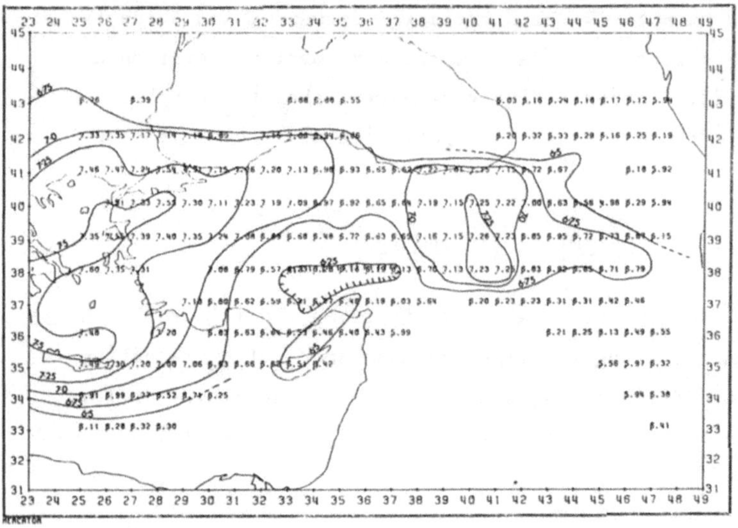

Fig. 10 Preliminary contour map of the largest earthquake magnitude M_{75} expected over an interval of seventy-five years. Point value forecasts for 4° cells are contoured in one quarter magnitude intervals and associated forecasting uncertainties can be assessed through associated covariance matrices.

release, particularly variations from the mean rate, and relate this to earthquake magnitudes which are thereby characteristic of the seismic risk. This approach is appropriate because it is the rate of strain energy release which governs seismic risk, rather than individual and isolated occurrences, and it is easy to apply through graphical means.

Earthquake magnitude and strain energy release, E ergs, can be related through Båth's (1958) equation.

$$\log E = 12.24 + 1.44M_s \tag{9}$$

Cumulative strain energy release diagrams are well known, described by Båth (1973), and used by Galanopoulos (1972) and Makropoulos (1978) to discern annual and maximum possible strain accumulation in Greece. Cumulative strain energy release diagrams for the cell 39N, 41E is shown in Fig. 11. The middle line, of the three superimposed, joins the origin to the total accumulated strain energy released, and therefore represents the average rate of annual strain energy release, which can

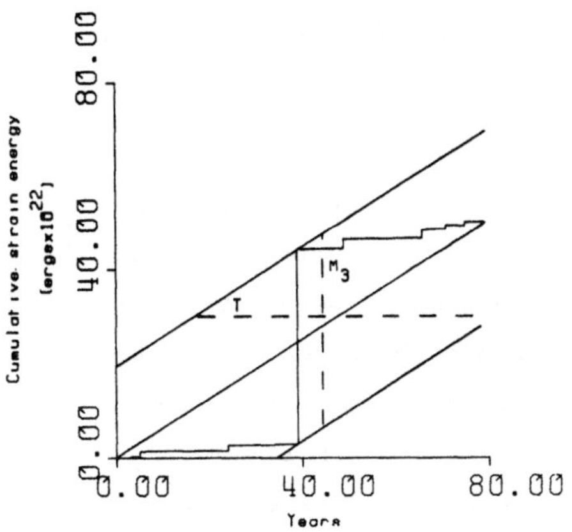

Fig. 11 Strain energy release. Cumulative strain energy release in ergs is plotted against time in years. The gradient of the middle line joining the origin to the total strain energy release measures the annual average strain energy release which can be related to an earthquake magnitude $\overline{M}_2$. Two outer lines, parallel to the middle line, envelope variations in the rate of release of strain energy. The vertical separation of the enveloping lines may be regarded as a measure of the maximum strain energy release which may be accumulated and released in a region, and may be interpreted as the large magnitude earthquake M_3. Horizontal separation of the enveloping lines may be interpreted as the waiting time T years required for the accumulation of strain energy corresponding to M_3 if no other earthquakes occur meanwhile.

be expressed as an equivalent magnitude $\overline{M}_2$ through equation (9). Lines parallel to $\overline{M}_2$ envelope variations in strain energy release from this mean rate. This type of cell diagram shows indications of periodicity in strain energy release, rather than just individual earthquake occurrence.

The projection of the vertical separation of the two enveloping lines, in ergs, is a measure of the total energy that might be accumulated and released in a region. If this energy was released in one earthquake then it would be the upper bounded earthquake for the region, and its magnitude M_3 can be assessed through equation (9). The horizontal projection of the enveloping lines represents the minimum time T years required to accumulate the energy equivalent to M_3 if no smaller earthquakes occur during this period.

This presentation is fundamentally different in that it does not imply a distribution of earthquake magnitude occurrence; it refers to rates of strain energy release which can be equally maintained by a few large earthquakes, or many smaller earthquakes, or a combination of the two. Fig. 11 possibly implies an environment where one major earthquake released most of the then available strain energy, which is now apparently re-accumulating. From Fig. 11 it is possible to deduce values of average strain energy release per annum, $\overline{M}_2$, M_3 and the associated waiting time, which for this cell are:

$\overline{M}_2$	6.64
M_3	7.90
T	64.60 years.

Values of M_3 and the upper bound to earthquake magnitude ω of Gumbel III should be comparable.

4.4 Perceptibility

Given we now have various estimates of the rate of earthquake occurrence available to us, particularly for large earthquakes, it is useful to be able to relate these occurrences to the levels of ground motion or vibration with which they are associated. This necessarily implies some combination between a statistical earthquake occurrence law, and a law describing the attenuation of ground motion with distance from an epicentre. Attenuation laws may describe ground acceleration or velocity, but until the current strong motion programme in Turkey is further developed, intensity attenuation laws, which relate directly to damage potential, seem more appropriate. Nevertheless, the method used leads to possible criteria for selection of engineering design suites of earthquake ground motion time histories.

$G^{III}(M)$ is the cumulative probability for extreme values, from this it is easy to calculate the probability density distribution, now denoted by P(M), for an earthquake magnitude M occurrence, where M now corresponds to the differential magnitude range $M - \frac{\delta M}{2} \leqslant M \leqslant M + \frac{\delta M}{2}$ (δM is, say, 0.2). There will be a set of intensities I_i associated with a magnitude M occurrence. Within a finite land area the probability $P_c(I)$ of perceiving an intensity I or greater at any point, given M, corresponds approximately to the land area fraction within the intensity I isoseismal.

Perceptibility, P_p, can now be defined as the probability of a magnitude M earthquake occuring and causing perceptible ground motion with intensity at least I, that is,

$$P_p = P(M|I) \approx P(M).P_c(I) \tag{10}$$

For small M the $P_c(I)$ tend to zero, and if focal depth is considered are truncated at zero for occurrences beyond particular depths: hence P_p tends to zero. Similarly, for large M the P(M) tend to zero, and for Gumbel III the P(M) will be truncated at zero for $M \geqslant \omega$: hence P_p tends to zero. The intermediate maximum of P_p defines a suite of most perceptible earthquakes (Burton, 1978c).

UNESCO (1976) and Yarar (1977) give expressions relating peak intensity to magnitude, and also the attenuation of that peak intensity with distance from the epicentre which for example can be reduced to,

$$I = 1.689M + 0.827 - 3.33 \log \Delta \tag{11}$$

Distance Δ may be redefined as $\Delta^2 = r^2 + h^2 + k^2$ where r is the epicentral distance, h the focal depth, and k an adjustment factor which allows for finite focal volume thereby eliminating spuriously high epicentral intensities. Equation (11) represents a relatively high degree of attenuation in the sense that it produces relatively small isoseismal areas for a given magnitude when compared to other attenuation laws; and this is probably most appropriate for seismicity near to the major active faults within Turkey. This attenuation law can be combined with Gumbel III cell parameters as indicated generally in equation (10) and perceptibility P_p calculated. The results of this process for the illustrative cell are shown in Fig. 12 at intensities VI, VII, VIII, and IX, for a notional uncorrected surface focus and with $h^2 + k^2$ set to 400 to allow for finite focal volume. The set of curves corresponding to a notional surface focus earthquake each give a maximum probability at the same magnitude value. This most perceptible earthquake magnitude is $M(P_p max)$; in Fig. 12 $M(P_p max)$ is 7.42.

This maximum for a cell corresponds to that earthquake magnitude which has the highest probability of being perceived at any intensity value, and earthquakes with magnitudes spanning this peak may be regarded as appropriate for engineering design purposes. This simplicity exhibited by the notional surface focus earthquake is slightly modified if corrections for focal volume or depth are incorporated. Positive values of $h^2 + k^2$ perturb the maximum of the curve corresponding to a particular intensity level in two ways: firstly, the maximum probability of

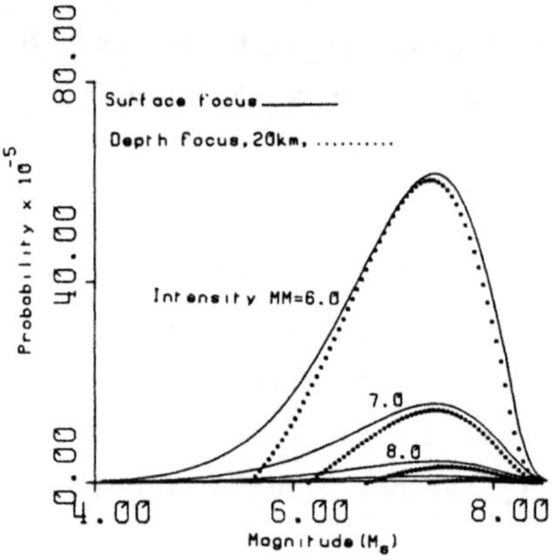

Fig. 12 Perceptibility. Cellular Gumbel III earthquake magnitude occurrence laws are combined with an intensity attenuation law (equation 11). Results give the probability of perceiving intensity VI, VII, VIII or IX from a particular magnitude of earthquake. The solid curves correspond to a notional surface focus earthquake; each such curve has a maximum at the same value of surface wave magnitude thus leading to the concept of a 'most perceptible' earthquake for engineering design purposes. Dotted curves give similar results but include a correction for finite earthquake volume which here may be interpreted as notional earthquakes at 20km depth.

perceiving a particular intensity is depressed, secondly, the magnitude $M(P_p max)$ at intensity I is slightly displaced to higher magnitude values. One other feature of these families of graphs is worth comment. For a surface focus earthquake the probability of perception is asymptotic to zero as magnitude decreases. With finite $h^2 + k^2$ the probability of perception of a particular intensity truncates at zero when the magnitude is insufficient to stimulate such an intensity for I_o over even an infinitesimally small area above the hypocentre.

5. Discussion and conclusions

The occurrence of upper bounded, large and damaging earthquakes by definition restricts us to a study of rare events in which statistical forecasting capability is vital. The problem of assigning return periods to earthquake occurrences, or estimating the occurrence of large magnitude earthquakes, depends fundamentally on available earthquake catalogues and histories. Two earthquake files for Turkey which span the years 1900 to 1970 have been examined in some detail, revealing that the file contents come from diverse catalogue and agency sources although basically describing the same earthquake epicentres and magnitudes, presumably originally available as phase data and amplitude data on a specific set of seismograms. The second of these files, TURK70, contains the catalogue of ATB which has attempted to refine the earthquake history of Turkey (1913 to 1970) by a systematic and homogeneous assessment of the available instrumental data, 'calibrated' using the wide time span of instrumental recordings on the Uppsala Wiechert instrument.

Both files show similar completeness properties, discrepancies between the two not surprisingly occurring at the smaller magnitudes. Except for the earlier decades of this century magnitudes down to $M_s = 4.5$ appear reasonably complete, although distinct differences may be expected between east and west Turkey.

Both files show similar forecasts of the largest earthquake magnitude to be expected in a seventy-five year period, that is point values of M_{75} obtained by cellular analysis of these two catalogues show good agreement usually within the uncertainties σ_M calculated on M_{75}. The forecasting capability of both catalogues is seen to be similar.

Throughout the file, attention has been centred on the surface wave magnitude M_s because this is less likely to saturate for large earthquakes than is the case for the now more readily available body wave magnitude m_b. Few values of M_s are likely to saturate (Kanamori, 1977; Abe & Kanamori, 1980) in the manner shown by some great earthquakes, particularly in the circum-Pacific belt. Extension of the earthquake file from 1971 onwards using ISC and PDE data requires a conversion formula from the readily available m_b to M_s, this, and a subsequent detailed analysis of seismic risk in Turkey obtained using the methods illustrated briefly here will be described elsewhere. These several methods have led to a comparison of upper bounded and other large magnitude earthquake occurrences estimated using intrinsically different processes. The 4° cell of seismicity centred on 39N, 41E, in Anatolia has been used to illustrate these methods, the main numerical results of which can be briefly tabulated as:

Gutenberg-Richter	Gumbel III		Strain energy release		Perceptibility
M_{75}	M_{75}	ω	M_3	T_{years}	$M(P_p max)$
7.51(0.62)	7.23(0.27)	8.60(1.59)	7.90	64.6	7.42

Seventy-five year forecasts M_{75} obtained using a and b parameters of the Gutenberg-Richter cumulative frequency law have been obtained using least squares, maximum likelihood, and indirectly through Gumbel I extreme value analysis. However, in all cases these estimates of M_{75} lie well within one standard deviation calculated on M_{75}; here the least squares result 7.51(0.62) is representative. It is apparent from Fig. 8 illustrating the whole process statistics of Gutenberg-Richter that values of a and b would usually be significantly influenced if the completeness analysis was not available; values of a and b obtained by least squares and maximum likelihood would be adversely affected. Values of a and b obtained here do not need adjustment in the manner described by Utsu (1966) because the 0.1 magnitude increment used in the cumulative earthquake magnitude count is sufficiently small.

Gutenberg-Richter cumulative frequency distributions often show curvature convex with respect to the origin implying observational roll-off of probability towards higher magnitudes which can be analysed using the part process Gumbel III distribution of extreme values. Forecasts of M_{75} using Gumbel III are thus expected to be lower than those from Gutenberg-Richter and this is certainly borne out by the mean value of 7.23. M_{75} from Gumbel III is invariably less than that from Gutenberg-Richter.

Point values of M_{75} from Gumbel III may be contoured producing the rudimentary risk map of Fig. 10. Using 4° cells without consideration of tectonics is most appropriate if the density of earthquake epicentral distribution changes gradually - 'linear' features such as the Anatolian fault are 'smoothed' by this process. However, these rudimentary contours do show major variations in seismic risk including an apparent low risk area in southern Turkey, and high risk areas in western Turkey extending along the Anatolian fault into eastern Turkey. The method of analysis used has also demonstrated that an assessment of uncertainty in the M_{75} point values from which these contours are constructed can be obtained from the complete covariance matrix between parameters of Gumbel III for one cell. Further, use of the entire covariance matrix includes the large and negative $\sigma^2_{\omega\lambda}$ which significantly reduces these forecasting uncertainties. In time the data base from which M_{75} can be obtained will expand and lengthen and refinement of the Gumbel III parameters should improve forecasts of M_{75}, however, actual values of M_{75} obtained using earthquake data spanning longer times will conservatively approach the true M_{75} from above.

During estimation of Gumbel III upper bounds to earthquake magnitude occurrence are obtained, typically with values of σ_ω around one or two magnitudes. The example above gives ω = 8.60(1.59). It should be recognised that this is only one parameter of a total distribution which is being estimated and should only be regarded as an uncertain indication of an upper bound. A second problem is that ω is an asymptotic upper bound and therefore theoretically corresponds to an unobtainable infinite return period, whereas observation of actual strain energy release indicates that the entire seismicity process may be cyclic thereby showing periods of relative quiescence and activity, with finite period, and a specific maximum to the range of strain energy release. Examination of cellular strain energy release diagrams like Fig. 11 leads to estimates of large magnitudes M_3 which correspond to the maximum strain energy which apparently may be accumulated and released in a cell. The summary Table above shows an M_3 of 7.90, less than ω and greater than M_{75} of Gumbel III, which is consistent with overall views of the seismicity process. In addition cell 39N, 41E, shows low activity followed by a major earthquake equivalent to M_3, intuitively this is in agreement with the long waiting period of about 65 years required for uninterrupted accumulation of strain energy equivalent to M_3. In this example M_3 = 7.9 is not particularly close to ω = 8.6; but it is within the bounds of $\sigma_\omega = \pm$ 1.59 on ω, and its waiting time approaches the time span of the present earthquake file data base. At this stage it would be speculation to consider if refinement in due course of these parameters by extending the observational time span will reduce ω conservatively towards M_3, or if an event larger than M_3 will occur.

Finally, it has been demonstrated how combination of recurrence statistics with an inferred intensity attenuation law for Turkey allows calculation of the probability of perceiving intensities generated by particular magnitude earthquakes. Maximum probability peaks of these families of intensity curves will show a range of most perceptible earthquakes throughout Turkey. Given this regional variation in these most perceptible earthquakes specific values may be assigned for individual areas; the example used here leads to $M(P_p max)$ of 7.42. Standard zoning practice would indicate which maximum intensities have occurred in an area, and perhaps, how often; this approach allows quantification of intensity occurrence in an area (by integration of the local perceptibility curve), but more importantly leads directly to the magnitude range which is likely to cause damage within that locality. It is apparent that these earthquakes need not cause the maximum possible damage, the larger and rarer earthquakes within the range $M(P_p max) \leqslant M_s \leqslant \omega$ may cause this, however, when making an economic evaluation it should be noted that magnitude $M(P_p max)$ will cause damage corresponding to any intensity level on the greater number of occasions than magnitudes nearer to ω.

Assessments of statistical seismic risk in a region can be founded on a range of determinations of earthquake magnitude corresponding to discrete return periods, to upper bounded, and to other large magnitude earthquakes, which for earthquake engineering design criteria may invoke the perceptibility of different levels of ground motion produced by earthquakes.

Acknowledgements

We wish to thank David Redmayne who helped eliminate faults in and compile the earthquake file TURK70. We are grateful to Dr Dizer of Kandilli Observatory, Istanbul, and to Professor Drakopoulos of the National Observatory, and the Seismological Laboratory of the University of Athens, for encouraging these cooperative studies.

The work of PWB and RWM has been supported by the Natural Environment Research Council and is published with permission of the Director of the Institute of Geological Sciences.

References

Abe, K. & Kanamori, H., 1980. Magnitudes of great shallow earthquakes from 1953 to 1977, Tectonophysics, 62, 191-203.

Aki, K., 1965. Maximum likelihood estimate of b in the formula logN=a-bM and its confidence limits, Bull. Earth.Res.Inst., Tokyo Univ., 43, 237-239.

Alsan, E., Tezuçan, L. & Båth, M., 1975. An earthquake catalogue for Turkey for the interval 1913-1970, Kandilli Obs., Istanbul, and Seismol. Inst., Uppsala, Rep. No. 7-75, 166pp. Also in: Tectonophysics, 1976, 31, T13-T19.

Ambraseys, N.N., 1971. Value of historical records of earthquakes, Nature, 232, 375-379.

Båth, M., 1958. The energies of seismic body waves and surface waves, In: H. Benioff, M. Ewing, B.F. Howell Jr., and Press (Editors), Contributions in Geophysics, Perganon, London, 1, 1-16.

Båth, M., 1973. Introduction to Seismology, Birkhauser Verlag Basel, 395pp.

Båth, M., 1979a. Seismic risk in Turkey - a preliminary approach, Tectonophysics, 54, T9-T16.

Båth, M., 1979b. Seismic risk in Fennoscandia, Tectonophysics, 57, 285-295.

Burton, P.W., 1978a. The IGS file of seismic activity and its use for hazard assessment, Inst. Geol. Sci., seism. Bull. No. 6, HMSO.

Burton, P.W., 1978b. The application of extreme value statistics to seismic hazard assessment in the European area, Proc. Symp. Anal. Seismicity and on Seismic Risk, Liblice, 1977 October 17-22, pp. 323-334; Academia, Prague.

Burton, P.W., 1978c. Perceptible earthquakes in the United Kingdom, Geophys. J.R. astr. Soc., 54, 475-479.

Burton, P.W., 1979. Seismic risk in Southern Europe through to India examined using Gumbel's third distribution of extreme values, Geophys. J.R. astr. Soc., 59, 249-280.

Epstein, B. & Lomnitz, C., 1966. A model for the occurrence of large earthquakes, Nature, 211, 954-956.

Erdik, M. & Özselçuk, A., 1978. Regional rearrangement of earthquake catalogues of Turkey, Middle East Technical University, Earthquake Engineering Research Institute, Ankara.

Ergin, K., Güçlü, U. & Uz, Z., 1967. A catalogue of earthquakes for Turkey and surrounding area (11AD to 1964AD), Tech. Univ. Istanbul Fac. Mining Eng., Publ. No. 24.

Galanopoulos, A.G., 1972. Annual and maximum possible strain accumulation in the major area of Greece, Ann. Géol. d. pays Hellén, 24, 467-480.

Gençoğlu, S. & Tabban, A., 1973. Statistical analysis of earthquake occurrence in Turkey (1900-1970), Earthquake statistic and risk symposium, Balkan Project, 26 May 1973. (Also report of Turkish Republic Ministry of Reconstruction and Resettlement, Earthquake Research Institute, Ankara).

Gringorten, I.I., 1963. A plotting rule for extreme probability paper, J. geophys. Res., 68, 813-814.

Gumbel, E.J., 1958. Statistics of Extremes, Columbia University Press, New York.

Kanamori, H., 1977. The energy release in great earthquakes, J. geophys. Res., 82, 2981-2987.

Kárník, V., 1968. Seismicity of the European Area, Part I, Academia, Praha.

Kárník, V., 1971. Seismicity of the European Area, Part II, Academia, Praha.

Makropoulos, K.C., 1978. The statistics of large earthquake magnitude and an evaluation of Greek seismicity, PhD thesis, University of Edinburgh, 193pp.

Makropoulos, K.C. & Burton, P.W., 1980. A catalogue of seismicity in Greece and adjacent areas, Geophys. J.R. astr. Soc., in press.

Pinar, N. & Lahn, E., 1952. Descriptive catalogue of Turkish earthquakes (in Turkish, Bayindirlik Bakanliği (Ministry of Construction), Publ. Ser. 6, No. 36.

Stepp, J.C., 1971. An investigation of earthquake risk in the Puget Sound area by use of the type I distribution of largest extremes, PhD thesis, Pennsylvania State University, 131 pp.

Üçer, S.B., Ayhan, E. & Alsan, E., 1975. Preliminary statistical results for the preparation of a seismic zoning map of Anatolia, Report of Kandilli Obs., Seismol. Dept., Istanbul.

UNESCO, 1976. Country Monograph of Turkey. Intergovernmental Conference on Assessment and Mitigation of Earthquake Risk, Paris, February 10-19, 1976.

Utsu, T., 1966. A statistical significance test of the difference in b-value between two earthquake groups, J.Phys. Earth, 14, 37-40.

Yarar, R., 1977. Seismic activities of Turkey and surrounding Balkan region, Bull. Int. Inst. Seis. Earthquake Eng., 15, 83-104.

Yegulalp, T.M. & Kuo, J.T., 1974. Statistical prediction of the occurrence of maximum magnitude earthquakes, Bull. seism. Soc. Am., 64, 393-414.

Faults and Recently Active Breaks Along the North Anatolian Fault Zone Between Gerede and Ilgaz

Melih Tokay

Middle East Technical University, Ankara, Turkey

Introduction

The main purpose of the paper is to present the results pertaining to the seismo-tectonic studies and field mapping carried out at the scale of 1:25 000 of the Gerede-Ilgaz portion of the North Anatolian Fault Zone (NAFZ) undertaken some years ago as a TBTAK Research Project (TBAG No. 43) (Fig. 1).

The studied portion represents 1/7 of the total length of this major fault zone and covers an area of 150 km length and 15-17 km width.

During the field work a detailed map was prepared including ancient and active faults resulting from the 1943, 1944, and 1951 earthquakes. These are plotted upon field observation and information obtained from literature or from eye witnesses.

Stratigraphy

Sedimentary and volcanic rocks of Mesozoic, Tertiary and Quaternary ages crop out in the area (Figs. 1-5). The following stratigraphic units have been differentiated along the fault zone:
Unit J on the map ranges from Liassic to Portlandian and mainly consists of fine grained and pseudo-oolitic limestone and yellow marl. Upwards it grades in to Lower Cretaceous in some localities, to form unit JK which consists of argillaceous, fine grained or pseudo-oolitic limestone.

KL unit of Albian-Cenomanian age is a flysch consisting of sandstone, marl, and occasional limestone. Huge olistoliths are noted when the unit bears a Wildflysch character.

Upper Cretaceous is represented either by Senonian sediments such as red argillaceous limestone, sandstone and conglomera-

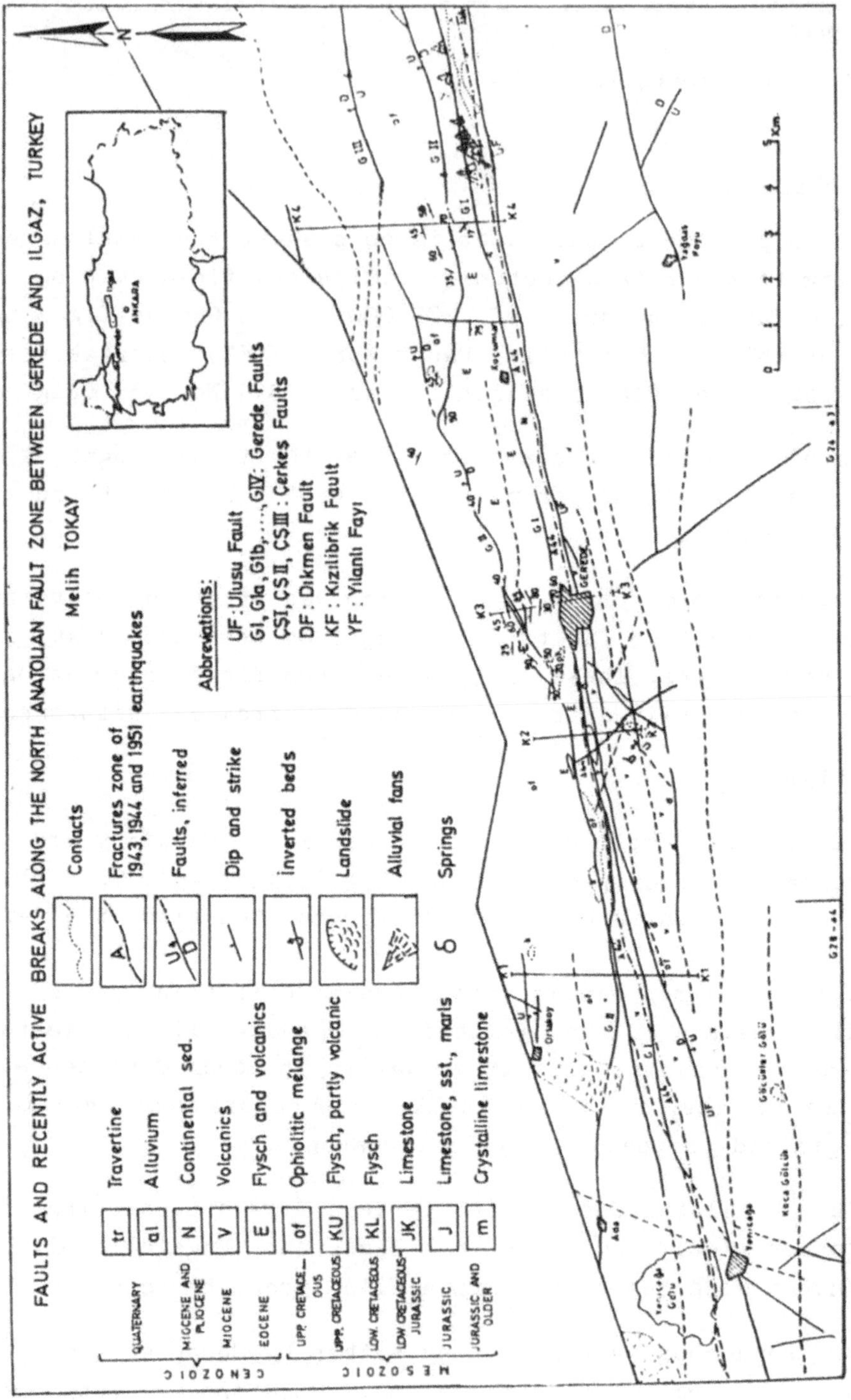

FAULTS AND RECENTLY ACTIVE BREAKS ALONG THE NORTH ANATOLIAN FAULT ZONE BETWEEN GEREDE AND ILGAZ, TURKEY
Melih TOKAY
CENOZOIC
QUATERNARY
tr Travertine
al Alluvium
MIOCENE AND PLIOCENE
N Continental sed.
MIOCENE
V Volcanics
EOCENE
E Flysch and volcanics
MESOZOIC
UPP. CRETACEOUS
of Ophiolitic mélange
UPP. CRETACEOUS
KU Flysch, partly volcanic
LOW. CRETACEOUS
KL Flysch
LOW CRETACEOUS-JURASSIC
JK Limestone
JURASSIC
J Limestone, sst., marls
JURASSIC AND OLDER
m Crystalline limestone
Contacts
Fractures zone of 1943, 1944 and 1951 earthquakes
Faults, inferred
Dip and strike
Inverted beds
Landslide
Alluvial fans
Springs
Abbreviations:
UF: Ulusu Fault
GI, GIa, GIb, ..., GIV: Gerede Faults
ÇSI, ÇSII, ÇSIII : Çerkes Faults
DF: Dikmen Fault
KF: Kizilibrik Fault
YF: Yilanli Fayi
ANKARA
Gerede
Yeniçağa
Yeniçağa Gölü
Koca Gölcük
Gölcükler Gölü
Örtülüköy
Ada
Yağsaş Fayı
0 1 2 3 4 5 Km

Fig. 1

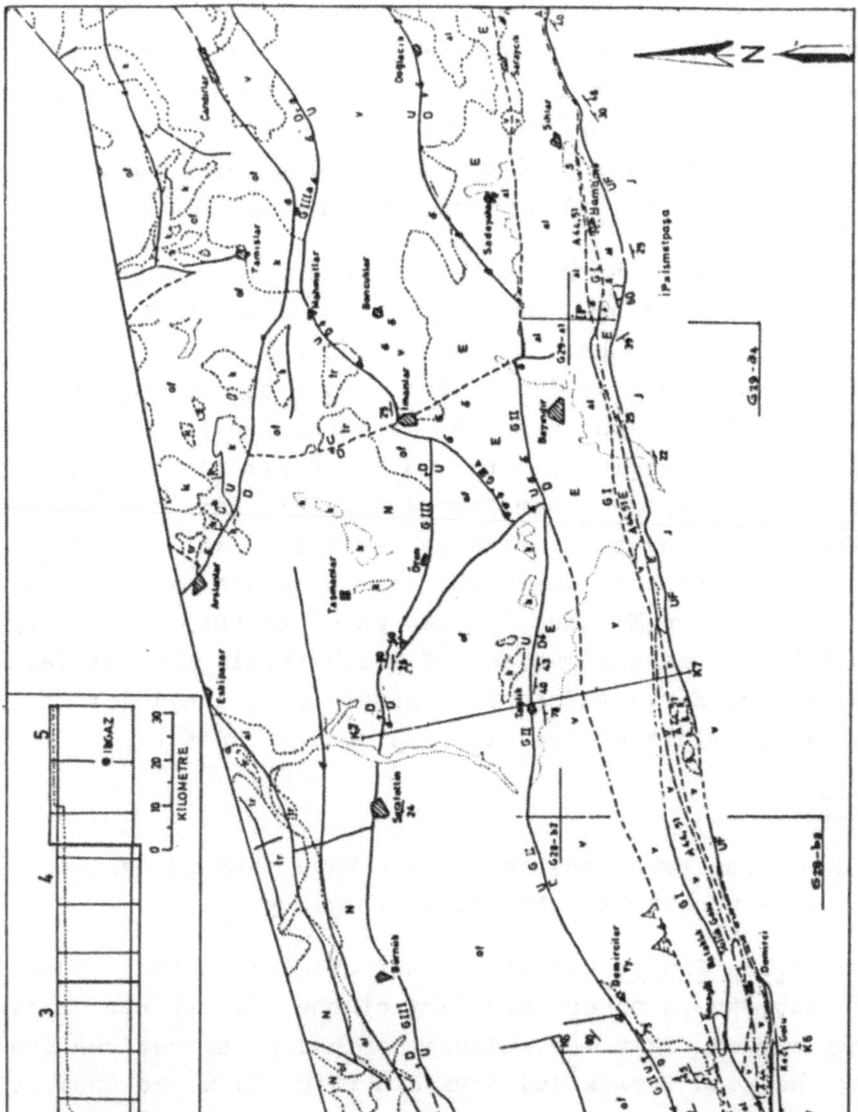

Fig. 2

tes (unit KU), or by an ophiologic mélange (unit 'of') includi serpentinite (s) and limestone (k) lying between Albian-Cenomanian flysch and Upper Campanian-Maastrichtian layers. Unit E is mainly of Lutetian age and consists of a lower yellow fossiliferous limestone and an upper brown sandstone with abundant plant remains. The unit E is overlain by conglomerates sandstones and marls of Late Eocene-Oligocene? age. Eocene sediments may occur along with volcanic rocks (Ev).

Unit N represents both sedimentary and volcanic rocks of Miocene and Pliocene age. Brackish marls of Aquitanian age, tuffaceous sandstones, colored tuffs, volcanic breccia, and small scale andesitic lava flows of Vindobonian age have been observed. This volcanic series is in turn overlain by extrusive rocks of the Köroğlu volcanic massif (unit v). An Upper Miocene lignite seam exposed near Yıprak (Fig. 3), and diatomaceous sediments bearing plant remains associated with fishes and reptiles have also been observed. All these are overlain by volcanic rocks Late Miocene in age, which in turn are cut by fault of the NAFZ, the maximum age of which should then be later Miocene, if not Early Pliocene. Younger loose conglomerates, presumably of Pliocene, are also included in unit N.

Faults

The following fault sets or faults have been recognized from north to south, within the Gerede-Ilgaz portion:

1 The Gerede faults (GI-GIV) a set of minimum four faults, constitute the northern boundary of the rift. Their traces are mostly curved. They are steeply north dipping reverse faults, except between Taşlik and Saraycik (Fig. 2) where they become south dipping normal faults. They displace Neogene (Lower Pliocene) sediments and elongated ridges and sag ponds are situated along them.

The GII fault seems to be the youngest of this set since it intersects GIb and GIII west of Bayındır (Fig. 2), GIV west of Sadeyaka (Fig. 2). GIV fault intersects east of Taşhanlar (Fig. 3) a W.NW-E.NE trending fault which offsets GIIIa.

2 Ulusu Fault (UF) forms the southern boundary of the central rift and constitutes a major element which may be traced along great distances. Its trace is straighter than those of the Gerede faults. It is better reflected in the topography to the east of İsmetpaşa (Fig. 2); its general trend is N 80 E. Its vertical displacement is estimated to exceed several hundred meters. Transversal faults with negligible lateral displacement cuts it at some localities.

3 Active faults of the 1943 Kastamonu-Lâdik-Bayramören, 1944 Bolu-Gerede, and 1951 Kurşunlu earthquakes are identified and plotted respectively as A43, A44 and A51 on the maps. Although these recent fractures may join, follow closely, trend transverse or intersect ancient faults, they generally remain within the fault zone. However, A44 leaves this zone at Oymaağaç (Fig. 1) as it may be traced north of GII Fault. The 1951 earthquake fracture A51 which normally follows A44, extends eastward, out of the fault zone and south of the Ulusu Fault, as seen near Bucura and Ömerli (Fig. 5).

In addition to sudden slip associated with the earthquakes, there are also evidence for subsequent aseismic tectonic creep along the same lines. The amount of creep is measured at regular intervals at İsmetpaşa, near the buildings of the Highway Department, the eastern wall of which is offset due to dextral strike-slipping along the active fault.

4 A group of north-dipping reverse faults called ÇŞI, ÇŞII, and ÇŞIII are present south of the Ulusu Fault. These faults may run parallel to each other, or intersect to form lens-shaped blocks. They constitute mainly the boundary between the southern Neogene deposits and older formations to the north, although they may also separate formations fo differing ages such as Upper Cretaceous, Jurassic, Ophiolitic Mélange etc. They form an imbricated structure with a south directed tectonic transport.

More to the south, Kızılıbrik (KF) and Yılanlı (YF) Faults represent other fractures mainly in Neogene deposits where beds may be overturned and and displaced southward.

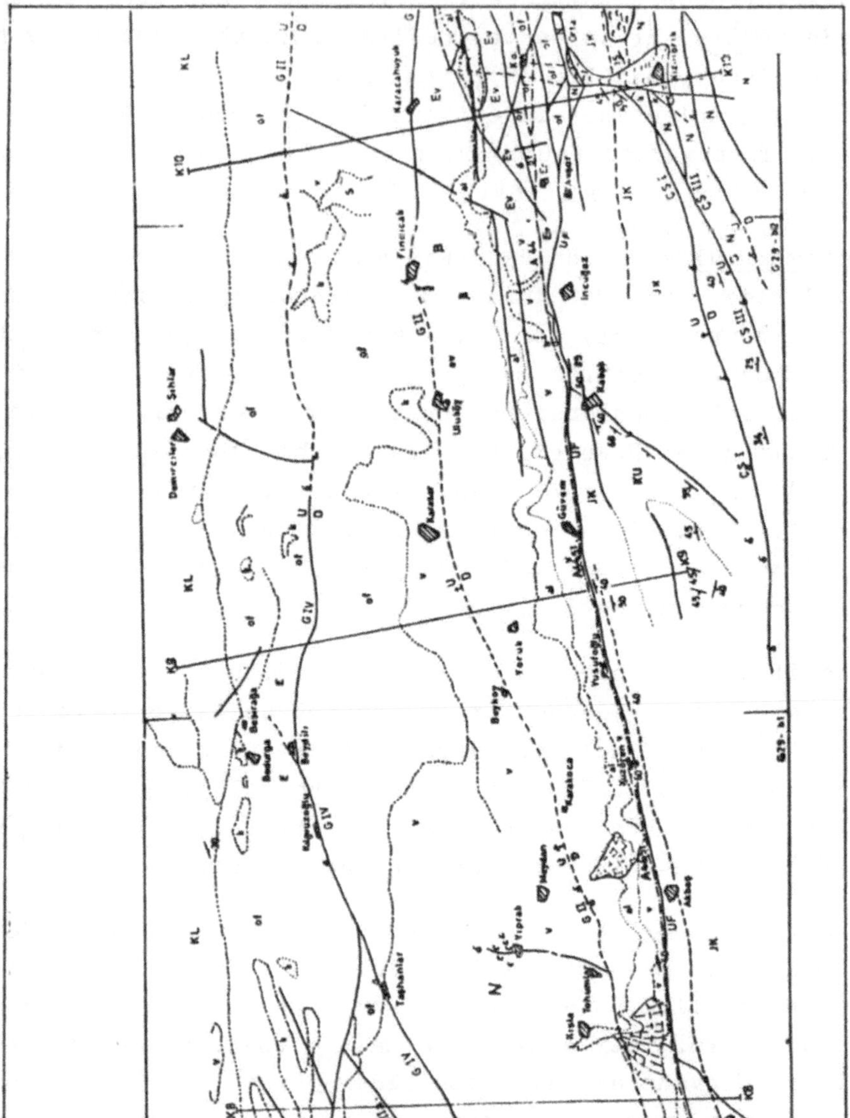

Fig. 3

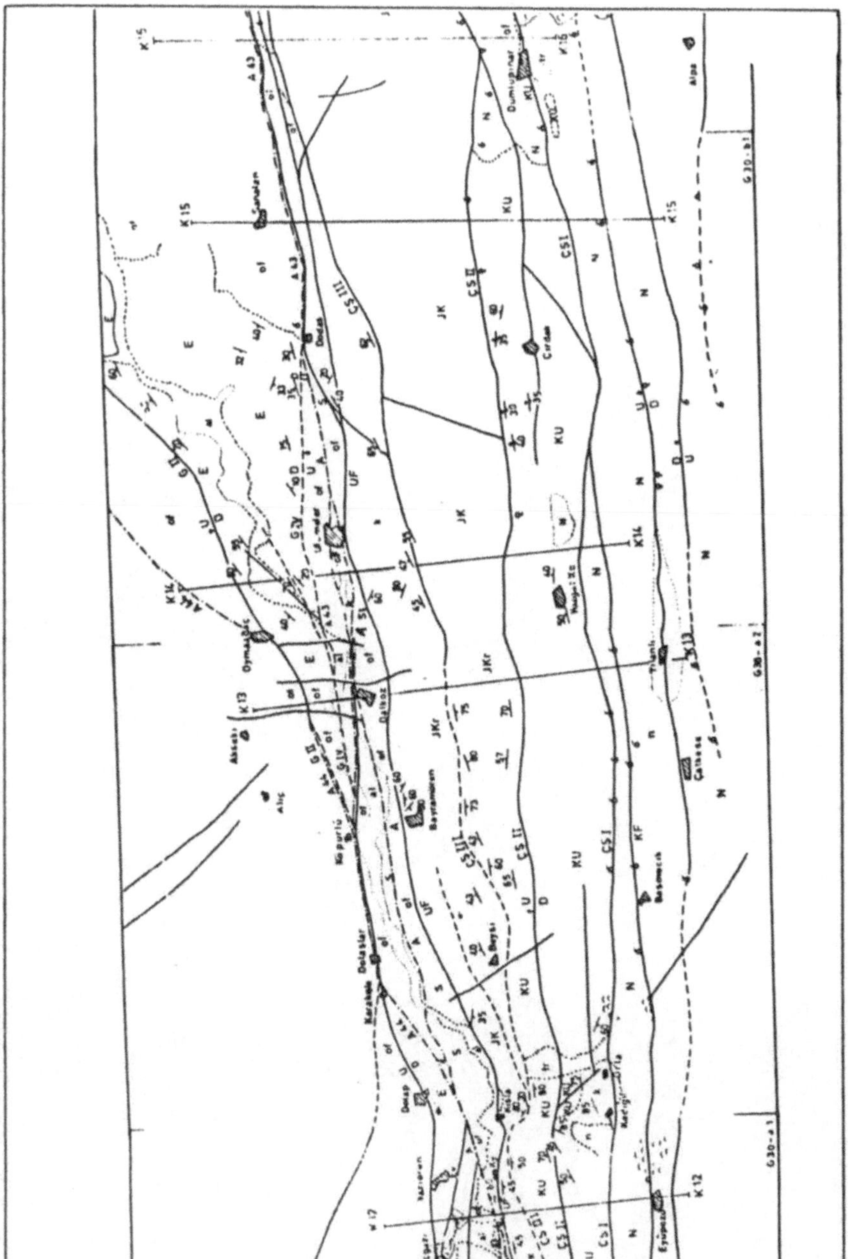

Fig. 4

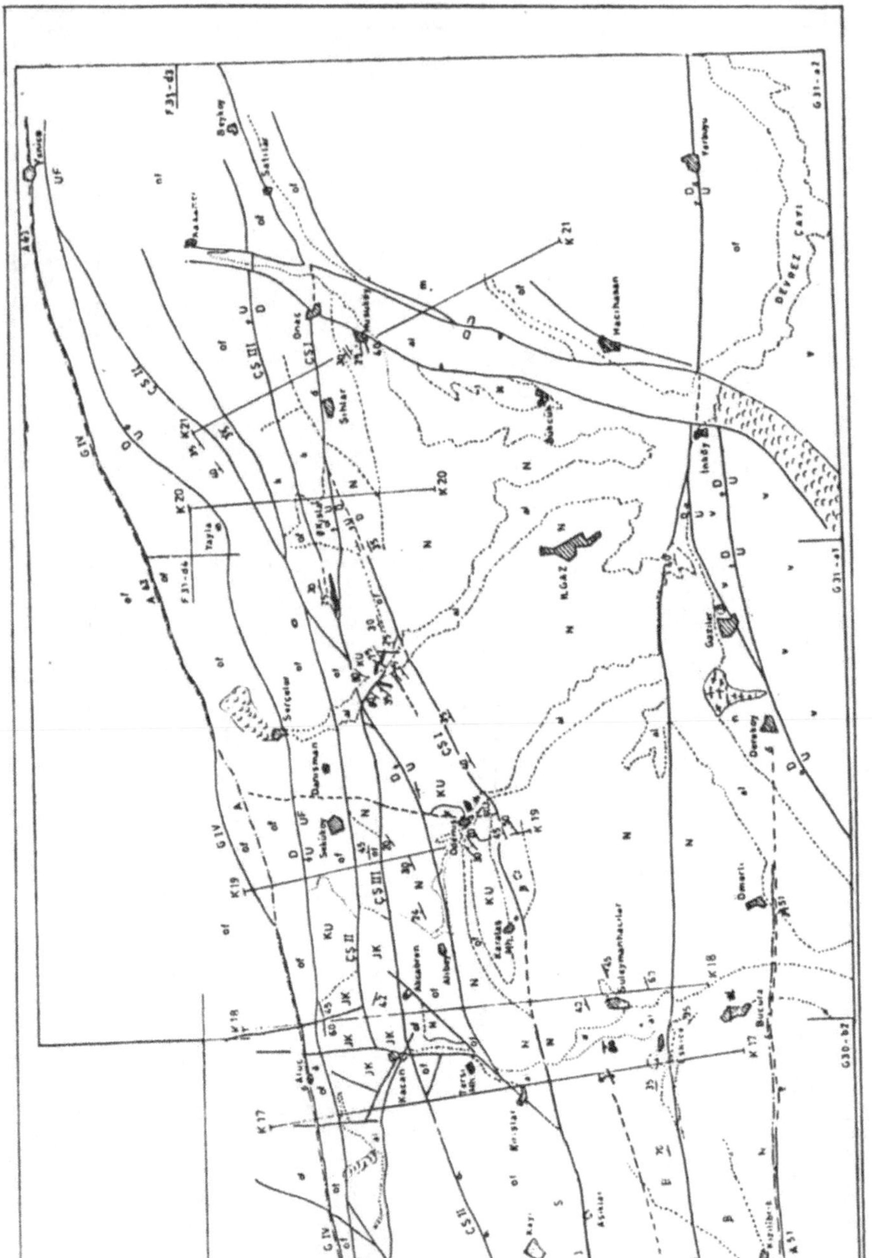
Yenice
F 31-d3
F 31-d4
G 31-e1
G 31-e2
G 30-b2
K 17
K 18
K 19
K 20
K 21
DEVREZ ÇAYI
K.GAZ
Sarıçalar
Danışman
Sekiköy
Gaziler
Bucura
Süleymanhacılar
Kızılibrik
GIV
CS I
CS II
CS III

Fig. 5

Few normal faults are also present in the area like the one that cuts Neogene sediments and volcanics in the İnköy-Dereköy region (Fig. 5) denoting extension. Transversal, vertical, some of them of gravity (Kızılıbrik, Demüzler etc.) intersecting the Fault Zone also denote younger extensional periods. Besides, thrusts and small recumbent faults with a S or SE vergence in Pliocene or Quaternary deposits at a place between Ankara and the Fault Zone should have resulted of compressional stresses which have been active in a more recent period.

Creep measured immediately to the North of İsmetpaşa railway station amounts to 1.5 cm per year horizontally and 0.6 cm per year vertically on the average, and it is intermittent (Aytun, 1973).

Conclusions

The structure of the zone is mainly one that has resulted from collision followed by dextral strike slipping. Dextral strike slipping was evident during the 1943, 1944, and 1951 earthquakes, and it is still at İsmetpaşa. Till now no geologic evidence has been found to measure the total displacement within the study area. However, assuming an average annual displacement of 1-1.5 cm, a total displacement of 60-80 km can be estimated since Middle Pliocene (Tokay, 1973).

Although the recent displacement within the NAFZ is definitely of right lateral strike-slip type, the structure of the zone can not be fully explained by east-west shear couple. The NAFZ, in a great number of localities, displays evidence for compressional forces oriented in N N30-S S 30 E direction, in the form of thrust faults on either side of the rift, imbrications, asymmetric folds with overturned flanks upthrusting younger formations, faulted tight overturned anticlines within the rift etc., the vergence generally being southward. That strike slipping is a relatively recent event which has followed compression (Tokay, 1973) and may be explained very briefly in the following way:
The North Anatolian Fault Zone was the site of a trench in Jurassic and Cretaceous times. To the north was an island

arc (Pontia) and to the south a microplate (Anatolia). Their convergence which was strongly evident during Late Cretaceous as a result of the advance of the Arabo-African block has lead to the formation of a subduction zone where Anatolia was plunging beneath Pontia. After different events such as formation of the Ophiolitic Mélange (prior to Maestrichtian), retreat of the sea, the Mediterranean Tethys (Late Eocene), volcanism (Late Cretaceous, Eocene, Neogene), continental regime (Miocene), collision has been completed, leaving behind evidence for intense compression in Uppermost Miocene (Rhodanian phase), or even Early Pliocene times. The North Anatolian right lateral strike-slip fault must have formed at the Miocene-Pliocene boundary, or in Early Pliocene during which orogenic movements have been noted in Turkey, whereas Late Pliocene was a period mainly characterized by tensional forces.

Results similar to those of the present paper were obtained by authors who carried out field work in more eastern portions of the NAFZ, Öztürk (1976) in Lâdik-Destek area, Seymen (1975) along the Kelkit valley (Reşadiye, Tokat), Tatar (1978) in Refahiye, Erzincan area, and Bergougnan (1975) in Erzincan area. The following conditions are independently and unanimously reported by these authors in their areas showing similar stratigraphic and structure, with some local differences: (a) closure of an ancient sea - Mediterranean Tethys -; (b) collision in early or middle Miocene, as a result of compressional forces roughly oriented NS, of two blocks - the northern one being an island arc -; followed by a dextral strike slipping movement which has started in late (? middle) Miocene - Pliocene times; (c) a total lateral displacement along the Fault Zone, estimated between 60 and 105 km; (d) westward displacement of the southernblock (Anatolia, or Turkish Plate of the authors) due to the northward advance of the Arabe block, etc.

Valuable contributions concerning the NAFZ have been made particularly using seismic data among others, by Mackenzie (1970, 1972, 1978), Dewey et al. (1973), and Dewey and Sengör (1979), who all have the merit of considering the Zone in the plate tectonics concept; and also by Dewey (1976). A review of previous studies concerning the North Anatolian Fault Zone is presented by Şengör (1979).

References

Abdüsselâmoglu, Ş., 1959. Almacıkdagı ile Mudurnu ve Göynük civarının Jeolojisi. İstanbul Üniv. Fen Fak. Monogr. 14, 94 pp.

Allen, C.R., 1969. Active Faulting in Northern Turkey. Contribution No.1.577, Div. of Geol. Sci., Calif. Inst. of Tech. 32p.

Ambraseys, N.N., 1970. Some characteristic features of the Anatolian Fault Zone. Tectonophysics, 9, No.2/3, 143-164.

Ataman, G., Buket, E., and Çapan, U.Z., 1975. Could North Anatolian Fault Zone be a paleo-Benioff Zone ?. Bull, Min. Res. Explor. Inst., Ankara, 84, 97-102.

Aytun, A., 1973. İsmetpaşa istasyonu civarında krip ölçmeleri. Kuzey Anadolu Fayı ve Deprem Kuşagı Simpozyumu, Ankara.

Bergougnan, H., 1975. Affrontement des blocs européen et arabique dans l'Est-Anatolien. Ann. Univ. A.R.E.R.S. Reims, 13, fasc. 1 et 2.

Bumenthal, M., 1948. Un aperçu de la géologie des châînes nordanatoliennes entre l'Ova de Bolu et le Kızılırmak inférieur. Inst. MTA, Monogr. Série B, 13, 265 pp.

Dewey, J.F. and Pitman, W.C. III, Ryan, W.B.F. and Bonnin, J. 1973. Plate tectonics and the evolution of the alpine system. Bull. geol. Soc. Am. 84, 3137-3180.

Dewey, J.F. and Şengör, A.M.C., 1979. Aegean and surrounding regions : complex multiplate and continuum tectonics in a convergent zone, Bull. geol. Soc Am. 90, 89-92.

Dewey, J.W., 1976. Seismicity of northern Anatolia. Bull. Soc. Am. 66, 843-868.

Horstink, J., 1971. The Late Cretaceous and Tertiary Geological Evolution of Eastern Turkey. First Petroleum Congress, Turkey, Proceedings, 25-41.

Ketin,İ.,1969. Kuzey Anadolu Fayı hakkında.Bull.Min.Res. Explor.Inst,Ankara,72, 1-27.

Mackenzie,D.P.,1970. Plate Tectonics of the Mediterranean Region. Nature. 226, 239-243.

-------------- ,1972. Active tectonics of the Mediterranean Region. Geophys.J.R.Astr.Soc.,30, 109-185.

-------------- ,1978. Active tectonics of the Alpine-Himalalayan belt : the Aegean Sea and surrounding regions.Geophys.J.R.Soc.55, 217-254.

Öztürk,A.,1968. Çerkeş-Eskipazar-Gerede Bölgesinin Jeolojisi Ph.D. thesis,Ankara Univ.Fac Sci.(unpublished).

-------- ,1976. Lâdik-Destek arasının jeolojisi ve faylarının özellikleri.Ankara Üniv.Fen Fak.Doç.tezi,79 s.

Şengör, A.M.C., 1979. The North Anatolian transform kult: its ag offset, and tectonic significance. Geol. Soc. London Jour. 136, 269-282.

Seymen,İ.,1975. Kelkit vadisi kesiminde Kuzey Anadolu Fay Zonununtektonik özelligi.İTÜ,Maden Fak.Yay.İstanbul, 192 pp.

Tatar, Y.,1975. Tectonic structures along the North Anatolian Fault Zone,north-east of Refahiye(Erzincan).In N.Pavoni and R.Green (Editors) Recent Crustal Movements Tectonophysics, 29,1-4, 401-410.

Tokay,M.,1973. Kuzey Anadolu Fay Zonunun Gerede ile Ilgaz arasındaki kısmında Jeolojik Gözlemler (Geologic observations along the North Anatolian Fault Zone,between Gerede and Ilgaz). Symposium on the North Anatolian Fault and Earthquake Belt,March 29-31,1972,Ankara , 12-29 (text in turkish,with an english summary).

Some Aspects of the Seismicity in the North Anatolian Fault Zone

T. Ahjos, H. Korhonen and J. Saari

Institute of Seismology, University of Helsinki Et. Hesperiankatu 4, SF-00100 Helsinki 10, Finland

Summary

A data file is compiled for studies of seismicity, seismic risk and earthquake prediction in the North Anatolian fault zone region. It contains 522 earthquakes occurred in and near the Anatolian fault zone in Turkey during the last 60 years. This file is used for statistical estimation of magnitude-frequency relationship, the upper magnitude limit and return periods for various magnitudes. The cumulative strain release pattern based on the file is also presented in time and space. In addition, the intensity distribution of earthquakes felt in the Anatolian fault zone and adjacent areas is considered. Maximum intensity map is computed on the basis of observed and estimated intensities.

Data file

For this study a computer-readable data file was compiled using NOAA's event tape and the earthquake catalogue for Turkey 1918-1970 (Alsan et al. 1975). The data file consists of 522 events occurred inside the area 38° - 42°N and 30° - 43°E in Turkey after the first World War. For each event the basic parameters - date, time, epicentral coordinates and depth - are given. Most of the events were shallow, the focal depth being less than 50 km and never over 160 km. The magnitude values are reported only for 460 events. The statistics worked out from the data file in Table 1 indicates that the frequency of observed earthquakes increases remarkably in the late 50's. This might be due to the improved instrumentation since the IGY 1956-57. One can notice that earthquakes with magnitudes smaller than four exist in the file only during the last twenty years.

Table 1. Distribution of magnitudes in the file

M	$T_{(1)}$	$T_{(2)}$	$T_{(3)}$
4.5	0.3	1.1	1.7
5.5	1.3	1.2	1.8
6.5	5.4	8.1	7.0
7.5	22.7	32.2	39.0

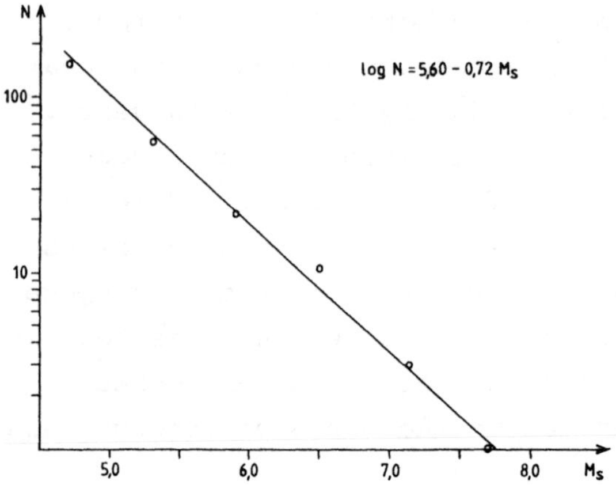

Fig. 1. Magnitude-frequency relationship.

Magnitude-frequency relationship

The magnitude-frequency relationship expressed by the well-known equation

$$\log N = a - bM \tag{1}$$

was used for earthquake frequency determination. Ignoring the earthquakes with magnitude smaller than 4.5 we computed the curve in Fig. 1. Least square fit yielded for constant b the value 0.72 ± 0.03. In agreement with this Alsan and others (1975) report for earthquakes in Turkey b-values ranging 0.68 to 0.78. Return periods obtained by this relationship are shown in Table 2, column 1.

Table 2. Estimates of return periods for various magnitudes

Period	$M<3$	$3\leq M<4$	$4\leq M<5$	$5\leq M<6$	$6\leq M<7$	$M\geq 7$	combined
1919-1928	-	-	26	21	1	-	48
1929-1938	-	-	37	12	2	-	51
1939-1948	-	-	20	11	4	4	39
1949-1958	-	-	32	16	4	2	54
1959-1968	6	63	60	8	3	-	140
1969-1978	16	86	23	1	2	-	128
Total	22	149	198	69	16	6	460

Upper magnitude limit and return period

The upper magnitude limit is estimated by using the analysis of extremes. The extremes of the series are obtained by dividing the time period covered into equal sections and extracting the maximum value for each interval. It has been shown (Gumbel 1958) that for a given time interval the probability P of the largest value having a value M or less, is given by one of three distributions. Two of these are used here. The propability distribution of P_i for unlimited upper values of M is

$$P = \exp[-X \exp(-YM)] \quad (2)$$

where X and Y are constants. When the upper limit W is imposed

$$P = \exp-[(W - M)/(W - U)]^K \quad (3)$$

where U and K also are constants.

If the extremes for n equal time intervals of t years are ranked in the order of size then the i:th largest value of M has a propability P_i of being an extreme in any t years where P_i is given by

$$P_i = i/(n + 1) \quad (4)$$

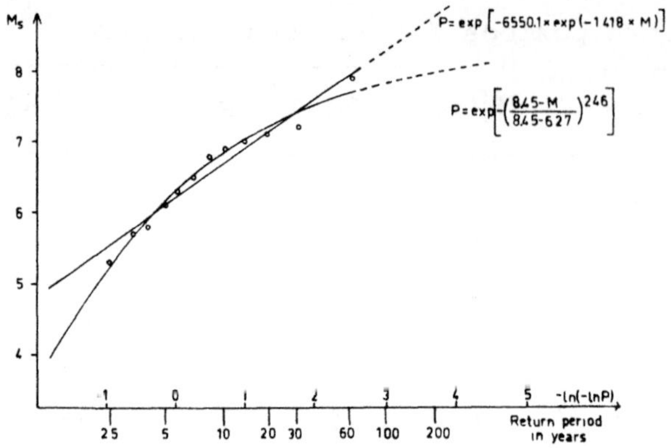

Fig. 2. Return periods for various magnitudes.

and the return period T for this event of magnitude M or larger is

$$T_i = 1/(1 - P_i). \qquad (5)$$

We applied the foregoing method to the data file of earthquakes in the North Anatolian fault zone by grouping events into intervals of five years. The results are shown in Fig. 2. The upper magnitude limit according to the equation (3) is 8.4. The method, however, is sensitive to the amount of data and grouping of it. The time interval concerned is too short for more accurate estimations. By testing with different groupings we got values ranging from 7.7 to 8.7.

Table 2 gives estimates of return periods for various magnitudes computed from equations (1), (2) and (3). The values in Table 2 are rather similar depending on the short time period. From the physical point of view we prefer the solution based on Gumbel's third equation.

In Table 2 column $T_{(1)}$ the return period is for magnitude $>M$ and in columns $T_{(2)}$ and $T_{(3)}$ for magnitude $>M$ as the yearly maximum.

<u>Cumulative strain release</u>

In order to consider the cumulative strain release in time the curve shown in Fig. 3 was computed by summing up the square

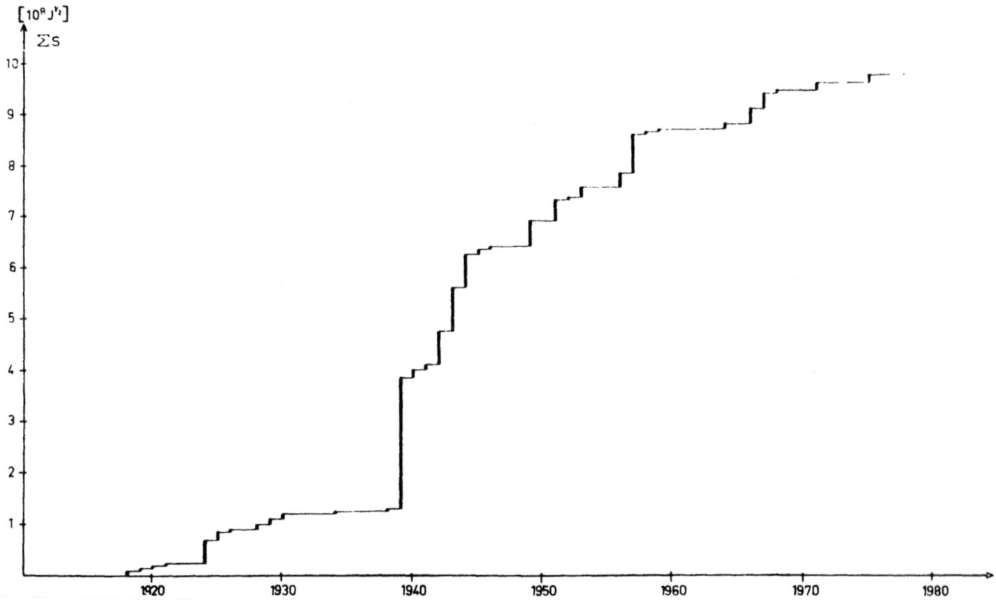

Fig. 3. Cumulative strain release.

roots of seismic energy E for all events with $M > 5.0$. This covers about 87% of the whole strain release in the area. In the evaluation of the energy E the following formulae were used (Gutenberg and Richter 1956)

$$\log E = 11.8 + 1.5\, M_S \qquad (h < 50 \text{ km}) \qquad (6)$$

$$\log E = 5.8 + 2.4\, m_b \qquad (h > 50 \text{ km}) \qquad (7)$$

For magnitude conversions IASPEI's and Romney's formulae (Kárník 1973) were used. The dominant feature in Fig. 3 is the increase of strain release in the period 1939-1943. In order to illustrate the regional distribution of strain release the whole area investigated was divided into squares $0.5^\circ \times 0.5^\circ$. The total strain inside each of the squares is illustrated by six categories in Fig. 4. The dashed lines indicate the location of main faults.

As expected the strain maxima can be found in the vicinity of the main faults. But it is interesting to note that regions with low strain release do exist along the fault as well.

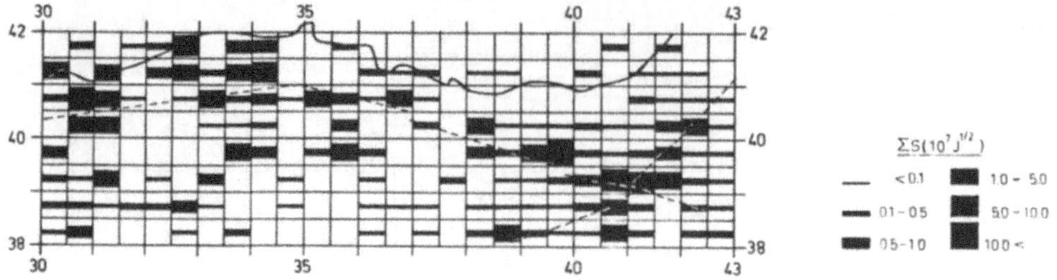

Fig. 4. Distribution of cumulative strain release.

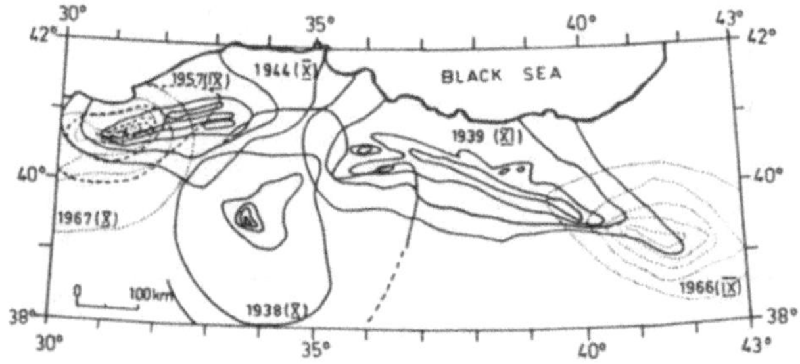

Fig. 5. Isoseismals selected from the catalogues of earthquakes for Turkey and surrounding area 11 A.D. - 1970 (Ergin et al. 1967, 1971).

Distribution of intensity

The isoseismals of earthquakes with maximum intensity I_O over IX felt in the North Anatolian fault zone are presented in Fig. 5. This map is based on the catalogues of earthquakes for Turkey and surrounding areas (Ergin et al., 1967, 1971). It is interesting to observe how well the shape and orientation of the isoseismals follow the North Anatolian fault. In addition, this figure gives an idea, how large earthquakes migrate in North Anatolian fault.

For estimating the isoseismals of the earthquakes with no macroseismic observations available we computed an intensity attenuation model in terms of earthquakes with known isoseismals. The intensity pattern obtained was applied to the events with the magnitude greater than 4.0 in our data file. By completing the observed maximum intensities with the estimated ones we got the maximum intensity map presented in

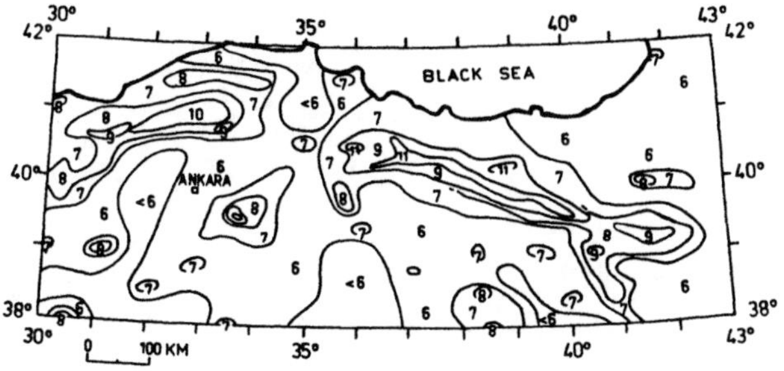

Fig. 6. Maximum intensity map based on observed and estimated intensities.

Fig. 6. An interesting low intensity area ($I_0 \leq$ VII) running NE from Ankara divides the fault zone into two high intensity sections. The highest maximum intensity is XI on the eastern section and X on the western section.

References

Alsan, E., Tezuçan, L. and Båth, M., 1975, An earthquake catalogue for Turkey for the interval 1913-1970. Seismological Institute, Uppsala, Report 7-75, 166 p.

Ergin, K., Güclü, U. and Uz, Z., 1967, A catalog of earthquakes for Turkey and surrounding area (11 A.D.-1964 A.D.). Technical Univ. Istanbul, Inst. Phys. Earth, Publ.no. 24, 169 p. + maps.

Ergin, K., Güclü, U. and Aksay, G., 1971, A catalogue of earthquakes of Turkey and surrounding area (1965-1970). Technical Univ. Istanbul, Inst.Phys. Earth, Publ. no. 28, 91 p.

Gumbel, E.J., 1958, Statistics of extremes. London & New York, Columbia univ. press, 375 p.

Gutenberg, B. and Richter, C.F., 1956, Magnitude and energy of earthquakes. Ann. Geofis., 9, 1-15.

Kárník, V., 1973, Magnitude differences. Pure appl. Geophys., 103, 362-369.

NOAA, 1972, Seismic data (magnetic tape) in the world. World Data Center A. U.S. Dept of Commerce, National Oceanic and Atmospheric Administration, Environmental Data Service.

A Comparative Analysis of the Seismotectonic Characteristics of the North Anatolian Fracture Zone and the Fracture System of Inner Dinarides

Boris Sikošek

Seismological Institute, Belgrad, Yugoslavia

ABSTRACT

The features of the territories of Turkey and Yugoslavia are similar in geotectonic characteristics. The Inner Dinaric fault belt and the North Anatolian fault system have an identical geotectonic position, but completely different seismic characteristics. Contrary to the Anatolian fault system where the seismic energy is continuously generated along a linear system, the Inner Dinaric belt represents a striking of separate seismogenic blocks.

INTRODUCTION

The territories of Turkey and Yugoslavia are situated on the seismically very active southern margin of the Eurasian continent. The origin of earthquakes in these areas is tectonic.

In this attempt the main task is to analyse two seismically active areas and to define their geotectonic position relative to the basic energy sources; this energy is transmitted to the particular seismotectonic motions, i.e. collision between Eurasia and Arabo-Africa. Such definition permits the identification of their contemporary seismic activities and eventually can be used for the earthquake prediction in these areas(Fig. 1).

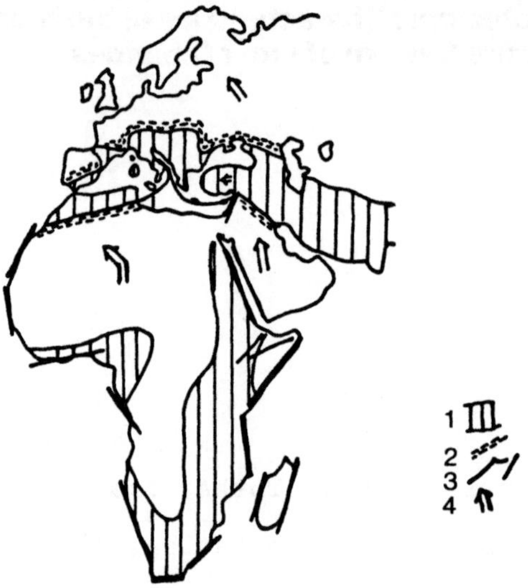

Fig. 1. RECENT GEOTECTONIC POSITION OF EURO-ASIA AND ARABO-AFRICA
1. Seismic active areas, 2. Variscian boundaries of Euro-Asia and Arabo-Africa, 3. Seismoactive boundary faults, 4. Direction of drift

Recent geotectonic interaction of Eurasia and Arabo-Africa

Recent geotectonic interaction, i.e. collision between Eurasia and Arabo-Africa is expressed by the seismoactive areas of the southern margin of Arabo-Africa (Alpine zone). Seismicity of these areas particularly indicates the localities of the active contacts between the sialic continental blocks and the oscillatory tidal movements of the simatic basement. Depths of the seismoactive levels are indicated by the earthquake foci (0-150 km). The majority of foci occurs at the depths of 0-60-80 km, i.e. within the compact crust (Algermissen et al. 1975).

The areas of Italy, Yugoslavia, Albania and Greece are situated on the contemporary active contact zone between the Alpine margin of the European plate and the Variscan margin of the African plate, with the subduction of the Variscan-African margin under the European Alpides. This presents the main source of seismic energy in these areas.

In Turkey, the Anatolian plate faces the African and Arabian plates. The Arabian plate - after the opening of the Red Sea - during the Miocene, progressed more than 100 km northwards from Africa along the leftside transcurrent fault of Levant, causing a westward lateral sliding motion of Anatolia between the right North Anatolian and left East Anatolian faults, where the pressures are smaller (Bergougnan, Fourquin, Ricou, 1978).

Geotectonic position of studied areas

The position of Inner Dinaric fault system and the North Anatolian fault is presented in Fig. 2 where the arrows show the motion of the blocks: Adriatic mass as a protruding part of the Variscan northern margin of Afro-Arabian plate and Anatolian block (Brunn, Burollet, 1979, Bingol, 1975, Izdar, 1975, Gorškov, Karnik, Sikošek, 1974).

The Inner Dinaric fault system

Yugoslavia is situated in the areas of the Alpine orogene zone where all of its categories are present: the northern branch in Pohorje block in the west and a part of the Carpatho-Balkan in Eastern Serbia. Intermountain elements are present in Alpine-Dinaric boundary zone in the west, in Slovenia (Karnian Alps - Karavanke) and in the Pannonian and Serbo-Macedonian mass in the east. To the southern branch of the Alpine orogene belongs the Dinarides which form the south-western parts of Yugoslavia adjacent to the Adriatic Sea (Petković, 1969, Kober, 1952, Medwenitsch, Beck-Mannagetta, 1979).

The symmetry of the older pre-Miocene tectonic structures was caused by the configuration of old consolidated margins of Europe and Africa (Adriatic mass).

The neotectonic and contemporary motions as well as the structures are reflected in the differential movements "en block", representing the response of the entire consolidated structure of lithosphere in this area to the tensions existing in the contemporary SIAL-SIMA system.

Fig. 3. presents the distribution and directions of the neotectonic faults over the Yugoslav territory with their clear concentration at boundary of the Adriatic mass and the Dinarides, where they present reactivated longitudinal faults of the Alpine structures intersected by transversal faults.

In the Inner Dinarides another system occurs composed of a comparatively narrow belt dominated by crossed fault. The deep structure of Yugoslave Dinarides is presented in Fig. 4 (Dragašević, Andrić, 1974).

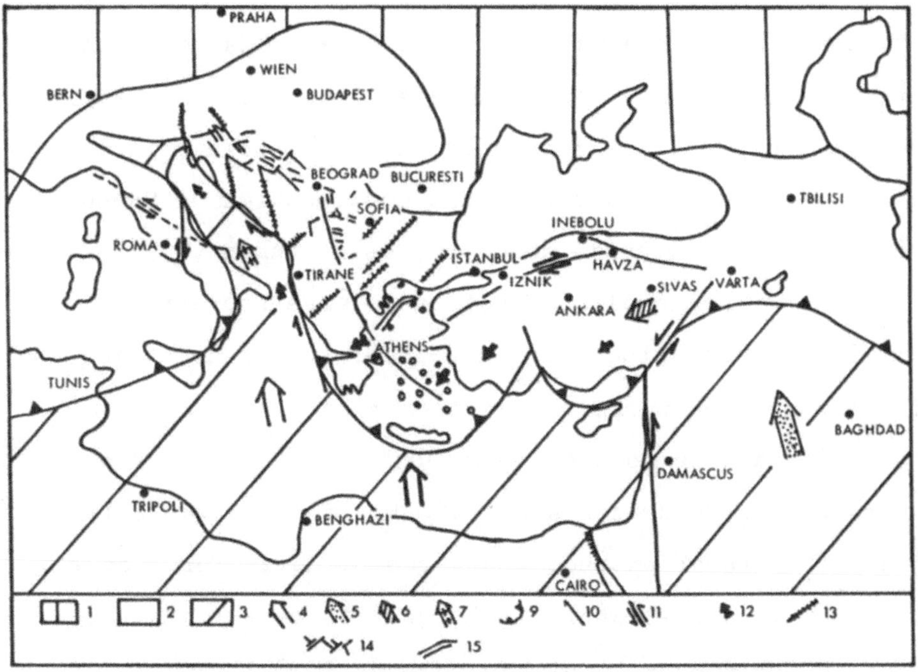

Fig. 2. GEOTECTONIC POSITIONS

1. Euroasian plate, 2. Alpine belt, 3. African-Arabian plate, 4. Main motions of African plate, 5. Main motions of Arabian plate, 6. Main motions of Anatolian block, 7. Main motions of Adriatic block, 8. Zone of subduction, 9. Longitudinal deep faults, 10. Strike-slip deep faults, 11. Motions of the secondary blocks, 12. Transversal faults, 13. System of Inner Dinaric neotectonic faults, 14. North Aegean tectonic graben

North Anatolian fault system

In Turkey, similarly as in Yugoslavia, all parts of Alpine orogene are present. The northern branch consists of Pontides, situated on the southern margin of the Black Sea and the southern continuation of the Carpatho-Balkanian range, while the Taurides are the eastern continuation of Dinarides (i.e. Hellenides). The Anatolides are the continuation of the Pelagonides. The North-Anatolian massif is the continuation of Serbo-Macedonian mass (Gutnic, Monod, Poisson, Dumont, 1979).

The North Anatolian fault is a large neotectonic right lateral longitudinal fault intersecting sometimes the older Alpine structures. This system consists of two parts which are in contact near Havza (Fig. 2). The western part extends for about 450 km from Havza to Iznik. The eastern branch extends eastwards of Havza for about 500 km up to Varto where it loses its seismic characteristics (Bergognan, Fourquin, Ricou, 1978).

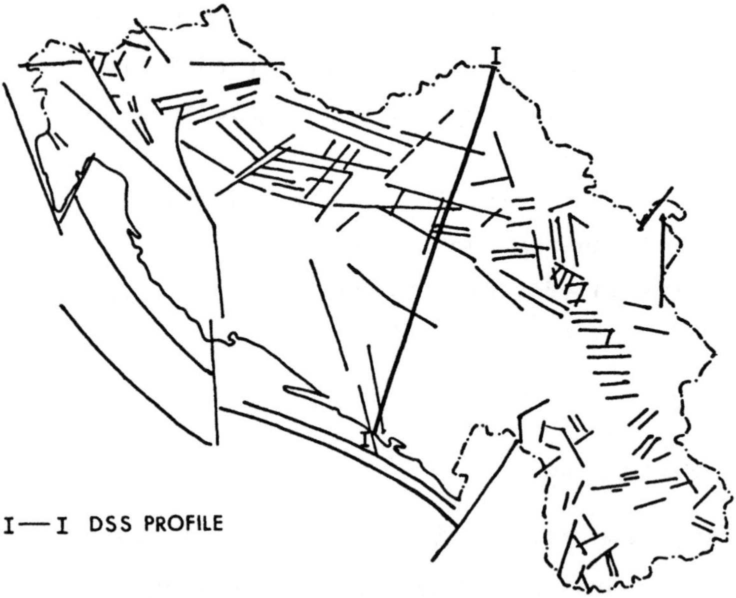

Fig. 3. THE MAIN NEOTECTONIC FAULT SYSTEM IN YUGOSLAVIA

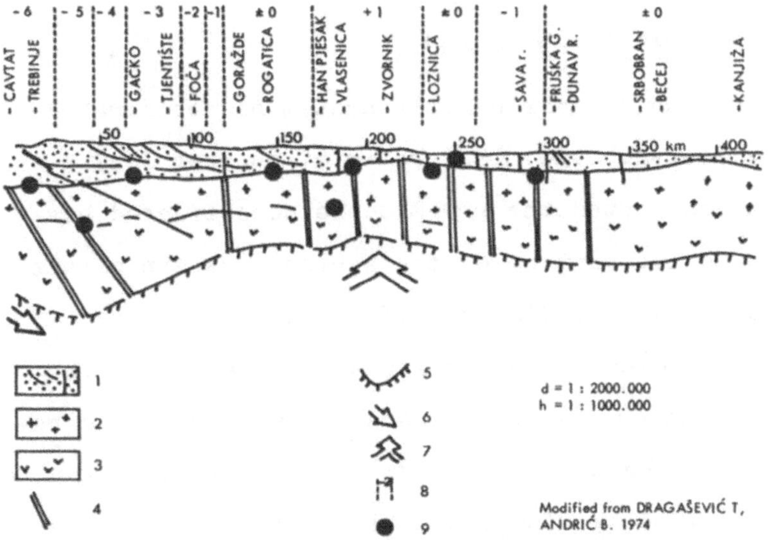

Fig. 4. THE DEEP SEISMIC SOUNDING CAVTAT-KANJIŽA PROFILE
1. Epidermal part with Alpine and neotectonic structures, 2."Granitoide belt", 3 "Basaltic belt", 4. Deep faults, 5. Earthquake foci, 6. Mohorovičić's discontinuity, 7. Direction of pressure of Adriatic mass, 8. Tidal effect of simatic zone, 9. Velocity of vertical motions (mm/y)

The system of the North Anatolian faults obtained its contemporary tectonic and the seismical function after the Miocene. The opening of the Red Sea and the faster northwards motion of the Arabian plate caused the westward movement of Anatolia. The continuation of the eastern segment from Havza to Inebolu which cuts the Pontides orogene belt toward the Black Sea, now is inactive seismically (Fig. 2).

Seismotectonic characteristics of studied areas

The Inner Dinaric fault system

The Inner Dinaric fault system has originated during the post-Miocene. The system of shorter faults which have destructed the pre-Miocene consolidated base follow the active contact of the southern margin of the European plate at the distance of about 250-300 km, from the NW (Slovenia) to the Aegean Sea, at a length of about 100 km. These faults are situated within a relatively narrow belt (150-200 km), and individual faults are usually 10-30 km long. Seismic activity in this belt is expressed in the segments, i.e. with some blocks being seismically active and other blocks inactive.

Seismotectonic characteristics of North Anatolian faults

In distinction from the Inner Dinaric fault system, which represents a string of blocks, the North Anatolian fault system is linear, i.e. over this system seismic energy is generated at the particular places of a unique, continual fault. The line: Aegean Sea- Iznik - - Havza - Varto presents the northern boundary of the block of the Anatolides, along which occurs the contemporary motion of the Anatolian block toward the west, relative to the Pontides (Fig. 5).

Conclusion

1. The geotectonic position of two compared areas are almost identical. The Inner Dinaric fault belt is located in the areas where the geotectonic elements of

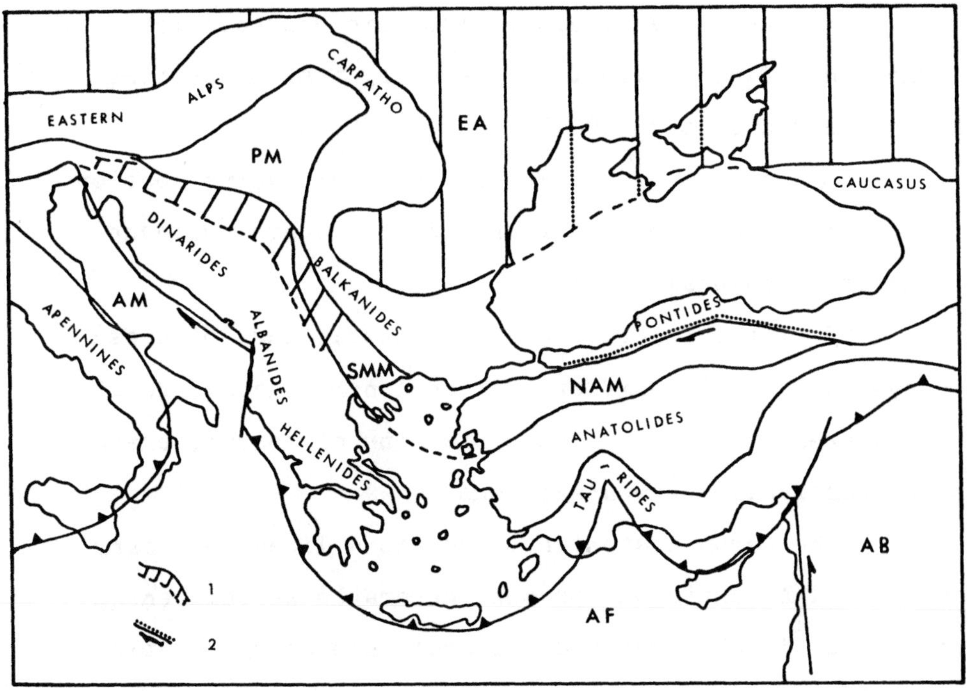

Fig. 5. SEISMOTECTONIC POSITIONS OF NORTH-ANATOLIAN FAULT SYSTEM, THE FAULT SYSTEM OF INNER DINARIDES AND THE NE BORDER OF ADRIATIC MASS

1. Inner Dinaric fault system, 2. North-Anatolian fault system AM-Adriatic mass, PM-Pannonian mass, SMM -Serbo-Macedonian mass, NAM - North-Anatolian mass, EA - Euro-Asiatic plate, AF - African plate, AB - Arabian plate

Alpine orogene have a common boundary: Dinarides - Serbo--Macedonian mass and Carpatho-Balkanian mountain range. The system of North Anatolian faults is also located in the boundary area between the Pontides and Anatolides. Both systems belong to the structural system of the central areas of the Alpine orogene zone. Both are about 1000 km long, and were formed by neotectonic post-Miocene movement.

2. The contemporary geotectonic characteristics of the Inner Dinaric fault zone are: systems of shorter faults grouped into a 100-150 km wide belt.

The system of the North Anatolian faults consists of several segments of the right linear faults without secondary orthogonal swarms.

3. The seismotectonic characteristics of these two systems are also different: the Inner Dinaric faults are featured by selective accumulations of seismic energy in the particular segments-blocks.

The North Anatolian system is a linear boundary fault zone. A continuous process generates seismic energy within this linear area and releases this energy at places of critical tension.

4. According to the seismotectonic characteristics, the system of the North Anatolian faults can be compared with the linear boundary fault system between the Dinarides and the Adriatic mass of similar lengths. Here the seismic activity is also related to the continuous progressive motion of the Adriatic mass via this fault system towards the NW (Fig. 2, fig. 5).

The study and identification of common characteristics in these two areas can be used eventually in future studies for the short-range earthquake prediction.

REFERENCES

ALGERMISSEN, S.T., D. PERKINS, G. ISHERWOOD, G. REAGOR, C. HOWARD (1975): Seismic risk evaluation. UNESCO - Surv. of the seism. of Balkan reg.

ALIAJ SH., S.KOCIAJ, E. SULTRATOVA (1974): Seismotectonic analysis and map of Albania. Proc. of the Seminar on the seismot. map (cum lit.) UNESCO/ UNDP, Skopje

BELOUSSOV,V.V., V.N. SCHOLPO (1976): Geodynamics of the eastern part of the mediterranean Alpine Belt. Tectonophysics 35, Amsterdam.

BELOUSSOV, V.V. (1978): Endogennye rezhimi materikov. "Nedra", Moskva

BEMMELEN, R.W. van (1964): The importance of the geonomic dimensions of lenght, time, pressure and temperature. Earth Sci. Rev. 3.

BERGOUGNAN, H.,C.FOURQUIN and L.E. RICOU (1978): Les deux tronçons et le double jeu de la faille nord-anatolienne dans le tectonique recente du Moyen-Orient.
C.R.Acad., Sc. Paris, t.287,Sér.D.

BINGOL,E.(1975): Evolution géotectonique de l'Anatolie de l'Ouest.
Veme Colloque sur la géologie des Régions Egeennes, Orsay

BRUNN,J., P.F. BUROLLET (1979): Island arcs and the origin of folded ranges.
Geologie en Minopuw, Vol. 58 (2)

DRAGAŠEVIĆ, T., and B. ANDRIĆ (1974): Results of the earth crust investigations obtained to the present day by deep seismic sounding.
Acta seismologica Yugoslavica, No 2-3, Belgrade

DUBOURDIEN, G.(1980): La théorie des pulsations internes de la terre.
Lab. de geologie du Collège de France Meudon

GORSKOV, G.P., V. KARNIK and B. SIKOŠEK (edit.) (1974): Proceedings of the seminar on the seismotectonic map of the Balkan region. UNESCO/UNDP, Dubrovnik-Skopje

GUTNIC, M., O. MONOD, A. POUSSON, D.F. DUMONT (1979): Géologie des Taurides occidentales (Turquie). Memoires de la Soc. geol. de France (Nouvelle série, No 137)

IZDAR,E.(1975): Géotectonic developments of Western Anatolia and the comparison with geological units of western part of Aegean Region Veme Coll. sur la géol. des Regions Egeennes, Orsay

ISIKARA, A.M., N. ORBAY, M. PANTOVIĆ (1979): An attempt for elimination of non local changes of the geomagnetic field for earthquake prediction. Proc. ESA-Counc. SO-149, Strassbourg

KOBER, L.(1952): Leitlinien der Tektonik Jugoslawiens. Po.izd. SAN T. 189, Geol. Inst. Beograd

KRAUS, E. (1968): Ueber die geotektonische Detungsmöglichkeiten der Undulationtheorie Geolog. Jahrb. 86, Hannover

MAKSIMOVIC, B., and B. SIKOSEK (1971): Importance of geotectonic regionalisation for the oil explorations in Adriatic Sea. "Nafta", No 4-5, Zagreb.

MONOD, O.(1977): Towards a tectonic connection between Hellenides and Taurides. Veme Colloque sur la géologie des Régions Egeennes, Orsay

OLUJIĆ, M., N. BRDAREVIĆ (1979): Contribution to the knowledge of tectonical structure of the Adriatic Sea bottom.
Zbornik radova rud. fak. Zagreb

JOVANOVIĆ,P. (1968): Shematic map of vertical movements of earth crust in Yugoslavia
IV Congr. geodes. and geomet., Sarajevo

PETERSCHMITT, E.(1979): Notes sur l'activité seismique de la région SE adriatique - Herzegovina, mars-avril, Centre séismologique Europeo-Mediterranean C.S.

PETKOVIĆ, K. (1961): Tectonical map of Yugoslavia.
Glas SAN knj. 22, odelj. prir. mat. nauka Beograd

SEMENENKO, N.P. (1975): Kontinentalnaja kora.
"Naukovaja dumka", Kiev

SIKOŠEK, B. and W. MEDWENITSCH (1965): Neue Daten über die Fazien und Tektonik der Dinariden.
Geol. Glasnik 13, Sarajevo.

MEDWENITSCH, W. and P. BECK-MANNAGETTA (1979): Geologische Karte der Donauländer mit Tektonik 1:2 000 000.
Ost und Südosteurop. Inst. Wien

SIKOŠEK, B.(1974): Geotectonics of Yugoslavia in the light of Neotectonics and seismic movements.
Faculty of mining and geology, Beograd

SIPAHIOGLU, S.(1974): Explanatory notes on some geological maps. Proc. sem. seismotec. map UNESCO/ UNDP, Skopje

SKOKO, D., M. ARSOVSKI, D. HADŽIJEVSKI (1976): Odredjivanje zona mogućih žarišta na području Jugoslavije.
Acta seismol. Iugoslavica, No 2-3, Beograd

SUBBOTIN, S., V. SOLLOGUB, D. PROSEN, T. DRAGAŠEVIĆ, E. MITUCH, K. POSGAY (1968): Junction of deep structures of the Carpatho-Balkan region with those of the Black and Adriatic Seas.
Canadian J. of Earth Sciences

VUKAŠINOVIĆ, M. (1975): Some problems related to seismotectonic features of the Southern Adriatic.
Acta seism. Iugoslavica, No 2-3, Beograd

Fatigue Crack Propagation Along the Anatolian Faults and Levant Coast and Earthquake Prediction

Uğur Kuran

Earthquake Research Institute, Ministry of Reconstruction and Resettlement, Ankara, Turkey

SUMMARY

Prediction of a destructive earthquake is an exciting problem and one of the goals of geophysicists in Turkey. For the last five years the problem of the earthquake mechanism (or fatigue mechanism) and earthquake prediction (or fatigue life prediction of crustal material) along the San Andreas, the North Anatolian and the East Anatolian faults has been under investigation (Kuran 1975, 1979). A procedure has been developed for the prediction of the length, magnitude and fatigue-crack-growth rate along the above mentioned historic fault breaks. This technique was also applied to the Levant Coast in order to achieve a better magnitude and time prediction of future destructive earthquakes. Three elliptically shaped macroseismically active areas, which are believed to be the terminations of four active faults, have been determined in the Levant. The terminations of four active faults (from north to the south) are: Kırıkhan (in Turkey) Baalbek, Ajlun, ElMazra and the junction of the Gulf of Suez and Gulf of Akaba. These active fault segments may produce at least four destructive shocks in the Levant with magnitude values ranging between M = 7.4 to 8.

Other three elliptically shaped macroseismically active areas which are believed to be terminations of at least five active faults, have been determined along the North Anatolian and East Anatolian fault zones. Since the Keban Dam is founded on two potentially active fault segments it will be extremely important to make time predictions for future earthquakes (M = 7.4 - 8.1) in order to control the water level in the reservoir behind the dam.

If the Kaiser effect can be applied to the loading arising from the lunar cycle of the 19 years between 1969 - 1987, there could be a strong possibility of a destructive earthquake between Palu and Karliova in 1982/3 because during this time the tidal load will exceed the maximum level reached in the previous lunar cycle (Kuran 1979).

I FATIGUE CRACK PROPAGATION ALONG THE NORTH ANATOLIAN FAULTS AND SOME SIGNIFICANT OBSERVATIONS

The following are important field observations and the main seismotectonic characteristics of the North Anatolian fault zone:

1) Almost all macro-seismic activity was concentrated within elliptically shaped areas along the East and North Anatolian fault zones. In other words macro-earthquakes show activation within elliptically shaped areas on opposite side of the earthquake-prone areas.

2) Each of these ellipses along the North Anatolian fault not only corresponds to the termination of two main fault breaks but at the same time they cover another fault break which is associated with relatively low-magnitude earthquakes. Each of the ellipses presented in Fig. 1 is responsible for at least two destructive earthquakes. As an example we shall look at the fault break which is associated with the 1942 earthquake in ellipse No. 2, and two fault breaks associated with the 1957 and 1967 earthquakes in ellipse No. 4.

3) Some macro-earthquake activity occurs at both ends of the fault zone before the main destructive shock. An example is the fault break associated with the great Erzincan earthquake of December 26, 1939 in ellipse No. 1. The eastern end of this fault break showed macro-seismic activity in 1906 and 1907. Similarly the western end of this fault segment showed activity in 1901 and 1908.

4) Both long and short axes of the macro-seismically active elliptically shaped areas vary in length and direction. However, the direction of the major axis strongly depends on the orientation of the zone of maximum shear strain.

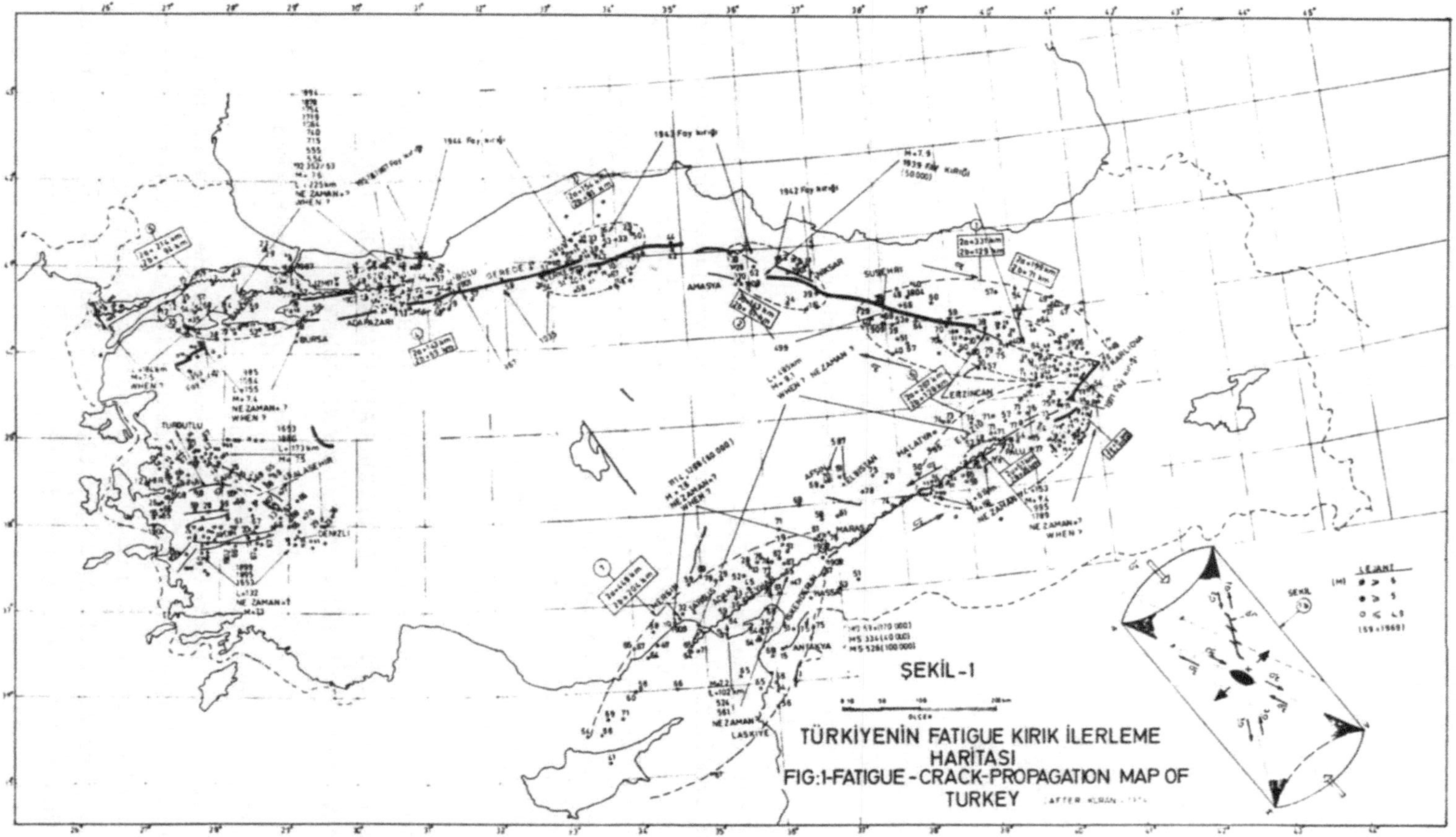

1. Fatigue-crack-propagation map of Turkey

5) The instrumentally determined epicentres of the main shocks which took place in 1939, 1942 and 1943 coincide with the centres of the ellipses numbered 1, 2 and 3 respectively in Fig. 1.

6) Discontinuous type of crack growth is one of the most important features of fatigue processes in crustal material. This phenomenon can be seen under both laboratory test conditions and under actual crustal conditions (see Fig. 1 and Kuran, 1979). As a result of this process there is a gap between the fault breaks which were associated with the 1944 and 1953 earthquakes.

7) If an earthquake occurs far from the North Anatolian fault zone the direction of the new fault break will be parallel to the North Anatolian fault zone. Examples are the fault break associated with the Caldiran earthquake of 1976 and also the 1957 and 1967 fault breaks within the ellipse No. 4 in Fig. 1.

8) It can be seen from Fig. 1 that the length of the fault break produced by an earthquake roughly equals the distance between the centre points of adjacent ellipses.
These are the most important seismotectonic characteristics of the North Anatolian faults and will be very useful in recognizing other earthquake-prone areas along the Anatolian faults (see the ellipse numbered 5, 6 and 7 in Fig. 1) and on the Levant Coast (see Fig. 2).

II PROCEDURES FOR PREDICTING THE LENGTH, MAGNITUDE AND TIME OF FUTURE DESTRUCTIVE EARTHQUAKES ALONG THE EAST ANATOLIAN FAULTS AND THE LEVANT COAST

Earthquake-prediction studies should be concentrated on the following questions (see Fig. 3):
a) Where is the near-future earthquake-prone area along a particular fault zone? As an example let us consider that four cities (A, B, C and D) were destroyed in 69 B.C. and 170000 persons killed. Collection and analysis of historical earthquake data is stage 1 in this type of study.

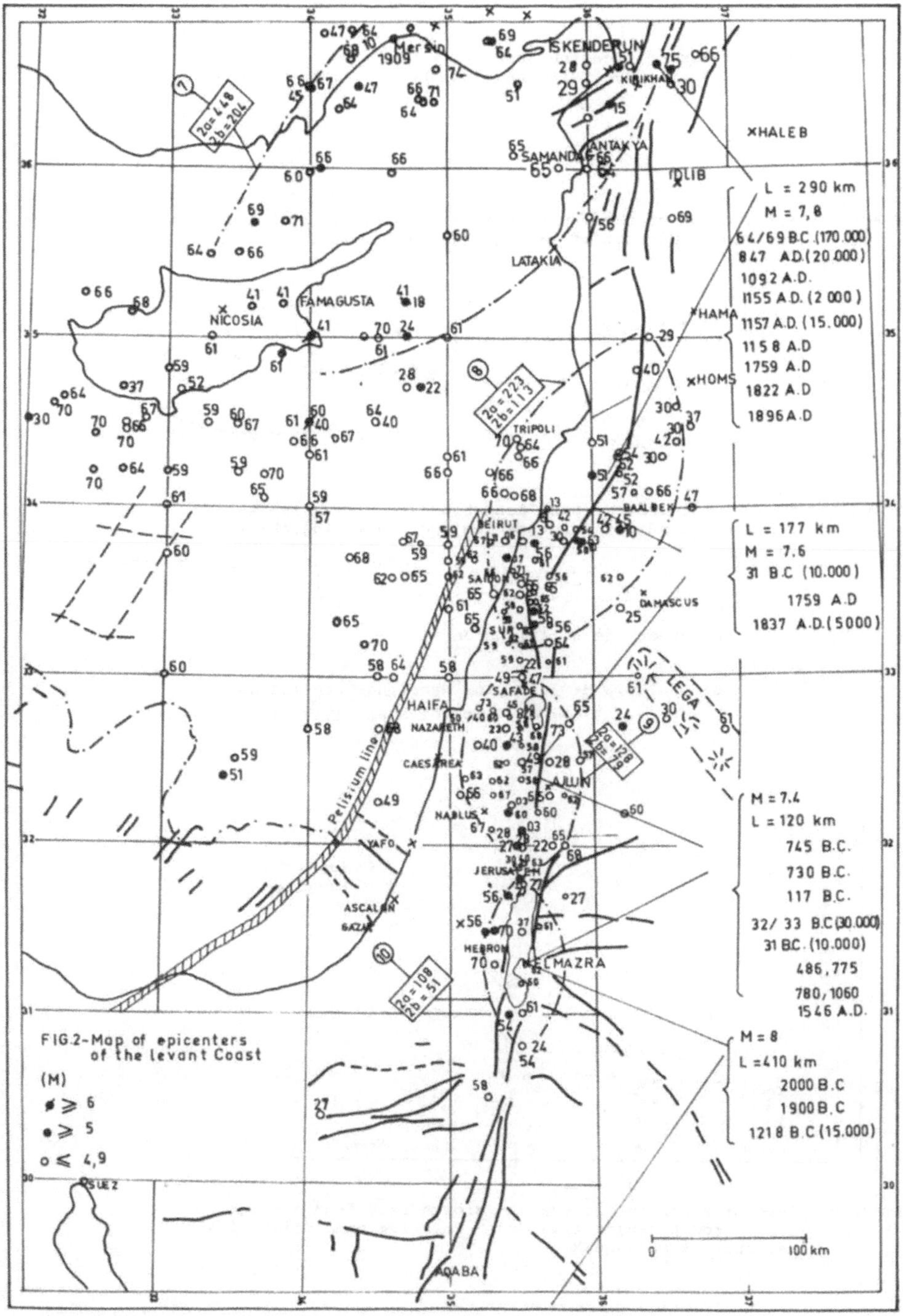

2. Map of epicentres of the Levant Coast

STAGE-I : Preparation of the historic earthquake data for place prediction.

Example : Date of historic earthquake : 69.B.C.
Casulties = 170000 persons killed and four city destroyed.

STAGE-II: Preparation of the map of epicenters between 1900 and 1980 for the prediction of earthquake prone segment.

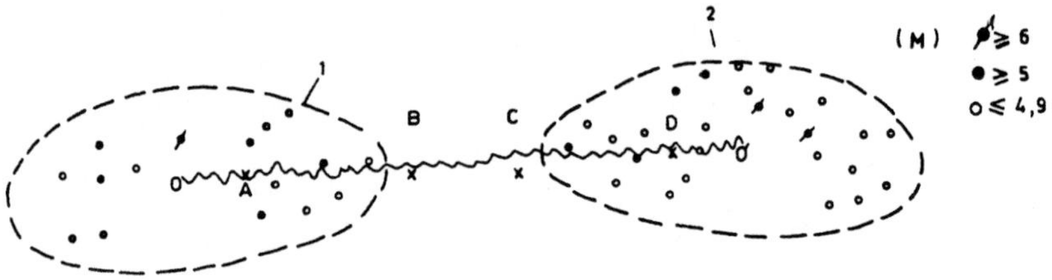

STAGE-III & IV Prediction of both length of the active fault and Magnitude of a future destructive earthquake

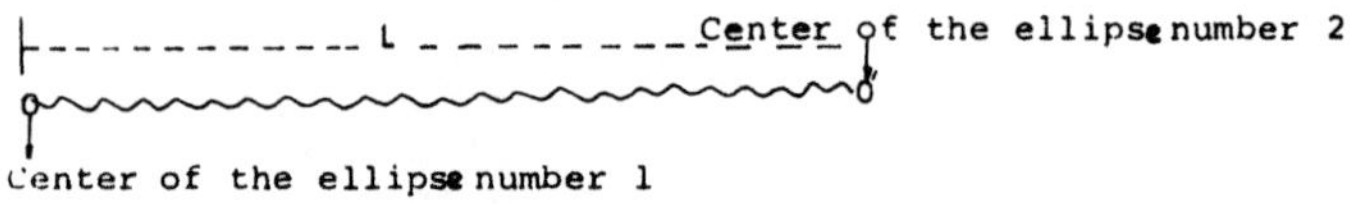

Log L = 0.78 M- 3.62 for Magnitude prediction.
(Toksöz N.et.al.1979)

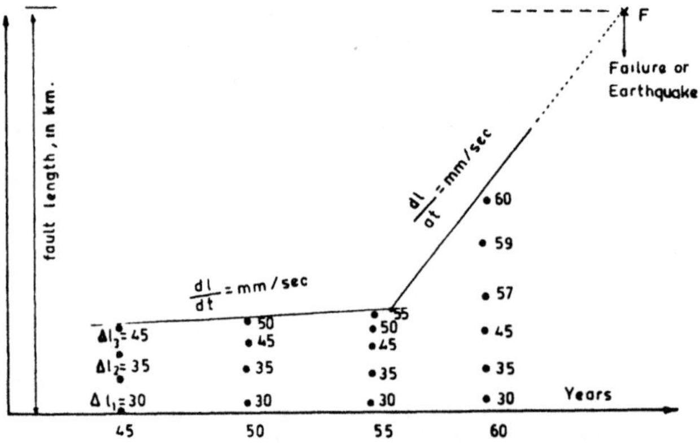

STAGE-V : Determination of the crack - growth rate from cumulative crack growth versus time graph for time prediction of a destructive earthquake.

3. Place, magnitude and time prediction of a future destructive earthquake

b) What was the length of the historic fault break and the magnitude of that earthquake?
In order to predict earthquake-prone segments all available macro- and micro-earthquke epicentres between 1900 and 1980 should be plotted on a map. Because all macro- and micro-earthquake epicentres were concentrated within elliptically shaped areas as indicated in Fig. 3, and because the main fracture associated with the destructive shock was located half-way between the centre points of the two ellipses O and O′ the fault length can be measured directly from the map. Stage 2 therefore is the analysis of macro- and micro-earthquake data.

c) The magnitude of a near-future earthquake can be determined by the following expression taken from Toksöz et al. (1979).

$$\text{Log } L = 0.78M - 3.62$$

Stages 3 and 4 are the determination of both length of the fault and the magnitude value of the impending earthquake (Fig. 3).

d) Which part of this historic fault break had already been fatigued (or damaged?), and how long does it take for a fatigue break to occur along a historic fault break? Laboratory experiments on rock specimens clearly indicate that, in order to have catastrophic fracture (or an earthquake) both crack initiation and macro-crack propagation processes along the maximum shear zone must be completed (Refs. 7 and 8). All the photographs which are taken parallel to the maximum shear zone of the centre notch in granite indicate that crack extension is a discontinuous process. The discontinuous nature of crack growth is also seen along the San Andreas and North Anatolian fault systems. In making an analysis of the crack propagation period for any seismically active part of a fault zone, it is necessary to consider the fatigue-crack growth process in detail. In order to obtain more accurate crack-growth results from micro (or macro) earthquakes it is essential to plot cumulative-fatigue-crack-growth versus number of years (Fig. 3). Stage 5 then is the prediction of the time of occurrence of a destructive earthquake.

III DISCUSSION AND CONCLUSIONS

In order to illustrate the basic concepts underlying fatigue-crack-propagation processes along the Levant Coast, it is necessary to consider macro- and micro-seismic developments in the vicinity of an active fault segment. Continuous epicentre determination of micro- (and macro) earthquakes within a fault zone may also provide a significant clue to the propagation of a fatigue crack into an inactive area. For the Levant Coast shocks of M = 3 1/2 - 6 1/4, epicentres and time of occurrence were taken from Arieh (1967) and Maamoun (1976). Historical data were taken from Calvi (1941), Maamoun (1976), Alsinawi et al. (1975), and Amiran and Kallner (1951). All of the instrumentally located epicentres of other large shocks are closely associated with major fatigue breaks and they do follow earlier breaks and mapped faults (Fig. 4). It is interesting to note that both the Dead Sea and the Sea of Galilee (Tiberias) are covered by ellipses, numbered 10 and 9 respectively. The centre point of ellipse No. 10 corresponds to the intersection between the Vadi el Araba and the Kerak-El-Fina Graben. Centre point of ellipse No. 9 corresponds to the intersection between the Jordan Valley and the Qishon Graben. The direction of the principal stresses which have been predominant during the course of geological time was determined. It was concluded that master faults could originate in the following stress regime: the principal cyclic stresses in a N 37°W-N 60°W direction, causing fatigue failure and producing zig-zag fault pattern in Israel.

An estimate of the magnitude of historic earthquakes produced the following results (see Fig. 2): The active fault which lies between Kirikhan and Baalbek may produce earthquakes of M = 7.8 (see the historic earthquakes which occurred in 64/69 B.C., 847, 1092, 1115, 1157, 1158, 1759, 1822 and 1896 A.D. in this region). The active fault which lies between Baalbek and Ajlun may produce earthquakes of M = 7.6 (see the historic earthquakes which occurred in 31 B.C., 1759 and 1837 A.D. in this region). The active fault which lies between Ajlun and El Mazra may produce earthquakes of M = 7.4 (see the historic earthquakes which occurred in 745, 730, 117, 32/33, 31 B.C.

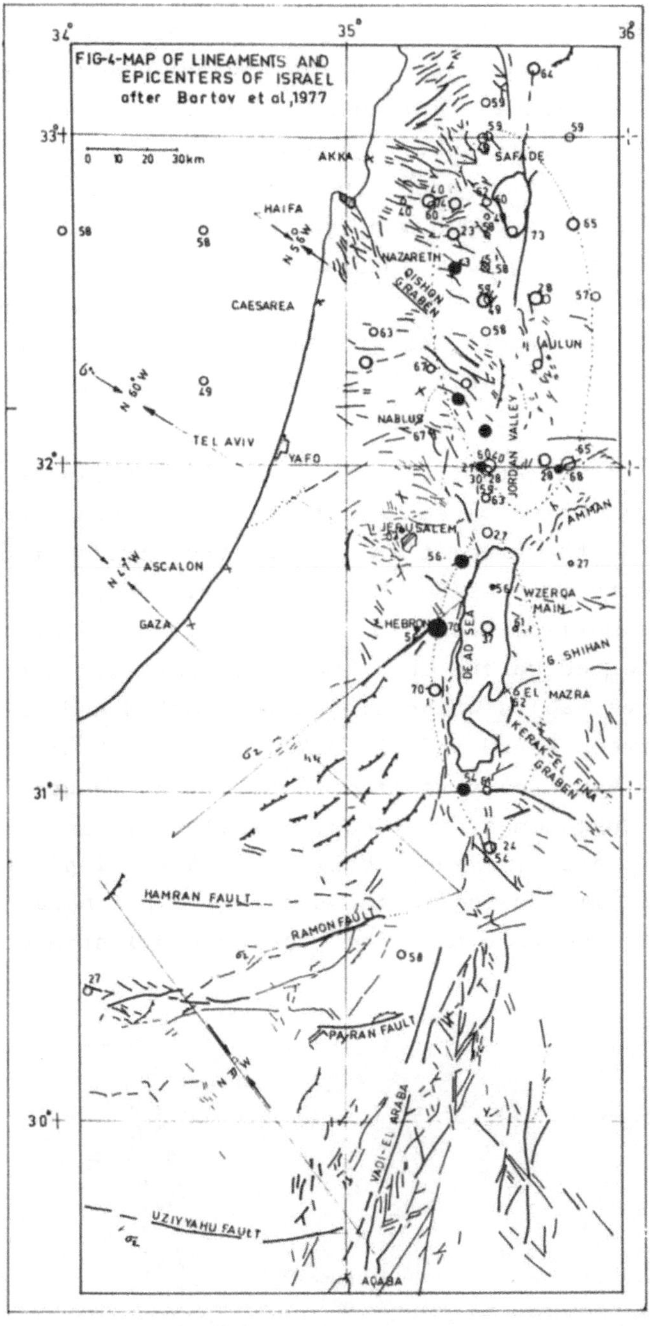

4. Map of lineaments and epicentres of Israel
(after Bartov et al. 1977)

and 486, 775, 780, 1060, 1546 A.D. in this region). Finally the active fault which lies between El Mazra and junction of the Gulf of Suez and Gulf of Akaba may produce earthquakes of M = 8 (see the historic earthquakes which occurred in 2000, 1900 and 1218 B.C. in this region).
It is believed that these active faults along the maximum shear zone have destroyed many countries in the Levant in early times (see Fig. 2). According to historical seismicity records of the last 3980 years twenty-five "major" earthquakes with Richter Magnitude greater than 6, and more than one hundred and fifty "moderate" earthquakes occurred in the Levant (Refs. 1, 2, 3, 4 and 5).

In order to illustrate the basic concepts underlying fatigue-crack-propagation processes along the Levant Coast and their application in the prediction of a destructive earthquake, it is necessary to consider macro- and micro-seismic developments in the maximum shear zone. The cumulative-fatigue-crack-growth is plotted versus the number of years (or tidal cycles) for two active faults which lie between AJLUN and EL MAZRA and also between AJLUN and BAALBEK (see Figs. 5 and 6). The results show that the rate of discontinuous fatigue-crack-growth as a function of the yearly tidal cycle is not linear and shows dramatic increases between 1958 and 1962, reaching 0.261 - 0.651 mm/sec (see Figs. 5 and 6). This behaviour is attributed to the fact that the greater tidal stress in those

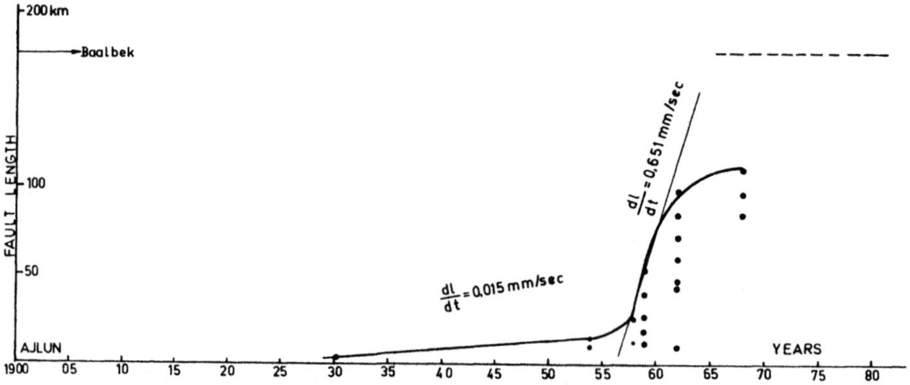

5. Growth of fatigue-crack between Ajlun and Baalbek
as a function of time (in years) or cycles

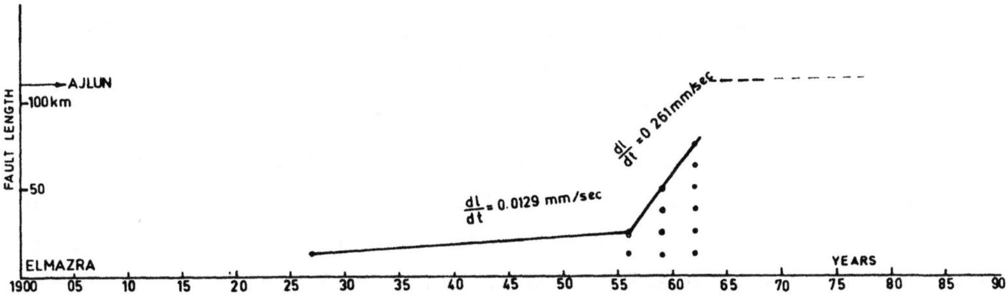

6. Growth of fatigue-crack between Ajlun and El Mezra as a function of time (in years)

years has a large influence upon the overall crack-growth rate.

The rate of fatigue-crack growth along the North Anatolian faults was calculated to range between 0.265 to 0.753 mm per second (Ref. 7). These two values correspond to the crack-growth rate just before the destructive earthquakes which occurred in 1939 (the great Erzincan earthquake), and 1944 (the Gerede-Bolu earthquake). It is also important to bear in mind that more than 63 percent of the total length of the fault which lies between Baalbek and Ajlun have been damaged. According to micro- and macro-earthquake results more than 68 percent of the total length of the fault between AJLUN and EL MAZRA have been damaged.

Because of large population growth and land use in recent years the potential for damage and casualties in the Levant is even more enormous today than compared with that existing in historic time. The numbers shown within the brackets in Fig. 2 represent the loss of life in this region. Because of lack of systematic macro- and micro-earthquake measurements along the other active faults between Hatay and Baalbek, and also between El Mazra and the junction of Gulf of Aqaba and Gulf of Suez, it is impossible to make satisfactory predictions of future destructive earthquakes. Although four earthquake-prone areas have already been determined in the Levant, much more research is needed to calculate average-crack-growth rate before we are able to predict future destructive earthquakes in this region. If there were a sufficiently long (from 1969 to 1980) micro-earthquake record to establish average slopes of the curves of Figs. 5 and 6, one might be able to predict the time of future destructive earthquakes.

It is important to emphasize that, if the Kaiser effect influences the loading of the next 19-year lunar cycle (between 1969-1987) there could be a strong possibility of destructive earthquakes between AJLUN and EL MAZRA and also between BAALBEK and AJLUN in 1982/3, as by this time the tidal stress will exceed the maximum level reached in the previous lunar cycle of 1963/4 (Refs. 7 and 10).

APPENDIX I - STRESS STRAIN TIME PHENOMENA WITHIN THE EARTH'S CRUST DUE TO CYCLIC TIDAL LOADING AND FATIGUE-FAILURE PROCESS

The effect of tidal loading on crustal strain has been investigated by many authors, but no attempt has been made to dive stress-strain data from existing strain-time data (see Ref. 7). During the interpretation of the strain-time recording a new conversion technique was used (Ref. 11). The graphical method which was used in these studies has been developed by the present author.

Our investigations indicate that earthquakes occur as a result of cyclic stress-strain variations (fatigue loading) rather than monotonically increasing loading within the earth's crust.

Fracture because of time dependent fluctuation rather than monotonically increasing uniaxial load is known as fatigue. Fatigue failure can occur at stress levels well below the compressive strength of crustal material and thereby produce en-echelon type of cracks (Figs. 1 and 1b, and Ref. 7). It is important to emphasize that the en-echelon-crack-growth rates decrease markedly with decreasing stress. This relationship indicates that crack-growth rates are proportional to the twentieth power of the applied cyclic stress (Ref. 8).

In the fatigue-crack-propagation tests, small cracks branching from the main cracks were observed. The majority of the branching observed in granite specimens are parallel to the conjugating shear zone (Fig. 1b, and Ref. 11).

This significant property of the en-echelon cracks enables us to determine the direction of the principal stress which predominated in crustal material during geological time (Fig. 4).

APPENDIX II - THE EFFECT OF THE TIDAL STRESSES ON EARTHQUAKE GENERATING PROCESSES

Most scientists would agree today that tidal stresses can only play a role as a triggering stress during the earthquake-generating process. As can be seen from Ref. 7 the amplitude of cyclic elastic tidal strain ranges from 10^{-4} to 10^{-7}. It can be assumed that the stress-strain curve varies linearly in the stress-strain plane, and the peak stress associated with the diurnal cyclic load is about 1 kg/cm^2 at a given depth. By using a linear relationship between stress and strain, the stress range would be about 100 kg/cm^2 as a result of secular tidal loading, and 1000 kg/cm^2 as a result of 19 years of lunar cyclic loading. It is important to remember now that the 1000 kg/cm^2 stress range is very near the compressive strength of a massive granite specimen under uniaxial load condition.

Stress measurements made 1977 by Kuran in the Zigana Highway tunnel showed that the stress range occurring during daily tidal cycles was between 2 and 14 kg/cm^2. In addition the daily horizontal stresses were three times greater than the vertical stresses.

Laboratory experiments (Ref. 7) indicate that fatigue failure can occur when the applied stress exceeds the endurance stress of the crustal material. It is evident from (Ref. 7) that peak tidal-strain (or stress) values are associated with a large number of earthquake counts. Laboratory experimental results also show clearly that peak compressive cyclic stresses are more damaging in the fatigue life of crustal materials, and are associated with a large number of acoustic emissions (Ref. 7). It is therefore concluded that the presence of cyclic stress in the earth's crust due to the tide-generation force is responsible for the triggering of earthquakes (or fatigue failures).

ACKNOWLEDGEMENTS

Much of this experimental investigation of earthquake mechanisms was conducted in the Rock Mechanics and Mechanical Engineering Department as part of the studies of the Geophysics Department at Imperial College, London.

I am greatly indebted to Professor R.G. Mason for supervision. I am particularly grateful to Professor E. Hoek for his personal interest, as well as the technical and laboratory supports. I am indebted to Dr. J. Rogan for the use of his Line-scale-fatigue machine, to Dr. B.P. Dash for the use of seismic instruments and to Dr. E.H. Rutter for the use of accelerometers.

Also, I am indebted to many people in Turkey, particularly Prof. Nezihi Canitez and Dr. Aydoğan Boray. Also, I wish to express my gratitude to Prof. Andreas Vogel for his valuable editorial comments and suggestions.

This work was supported by the CENTO.

REFERENCES

1. CALVI, WILHELM S., 1941: Erdbebenkatalog der Türkei und einiger benachbarter Gebiete. M.T.A.Rep. 276

2. MAAMOUN, MOHAMED M.EL SAİD, 1976: La séismicité du moyen et du proche-Orient dans le cadre de la séismotectonique mondiale. Ph. D. Thesis.

3. ALSINAWI, SAHİL., HAFİDY A.A. GHALİB, 1975: Historical seismicity of Iraq. Bulletin of the Seismological Society of America, Vol. 65, No. 5 pp. 541-547

4. AMIRAN, D.H. , Kallner , 1951: A revised earthquake catalogue of Palestine. The Israel Exploration Journal. Vol. I.No. 4

5. ARIEH, E., 1967: Seismicity of Israel and adjacent areas. Geol. Survey of Israel Bull. 43

6. BARTOV, Y. and Y. ARKINS, 1977: Geological photomap and lineament map of Israel, 1/500 000. Geol. Survey of Israel

7. KURAN, UĞUR, 1979: Fatigue-crack-propagation within the earth's crust due to cyclic loading and earthquake prediction along the San Andreas and North-Anatolian Fault Zones. JOFİZİK, Vol. VIII, No. 3, pp 75-123

8. KURAN, UĞUR, 1975: An experimental investigation of cyclic stress strain relation and fatigue crack propagation in crustal material. Ph.D. Thesis, Imperial College.

9. TOKSÖZ, M.N., SHAKAL, A.F., MICHEAL, A.J., 1979: Space time migration of earthquakes along the North Anatolian Fault Zone and seismic gaps. Pageoph., Vol. 117

10. KAISER, J., 1950: Untersuchungen über auftretende Geräusche beim Zugversuch. Ph.D. Thesis. Technische Hochschule München

11. KURAN, UĞUR, 1977: Stress/deformation - deformation/time phenomena in the Zigana Highway Tunnel and Landslide Area. JEOFIZIK, Publication of Geoph. Assoc. of Turkey. No. 2,3, Vol. VI, pp 5-28

Some Aspects of the North Anatolian Fault Zone Derived from the Comparison of its Instrumental Data with Historical Information

H. Soysal, D. Kolçak and S. Sipahioğlu

Geophysical Department of the Faculty of Earth Sciences, Istanbul University, Istanbul, Turkey

ABSTRACT

The main goal of this study is to point out some aspects of the North Anatolian Fault Zone (NAFZ) excluding the well known features.

In addition to the instrumental records, a great deal of historical information obtained in order to prepare the "Historical Earthquake Catalogue for Turkey and the Surrounding Area", was used for a better understanding of the fault zone. The analysis of all the data covering approximately 2000 years showed that there are no regular and visible seismic activity periods or dormancies in the history of the zone when it is considered as a whole. This is also applicable to each of the segments. However, various segments within the zone obviously show different characteristics. These peculiarities may help us for prediction and risk studies.

INTRODUCTION

For the past two years, a group of seismologists at the Geophysical Department has been involved in a project called "Historical Earthquake Catalogue for Turkey and the Surrounding Area". Therefore, a great deal of historical information has been collected, classified and listed. We believe that this catalogue, which is in preparation, will be very helpful to obtain real and dependable information for further earthquake prediction and risk estimation studies.

In this study, we exhibit some aspects of the North Anatolian Fault Zone (NAFZ) using both historical and instrumental earthquake data. Of course, this is a preliminary estimation which could be developed and completed later. Nevertheless, our knowledge of historical earthquakes is now much more comprehensive because many confirmative sources have been found related to them.

Before going into the details of our work we have to consider some aspects of the NAFZ which may be summarized as follows:

- Although the NAFZ has been producing large and destructive earthquakes, its features are not well mapped nor have relevant studies yet been completed. Probably, the efforts which have been spent recently to understand the geology of the zone, may put some of the results of previous geologic studies out of date (Ref. 8).
- Many of the investigators from the various disciplines were concerned with the NAFZ only when an earthquake devastated an area. As a result of this, most of the reports and papers, etc. available today describe one particular destructive earthquake, or specific studies were carried out to understand the geology or seismicity of a limited area of it (Ref. 4,5,12,15,18,23).
- There are some studies in which the NAFZ has been investigated as a whole, but they do not yield sufficient information for the purpose of "Earthquake Prediction" (Ref. 1,6, 10,11,16,17,22).
- Some papers contain very speculative results (Ref. 6), which were later uncritically used by other authors (Ref. 2). The lack of historical records may be the cause for this kind of speculative work.
- Unfortunately, in or around the zone, the permanent seismograph networks are too poor to record and evaluate current seismic activity (Fig. 1).
- To date, the NAFZ has contributed towards understanding how similar earthquakes and active fault movements have occurred in other parts of the world. In the future we have to confine ourselves to the NAFZ, and by intensified studies we should try to increase our understanding and knowledge of it.

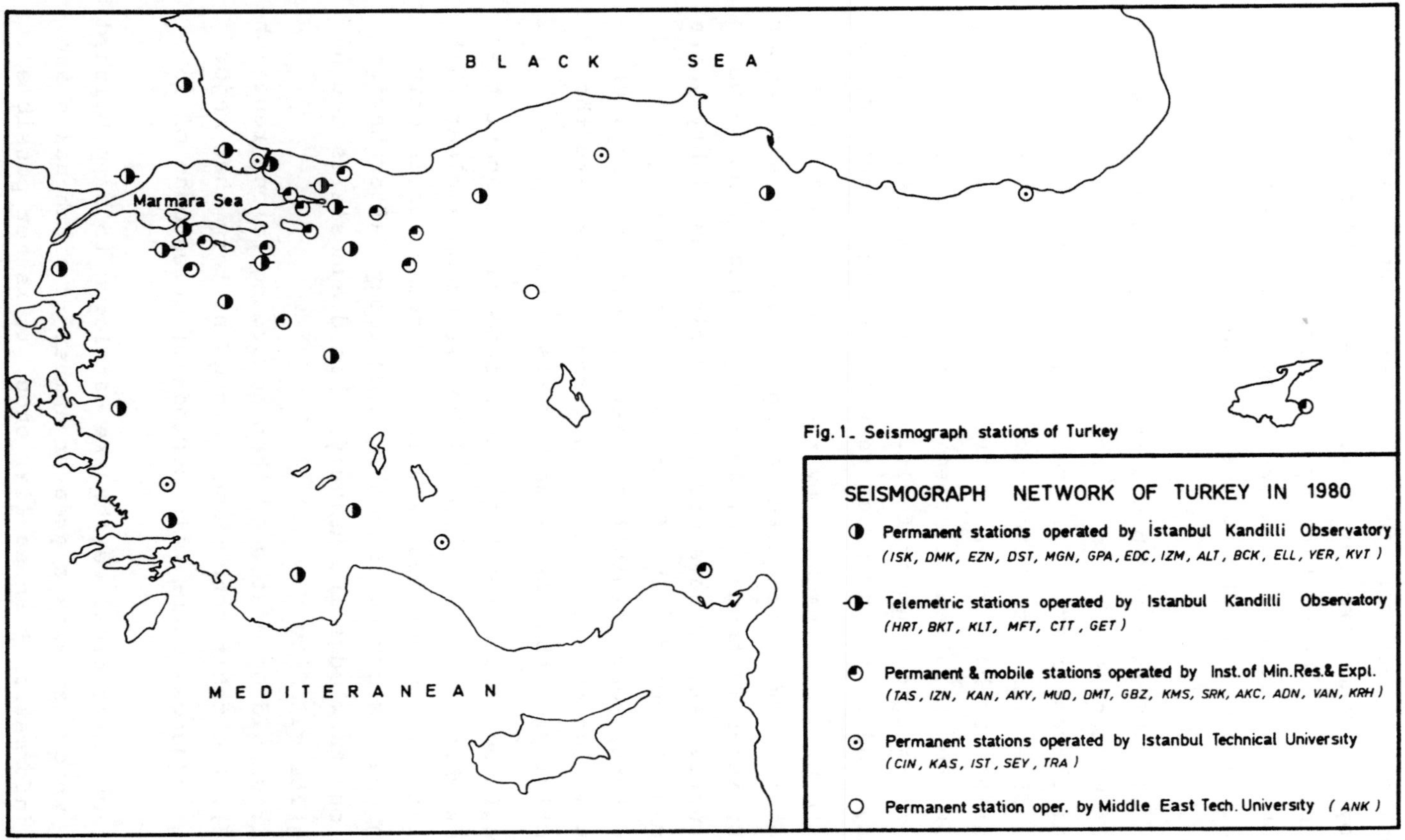

Figure 1. Seismograph Network of Turkey in 1980.

THE PROBLEM AND WAY OF APPROACH

The purpose of our study is to gain a better understanding of the NAFZ by all means presently available. For this purpose we try to approach the problem by analysing both historical information and instrumental data of the earthquakes which occurred along the zone during the last 2000 years.

The length of NAFZ is more than a thousand kilometers from east to west. If we consider it as a whole, we can identify only the general characteristics of the zone, such as its strike, slip type, direction, relative block movements, etc. However, some investigators have tried to evaluate the seismicity along the zone or the migration of seismic activity by considering the zone as a single segment fault. In our opinion this conception may lead us to incorrect results or misunderstandings.

According to previous research by the authors, the largest and most destructive earthquakes along the zone began to occur with the Erzincan Earthquake of 1939. Since that time, nine very destructive shocks have occurred. During the eighty years of instrumental records, the first forty years were quiet and the second forty years exhibit a great seismic activity. On this basis we can do little more than roughly evaluate the seismic activity of the last eighty years.

When we begin searching for historical information about the earthquakes which occurred before 1900, it becomes clear that the existing data do not show any regular active or inactive periods in the history of the NAFZ. Even today, after having increased our knowledge intensively, this diagnosis is roughly keeping its validity.
However, it should be mentioned that according to one researcher (Ref. 6), the fault zone shows through history that major earthquake sequences occur with periods of quiescence of 150 years.
In view of our historical earthquake catalogue to be published next year, which contains a greater number of earthquakes and much more information than earlier ones, it is not possible to make such definite statements.

It is very well known that the NAFZ is exposed as a complex of many en-echelon segments. Generally, the formation of each of the fault segments is closely related to individual earthquakes. The slip values vary from place to place along the zone. In some segments, the horizontal displacement associated with earthquakes reaches 4-5 metres, while others only show creep movements. The segments have produced different seismic activity. According to historical and instrumental records some segments of the region show very high seismic activity while other ones are of low activity.

Undoubtedly, all of these features are parts of the NAFZ and represent considerable danger for the millions and millions of people who are living within this zone or the surrounding area. The zone as a whole is much more hazardous than other regions in Turkey. However, the rapidly increasing problems of population and industrialization necessitate a more detailed identification of the earthquake hazard which is a challenging task to earth scientists. Which segments of the NAFZ are more hazardous than others? These questions and similar ones are still waiting for reasonable answers before we are able to protect human life and reduce the damage to man-made structures.

EVALUATION OF SEISMIC ACTIVITY

Instrumental records are much more precise than historical information. To begin with we try to describe the character of seismic activity utilizing relevant data of the last eighty years.

Generally, the earthquakes, having magnitudes of equal to or greater than 6.0 and/or intensity degree VIII and over, are related to surface faultings and therefore cause great damage. In this category, including nineteen of magnitude 6.5 and over, and eleven shocks of magnitude 7.0 and greater, thirty-five destructive earthquakes have directly been related to the NAFZ. As shown in Table 1, at least five destructive shocks occurred in that region before the Erzincan Earthquake of 1939. Taking into consideration the earthquakes of magnitude 7.0 and over, it can be seen that there are two shocks which occurred before the 1939 event.

TABLE 1

Earthquakes of M > 6.0 or $I_o \geqslant$ VIII Which Occurred Along the NAFZ (1900 - 1980)

No.	Date	Epicenter N	Epicenter E	M	h (km)	I_o	Max. Affected Locality	Remarks	References
1	19.04.1903	39.1	42.4	6.5		IX	Malazgirt-Bulanık		13,9
2	09.02.1909	40.0	38.0	6.75	60	VII	Şebinkarahisar		13
3	09.08.1912	40.6	27.2	7.3	16	X	Mürefte-Şarköy	M=7.75(13)	21,13,7,9
4	24.01.1916	40.3	36.8	7.1	10	?	Demirözü (Sivas)		3
5	13.09.1924	39.95	41.95	6.8	10	VIII	Pasinler (Erzurum)		3,13,9
6	18.05.1929	40.2	37.9	6.1	10	VIII	Suşehri (Sivas)		3,9,13
7	10.12.1930	39.75	39.25	5.6	30	VIII	Kemah (Erzincan)		3,9
8	04.01.1935	40.4	27.5	6.7	30	IX	Erdek-Marmara Islands		3,13,9
9	04.01.1935	40.3	27.45	6.6	20	?	Erdek-Marmara Islands		3,13
10	21.11.1939	39.8	39.7	5.9	80	VII	Tercan (Erzincan)		3,13,9
11	26.12.1939	39.8	39.5	7.9	20	XI	Erzincan	M=8(13)	7,3,13,9
12	11.12.1942	40.76	34.83	6.1	40	VIII	Osmancik (Çorum)		3,13
13	20.12.1942	40.87	36.47	7.0	10	IX	Niksar-Erbaa	M=7.3(13,7)	3,13,7,9
14	20.06.1943	40.85	30.51	6.6	10	IX	Adapazarı		3,13,7
15	26.11.1943	41.05	33.72	7.2	10	X	Ilgaz-Vezirköprü-Lâdik	M=7.4(13)	3,13,7,9
16	01.02.1944	41.00	33.00	7.2	10	X	Çerkeş-Gerede-Bolu	M=7.4(13) M=7.5(7)	3.13.7.9
17	26.10.1945	41.54	33.29	6.0	50	VI	Ilgaz and surroundings		3,13
18	20.11.1945	38.63	43.33	5.5	10	VIII	Van		3,13
19	31.05.1946	39.29	41.21	5.7	60	VIII	Varto-Hınis	M=6.0(7,13)	3,13,7,9
20	17.08.1949	39.57	40.62	7.0	40	IX	Kiği-Karlıova		3,13,9
21	13.08.1951	40.88	32.87	6.9	10	IX	Kurşunlu-Ilgaz	M=6.5(7)	3,13,9,7
22	03.01.1952	39.95	41.67	5.8	40	VIII	Pasinler (Erzurum)		3,13,7
23	18.03.1953	39.99	27.36	7.4	10	IX	Yenice-Gönen	M=7.2(7)	3,13,7,9
24	07.09.1953	41.09	33.01	6.4	40	VIII	Kurşunlu (Çankırı)		3,13,9
25	24.10.1954	40.	40.	6.0	10	VII	NE of Erzincan	M=4.6(3)	13,3,9
26	26.05.1957	40.67	31.0	7.1	10	IX	Bolu-Abant		3,13,9,7
27	25.10.1959	39.25	41.63	5.0	50	VIII	Hinıs-Varto	M=6.2(13)	3,13,9

TABLE 1 (Continued)

No.	Date	Epicenter N	E	M	h (km)	I_o	Max. Affected Locality	Remarks	References
28	26.01.1960	40.19	38.75	6.1	20	?	Kelkit Valley (North of Refahiye)	M=5.9(3)	3,13
29	18.09.1963	40.77	29.12	6.3	40	VIII	Yalova-Çınarcık		3,13,9
30	06.10.1964	40.30	28.23	7.0	35	IX	Manyas-Karacabey		3,13,7,9
31	31.08.1965	39.36	40.79	5.6	15	VIII	Karlıova		3,14
32	07.03.1966	39.2	41.6	5.6	30	VIII	Varto-Hınıs		3,14,9
33	19.08.1966	39.17	41.56	6.9	25	IX	Varto-Hınıs-Karlıova	M=6.6(7) M=5.8(14)	3,14,7,9
34	22.07.1967	40.67	30.69	7.2	35	IX	Adapazarı-Mudurnu		3,14,7,9
35	26.07.1967	39.54	40.38	6.2	30	VIII	Pülümür		3,14,7,9

TABLE 2

Earthquakes of M = 7.0 or Greater Which Occurred Along the NAFZ (1900 - 1980)

No.	Date	Epicenter N	E	M	h (km)	I_o	Max. Affected Locality	Remarks	References
1	09.08.1912	40.6	27.2	7.3	16	X	Mürefte-Şarköy	M=7.75(13)	21,13,7,9
2	24.01.1916	40.3	36.8	7.1	10	?	Demirözü (Sivas)		3
3	26.12.1939	39.8	39.5	7.9	20	XI	Erzincan	M=8(13)	7,3,13,9
4	20.12.1942	40.87	36.47	7.0	10	IX	Niksar-Erbaa	M=7.3(13,7)	3,13,7,9
5	26.11.1943	41.05	33.72	7.2	10	X	Ilgaz-Vezirköprü-Lâdik	M=7.4(13)	3,13,7,9
6	01.02.1944	41.0	33.0	7.2	10	X	Çerkez-Gerede-Bolu	M=7.4(13) M=7.5(7)	3,13,7,9
7	17.08.1949	39.57	40.62	7.0	40	IX	Kiği-Karlıova		3,13,9
8	18.03.1953	39.99	27.36	7.4	10	IX	Yenice-Gönen	M=7.2(7)	3,13,7,9
9	26.05.1957	40.67	31.0	7.1	10	IX	Bolu-Abant		3,13,9,7
10	06.10.1964	40.30	28.23	7.0	35	IX	Manyas-Karacabey		3,13,7,9
11	22.07.1967	40.67	30.69	7.2	35	IX	Adapazarı-Mudurnu		3,14,7,9

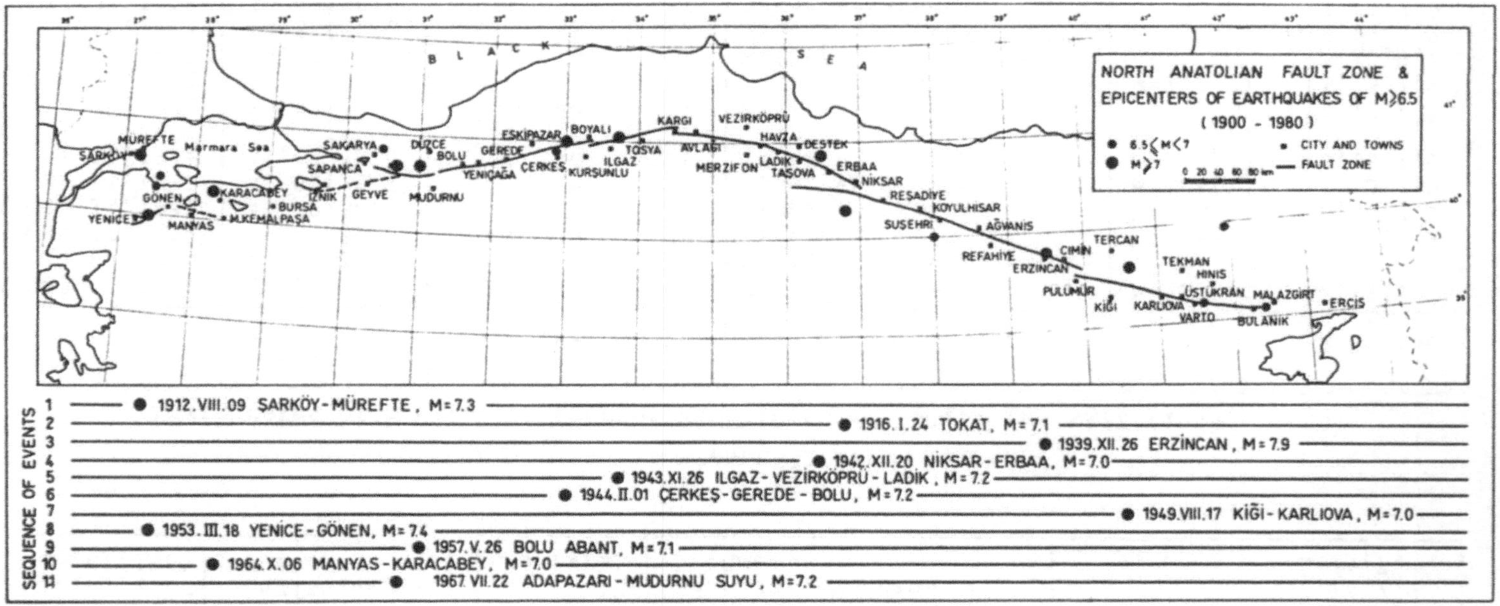

Figure 2. The North Anatolian Fault Zone and the Epicenters of Destructive Earthquakes of M = 6.5 and Greater. The Horizontal Lines Indicate the Sequence of Earthquakes of M = 7.0 and Greater.

At the beginning, the epicentres of the earthquakes of magnitude 7.0 and over were plotted on a map to see whether any relationship exists between the epicentres and the NAFZ system. The epicentres of the earthquakes which are tabulated in Table 2 and plotted in Fig. 2, are located just on the fault lines, forming some sub-groups.

Later, the epicentres of earthquakes between magnitudes 6.5 and 7.0 were plotted on the same map (Fig. 2). We found that these epicentres also were more or less located within the above-mentioned sub-groups. This grouping of epicentres may lead us to define some seismogenic zones or hazardous segments of the NAFZ. For this reason, we decided to take into consideration the preliminary results of our studies on the historical earthquakes which occurred before 1900.

Today, we have reasonably well-documented information on 152 historical earthquakes which occurred between 330 BC and 1900 AD. Seventy-nine of them had VIII and greater intensity degrees. The chart in Fig. 3 was compiled to clarify the pattern of seismic activity in relation to time and location. The data collected in Fig. 3 permit us to draw the following conclusions:

- There is indeed a clear accumulation of earthquake epicentres around segments which also produced intensive earthquakes during the last eighty years. In Fig. 3 those are defined by the letters A-E and related to the longitude scale.
- According to the presently available information there are no regular activity periods of dormancies visible for the NAFZ as a whole.
- None of the segments shows any regularity as to periods of quiescence. Fig. 4 shows the time distribution of all historical earthquakes. It is a further proof for this statement.
- Generally, the sub-groups of earthquakes which occurred during the last eighty years correspond to the segments which have been defined as A,B,C,D and E in Fig. 3. Sometimes there was earthquake activity in some of these segments during historical periods and none during instrumental pe-

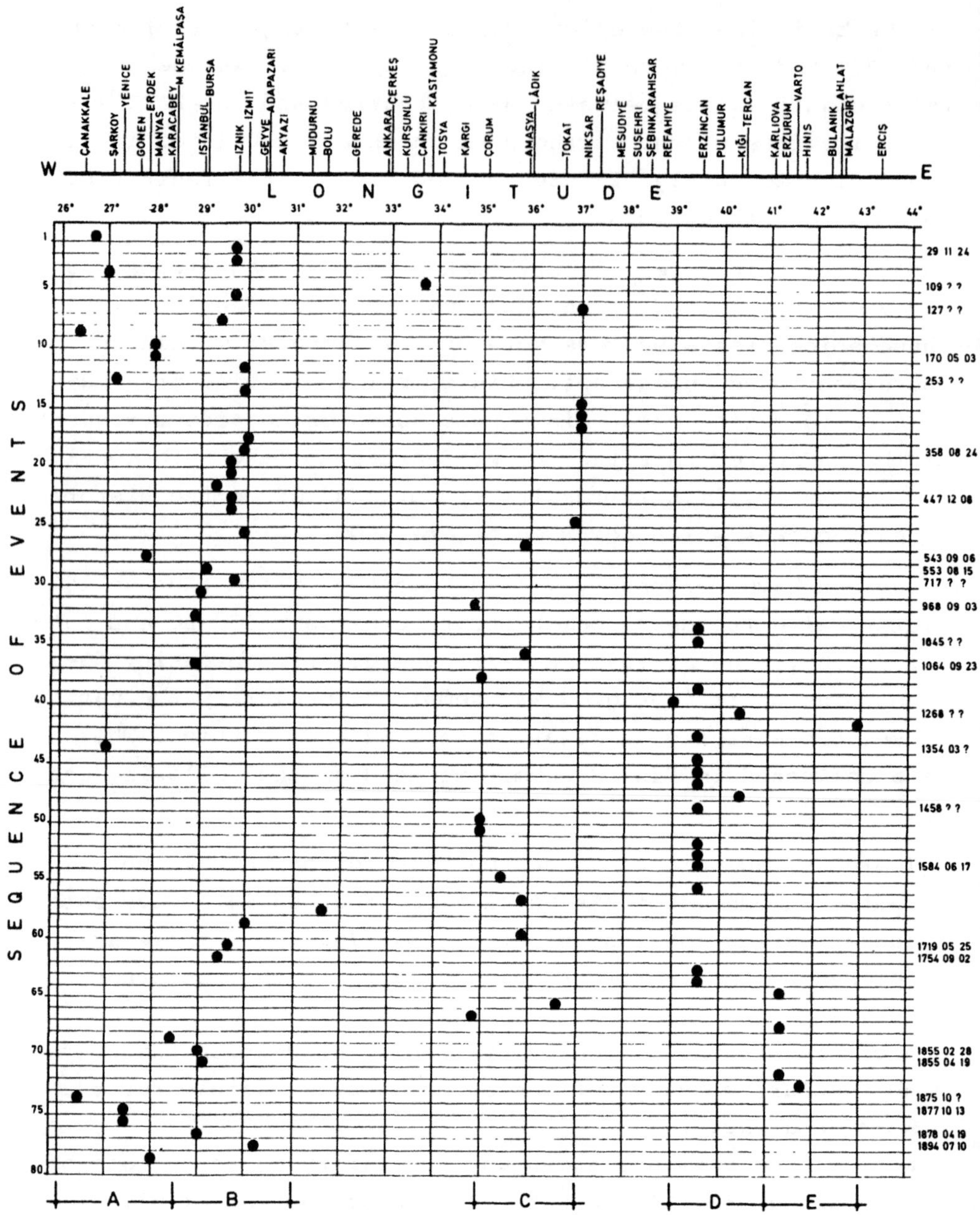

Figure 3. Spatial Distribution of Historical Earthquakes. Date Indication on the Right Refers to Earthquakes of Intensity IX and Greater.

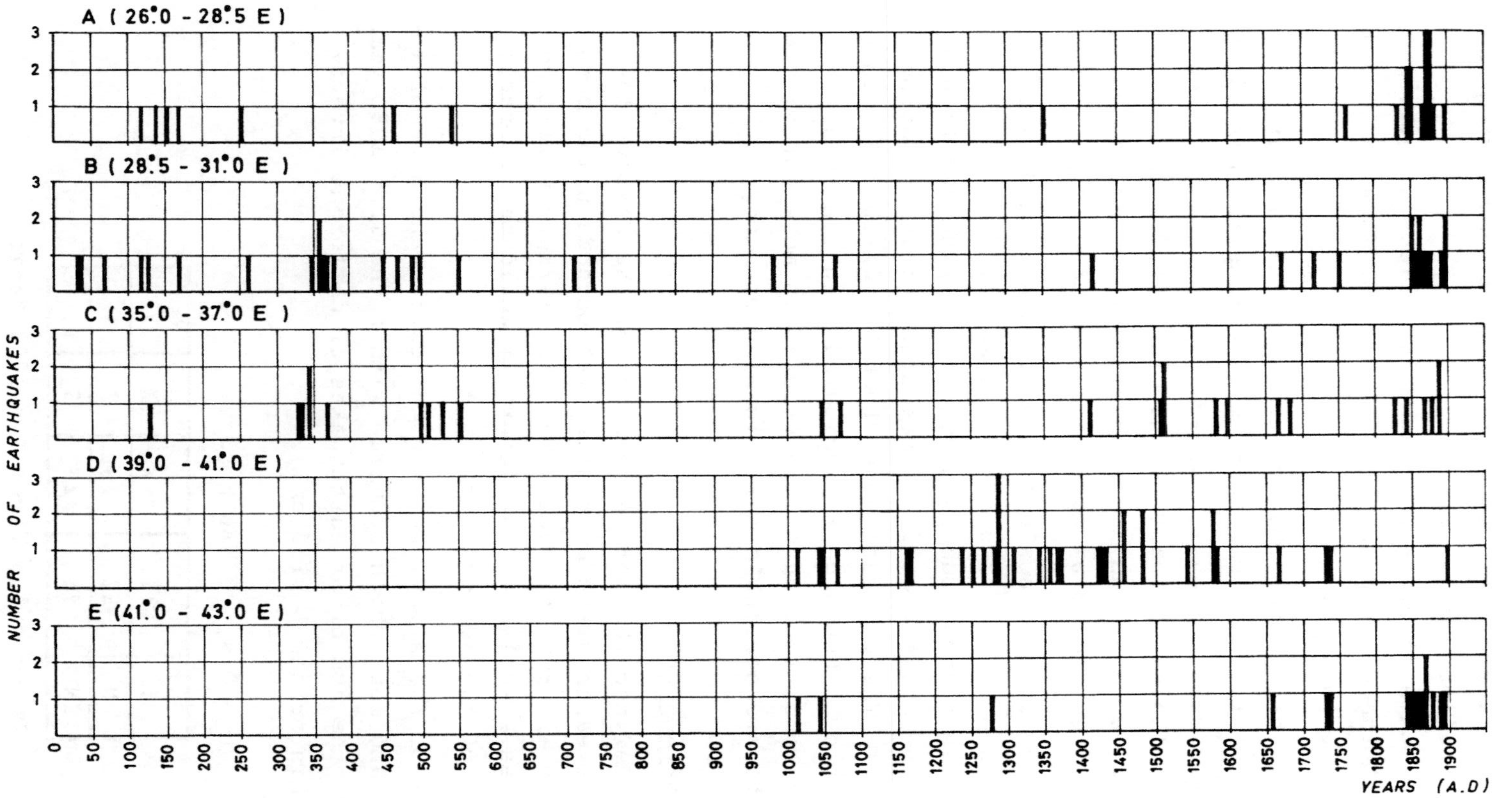

Figure 4. Time Distribution of Historical Earthquakes Within Various Longitude Sections.

riods, or vice versa. However, most of them show earthquakes during both periods continuously.

- For some segments, there are remarkable periods of quiescence lasting a thousand years as shown in Fig. 3. These dormancies may be real features of seismic history.

As a result, it has become obvious that the existing historical and instrumental data can help us to define the hazardous segments of the NAFZ. However, the evaluation made by using only the seismological records is insufficient for the understanding of those hazardous segments. Therefore the following geological studies should be carried out as soon as possible:

1. Strip maps at 1:25000 scale should be made to understand the physiographic features of the segments.
2. Comprehensive trench studies should be carried out to understand how the seismic activity has affected local geology (Ref. 19,20). The destructive earthquakes which are tabulated in Table 2 should be reconsidered when the local geological conditions have been studied in more detail and the results are available for this kind of study.

STATISTICAL APPROACHES

After having defined the active segments of the NAFZ we investigated the earthquakes by statistical methods using both historical and instrumental data. The calculations were made for the NAFZ as a whole. In the future it would be desirable to treat each of the segments separately.

In this stage of the study, we used the earthquake data of intensity VIII and more which occurred within the last 1980 years. Table 3 shows the number of earthquakes of different intensity for the historical period only.

TABLE 3

I	VIII	IX	X	XI
N	52	20	2	1
ΣN	75	23	3	1

Observation Time: T = 1900 Years

Applying the method of least squares to these data the following relationship was obtained

$$\log (\Sigma N/T) = 3.834 - 0.651\, I \qquad (1)$$

Adding the number of earthquakes occurring within the last 80 years to the values tabulated in Table 3, we found the shock numbers for the whole 1980 years as shown in Table 4.

TABLE 4

I	VIII	IX	X	XI
N	62	33	2	2
ΣN	99	37	4	2

Observation Time: T = 1980 Years

In the same way, the following relationship was obtained:

$$\log (\Sigma N/T) = 3.568 - 0.605\, I \qquad (2)$$

Then, using formulae (1) and (2) separately, the recurrence periods were found as tabulated in Table 5.

There is a significant difference between the recurrence times as shown in Table 5. The regime of seismic activity may have changed within the instrumental period. But it is very probable that these differences are due to incomplete information from the historical period.

TABLE 5

Intensity equal to or greater than	Recurrence Periods (Years)	
	Formula (1)	Formula (2)
VIII	23.7	18.7
IX	105.9	75.3
X	474.2	303.4
XI	2123.2	1221.8

CONCLUSION

The following results, which were obtained from this study, may not be of great benefit for earthquake prediction, but we believe that they will require new approaches to most of the problems related with the NAFZ:

1. There are no regular dormancies within the seismic history of the NAFZ when the zone is considered as a whole. This is also true for each of the segments of the NAFZ.
2. After a preliminary estimation of the zone we believe that each of the segments of the zone has different characteristics. The same methods of research have to be applied to study each of the segments separately.
3. The seismological data which are available today are insufficient to evaluate the seismic activity of the past. It is our hope that geological studies may reveal further information which may help us to gain a better understanding of the zone.

For this purpose the following work is required:

a) Strip mapping studies at 1:25000 scale.
b) Dating by Carbon 14 methods of rock - samples which are collected from trenches cutting segments of the NAFZ where the local geologic conditions are discernible.

ACKNOWLEDGEMENT

The authors are especially grateful to the Scientific and Technical Research Council of Turkey (TUBITAK) for supporting the Project TBAG 341 on the compilation of the "Historical Earthquake Catalogue for Turkey and Surrounding Areas".

The authors would also like to thank the colleagues for their guidance, in particular Dr. Clarence R. Allen from the California Institute of Technology.

The authors express their gratitude to Prof. Vogel and Dr. Işıkara for a critical review of the article.

REFERENCES

1. Allen, C.R. (1969) - Active Faulting in Northern Turkey. California Inst. Tech. Contribution No. 1577, Division of Geol. Sci.

2. Allen, C.R. (1975) - Geological Criteria for Evaluating Seismicity. Geol. Soc. Amer. Bull. 86, 1041-1057.

3. Alsan, E., Tezuç, L., Båth, M. (1975) - An Earthquake Catalogue for Turkey for the Interval 1913-1970 Report No. 7-75, Uppsala - Sweden.

4. Ambraseys, N.N., Zatopek, A., Taşdemiroğlu, M., Aytun, A. (1968) - The Mudurnu Valley Earthquake of 22 July 1967. Unesco Report, Serial No. 622/BMS.RD/AVS, Paris.

5. Ambraseys, N.N., Zatopek, A. (1968) - The Varto-Üstükan (Anatolia) Earthquake of 19 August 1966, Summary of a Field Report. Bull. Seismol. Soc. Amer. Vol. 58, 47-102.

6. Ambraseys, N.N. (1970) - Some Characteristic Features of the Anatolian Fault. Tectonophysics 9, 143-165.

7. Ambraseys, N.N. (1975) - Studies in Historical Seismicity and Tectonics. Geodynamics Today. Chapter 2. The Royal Soc. London.

8. Arpat, E., Şaroğlu, F. (1975) - Türkiye'deki Bazı Önemli Genç Tektonik Olaylar. Bull. Geol. Soc. Turkey, Vol. 18, No. 7, 91-101.

9. Ataman, O., Tabban, A. (1977) - Türkiye'de Yerleşme Alanlarının Doğal Afetlerle İlişkileri. Mimarlık Dergisi No. 77/4, Istanbul.

10. Canıtez, N. (1962) - Gravite Anomalileri ve Sismolojiye Göre Kuzey Anadolu'da Arz Kabuğu'nun Yapısı. Doktora Tezi, İ.T.Ü. Maden Fakültesi.

11. Canıtez, N. (1973) - Yeni Kabuk Hareketlerine İlişkin Çalışmalar ve Kuzey Anadolu Fay Problemi. Kuzey Anadolu Fayı ve Deprem Kuşağı Simpozyumu Tebliğleri. M.T.A. Enstitüsü, 35-38.

12. Crampin, S., Üçer, S.B. (1975) - The Seismicity of the Marmara Sea Region of Turkey. Geophys. J.R. Astr. Soc., 40, 269-288.

13. Ergin, K., Güçlü, U., Uz, Z. (1967) - Türkiye ve Civārının Deprem Kataloğu (11-1964). İ.T.Ü. Maden Fak. Arz Fiziği Enst. Yayınları No. 24, Istanbul.
14. Ergin, K., Güçlü, U., Aksay, G. (1971) - Türkiye ve Dolaylarının Deprem Kataloğu (1965-1970). İ.T.Ü. Maden Fak. Arz Fiziği Enst. Yay. No. 28, Istanbul.
15. Güçlü, U., Uz, Z., Canıtez, N. (1969) - 22 Temmuz 1967, Adapazarı-Mudurnu Suyu Vadisi Depremi Üzerine İncelemeler. İ.T.Ü. Maden Fak. Arz Fiziği Enst. Yayınları No. 27, Istanbul.
16. Ketin, İ. (1969) - Kuzey Anadolu Fayı Hakkında. M.T.A. Enst. Dergisi No. 72, 1-25.
17. Kuran, U. (1979) - Fatigue-Crack-Propagation Within the Earth's Crust Due to Cyclic Tidal Loading and Earthquake Prediction Along the San Andreas and North Anatolian Fault Zone. Publ. Geophysicists Ass. Turkey Vol. III, 73-122.
18. Omote, S., Çöloğlu, M.G. (1968) - Seismic Activity in Marmara Region, Turkey. Bull. IISEE Vol. 5, 17-30.
19. Sieh, K.E. (1978) - Slip Along the San Andreas Fault Associated With the Great 1857 Earthquake. Bull. Seismol. Soc. Amer. Vol. 68, 1421-1448.
20. Sieh, K.E. (1978) - Prehistoric Large Earthquakes Produced by Slip on the San Andreas Fault at Pallett Creek, California. Jour. Geoph. Res. Vol. 83, 3907-3939.
21. Shebalin, N.N., Karnik, V., Hadzievski, D. (1974) - Catalogue of Earthquakes (of Balkan Region). Unesco. Skopje-Yugoslavia.
22. Tabban, A., Aytun, A. (1973) - Kuzey Anadolu Fay Zonu Boyunca Depremlerin Şiddet-Tekerrür Kümelenmeleri ve Bunun Yerkabuğu Deformasyonlarıyla Muhtemel İlişkisi. Kuzey Anadolu Fayı ve Deprem Kuşağı Simpozyumu Tebliğleri. M.T.A. Enstitüsü, 59-67.
23. Wallace, R.E. (1968) - Earthquake of August 19, 1966, Varto Area, Eastern Turkey. Bull. Seismol. Soc. Amer. Vol. 58, 11-45.

A Statistical Analysis of the Major and Microearthquakes Along the East-Anatolian Fault

Ahmet Ercan

Technical University of Istanbul, Geophysics Department, Teşvikiye, Istanbul, Turkey

Abstract

In order to evaluate past and present earthquake activity along the East Anatolian Fault between Karliova and Maras, the distribution in time and space of earthquakes was investigated. The investigation dealt with an interval of 75 years for large-magnitude earthquakes, and with a period of 105 days for micro-earthquakes occurring in a limited area. The most active regions of the fault zone are around Bingöl in the north, the Lake Hazar in the middle, and Adiyaman in the south. Along the fault zone the average Richter magnitude of earthquakes is about 5, the hypocentral depth about 25 kilometers, the period of occurrence (determined by spectral analysis) about 11 years, and the b-value for the magnitude range of $7 > M > 4$ approximately 0.5.

The microearthquake recordings around Lake Hazar indicated a daily average of 5 events with magnitudes less than 3. The composite fault plane solutions confirmed a lefthand strike-slip motion for most of the secondary faults. The average daily energy release of 10^{16} erg and explicit alignment of strain release contours with the fault are some of the indications of the continuous tectonic activity in this region even in this quiet period. An 11 and 27 days periodicity of microearthquake activities has been obtained from spectral analysis. A close correspondence in epicentral distribution exists between major earthquakes and the present microearthquakes.

INTRODUCTION

This study was not planned to predict earthquakes in the eastern region of Turkey. However, this task actually had become inevitable after the publication of a paper by Arpat (1977), in which the author strongly claimed that, as the title indicates "the Karakaya dam site will in the future undergo ground-shaking by major earthquakes". His assumptions were strictly based on the geological features along the fault and on collected documents concerning historical earthquakes of the East Anatolian fault zone. In addition, information exists of numbers of completely ruined towns and on the migration of large groups of people, who used to inhabit these towns. The main concern of my colleague is that the Karakaya dam was planned to be built only 14 kilometers away from one of the major fracture zones of Turkey. The East Anatolian Fault borders the Arabic Arc in the south-east and the Mid-Anatolian sub-plate in the west. Although the average magnitude of the earthquakes recorded since 1903 is about 5 on the Richter scale, Arpat (1977) strongly points out and demonstrates a degree of destruction and slip in the region which can only be the result of at least magnitude 7 or 8 earthquakes. This information was sufficient to draw the attention of earth scientists to this problem and to study available present-day evidence. For this purpose we cooperated with Prof. N. Canitez of the Technical University of Istanbul, Prof.Dr. M. Nafi Toksöz of M.I.T. and Esen Arpat of the M.T.A.

The aim of the field work was to record microearthquakes in order to determine present activity and the range of energy release characteristics of the secondary faults, to analyse earthquake risk by considering the magnitudes over 4, and to see if there is any correlation in epicentral distribution of microearthquake activities with that of large ones recorded in the past. We were hoping that this kind of information would provide a basis for the derivation of engineering parameters which would lead to an estimate of the acceleration of the biggest shake expected during the life-time of the dam.

DATA RECORDING AND COMPILATION

The field work took place in a highly rugged site which is close to the junction of four provinces of Turkey, namely Adiyaman, Diyarbakir, Elazig and Malatya. Presumably, the fault begins at Karliova in the north, which is also a crossing point with the North Anatolian fault, goes south from there through the recording area and continues over the Hatay province of Turkey, to Lebanon, the Dead Sea and most probably into Ethiopia. In contrast to the North Anatolian fault, the East-Anatolian fault shows a left-lateral strike-slip motion. This is explained by a relative movement resulting from the continuous opening of the Dead Sea and the resultant pushing of the Arabian and Persian block to the North-East, and by its relation with the Van and other Turkish sub-plates.

A six-station radio-link recording system was distributed around a circle with a radius of 18 kilometers. The receiving centre of the network was about 5 kilometers west of Pütürge (Fig. 1). The circumference of the circle included a 36 kilometer long part of the fault which lies in the Şiro Valley. Therefore, it was even possible to monitor extremely

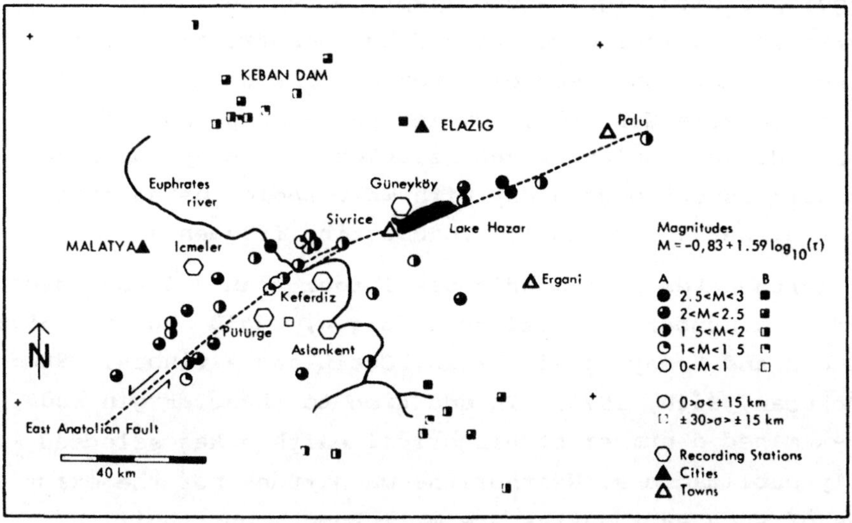

Fig. 1: Epicentral distribution of micro-earthquakes in the vicinity of Lake Hazar on the East Anatolian fault of a recording period of 105 days in 1977. Shaded circles (A) and squares (B) indicate magnitudes as well as expected standard deviations in epicentral locations, 0 to ± 15 km and ± 15 to ± 30 km respectively.

small micro-earthquakes occurring in the local region. Each receiving-transmitting station was equipped with a high-frequency and high-sensitivity Kinemetrics seismometer, a modulater, a transmitter and a Haki-type horizontal antenna. The carrier radio waves from each station having a fraction of 460 megahertz were transmitted to corresponding receivers operated in the centre of the network. Earthquake-singnals transmitted by the carrier were discriminated and then recorded on the twelve channels of the tape recording system. We had many problems particularly with breakdowns of the transmitting systems which are quite sensitive to humidity change in high mountains, in addition to timing and some general maintenance troubles.

In addition, three high-frequency, high-gain, and portable seismographs were operated at the latitudes of 38° - 39° and at the longitudes of 38° - 40° around the city of Pütürge in Malatya province. The instruments which we loaned from Massachusetts Institute of Technology were MEQ-800's from Sprengnether equipped with light and well-protected seismometers. These portable stations were shifted between two triangles during the 105 day-long recording period. The stations of the first triangle were Güneyköy, İçmeler and Pütürge and of the second one Pütürge, Keferdiz and Aslankent. Time synchronization was carried out once a day with WWVT radios using time signals at either 5, 10 and 15 broadcasting frequency.

Lists of earthquakes with magnitudes larger than 3.5 have been published in a number of catalogues (Ergin, Güçlü and Uz, 1967; Ergin, Güçlü and Aksay, 1971; Ergin, Güçlü and Altinbaş, 1980; Alsan, Tezuçan, Båth, 1975). In addition to these, Ergin and Soysal* examined a number of historical earthquakes selected from early publications. Neither the magnitudes nor the exact locations of such earthquakes are well-known, but their intensities are. From 1905 to 1979, 41 large-magnitudes earthquakes occurred whose epicentres were reported to lie between the latitudes of 37° - 41° E and longitudes of 37.5^{o} - 39.5^{o} N (Fig. 2 and Appendix 1). Numbers of other earthquakes with

* H. Soysal, Istanbul University-Personal communication

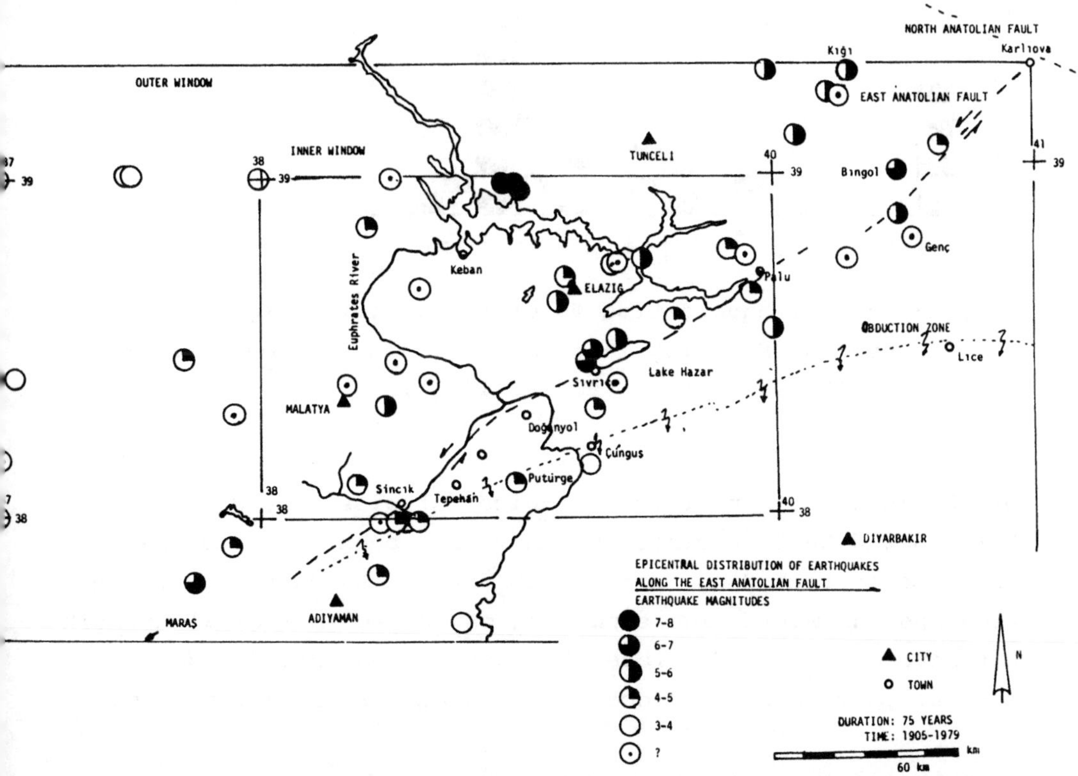

Fig. 2: Epicentral distributions of large-magnitude earthquakes along the East-Anatolian fault zone near Lake Hazar for a period of 75 years (from 1905 to 1979). Richter magnitudes are indicated. The two windows, drawn with thin lines indicate the study areas.

epicentres in the same limited region were excluded because their sources of energy are considered to be the North Anatolian Fault or the obduction zone of the Arabic Arc in the south-east. The earthquakes along the arc have mostly deep hypocentres, while the East-Anatolian Fault is characterized by shallow earthquakes with hypocentres less than 25 kilometres deep. Focal mechanisms were some other indications to help in discriminating different origins.

During the time period of 75 years, there were no earthquakes recorded between 1910 and 1925 (Fig. 3). This may be a result

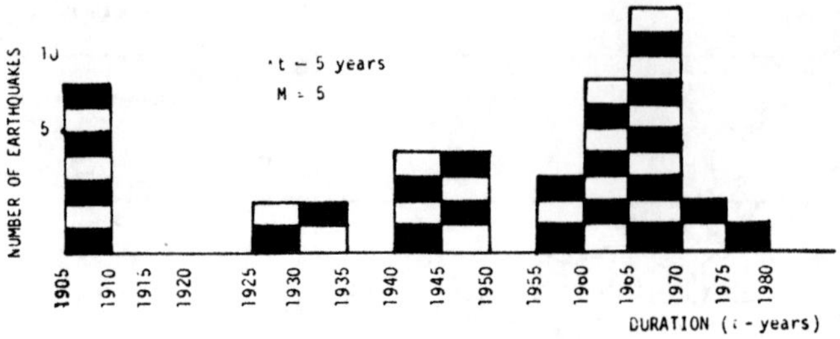

Fig. 3: The earthquake-time function for large magnitude events. Sampling interval is 5 years. Time interval is 75 years (from 1905 to 1980).

of the interruption of seismological studies caused by the First World War. In contrary, there has been a substantial increase in recordings after 1940 due to the denser distribution and the world-wide expansion of the seismological network. Earthquakes reported before 1940 have to be considered with care as to magnitudes, locations and numbers of occurrences.

MAGNITUDES AND NUMBERS OF OCCURRENCES

There are about 40 earthquakes recorded in this region, with Richter magnitudes varying from 3.2 to 6.8 (Fig. 3). The most recent one of these is the Palu earthquake which occurred on March 24, 1977 with an epicentral location at 38.562 N - 40.024 E, and a hypocentral depth of about 21 kilometers. This event appears to characterize the average properties of all occurrences in the region with respect to both magnitudes and focal depths (Fig. 4). The arithmetic averages of earthquake magnitudes and focal depths were calculated to be 5.1 and 25 kilometers, respectively. It is obvious that the hypocentres are shallow in the Pütürge-Hazar area and that they are becoming deeper in both NE and SW directions. The largest magnitude reported in the time period is 6.8, the magnitude of the Tunceli earthquake (Memişoğlu, 1978).

The daily average of occurrence of micro-earthquakes is about 5 (Fig. 5a). The Pütürge-Hazar region seems to be active with continuous small energy release which approximately amounts to 10^{15} ergs per day with magnitudes varying from 0.5 to 3.3.

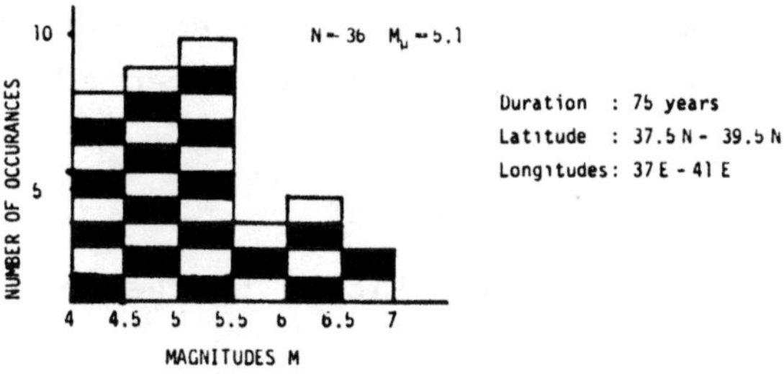

Fig. 4: Number of occurrences of large earthquakes in the vicinity of Lake Hazar as a function of magnitudes. Time period is 75 years. Latitudes 37.5N - 39.5N, longitudes 37E - 41E.

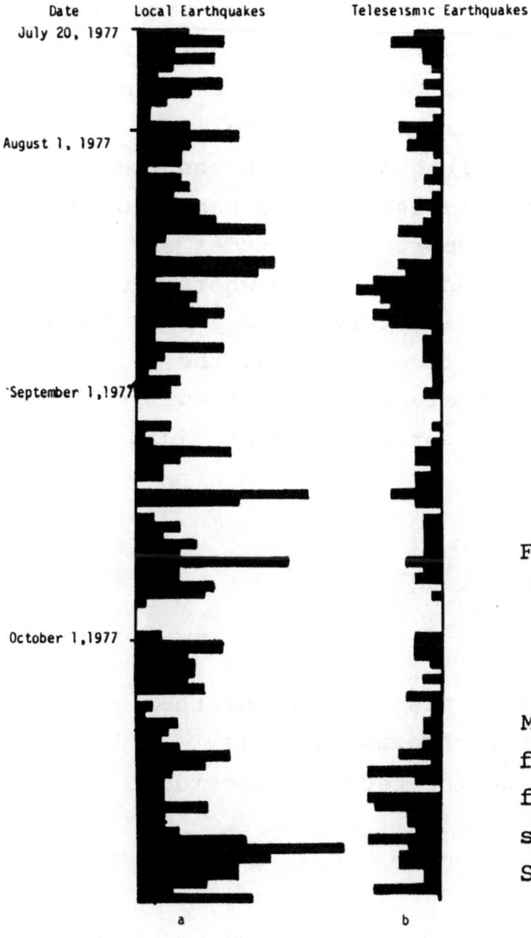

Fig. 5:

Micro-earthquake time functions for local (a), and for tele-seismic ($t_s - t_p \geq 12$ seconds) events (b). Sampling interval is one day.

The magnitudes were calculated from the following established empirical magnitude-duration relationship which was accepted for this particular region

$$M = -0.83 + 1.59 \log_{10} \text{(duration)}$$

The energy release estimates were made in relation to the magnitudes according to

$$\log_{10} E = 9.4 + 2.14\ M - 0.054\ M^2$$

As to the seasonal distrubution of micro-earthquake occurrences they appear to increase from summer to autumn, (i.e., 3.3 to 6.6) (Fig. 6). In order to analyse the day by day activity change, the micro-earthquakes were discriminated in terms of (t_S - t_P) differences into two groups, namely local (t_{SP} less than 12 seconds) and tele-seismic (t_{SP} larger than 12 seconds). Parallel plot of these groups versus time indicates whether or not the local and regional micro-earthquakes are activated by or influenced simultaneously by the same source (Fig. 5a,b). A close correlation seems to exist. In addition, the large magnitude earthquakes which occurred in neighbouring countries were superimposed on the same plot according to their approximately reported magnitudes and locations. This comparison was carried out to see whether or not there is a relationship with global activity. Although there is a similarity in the time-rate of activities, the evidence is not clear. The duration of observation is not sufficient to conclude whether the observed correlations are just a few

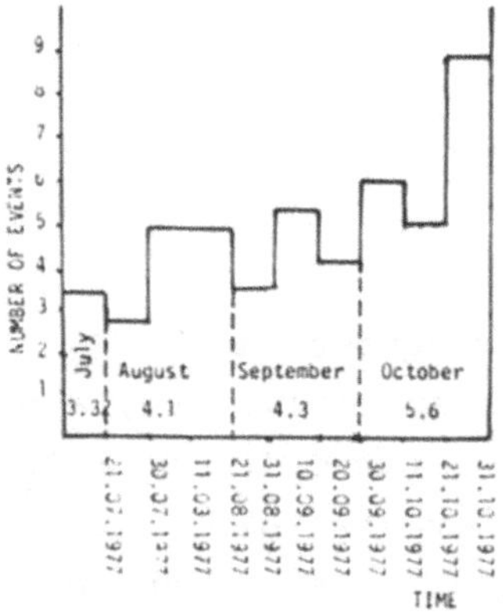

Fig. 6:

Monthly distribution of the microearthquake activity during the recording period.

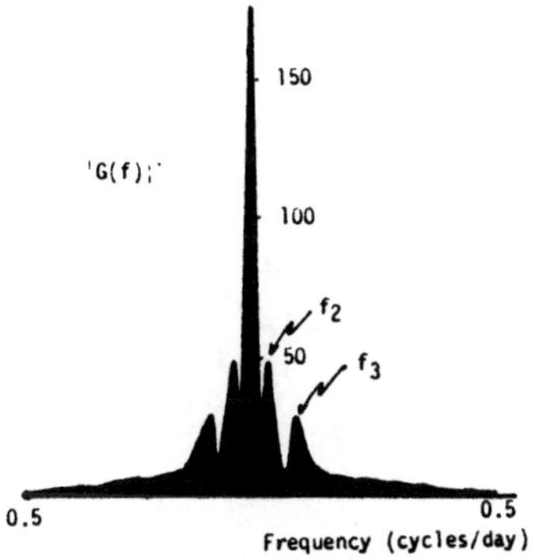

Fig. 7: The Fourier amplitude spectrum of the local micro-earthquake time function of the East-Anatolian fault zone. The sampling rate f, is 0.12 cycles per day, and the Nyquist frequency f_k = 0.5 cycles per day. The main frequency components obtained from the two local peaks of the spectrum are f_2 = 0.036 cycles per day and f_3 = 0.096 cycles per day which correspond to periodicities of 27.7 days and 10.4 days, respectively.

coincidences or are the result of a common mechanism. The spectral analysis of the micro-earthquake-time function shows two peaks at two different frequencies (Fig. 7). One of these, f_2, is at the frequency of 0.036 cycles per day, and the other, f_3, is at about 0.096 cycles per day. The Nyquist frequency is 0.5 cycles per day and the sampling interval is 1 day. The first frequency implies the existence of a periodicity of 27.7 days which may be correlated with sunspot activity and the second one is a periodical variation of 11 days. These correlations should not be taken too seriously into consideration before further recordings are available and analysed.

In a similar way we studied the earthquake-time function based on occurrences of earthquakes larger than magnitude 3.3 during the time interval of 75 years. We recognized two peaks corresponding to two frequencies, namely f_1 = 0.057 cycles per year and f_2 = 0.085 cycles per year (Fig. 8). The first of these two peaks indicates a periodicity of 18 years, and the

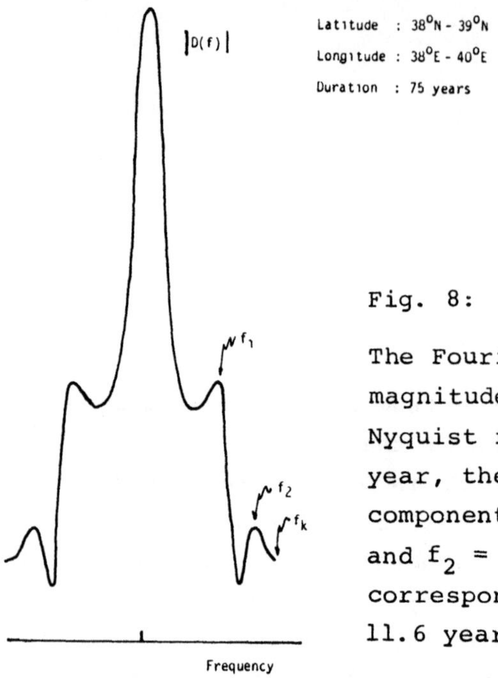

Fig. 8:

The Fourier amplitude spectrum of large-magnitude earthquakes in the region. The Nyquist frequency is f_k = 0.1 cycles per year, the two anomalous frequency components are f_1 = 0.057 cycles per year and f_2 = 0.085 cycles per year which corresponds to periodicities of 17.5 and 11.6 years respectively.

second one shows a periodicity of 11 years. Remembering that there is a gap in the earthquake reports between 1910 and 1940, the 11-year periodicity seems relatively reliable compared with the longer periodicity. However, we have to remember that the sampling interval is 5 years. If the above periodicity is nevertheless true, one has to expect an earthquake of about magnitude 5 once every eleven years. Whether or not there exists a correlation with the 11-year-periodicity of the external magnetic field variations has to be checked.

MAGNITUDE - FREQUENCY RELATIONSHIP

In order to arrive at conclusions about earthquakes which are to be expected in the area, one of the procedures is to apply the first-degree regression analysis to the data collected in a specified length of time, in a limited region and in a definite range of magnitudes.

Hypocentral depths and magnitudes of earthquakes in the Pütürge-Hazar region are somewhat different from those which originate outside this region. Considering that such earthquakes which do not apply to this specific field of interest, may originate from, or their mechanisms interrelate with, motions related to some other tectonic features, the available data in two areas were collected (Fig. 2). One is a rectangle circumscribing the inner Pütürge-Hazar area.

The earthquakes in the inner region were also taken into account while studying the enclosing outer rectangle. The geographical coordinates of the large rectangle are 37.5° - 39.5°N and 37°E - 41°E and of the small one 38°N - 39°N and 38°E - 40°E.

The epicentral distribution of the inner rectangle is particularly dense in the Elaziğ-Doğanyol-Hazar triangle and in the Sincik-Tepehan region. According to the magnitude-energy release calculations of both macro- and micro-earthquakes the most active part is the bending point of the fault where the Firat (the Euphrates) river crosses and flows approximately five kilometers along the fault line (Fig. 2). The first degree least-squares analysis of the 75 year-long data collection leads to the following magnitude-frequency relation

$$\log_{10} N = 3.85 - 0.55\ M,$$

where M represents the magnitude and N is the frequency of occurrence of earthquakes larger than M (Fig. 9).

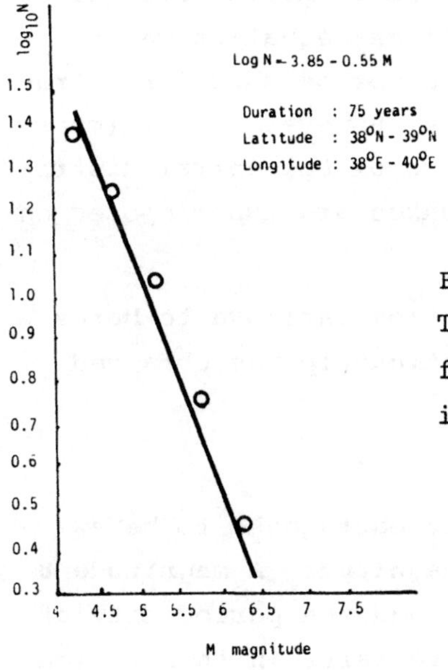

Fig. 9:
The magnitude - frequency relationship for largemagnitude earthquakes of the inner window (see Fig. 2).

This relation enables us to estimate the number of occurrences, N, in the time duration, D (which is 75 years for this case), for magnitudes larger than M for the purpose of interpolation as well as extrapolation. Accordingly a magnitude 8 earthquake is expected once in 250 years. The recurrence time period for a magnitude of 7.5 is 133 years, 75 years for a magnitude of 7, 28 years for a magnitude of 6, and 5.5 years for the magnitude of 5. This periodicity is in particularly good agreement with the 5.5 Richter magnitude, which was calculated by spectral analysis. If we neglect the 1905 Tunceli earthquake which has only been referenced by Memişoğlu (1978), there is in 75 years no earthquake reported with magnitude > 7. If this is true, do we have to expect one in the near future? It is risky to answer this question without knowing the reliability of these statistical measures and without having more information about the fault mechanisms. However, if such an earthquake occurs it will create an acceleration of at least 0.75g to a distance of 15 kilometers from the epicentre.

From the present-day micro-earthquakes the following magnitude-frequency relationship was obtained

$$\log_{10} N = 4.4 - 1.7\ M$$

using the events recorded in 105 days with magnitudes between 1.5 and 3.3. This shows that extremely small micro-earthquakes dominate in the region. Because this linear equation was derived from microearthquakes it should not be used for extrapolation to large magnitude earthquakes. However, it is important to state that there is a good match of epicentral distributions when present-day micro-earthquakes are superimposed on the large earthquakes of the 75 years.

For the large rectangle which extends from Karliova to Maras the following magnitude frequency relationship was obtained (Fig. 10).

$$\log_{10} N = 3.8 - 0.5\ M$$

According to this equation, the largest earthquake to be expected in 75 years is of 7.6 Richter magnitude. A magnitude 8 earthquake may occur once in 120 years and the periodicity of magnitude 6 is 12 years. The most active parts in this region

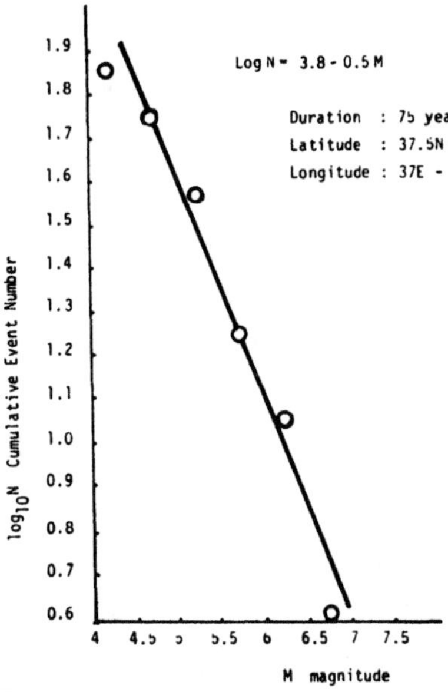

Fig. 10:

The magnitude - frequency relationship for largemagnitude earthquakes of the outer window (see Fig. 2).

are Bingöl in the North and the Adiyaman-Maraş province in the south. These sections have been experiencing earthquakes with magnitudes larger than 5 on the Richter scale.

CONCLUSIONS

A close correlation exists between the major faults and the epicentral distribution of earthquakes as well as micro-earthquakes (Fig. 11 and 12). The composite fault-plane solution of micro-earthquake activities confirmed a left-hand strike slip motion for most of the secondary faults. The average b value is 0.5, the focal depth is 25 kilometers and the magnitude is 5 on the Richter scale. The expectation of an earthquake with a magnitude of about 5.5 is approximately every 12 years, estimated both by spectral analysis of the earthquake-time function and by regression analysis. The most active parts of the investigated area are between Pütürge-Hazar and Hazar-Palu. The micro-earthquakes may be divided into two groups of different origin. One of these has a periodicity of about 11, and the other one of about 28 days. Although there are some gaps of information, an 11-year periodicity seems to exist. Because one magnitude 5 earthquake

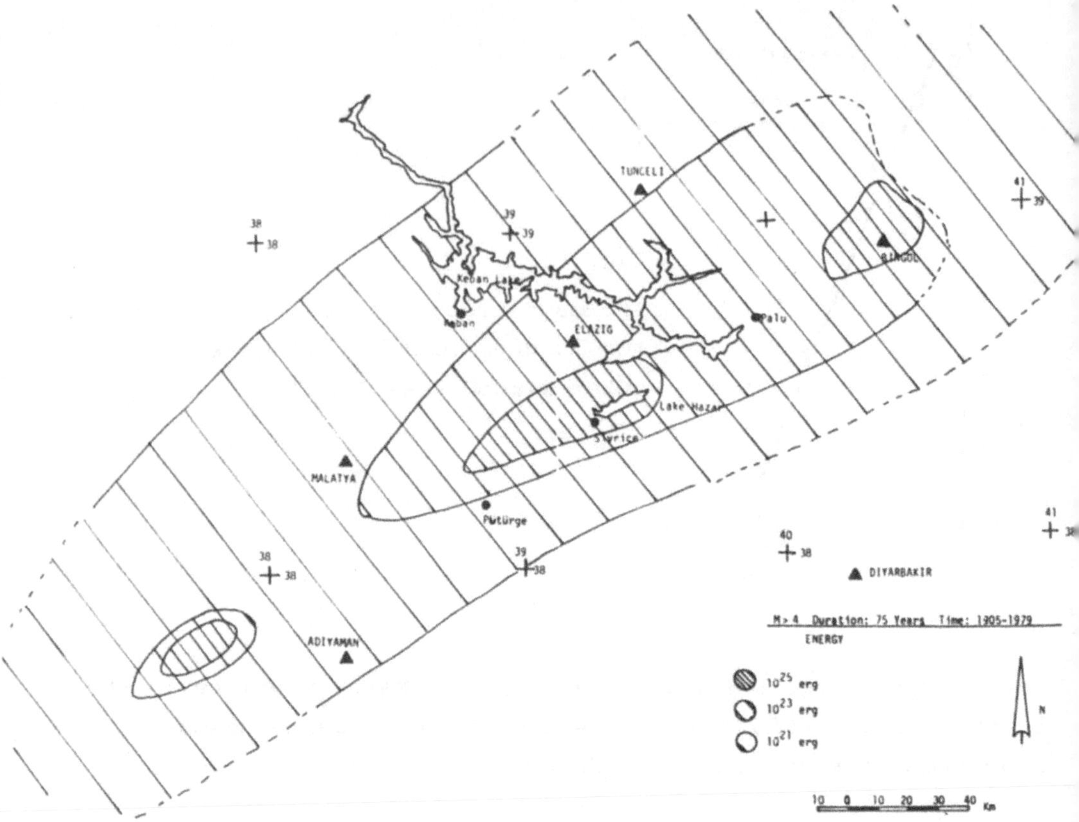

Fig. 11: Energy release map, derived from large-magnitude earthquakes along the East-Anatolian fault in the vicinity of Lake Hazar.

is observed about once in five years, the observed high microactivity does not prevent energy-build up for a major earthquake in the region. A more detailed study of micro-earthquake activity may point out which part of the fracture zone will be subject to a future shake. To confirm these first conclusions and to investigate the mechanisms along the East-Anatolian Fault, extension of the field work is recommended. The network should however be improved and operated during a much longer period of time.

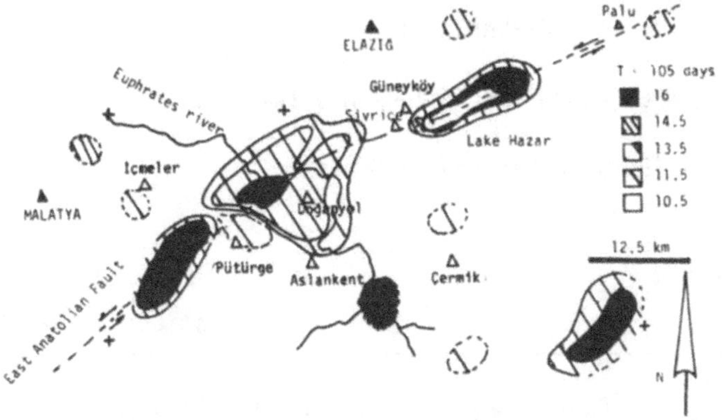

Fig. 12: Energy release map, derived from micro-earthquake magnitudes along the East-Anatolian fault zone for a record period of 105 days. Different densities in shading indicate the logarithmic degree of energy release in erg.

ACKNOWLEDGEMENTS

I am thankful to Prof. N. Canitez, Prof. M.N. Toksöz, and Esen Arpat for their cooperation and valuable discussions. Also I am grateful to the Massachusetts Institute of Technology and Kandilli Observatory of the Ministry of Education of Turkey for the loan of their instruments, and to the Turkish Mineral Research and Exploration Institute whose support made it possible to carry out the field work. I thank Fuat Şaroğlu, Halit Sav of the M.T.A. and my colleagues, Besim Başlangiç, Orhan Alpay, Necati Saraç, İbrahim Türer and Zafer Berker of the Technical University and other crew members for their participation in the field work.

I am grateful to Prof. A. Vogel for his critical review of the manuscript.

APPENDIX I

EARTHQUAKES ALONG THE EAST-ANATOLIAN FAULT

(Between Karlıova - Maraş)

Latitude : 37.5 - 39.5 N

Longitude: 37 E - 41 E Duration: 1905-1979

Date (Year)	Coordinates lat.	long.	Depth	Magnitude	Intensity	Location	Source
1905	39	39		6.8		Tunceli Akçapınar	Memişoğlu
1905	39	39		5.		Tunceli Akçapınar	"
1905	39	39		5.8		Tunceli Akçapınar	"
1905	39	39	33	5.6		Tunceli Akçapınar	"
1908	38.68	39.2		5.2	6	Elazığ	TDIK
1909	39.32	40.35		5.2	6	Bingöl Kiği	TDIK
19.05.1929	39	37.5				Sivas Konakpınar	ISS
28.06.1929	39	37.5				Sivas Konakpınar	ISS
08.02.1930	38.52	39.40	100	5.1		Elazığ Sivrice	ISS
09.01.1931	38	38.5				Adıyaman Sincik	ISS
28.05.1940	38.89	40.5		5.2	6	Bingöl	TDIK
23.09.1940	38.9	39.4				Elazığ Harput	ISS
03.03.1941	39	38				Malatya Hasançelebi	BCIS
07.02.1942	38.9	39.4				Elazığ Harput	ISS
14.07.1947	38	37				Maraş Çardak	ISS
16.03.1948	37.8	37.7		6.3	6	Adıyaman Gölbaşı	TDIK

Date (Year)	Coordinates lat.	long.	Depth	Magnitude	Intensity	Location	Source
18.08.1945	38.48	39.23		6.3	6	Elazığ Sivrice	ISS
25.04.1949	38.68	39.25		6.3	6	Elazığ Sivrice	ISS
07.07.1957	39	40.5		6.5		Bingöl	BCIS
14.01.1958	39.25	40.25				Bingöl Kiği	BCIS
09.12.1959	38.2	38.2				Diyarbakır Çüngüş	BCIS
19.06.1960	39	38.5				Malatya Arapkir	BCIS
02.12.1960	38.7	40.3	127			Malatya Akdağ	USCGS
01.06.1961	37.9	36.8	60			Kayseri Pınarbaşı	USCGS
25.08.1963	38.9	38.4	33	4.8		Malatya Taşdelen	USCGS
14.03.1964	38.7	39.6		4.2	7	Elazığ Hazar	İTO
14.06.1964	38.4	38.4	8-33	5.5	8	Malatya	USCGS İTO
14.06.1964	38.1	38.3	31	4.7		Adıyaman Çelikhan	USCGS
16.05.1965	38.1	39	26	4.9		Malatya Pütürge	ISC
17.10.1965	38	38.5	41	4.6		Adıyaman Sincik	ISC
20.10.1966	38.4	38.3				Malatya	BCIS
20.10.1966	38.4	38.6				Malatya Kale	BCIS
24.09.1968	39.1	40.1	14	5.1		Malatya Çan	BCIS
25.09.1968	39.2	40.2	47	5.1		Bingöl Kiği	BCIS
30.10.1968	38.0	38.6	3	4.9		Adıyaman Sincik	BCIS
01.10.1969	39.4	40.5	39	4.8		Erzincan Yedisu	BCIS

Date (Year)	Coordinates lat.	long.	Depth	Magnitude	Intensity	Location	Source
24.08.1970	38.3	39.3	33	4.4		Elazığ Sivrice	CGS
13.10.1970	38.2	37.0				Maraş Afşin	BCIS
24.05.1971	38.8	39.5	33	4.2		Elazığ Harput	ISC
24.05.1971	38.8	39.9	62±24			Elazığ Palu	ISC
29.05.1971	39.3	40	33	4.4		Tunceli Büyükyurt	ISC
02.03.1972	38.37	39.3	41 ±5			Elazığ Sivrice	ISC
22.04.1973	38.7	38.6				Elazığ Keban	ISC
18.05.1973	38.3	37.9	56 ±35			Malatya Akçadağ	ISC
10.09.1973	38.48	37.6	39	4.7		Malatya Balaban	ISC
18.01.1974	37.9	37.9	30	4.4		Malatya Akdağ	ISC
31.03.1974	37.7	37.8		3.4		Adıyaman Kahta	DGIS
23.06.1974	38.75	39.17	74	4.5		Elazığ	ISC
29.07.1974	39.05	40.65	17	4.2		Bingöl	ISC
21.08.1974	38.8	38.8	11	4.4		Elazığ Palu	ISC
23.09.1975	38.44	37.03				Maraş Afşin	ISC
28.11.1975	39.56	39.69	32	3.7		Erzincan Yedisu	ISC
22.08.1976	38.8	40.5				Bingöl Genç	STR
25.03.1977	38.45	40.01	21	5.2		Elazığ Palu	STR
04.05.1977	38.66	39.92	10	4.		Elazığ Palu	STR
21.09.1978	37.88	38.41		4.6		Adıyaman	STR
04.05.1977	38.66	39.92	10	4.		Elazığ Palu	STR
21.09.1978	37.88	38.41		4.6		Adıyaman Kuyucak	STR

REFERENCES

Alsan, E., Tezuçan, L., Bàth, M., 1975, An earthquake catalogue for Turkey for the interval 1913-1970. Kandilli Observatory. Report No. 7-75. 166 pages.

Arpat, E., 1977, Karakaya baraji çok büyük depremlerle sinana-caktir: Yeryuvari ve İnsan, Şubat sayisi, s. 59-62 (The Karakaya dam will experience extremely large earthquakes).

Ergin, K., Güçlü, U., Uz, Z., 1967, Türkiye civarinin deprem kataloğu (M.S. 11-1964) İ.T.Ü. Arz Fiziği Enstitüsü Yayinlari No. 24 (Earthquake catalogue of the vicinity of Turkey).

Ergin, K., Güçlü, U., and Aksay, G., 1971, Türkiye ve civarinin deprem kataloğu: İ.T.Ü. Maden Fak., Arz Fiziği Enstitüsü Yayinlari No. 28, 92 sayfa (Earthquake catalogue of the vicinity of Turkey).

Ergin, K., Güçlü, U., and Altinbaş, G., 1980, Türkiye ve dolaylari deprem kataloğu 51975-1980), İ.T.Ü. Yerfiziği Enstitüsü Yayinlari (Unpublished).

Memişoğlu, S., 1978, Karakaya barajinin deprem riski. Ist. Üniv. Fen Fak. Jeofizik Böl. Dipl. Tezi. (Earthquake risk of the Karakaya dam).

3 Pre-earthquake strain and triggering effects from continuous records

Possible Use of Tilt and Tidal Measurements for Earthquake Prediction

Ugur Yaramanci

Technical University of Istanbul, Department of Geophysics, Istanbul, Turkey

Summary

Tilt changes preceeding earthquakes and the prospects for a possible use of earth tide measurements for earthquake prediction are shortly reviewed.

Encouraging results obtained so far convince us that earth tide measurements should be included into an extensive earthquake prediction program in the North Anatolian fault zone.

Introduction

It is well-known that many physical parameters and phenomena such as seismic wave velocities, the distribution and characteristics of seismic events, electrical resistivity, the ground-water level in wells etc. may show anomalous changes prior to an earthquake. Earthquake prediction research is therefore basically concentrated upon the observation and the geophysical explanation of such changes.

Anomalous changes of crustal tilt preceeding earthquakes have been observed many times. Particularly Japanese researchers investigated these phenomena very early and increased efforts have been made recently in the United States to install tiltmeter arrays along the San Andreas fault. Increasing capability and lower costs of the instruments made it worthwhile to establish tilt networks in major earthquake areas.

Problems of tilt measurements

Though there are remarkable examples of anomalous changes of tilt prior to an earthquake (e.g. Johnston and Mortensen 1974, Press 1975), there are also many examples of a mistaken association of tilt changes to earthquakes. The measured non-tidal changes in tilt are many times due to other disturbancies than tectonic movements related to earthquakes. Particularly in the case of tiltmeters when installed near to the surface, meteorological effects cause tilts. Along with variations of temperature and air pressure, rainfall may cause tilts (e.g. Gerard 1978) and their elimination by analysis of the records can be quite complicated and unsatisfactory since the modelling of such effects is difficult. Long period tilts with large amplitudes induced by air pressure (Zschau 1977) and by sea loading in coastal areas are also reported (Yaramanci 1978). In order to minimize disturbing signals the installation techniques of tiltmeters should be improved and the location of the recording sites should be chosen carefully.

Also the choice of the tiltmeter type is important when elimination of unwanted signals is considered. Long base length tiltmeters measuring the integrated tilt over a distance of 50 meters and more are less sensitive to very local tilts than tiltmeters of short base length such as the vertical or horizontal pendulums (Peters 1978). Vertical pendulums, however, placed in deep boreholes are less influenced by meteorological effects (Zschau 1977, Flach et al. 1975), problems of the cavity effect experienced by measurements in mines and tunnels are also avoided.

Further problems due to the thermoelastic effect caused mainly by daily and seasonal variation of the temperature (Harrison and Herbst 1977), due to the strain-tilt coupling caused by the cavity effect (Harrison 1976, Berger and Beaumont 1976) and due to the site topography and local geological structure point out the complexity of extracting the tilt component related to the earthquakes. This complexity is increased by the erratic drift behaviour of some instruments. In extreme cases the real tilt signal may be lost totally.

Signals of various origin mentioned above are also present in measurements of tilt, strain and gravity made specially for earth tide research. These are traditionally regarded as disturbing effects. They are eliminated by use of digital filtering techniques leaving only the tidal frequency bands for the further analysis. The non-tidal part of the signal, however, should always be examined for effects related to earthquake and for other phenomena mentioned above.

Tidal triggering of earthquakes

It has often been suggested that earthquakes are influenced and triggered off by the periodic tidal stresses in the earth. Tidal stresses are quite small in comparison to the tectonic stresses. However, the stress rates of both may be comparable. Attempts to find a simple correlation of earthquake events with the tidal phases usually do not yield positive results. But detailed investigations show, when factors such as the fault plane and slip vector and such as the tectonic and tidal stress tensor are considered, that correlations can be found for certain types of earthquakes. In a study by Heaton (1975) using earthquakes with magnitude greater than 5 such correlations were found for shallow oblique-slip and dip-slip events. His theoretical considerations, however, indicate that small earthquakes which undergo a dilatancy diffusion process cannot be tidally triggered.

For practical purposes it should be mentioned that studies on tidal triggering would be more realistic if they were based on the measured tides since effects of ocean loading tides and of local structure influence both the amplitude and the phase of tides. This is particularly true for tilt and strain components.

Modification of Earth tides by dilatancy

Tidal admittance, that is the normalized response of the earth to the tidal potential input in various components such asgravity, tilt and strain is basically a function of the elastic parameters of the concerned region. Therefore a change in the elastic properties of a region, for example due to dilatancy before an earthquake, will influence the tidal admittance. Consequently it will influence the actually measured tides.

Figure 1. shows some results of extensive model calculations made by Beaumont and Berger (1974). Here two layers are assumed with a boundary at 10 km,and from 26 km downward Gutenberg-Bullen-A model is used. The dilatant region included extends over 40 km in North/South direction, and it is 20 km deep. The actual extension of the finite element model for the calculation was 2200 km in North/South direction and 200 km in depth. Calculated are the amplitudes of the major tidal wave M2 for radial displacement and the North/South components of tilt and strain. The amplitudes of the M2 wave for the same model but without the dilatant inclusion is given with the normalizing curves (left axis), and they are used to normalize the amplitudes of the M2 wave for the model with dilatant inclusion. Normalized functions represent the fractional change in M2 amplitude as a function of the position, when the seismic P-wave velocity in the dilatant region is reduced by 5, 10, 15, and 20 percent. Fractional change of the M2 in the radial component is too small to be detected. Changes in tilt and strain, however, are easily detectable. They are

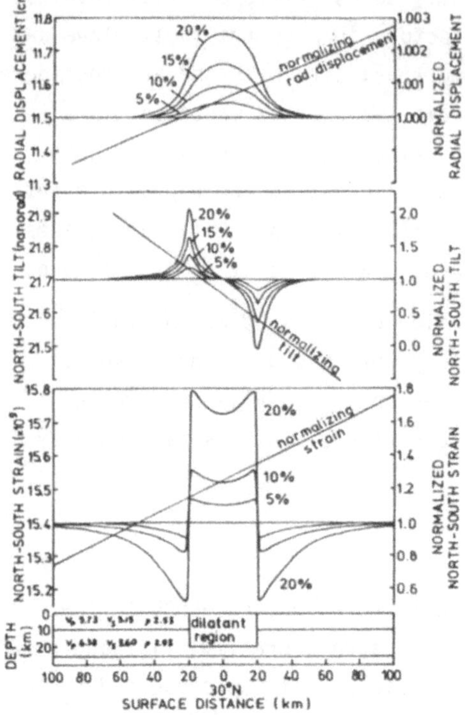

Fig. 1.
Modification of the amplitude of the M2 tidal wave in radial displacement, tilt and strain components for the model shown at the bottom of the figure. (From Beaumont and Berger, 1974)

basically edge effects and decrease towards the center of the dilatant region. In a modified form of the model in Figure 1. having an infinitely wide dilatant region from the edge with 15 percent decrease in P-wave velocity, the tilt anomaly is greater than 2 percent in a distance of 30 km from the edge. The tidal strain anomaly is about 2 percent even in a distance 120 km from the edge.

Model calculations shown so far are based on elastic dilatant crustal inclusion and represent a simplification since in laboratory crustal rocks stressed to more than 0.6 of their failure strength exhibit nonlinear elastic properties. But even for a possible nonlinear stress-strain relationship the tidal admittance should remain elastic since for the small range of the tidal stress the nonlinear stress-strain relation will be approximately linear. With changing tectonic stress ratė, however, tidal admittance will change depending on the slope of the nonlinear stress-strain relation (Beaumont 1974). Apart from the changes in the amplitude of the tidal admittance phase changes in the admittance are expected due to the existing anisotropy (Heaton 1975).

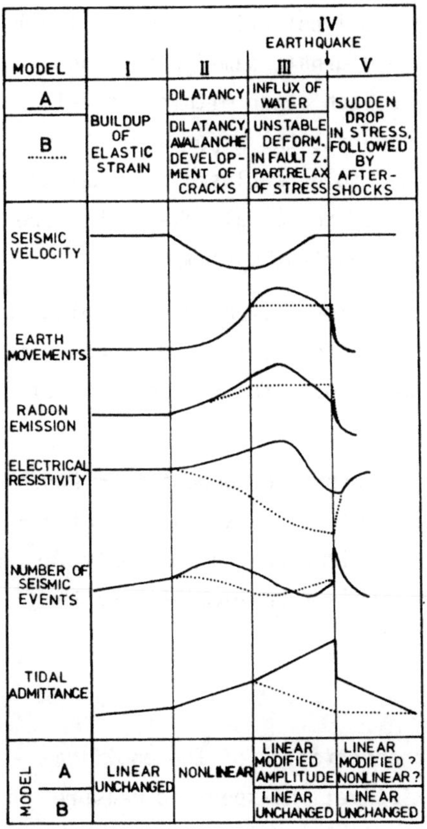

Fig. 2.

Variation of some parameters and of the tidal admittance during an earthquake cycle.
(From Press 1975, Beaumont 1979)

Model A : Dilatancy-Diffusion model

Model B : Dilatancy-Instability model

Nonlinear changes on the other hand for the tidal admittance are predicted in the case of crustal rocks exhibiting hysteresis when tidal and tectonic stress rates are approximately the same (Beaumont 1979).

Following these results possible variations in the tidal admittance for an earthquake cycle are summarized in Figure 2, along with the changes of other physical parameters. It should also be mentioned that modifications of the loading tide admittance of a dilatant region in coastal areas are expected (Tanaka 1976).

Tidal analysis for detecting tidal anomalies

Despite the long history of the tidal analysis, it is only in the last decade that methods of modern time series analysis found its way into tidal analysis which made possible to establish a modern framework for comparison and further development of so many seemingly different analysis techniques and for new fundamentals of the error analysis (Yaramanci 1978).

In the tidal analysis the main input is the tidal potential with the tidal frequencies. These tidal waves are grouped in diurnal, semidiurnal and higher order bands and they are very close to each other within a tidal band. Therefore long records are needed to separate the major waves. Short sequences of the record on the other hand have to be analysed in order to detect the changes in the amplitude and phase of the tidal waves. Thus one has to deal with the analysis problems of time variant systems.

Nonlinear tidal waves coincide sometimes with primary tidal waves adding to the frequency separation problem. Techniques for the analysis of ocean tides using the nonlinear waves in routine would be helpful for the analysis of earth tides as well (Baker 1978).

As common to the analysis of finite signals, longperiod components disturb tidal components due to the spectral leakage. Therefore careful use of digital filters is necessary. These and other problems are continuously investigated in order to optimize the tidal analysis techniques for earthquake prediction as well.

Conclusions

It should be emphasized that no significant observation of the modification of earth tides due to an earthquake exist yet. No specific measurements have been made for this purpose. Theoretical results, however, promise a valuable earthquake predictor which can be used with others. Particular advantages of this predictor would be that it is aseismic, therefore independent from background activity and allowing a continuous sampling of elastic parameters in a region. Modification of the tides due to the dilatancy are expected to be higher than the modification of the seismic velocities (as shown in Figure 1.), in fact to be higher than the modifications of many physical parameters of the earthquake region.

Considering the frame of this symposium it is suggested that any future program of tilt and strain measurements in the North Anatolian fault zone should not be limited to the long term changes but it should include earth tides as well.

References

BAKER, T.F.: A review of the objectives of tidal analysis. Bulletin d'Information, Marées Terrestres, Vol. 78, 1978

BEAUMONT, C., BERGER, J.: Earthquake prediction: Modification of earth tide tilt and strain by dilatancy. Geophys. J.R.Astron.Soc. 39, 111-121, 1974

BEAUMONT, C.: Linear and nonlinear interactions between the earth tide and a tectonically stressed earth. Proc. of 9. GEOP Conference, Dept. of Geodetic Science Rept. No. 280, The Ohio State Univ., 1979

BERGER, J., BEAUMONT, C.: An analysis of tidal strain observations from the United States of America: II. Laterally inhomogeneous tide, Bull. Seis. Soc. Am., 66, p. 1821, 1976

FLACH, D., JENTZSCH, G., ROSENBACH, O., YARAMANCI, U.: Untersuchungen von Gezeitenregistrierungen mit statistischen Verfahren, Deutsche Geod. Komm., Reihe B, 211, 1975

GERARD, V.B.: Earthquake precursors from earth tilt observations corrected for rainfall, Nature, Vol. 276, 1978

HARRISON, J.C.: Cavity and topographic effects in tilt and strain measurements, Geophys. J. R. Soc., 81, p. 319, 1976

HARRISON, J.C., HERBST, K.: Thermoelastic strain and tilts revisited, 8th Int. Symp. on Earth Tides, Bonn, Sept. 1977

HEATON, T.H.: Tidal triggering of earthquakes, Geophys. J.R. Soc., 43, 307-326, 1975

JONSTON, M.J.S., MORTENSEN, C.F.: Tilt precursors before earthquakes on the San Andreas Fault, California, Science, Vol. 186, 1031, 1974

PETERS, J.A.: Tidal measurements using a new long baselength tiltmeter. Ph.D. Thesis, University of Liverpool, 1978

PRESS, F.: Earthquake prediction. Scientific American, Vol. 232 No. 5, 14-23, 1975

TANAKA, T.: Effects of dilatancy on ocean load tides. Pure. Appl., Geophys. 114, 1976

YARAMANCI, U.: A unified approach to signal analysis in earth tides. Ph.D. Thesis, University of Liverpool, 1978

ZSCHAU, J.: Air pressure induced tilt in porous media. 8th Int. Symp. on Earth Tides, Bonn, Sept. 1977

Normal Earthquake Gravitational Precursors from Earth-Tide Data

E. Druta

Central Institute of Physics, Centre for Earth Physics and Seismology, Bucharest, P.O.Box 5206, Romania

ABSTRACT : Tidal gravity records from an Askania gravimeter (sensitivity about 2 μgal) are used to detect coseismic gravity drops. The non-tidal drift shows that normal earthquakes occurring in the vicinity of the earth-tide station (in an area R ∿ 100 Km) are accompanied by precursor anomalies. The duration (in days) and the amplitude (20 - 200 μgal) of the precursors depend on the magnitude and the epicentral distance to the station ; a best fitting magnitude-precursor duration relationship : $\log \tau = 0.6\ M - 1.75$ was established for 10 earthquakes ($2 \leq M \leq 4$) which occurred in Romania near Bucharest. It is worth - while to note that the earthquakes occur always after the extremum of the gravitational anomaly is attained and this particularity allows the prediction of occurrence time and magnitude of the incoming earthquake.

INTRODUCTION

Gravity measurements performed in epicentral areas show that earthquakes may be accompanied by gravity changes /1/, /2/. In order to emphasize such forerunning gravity anomalies, we used earth-tide records (vertical component) obtained at the Gravimetric Observatory Căldărușani /3/, with an Askania gravimeter operating at a 1.5 μgal sensitivity.The data was processed by standard tide-analysis programs / 4 /, which allows the removal of the tidal components so that the remaining long-period components (drift) contain gravity variations due to geophysical and instrumental causes. The drift analysis for a two years period (Jan. 1977 - Mar. 1979) led to the conclusion that normal earthquakes occurring within a ∿ 100 Km radius area are preceded by gravity changes. The earthquake data are taken from the Romanian Seismic Bulletins /7/.

1. MAGNITUDE-PRECURSOR DURATION RELATIONSHIP

Some gravitational precursor characteristics are derived from Table 1 and Fig. 1 :

- in all cases the gravitational anomaly is negative (gravity drop);
- the earthquakesoccur always after the minimum of the anomaly is attained;

TABLE 1

Seismic data and some precursory characteristics, M=earthquake magnitude; τ_o, τ, τ_d = precursor durations (Fig. 1) in days; Δg = gravity drop in μgal; Δ = epicentral distance to the earth-tide stations in Km; t_{o-c}= the difference between observed and computed time. For the two last cases, the reading of precursor durations was not possible due to an insufficient gravimeter pressure compensation.

No.	M	τ_o	τ	τ_d	Δg	Δ	t_{o-c}
1	3.0	2.5	1.0	3.0	60	60	2.4
2	3.9	20.0	4.5	21.5	250	20	-14.4
3	2.7	1.0	0.5	3.0	40	100	2.4
4	2.5	1.0	0.5	2.0	15	110	2.4
5	2.9	1.2	1.0	2.5	20	100	- 6.7
6	2.9	1.5	0.8	2.5	40	100	0.0
7	3.5	7.0	2.0	7.0	200	35	0.0
8	3.2	5.0	1.0	4.0	180	30	16.6
9	3.5	-	3.0	-	-	50	-
10	3.9	-	4.5	-	-	45	-

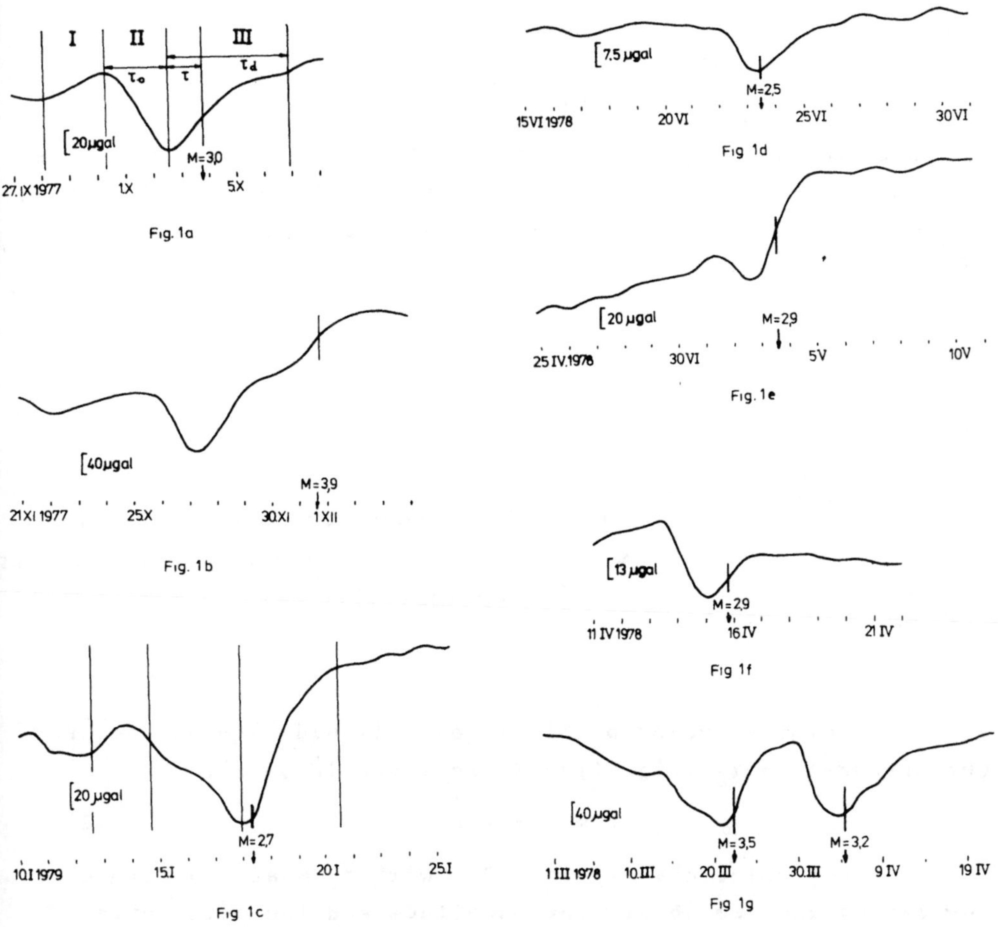

Fig. 1.a - 1.g Gravity precursors associated with event 1 - 8 from Table 1.

- the amplitude of the anomaly depends on the epicentral distance to the gravimetric station and on the earthquake's magnitude. The anomaly time-history may be divided in three states (identified in Fig. 1 by Roman numerals) with the durations τ_0, τ and τ_d respectively, which are related to the physical processes in the focal zone. We note the precursor duration τ (measured between the anomaly minimum and the event) lie on a line which satisfies the following equation (best - squares fit through 9 points), Fig. 2:

$$\log \tau = 0.6M-1.75 \qquad (1)$$

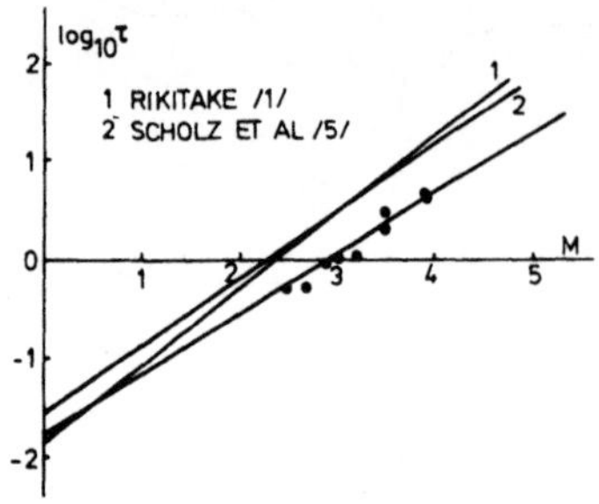

Fig. 2

Magnitude - precursor duration τ relationship: log τ = 0.6 M - 1.75 obtained for events 1 - 10 from Table 1.

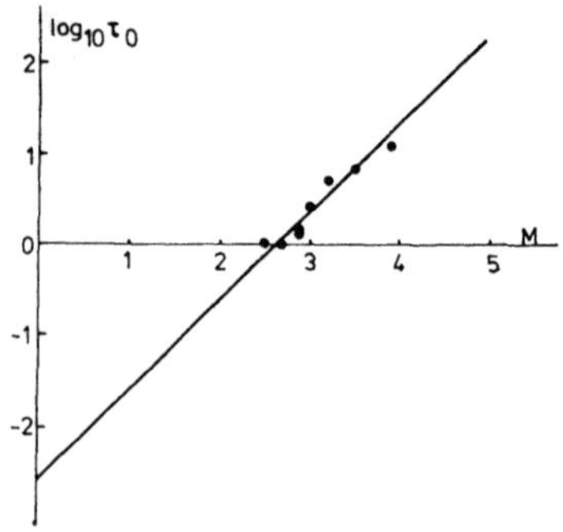

Fig. 3

Magnitude - precursor duration relationship: log τ_0= M - 2.6 obtained for events 1 - 8 from Table 1.

A similar relationship is established between the precursor duration τ_0 (Fig. 1) and magnitude (Fig. 3):

$$\log \tau_0 = M - 2.6 \qquad (2)$$

From relations (1) and (2), with τ_0 measured before the earthquake, we obtain the magnitude and the occurrence time of the seismic event:

$$M = \log \tau_0 + 2.6$$
$$\log\tau = 0.6 \log \tau_0 - 0.19 \qquad (3)$$

To illustrate the precision of the prediction we give in Table 1 the difference between the computed and observed occurrence time.

which is compatible with similar relations obtained by other authors /1/, for instance: log T=0.68M-1.57 due to Scholz et al./5/.

2. A PHYSICAL MODEL OF THE DILATANT ZONE

The resemblance of (1) to the statistical relation of Rikitake /1/ obtained for a wide variety of precursors, proves the existence of a unique explanation for all precursors based on the dilatancy wet model /5/. Stage I (Fig. 1) indicates a slight increase in gravity due to density increase as a consequence of simple accumulation of tectonic stress. As microcracks are produced for stage II, the gravity decreases because of the density decrease. At the same time, inflow of water into newly created pores from existing pores, would result in a decrease in pore pressure giving rise to the socalled dilatancy hardening which prevents further generation of microcracks.

As inflow of water keeps up and overcomes crack generation, in stage III the density tends to increase leading to a gravity increase.

The increase of pore pressure lowers the shear stress for rupture and therefore the earthquake occurs at moment τ after the beginning of water diffusion into the dilatant zone.

2.1 *The diffusion coefficient*

The diffusion of water into the dilatant zone is governed by the differential equation:

$$\frac{\partial p}{\partial t} = D\nabla^2 p \qquad (4)$$

where, p, t and D are the pore pressure, time, and diffusion coefficient, respectively. The diffusion coefficient is defined as the quotient of the squared length through the diffusion time /5/:

$$D = \frac{L_f^2}{\tau} = \frac{L_d^2}{\tau_d} \qquad (5)$$

where L_f, L_d are the typical length of the focal zone and dilatant zone, respectively; τ, τ_d are the needed time for diffusion of water into the whole focal zone and dilatant zone, respectively. The quantity L_f may be identified with the aftershock area related to the magnitude by the relationship due to Utsu and Seki /6/:

$$\log L_f^2 = M + 6 \qquad (6)$$

From (5) and (6) we obtain for the diffusion coefficient the relation:

$$\log D = M + 6 - \log \tau \qquad (7)$$

Using the τ and M values from Table 1 the mean diffusion coefficient is $D = 1.1 \cdot 10^4 cm^2/sec$. with the same order of magnitude as the value given in /5/.

2.2 *Ratio focal-dilatant zone*

From (5) we may derive a dependence between the typical length of the focal and the dilatant zone:

$$L_d = \sqrt{\frac{\tau_d}{\tau}} L_f \quad (8)$$

assuming that the water continues to diffuse into the remaining dilatant zone after the shock, for a duration τ_d. The data for τ and τ_d fit the following relationship (Fig. 4):

$$\tau = 0.20 \cdot \tau_d + 0.28 \quad (9)$$

From (8) and (9) we have:

$$L_d = \sqrt{5 - \frac{1.4}{\tau}} L_f \quad (10)$$

The graph of the function $F(\tau) = \sqrt{5 - \frac{1.4}{\tau}}$ (Fig.5) shows that for $\tau \geq 1$ ($M \geq 3$) the typical length of the dilatant zone is practically twice the focal zone length:

$$L_d \cong 2 L_f \quad (11)$$

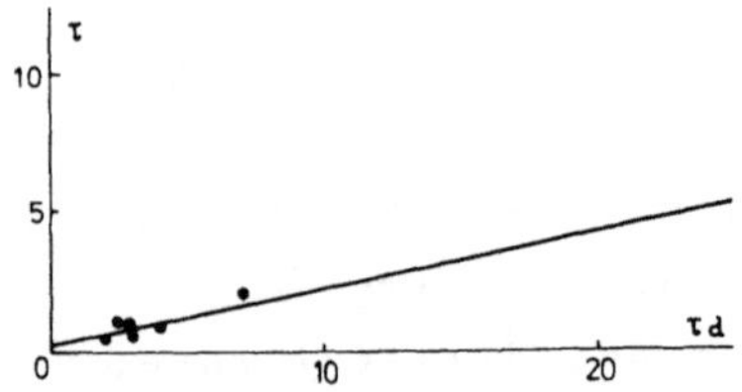

Fig. 4

The best - fitting straight line: $\tau = 0.2\ \tau_d + 0.28$ obtained for events 1 - 8 from Table 1.

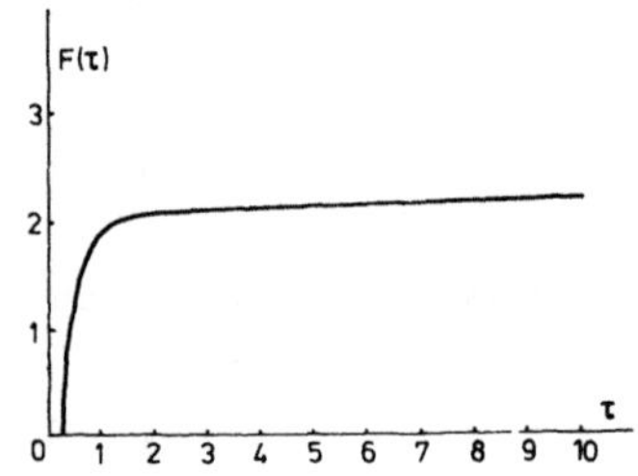

Fig. 5

Graph of the function $F(\tau) = \sqrt{5 - \frac{1.4}{\tau}}$

2.3 *Physical quantities of the dilatant zone*

The diffusion coefficient is determined by the porous medium and fluid characteristics:

$$D = \frac{k}{\eta \cdot P \cdot \beta} \tag{12}$$

where: η = water viscosity; P = porosity ; β = compresibility of water. Considering the model proposed in /2/ for a cylindrical dilatant zone with radius a and height H, the gravity change due to water inflow filling a fraction h/H of the initial volume, is:

$$\Delta g = 2 \cdot \pi \cdot G \cdot \rho (1 - F(H/a)) \cdot h \tag{13}$$

where: G = gravitational constant ($=6.67 \cdot 10^{-8}$ c.g.s.);ρ=water density ($=1.0$ g/cm^3); $F = ((1+(H/a)^2)^{1/2}-1) \cdot a/H$; h is related to the porosity P = h/H. For the particular case a = H, from (13) results: h(cm) = 4.Δg(μgal) and taking for H the typical length of the dilatant zone, we obtain the porosity:

$$P = \frac{4 \cdot \Delta g(\mu gal)}{L_d(cm)} \tag{14}$$

For M = 3.9 and Δg = 250 μgal one obtain from (14) $P = 6.0 \cdot 10^{-3}$ and for M = 7.3 and Δg= 350 μgal (the Haicheng earthquake /2/) results $P = 1.0 \cdot 10^{-5}$. The first value is compatible with the one obtained by Scholz et al. /5/ and the second value with the one given in /2/. The difference of two order of magnitude is probably due to the distance dependence of the gravity drop values used in (14) . For the following values: $\eta = 10^{-2}$ Poisse, $P = 6.0 \cdot 10^{-3} (10^{-5})$, $\beta = 3.2 \cdot 10^{-11}$ cm^2/sec we obtain the permeability ; the permeability is the same as that in /5/: $k = D \cdot \eta \cdot P \cdot \beta$ = 1.8 millidarcy (0.003).

2.4 *The amplitude-distance dependence*

From Table 1 we may conclude that the gravity drop Δg depends on the distance and this dependency may be written as (Fig. 6):

$$\log \Delta g = -1.571 \log \Delta + 4.9 \tag{15}$$

i.e. the gravity change depends on distance like $1/r^{1.57}$.

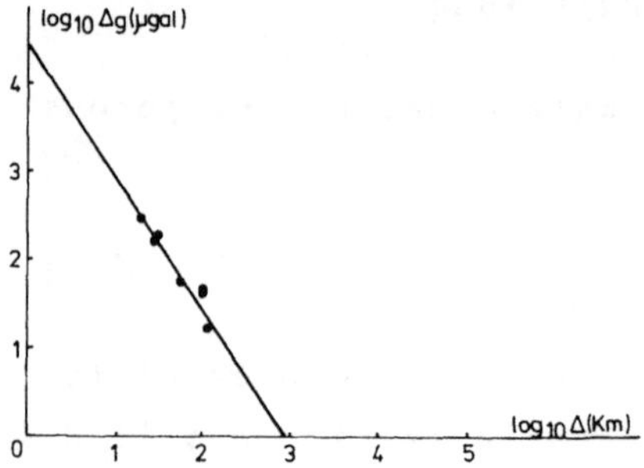

Fig. 6

Gravity drop - distance relationship $\log \Delta g = -1.57 \log\Delta + 4.9$ obtained for events 1 - 8 from Table 1.

The distance dependence of the gravity due to a dilatant zone located at a distance r from the observation point, should be $1/r^2$ (for $r >> L_d$). The discrepancy between the two formulas could be caused by the magnitude-dependence neglection in deducing (15).

This problem may be solved only by using at least two earth-tide stations located at different epicentral distances.

3. *CONCLUSIONS*

The processing of tidal gravity measurements revealed some short-period gravitational precursors, that allow the prediction of normal earthquakes occurring up to 100 Km around the earth-tide station. From the results which support the dilatancy wet model, we obtained some physical quantities of the dilatant zone at least concerning the order of magnitude.

As the earthquake's magnitude depends on the size of the diffusive zone, we can state that, for a constant diffusion coefficient, the magnitude is a function of the stress velocity. If this velocity is greater than the water diffusion velocity into the dilatant zone, the material strength is overcome only in a fraction of the dilatant zone volume giving rise to a small shock. The low seismicity areas are typical for local fast stresses, while the great earthquakes are generated by slow regional stresses.

Finally, we remark that for increasing magnitudes the precursor duration will increase also, therefore the gravitational precursors are of long-period type (i.e. for $M = 7.2$ the precursor duration τ (1) is about one year).

REFERENCES

/ 1 / Rikitake, T. 1976, "Earthquake prediction", Elsevier, Amsterdam.

/ 2 / Chen Yun-Tai, Gu Hao-Ding and Lu Zao-Xun, 1979, "Variations of gravity before and after the Haicheng earth - quake 1975 and the Tangshan earthquake, 1976," Phys. Earth. Planet. Inter. 18, 330 - 338.

/ 3 / Zugrăvescu, D. 1974. "Stations for Earthtide Recordings in the Socialist Republic of Romania", Geophysique, 18 pp, 55-58, Bucharest.

/ 4 / Chojnicki, T. 1975, "Ein Verfahren zur Erdgezeiten analyse in Anlehnung an das Prinzip der kleinsten Quadrate", Mitt. Inst. Theor. Geod. Univ. Bonn, 15.

/ 5 / Scholz, C.H., Sykes, L.R. and Aggarwal, Y.P., 1973, "Earthquake prediction : a physical basis", Science, 181 : 803 - 809.

/ 6 / Utsu, T. and Seki, A. 1954, "A relation between the Area of Aftershock region and the Energy of Mainshock" , Zishin (Earthquake), 7, 233.

/ 7 / Romanian Seismic Bulletins, 1977 - 1979, Seismological Lab. Center for Earthphysics and Seismology, Romania.

Creep Measurements in the Ismetpaşa Region of the North Anatolian Fault Zone

Alkut Aytun

The Scientific and Technical Research Council of Turkey (TÜBITAK), Ankara, Turkey

ABSTRACT

Cumulative creep effects in İsmetpaşa region are described with reference to the first awareness of, and the immediate preliminary investigations on, the phenomenon. After a brief review of the earthquake history of the region, it is concluded that the line, along which creep took place, is exactly the same fault trace that had been activated during the most recent major earthquakes.

The setting up of a simple surveying network by the author across the trace in 1969, and the way of evaluating the readings, are described.

Utilizing the measurement made on this network within the last 10 years, the trend of creep at the locality is presented instrumentally, without going into geological interpretations. Creep was of episodic character, with intermittent stagnant periods. Visual changes, meanwhile observed on natural and man made features in the locality, are given for documentation purposes.

Depending on the mentioned instrumental data, an average horizontal slip rate of 1.0 cm/year is obtained. The sense of movement is consistent with those observed in major earthquakes along the North Anatolian Fault Zone, i.e. right-handed. A throw is also accompanied, again consistent with those observed macroseismically, i.e. North side down.

The ratio of vertical and horizontal slips is again found similar, for both the creep and the major seismic events.

Finally, recommendations are made for future measurements of this nature, depending on faced problems.

FIRST AWARENESS OF THE PHENOMENON

Introduction

The first evidences of fault creep in İsmetpaşa region, have been observed during a field excursion in May 1969. The excursion was arranged following an international meeting, with the purpose of visiting the area of a recent (1951) fault activation and seeing the accumulated topographical modifications caused by past earthquakes.

The group started with Gerede, observing a well-known offset street in the shopping center of the town, then moved to the East. The fault trace, having preserved its features in certain segments, very easily lent itself for observation, although 18 years elapsed since the last activation. In some segments it went through cultivated land, so the trace was not followable. However, in general, one could mark the occasional clearer evidences of the fault, and could reconstruct the trace, by interconnecting those key points with imaginary lines.

In the process of following the trace, advancing from the point where the 1951 fault crossed the railroad track, towards Hamamlı village, in the vicinity of İsmetpaşa station (Fig. 1), a stone masonry wall, surrounding a one-story stone masonry building, was encountered. Clearly it did not exist at the time of the last activation. As a precise map of the 1951 fault trace was not available, the points where the trace passed through, in the locality, were not accurately known. However, considering that the wall stands across the line of marching, which is supposedly the recent trace, the shape and condition of the wall were carefully examined, with a suspect of deformation due to earthquakes.

The Wall

The surrounding wall had a rectangular layout of 79.50 m X 49.50 m in the plan (Fig. 2). The cross section of the wall was about 1.00 m high and 0.50 m wide, having concrete capping at the top.

Eastern side (i.e. B-C in Fig. 2) of the rectangle, which is perpendicular to the railroad track, was sheared by 18 centimeters right-laterally (Fig. 3). The Southern side (i.e. A-B in Fig. 2), which is almost parallel to the railroad track, was compressed, and at two distinct locations the masonry wall was crushed. The concrete capping was broken after buckling at those locations due to compressive action, and thrusted over the Western portion (Fig. 7).

This gave the impression that the trace intersects the rectangular surrounding wall-course at two points, i.e. points G' and H' in Fig. 2. No signs of ground deformation were observed along the zone interconnecting these two locations. The stone masonry building too, showed no signs of damage. Anyhow, it is a few meters off the zone.

On the Eastern extension of this zone, outside the wall, where it crosses a service road (Fig. 2), a local settlement was observed. Another similar settlement existed on the service road, at the point of intersection with the Western extension (Fig. 2 and 11). Going on West, a small pond was formed and a drastic change of color and density in the vegetation due to water, was clearly noticed. The existence of water and plants had attracted the ducks (F in Fig. 2).

The Railroad Tracks

At the time of excursion, a railroad upkeeping team was observed, busily working on a certain segment of the track. Although noticed at the beginning, it had been considered a common view of the stations, and attracted nobo -

Fig. 1 - View of the compressed railroad tracks and İsmetpaşa station. Looking towards East. Highway maintenance center is seen on the left. The building on extreme right is the repair shop of railways.

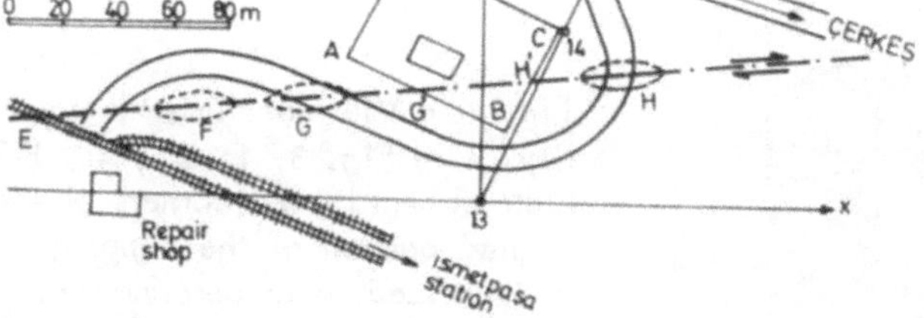

Fig. 2 - Unscaled sketch of the masonry wall and surrounding, where E: intersection of fault trace with railroad tracks, F: ponding, G: settlement on the service road, at intersection with fault trace, H: settlement on service road, at intersection with fault trace.

Fig. 3 - View of the sheared masonry wall B-C, from point B, looking towards N 15° E. Note the two transversal lines on concrete capping. The nearer (Southern) one is a construction joint, and the farther (Northern) one is a crack, due to shearing. The portion in between was rotated clockwise, due to shear deformation of 18 cm (October 14, 1969).

Fig. 4 - View of the same locality as in Fig. 3, in July 11, 1978. Note the crumbled corners of the rotated portion of the capping, due to increased shear deformation. Note the steel posts of rectangular section and the galvanized wire mesh. Also note the crushed stones outside the wall B-C and the water drained out from the disintegrated part of the wall.

Fig. 5 - View of the sheared wall B-C, along the fault trace. In the background, the buckled capping is seen. The cross-sign denotes the intersection of the service road with the railroad tracks, approximately the same location where the trace crosses the tracks (October 14, 1969).

Fig. 6 - Same locality as in Fig. 5, on wall B-C, in July 11, 1978. Note the wire mesh fence (as in Fig. 2), and the concrete U-channel for asphalt piping (upper-left of photo). Compare the corresponding stones in masonry, with Fig. 5.

Fig. 7 - View of wall A-B, looking towards South . Note the overthrusting of the Eastern side of the concrete capping, due to compression effects. The overthrusting segment has a hinge on the East, actually a bending crack, forming an inclined beam, cantilevering at Western end (March 20, 1972).

Fig. 8 - View of the wall A-B, as in Fig 7, in July 11, 1978. Overthrusting progressed due to increased compression, the cantilevering end was broken and the rest fell on the masonry again. The broken part still rests on the capping below. Note the change in the fence. Compare the corresponding stones in masonry, with Fig. 7 .

dy`s attention up to that time. However, the segment was just on the intersection with the Western extension of the supposed zone (Fig. 2), and this fact instantane - ously brought to mind the probability that the rails might have been deformed in the same manner as the wall. This prediction was strenghtened after seeing the irregularities due to compression, accompanying a just noticeable shear component, in the right-lateral sense (Fig. 9). The inner sides of the railtops had signs of impact, due to these irregularities (Fig. 10). A positive answer was received to the question as to whether maintenance work was frequently needed in this part. The Station Master spoke about the complaints that arose from lateral disturbances. But his opinion was that it was a usual scene and should be considered normal, near the stations, due to acceleration and deceleration of trains. However it was a small station, where the express trains don`t usually make stops, but continue their way. The stations of comparable conditions, along the railway, did not need such fre - quent maintenance.

The effects on the rails were similar, in a smaller scale, to those of the wall A-B (Fig. 2). The difference can easily be explained from the fact that the fragile wall B-C failed in the shear mode, at the point where the shear deformation was imposed, but the wall A-B, and the rails near the point E , exhibited irregular compression effects, not necesarrily at the point of application, as compression is readily transmitted along the rails and the wall. However, both the wall and the rails had a minor shear component, due to small angle of intersection, and it might have eased the compression effects to initiate at the intersection with the zone. One could also conclude, from the way of thrusting of the capping concrete on the wall A-B, that the Western portion of the wall (i.e. the Northern block of the fault) went down.

PRELIMINARY INVESTIGATIONS

It was clear that all these deformations (ponding, settlement, shear, compression etc.) were caused by the same movement, however, a question arose as to whether this slip was a result of a series of earthquakes that occured after construction of the wall, or was due to another reason, e.g. landslide etc. The land was flat, ruling out any possibility of landsliding. This led the author to in - vestigate the record of the construction. In fact it had been built and functioned as a highway maintenance station, to accomodate a number of maintenance personnel and material storage, for regional service. The station used to be inactive in winters, the personnel being transferred to a more central station. Depending on the information, provided by the builders (4) and by the General Directorate of Highways, the construction of the surrounding wall was completed just before the spring of 1957, and shortly after the team moved to Bolu for another job. Then, on May 26, 1957 , Abant-Bolu earthquake occured. But the fault activation did not reach this locality, and no observable harm was done to the wall. Mudurnu earthquake of July 22, 1967 was felt in the area, but caused no damage. It was understood that no fault activation nor a destructive earthquake motion was experienced in the locality since then (to the present, i.e. 1980).

Fig. 9 - View of railroad track, looking towards West. Note compression effects and buckling, near the intersection with the fault trace.

Fig. 10 - Railway officials pointing to the signs of impact, on the inner sides of railtops, due to compression. The buckling caused discomfort during the passage of express trains, that normally do not stop at this station.

Fig. II - Settlement caused by vehicles passing over the service road between the wall A-B and the railroad tracks. Looking towards West. Note the water and vegetation, possibly due to weakening of soil. Crushed stones indicate that filling is required occasionally.

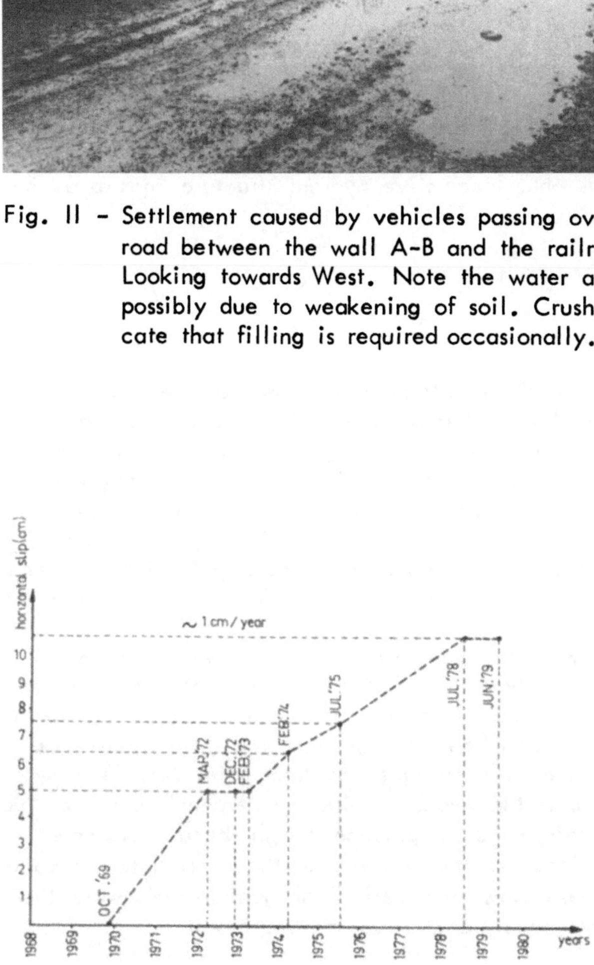

Fig. 12 - Time-history of creep, as revealed by conventional surveying.

RECENT PAST EARTHQUAKES IN THE REGION AND THEIR EFFECTS

The recent earthquakes, which were accompanied by known fault activation in the locality, are as follows :

1045 Hamamlı earthquake
February 1, 1944 earthquake
August 13, 1951 earthquake

It is known that 1944 earthquake was associated with a strike-slip of 1.50 m (1, 2), and 1951 earthquake again caused shear deformation forcing the railroad tracks to assume a Z-shape (3). As the fault traces were not precisely mapped on a medium or small scale (e.g. 1/25,000) basis, it is not possible to tell clearly today, where they exactly passed. However, it is believed, for at least two times (the two most recent earthquakes above), the trace crossed the railway at E region (Fig. 2). Although several earthquakes were felt in the locality, the most recent one, which was accompanied by a fault activation, occured in 1951. It is still possible to see the ruins of a locomotive hydrant structure, destroyed by that earthquake.

THE DEFORMATION RATE

The Horizontal Slip

It is understood that the shear deformation of the wall B-C, has accumulated within the period between 1957 and 1969, the latter denoting the observation for the first time. However, it is not known during which portions of this interval of 12 years, actual movement took place. Considering the observations in California (5,8 and 9) it seems reasonable that this shear deformation was achieved within a considerably shorter cumulative duration. In case of the assumption that the shear deformation took place gradually and uniformly (clearly leading to a smaller rate) throughout the time segment, an average rate of 1.5 cm/year is obtained.

A simple surveying network, consisting of three reference points (brass markers on concrete blocks) has been established by the author. As seen on Fig. 2, these points were so selected that they were approximately collinear at the beginning. Two of them were on the Northern block, while the third was on the South, with respect to the fault trace. The straight line from point No. 13, towards the corner of repair shop, constitutes the x-axis, while the perpendicular from No. 13 to this line constitutes the y-axis, so as to produce a right-hand coordinate system. A procedure is adopted to observe the variation of the coordinates of points No. 14 and 15, with respect to these axes, and periodical measurements were then taken by the author, using quality conventional surveying equipment.

The results are summarized in Fig. 12. The average horizontal slip, in the period between the years 1969 and 1978 is approximately 1.0 cm/year. If the period between the years 1957 and 1978 is considered, this rate is raised to 1.3 cm/year. The horizontal slip is in the same sense, as observed in major earthquakes.

The Vertical Slip

Depending on levelling measurements conducted in October 1969 and March 1972, North side went down by 13 mm . This gives a vertical slip rate of 5 mm/year, approximately. The direction of dip is again in consistence with those observed in major earthquakes.

VISUAL CHANGES MEANWHILE OBSERVED ON NATURAL AND MAN MADE FEATURES, IN THE LOCALITY

During the period of ten years (1969-1979), various visual changes have been observed. Some of them were directly or indirectly related to the ongoing creep phenomena, while the others were merely modifications or improvements, without any tectonic implications. Major changes from both categories are summarized below, where possible with chronological information, for documentation purposes.

Man Made Features Above The Level Of Concrete Capping Of The Masonry Wall

Month and Year of Observation	Condition, Modification or Improvements
May 1969	Nothing existed on the concrete capping, except hollows left in the concrete, with certain spacing, to fix the posts for fencing (Fig. 3 and 5). But the posts were not yet fixed.
May 1972	The posts were fixed and a sort of fence was formed, held in place by the former. The fence consisted of of vertical timber laths, laterally connected by iron wires at five levels (Fig. 7).
February 1974	The timber posts still stood in place, but the fence on the segment B-C completely separated from them, fell outside, and leaned to the external face of the wall. The fence on wall A-B fell out too, at the mid portion, while the ends were intact.
July 1978	Steel posts were fixed, cemented, and a galvanized wire mesh of diagonal configuration (Fig. 4 and 6) spanned the spaces between. On the segment A-B, near the corner B, a gate was constructed, consisting of steel sections and wire mesh. A line of barbed wire at the top, completed fencing.

The Locality Of Shear Failure In The Wall

The sheared portion of the wall has not been repaired as a matter of good luck, and thanks to this, one could follow the progression of the shear deformation (Figs. 5 and 6). Two lines on the concrete capping are seen. One on the left (i.e. Southern) is an expansion joint, while the other is a crack, formed by shearing. Due to the rotation of the portion between, the pertinent corners subjected to compression, and crumbled (Fig. 6). The wall is more distinctly separated and gave way the water to drain out.

The Locality Of Compressive Failure In The Wall

In 1969, the Eastern part of the capping on A-B, overthrusted the Western part due to compression. The overthrusting portion had a bending crack on the other side (East), thus assuming an inclined simple beam configuration (Fig. 7). Later on, the length of the overthrusting portion increased, thus increasing the slope and decreasing the distance between supports. Present condition is that, the cantilevered part is broken and rests on the capping below, while the rest fitted in the space left (Fig. 8). It is not known whether this occurence is natural or man made, although one may think that the broken cantilever should not still rest on the capping, if the latter was true.

Field Boundaries Crossing The Line From Ismetpaşa Station To Hamamlı Village

The field boundaries, generally in ridge form, were observed to assume S-shape, conforming the right-hand sense of the movement. The deformation should have accumulated in a longer time period than that of the masonry wall, surrounding the maintenance station, considering the larger amount of shear and the fact that they should be straight originally. It is very probable that they also included slips of recent earthquakes and subjected to readjustment, after events.

CONCLUSION AND RECOMMENDATIONS

Depending on instrumental data, an average horizontal slip rate of 1.0 cm/year is obtained, from an observing interval of 10 years window. The sense of the movement is consistent with those observed in major earthquakes along the Noth Anatolian Fault Zone, i.e. right-handed. The accompanied throw is also consistent with those observed macroseismically, i.e. North side down. The ratio of vertical and horizontal slips are again comparable for both the creep and the major seismic events.

It is not yet known whether the creep phenomenon reduces the magnitude of a potential shock, in the locality, by releasing the energy through gradual deformation, or is indicative of the forthcoming of an intensive one. Creep in one locality will probably release local stresses (if it takes place in the same way as known in material behavior), but the extent of reduction is not known and new areas of stress accumulation might be created at the ends of the locality. Presently, the distribution and dimension of creep locations along the North Anatolian Fault Zone, are not known.

The segments of the fault, crossing through such rocks as serpentine, which can be deformed even under constant stresses, constitute the most susceptible locations to creep. The author is in the opinion that various segments on the North Anatolian Fault Zone exist, which would lend themselves as suitable observation sites for creep phenomena.

Depending on the faced problems, the following are recommended:

1. Small networks are much more useful for preliminary creep and deformation observations, because the accuracy is adversely affected as the network gets larger.

2. Individual triangles or quadrilaterals, with a certain spacing, should be preferred. The relative movement is directly obtained, with a minimum of balancing effort. That means minimum influence from the surrounding. Comparison of the results of two or three individual (but not neighboring) networks, leads to more reliable interpretations, instead of depending on a single but combined result from a larger network.

3. Although the connection of several small networks to a larger network is desirable, to have a broader idea about the movement, it should be known that the relative deformations thus obtained along the fault line, will highly be affected by other measurements (and introduced errors) at a distance. Many points will be dragged in the process of balancing. In this case, more reliance should be placed on the results, obtained from small networks.

4. Utilization of a network of reference points and surveying equipment is uncomplicated and cheap, in revealing localities where creep takes place. However a detailed time-history can only be obtained by continuously recording instruments. Once creep phenomena is identified, then the recording instruments can be installed. It is recommended that these recording instruments be installed at certain distances along the trace (e.g. a few hundred meters or kilometers) to allow search on advancing of the movement. The time signals should be synchronized.

ACKNOWLEDGEMENT

Mr. Esen Arpat accompanied several of the measurements and related field excursions. The Ministry of Reconstruction and Resettlement and the General Directorate of Mineral Research and Exploration Institute are acknowledged for providing the author quality survey instruments.

REFERENCES

1. Pınar, N. and E. Lahn, "Catalog of Earthquakes in Turkey, With Explanations" (in Turkish), Bayındırlık Bakanlığı, Yapı ve İmar İşleri Reisliği, Seri: 6, Sayı: 36, Ankara, 1952.

2. Ergin, K., Z. Uz and U. Güçlü, "Earthquake Catalog of Turkey and Surrounding Areas" (in Turkish and English), Institute of Earth Physics, Technical University of İstanbul, 1967.

3. Pınar, N., "Etude Géologique et Macrosismique du Tremblement de Terre de Kurşunlu (Anatolie Septentrionale) du 13 Août 1951" (in French), Revue de la Faculté des Sciences de l'Université d'Istanbul, Série A, Tome XVIII, Fasc. 2, 1953.

4. Private communication with Mr. Süleyman Jaji, Executive Engineer of Construction, The General Directorate of Highways, Ankara, Turkey, 1969.

5. Nason, R.D., "Measurements and Theory of Fault Creep Slippage in Central California", Proceedings of Symposium on Recent Crustal Movements, Bulletin of Royal Society of New Zealand, pp. 181-187, 1971.

6. Aytun, A., "Creep Measurements in the Vicinity of İsmetpaşa Railway Station" (in Turkish), Proceedings of Symposium on the North Anatolian Fault and Earthquake Belt, Ankara, March 29-31, 1972 ; published by the Mineral Research and Exploration Institute, Ankara, 1973.

7. Lensen, G.J., "Phases, Nature and Rates of Earth Deformation", Proceedings of Symposium on Recent Crustal Movements, Bulletin of Royal Society of New Zealand, pp. 97-105, 1971.

8. Bonilla, M.G., "Deformation of Railroad Tracks by Slippage on the Hayward Fault in the Niles District of Fermont, California", Bull. Seismological Soc. of America, Vol. 56, No. 2, pp. 281-289, April 1966.

9. Brown, R.D. and R.E. Wallace, "Current and Historic Fault Movement Along the San Andreas Fault, Between Paicines and Camp Dix, California", Proceedings, Conf. on Problems of San Andreas Fault System, Stanford University publication, Geological Sciences, Vol. XI, 1968.

4 Pre-earthquake strain and deformation from repeated high-precision geodetic surveys

Possible Use of the Turkish National Triangulation Network for the Study of Crustal Movements in the North Anatolian Fault Zone

Ahmet Aksoy

Istanbul Teknik Universitesi, Insaat Fakültesi, Jeodezi ve Fotogrametri Bölümü, Istanbul, Turkey

Abstract

A short review is given on previous geodetic work in Turkey as related to crustal deformation studies. Plans for future measurements which may be of importance for earthquake prediction research in the North Anatolian Fault Zone are presented.

Introduction

If we know the movement of the crust according to its directions and velocity, then by separating of the movement vectors into their systematic and unsystematic parts, indications of the latter give references to tectonic strain and mass displacement of the crust. They serve the realizing of strain accumulation, therefore the earthquake prediction.

The goal of the geodetic part in the investigation of crustal motions lies in the significant determination of the movement vectors. They will help to clear up major tectonic questions posed by the neighbouring geo-sciences. The points to be cleared are the questions of tectonic plate movements, continental drift and changes of the position of the earth core and the resulting variations of the gravity field of the earth.

The non-periodic crustal movements are generally very slow and can be determined by repeated measurements. They attain magnitudes of some centimeters per year or less (Peschel, 1967). This demands a high resolution of the geodetic measuring technique for the determination of the movement vectors. In the classical solution of the problem, the movement vectors are separated into their horizontal and vertical components. Time is considered as an additional component, as the movement procedure is time-dependant.

Possible use of the Turkish Network for crustal deformation studies

More exact determination of crustal movements by terrestrial measurements are applied in limited areas with special geodetic nets. To cover areas of large extent such as the North Anatolian Fault Zone (about 1000 km) by classical means, if possible at all, is very difficult and expensive. In order to determine all the current motion vectors in such large areas, combined terrestrial and space techniques should be applied, in future. At the same time the national nets could be engaged and renewed. The earlier data would give very useful values on the crustal movement components over long time intervals.

Therefore, the Turkish National Network, specially the first order net, offers a good chance to identify the motions in the North Anatolian Fault Zone over a long time interval (Fig. 1).

The first order triangulation works were essentially started in 1940 and completed in 1953. The net is built up from chains in North-South and East-West directions with an average length of 180 km. It consists of well-shaped triangles, with the exception of 6 chains which are made up of complete quadrilaterals. On both ends of each chain and additionally at places where the length of the chains exceeds 200 km, a baseline with an accuracy of less than 1/1.800 000 has been measured. The sides of triangles are usually between 25 and 35 km, and there are no angles less than 40 degrees.

The angular observations were generally carried out using Wild T4 and partly the Max Hillebrand theodolites. Until 1947, Schreiber's combination method was applied, after which 24 direction sets were measured at every station. Laplace stations were established on one station of the junction figures and at one of the stations 70-100 km, along the chains, with the average accuracy of

$$m_\varphi = \pm 0''.191, \qquad m_\lambda = \pm 0''.240, \qquad m_\alpha = \pm 0''.350$$

The first order triangulation network of Turkey consists of 27 chains, 786 station points, 1769 sides, 3538 directions, 40 baselines, 98 Laplace stations and 1027 triangles. The only fundamental point is Meşedağ (near Ankara) with its astronomical coordinates (Ulsoy, 1979)

$$\varphi = 39^\circ\ 52'\ 10''.451 \pm 0''.139$$
$$\lambda = 32^\circ\ 54'\ 38''.430 \pm 0''.210$$

(General report of geodetic activities in Turkey, IGGC, Rome, 1954).

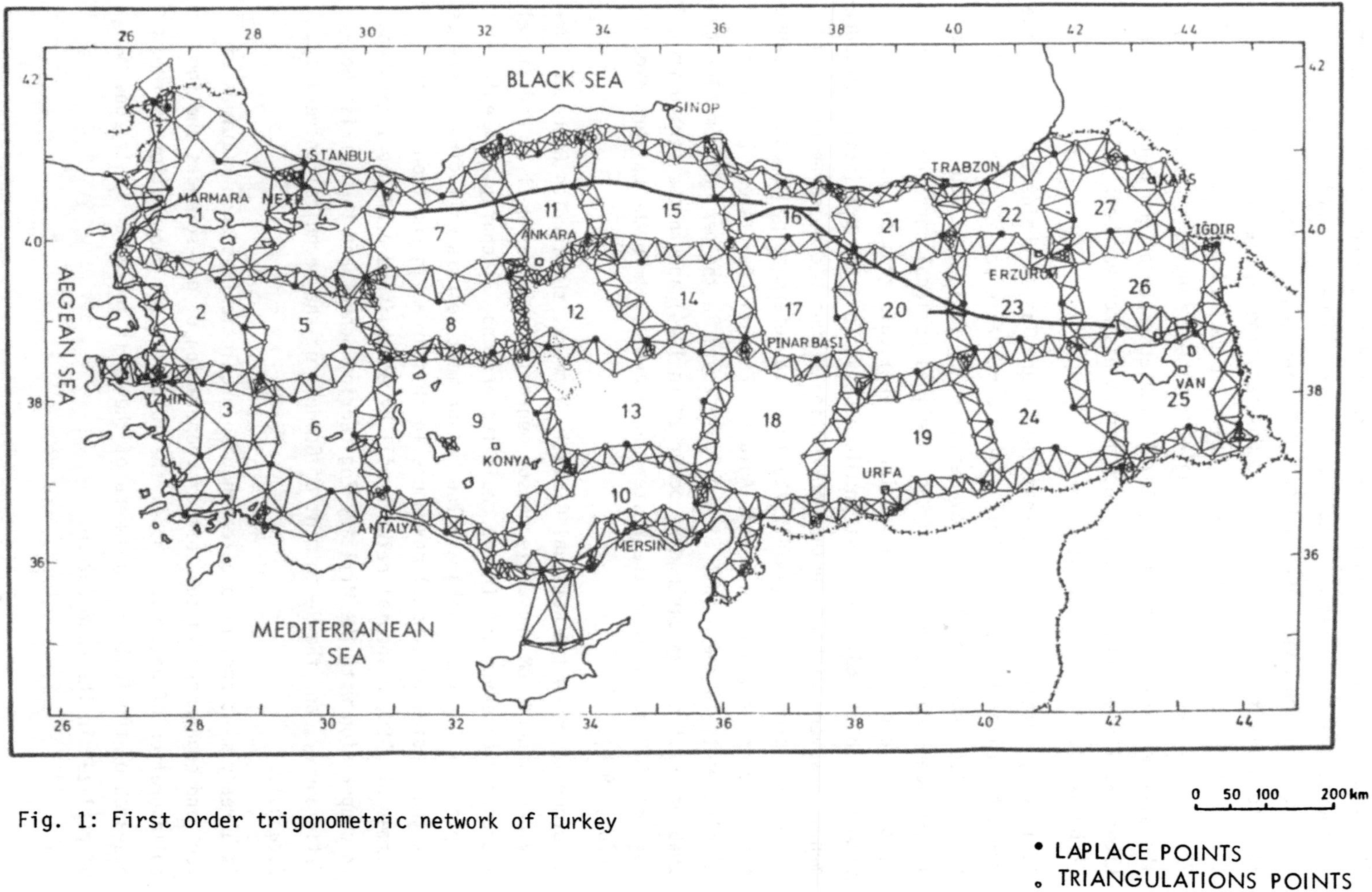

Fig. 1: First order trigonometric network of Turkey

The adjustment was performed entirely in 1954 by the method of variation of coordinates. The astronomical coordinates in Laplace points were only used for azimuth conditions and the direction angles were not corrected for the deflection of the vertical. However, the square mean error of a direction after adjustment is ± 0".681, and the maximum residual is 2".885. The residual sum of the angles in the triangles averages 0".907, the maximum being 4".046 (Uğur, 1976). The accuracy of coordinates is not known, as the variance - covariance matrix has not yet been calculated (Erbudak, 1966, Özgen, 1970).

It is very interesting to see that the triangulation chains cut across the North Anatolian Fault Zone. This offers a suitable possibility to study by repeated measurements the movements in the faults. It would be of use to consider the chains between the baselines as local nets and to remove systematic errors such as deflection of verticals from all measurements, with the aid of additional astronomic observations and gravimetric measurements. However, without further correction to the old measurements from systematic errors, Uğur (1974-1976) tried to determine by repeated measurements the crustal motions in the Ankara - Çaycuma chain area. He found displacements in the points of the Fault Zones in the period from 1942 to 1946 of 2-2.5 m (an earthquake occurred in 1944) and in the period from 1946 to 1972 of 20 cm in the Western area and 75 cm in the Eastern area (Uğur, 1974).

In order to renew the National Geodetic Network, the Turkish Geodetic Survey has planned EDM and angle measurements for 1980, as well as additional astronomic observations in the chain traverse no. 7 (chain: Eskisehir - Adapazari - Caycuma - Ankara). The measurement results will give references for the studies on crustal movements in this area. In addition, in the summer of 1980, a special network will be established near the Iznik Sea, in cooperation with the Military Engineering School,and the Turkish Geodetic Survey, and the Department of Geodesy of the Technical University, Istanbul.

As already mentioned before, earthquake prediction research requires effort and engagement in the form of cooperation of geo-sciences and the collaboration of nations. I hope, that a successful cooperation in the research of the tectonic phenomena of the North Anatolian Fault Zone may be realized in the near future.

References

Erbudak, M., 1966: Die geodätischen Arbeiten in der Türkei, Vortrag geh. im Institut für theoretische Geodäsie der Universität Bonn

Özgen, M.G., 1970: Geodätische Arbeiten in der Türkei, Inter. conference on methods and instruments for densification of geodetic network, Budapest

Peschel, H., 1967: Der geodätische Beitrag zur Erforschung rezenter Erdkrustenbewegungen. Festschrift zum 70. Geburtstag von Prof. Walter Grossmann, 1967

Uğur, E., 1974: Kuzey Anadolu Fay Kusağinin Gerede-Çerkeş-Bölümünde Yerkabuğu hareketlerinin Jeofizik Yöntemlerle Incelenmesi, Ankara, Harita Gn.Md., 1974

Uğur, E., 1976: Türkiye 1.inci derece triangulasyon ağinin dengelenmesi, Harita Dergisi, sayi 83, S. 30, Harita Gn.Md. Yayinlari, Ankara

Ulsoy, E., 1979: Ülke başlangiç noktasinin değerleri, Elektronik uzaklik ölcme semineri. IDMMA, Istanbul, Haziran 1979

Θ^2 Testing for Survey of Recent Crustal Deformation in Akyazi/Adapazari

N. K. Demirel, Z. Karahan, F. Pirselimoglu and K. Uysal

The Mineral Research and Exploration Institute of Turkey, Ankara, Turkey

E. Öztürk and E. Uğur

Keradeniz Teknik Universitesi, Trabzon, Turkey

Abstract:

Along a section of the North Anatolian Fault Zone near Akyazi/ Adapazari a triangulation and trilateration network was established in order to determine the principals of horizontal displacements. By comparison of two measurements taken at different times applying the Θ^2 testing criteria the horizontal displacement could be determined.

The various studies carried out in the Northern Anatolian Fault Zone showed that the largest value of displacement determined up to now was as much as 85 to 90 km. In connection with recent crustal deformation, the displacements were predicted by different research scientists to be between 0.5 cm and 2.0 cm per year, but in fact geodetic instrumental data could not be determined up to 1970. AYTUN (1972) noticed a creep movement in the area of ISMETPASA (Cerkeş) and UĞUR (1974) also determined a displacement reaching a value of 1 m per 30 years in the area between Gerede and Cerkes by comparing triangulation and trilateration measurements taken over a period of 30 years.

AKYAZI (Dokurcun Valley) is about 150 km away to the west of this district. In the east-west direction in the Dokurcun Valley the fault line can be easily seen on the surface.

The network was formed by 11 pillars in order to determine the principals of the horizontal displacements. These pillars were located on the top of the hills surrounding the north and the south sides of the valley, on the both sides of the

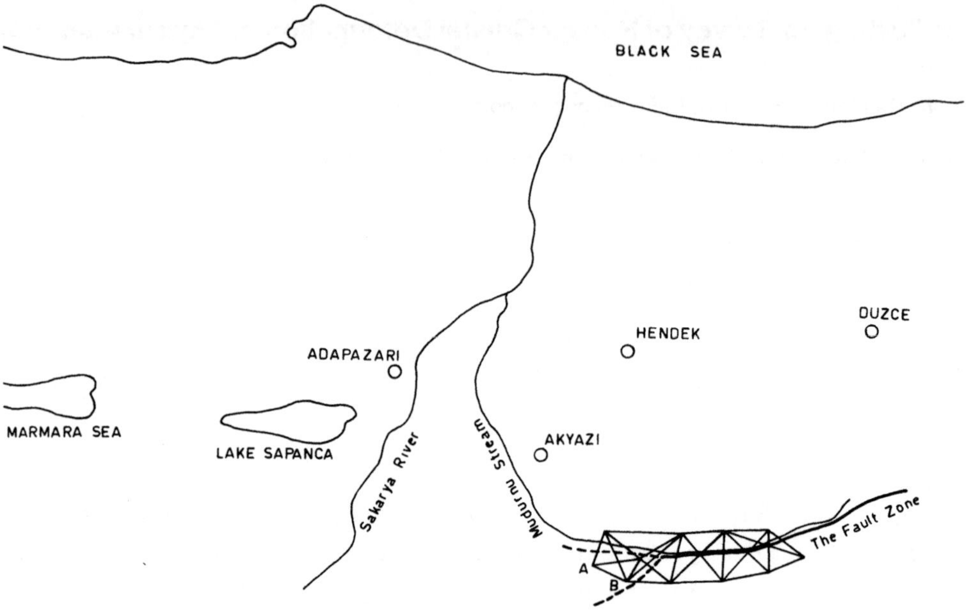

Fig. 1 Location of the triangulation and trilateration network

fault line, and especially outside the landslide area, if possible. This is a triangulation and trilateration network formed by a chain of quadrilaterals. Its stations are 5 to 10 km in distance from each other, Fig. 1.

The measurements in the first period were taken in 1976 with a Wild-T3 theodolite only. Measurements with the Rangemaster II (0.5 mm + 1.10^{-6}.D) were also included when taking the angular measurements in the second period in 1979.

A computer programme has also been developed at the KTÜ Dr.N. BULUT Computer Center by using the IBM 370-125/2 system to determine the horizontal crustal movements by making use of the geodetical network (Niemeier 1976). This programme is coded in Fortran IV and used to compare the measurements of two stations each obtained as a result of new measurements taken within certain time intervals to determine whether they are equivalent by applying the Θ^2 criteria (Pelzer 1971).

The initial measurements taken at the t_1 time (directions + distance or directions only) are reduced onto the ellipsoid.

The approximate coordinates of the stations of the net are taken from a 1/25000 scale map with a tolerance of 25 m. The coordinates of all the stations are assumed to be unknown and then the adjustment of the net is carried out as a free net, using the method of variation of coordinates. The cofactor matrix ($\underline{Q}_{x1}\underline{Q}_{x1}$) and the solution vector obtained as the results of this adjustment are stored in the disc files.

These operations are then repeated using the measurements taken at the time t_2 and the results are stored in the disc files.

The vector of the coordinate differences $\underline{d}$, can be written as: $\underline{d} = \underline{x}_2 - \underline{x}_1$, where $\underline{x}1$ and $\underline{x}_2$ are the solution vectors obtained as a result of the adjustment of the measurements taken at times t_1 and t_2 respectively.

$\underline{Q}_{dd} = \underline{Q}_{x_1x_1} + \underline{Q}_{x_2x_2}$ where $\underline{Q}_{dd}$ is the cofactor matrix.

m_1 and m_2 calculated from the measurements taken at times t1 and t_2 respectively are obtained as:

$$m_1 = \pm \sqrt{\frac{\underline{V}_1^T \ \underline{P}_1 \ \underline{V}_1}{f_1}}$$

$$m_2 = \pm \sqrt{\frac{\underline{V}_2^T \ \underline{P}_2 \ \underline{V}_2}{f_2}}$$

The weighted mean value m is written as :

$$m = \pm \sqrt{\frac{f_1 m_1^2 + f_2 m_2^2}{f_1 + f_2}}$$

The mean residual value Θ is obtained as:

$$\Theta = \pm \sqrt{\frac{\underline{d}^T \ \underline{Q}_{dd}^{-1} \ \underline{d}}{h}}$$

, where h is the total number of the independent elements (i.e. range of the $\underline{Q}_{dd}$ matrix).

A statistical test $F = \theta^2/m^2$ is performed. F_c is a value taken from the table of F distribution depending on the $h = f_1 + f_2$ degrees of freedom chosen at the c = 5 percent level of significance. If $F > F_c$, then it means that a horizontal movement exists with a reliability of 95% at one of the stations of the net at least. If $F < F_c$, then it means that the horizontal movement is so small that it requires much more precise measurements.

The stations where the crustal movements exist are determined by making use of the vectors of coordinate differences d_i and the value of the mean residuals θ_i are calculated at each station of the net. Where d_i and θ_i can be written as follows:

$$\underline{d}_i = \begin{bmatrix} x_{i_2} - x_{i_1} \\ y_{i_2} - y_{i_1} \end{bmatrix}$$

$$\theta_i = \pm\sqrt{\frac{\underline{d}_i^T \; \underline{Q}_{ii}^{-1} \; \underline{d}_i}{2}}$$

The station having the largest value θ_i is assumed to be one of the stations where the horizontal movement occurs. This station is then omitted and the "complete test" is repeated. If $F > F_c$, then it means that there is another station where the horizontal movement exists with a reliability of 95%, and the vectors of the coordinate differences d_i and the value of the mean residuals θ_i are calculated at each station of the net with the exception of the station that was omitted. The procedure continues until $F < F_c$. When $F < F_c$, the programme is terminated, and it is assumed that the remaining stations of the net become equivalent stations with a reliability of 95%.

The results of such a calculation are given in the following where:

	I	II
θ^2 (cc^2)	421	16836
m^2 (c^2)	256	81
F	1.64	207.85
Statistical boundary value	2.53	2.69
	F < 2.53	F > 2.69

In the first calculation's set, the 1976 period is formed with directions only, but the 1979 period is formed with directions and distances. It must be noted that the amount of the displacements in the first set of calculations stayed within the boundary of measurement errors, that is to say all of the stations become equivalent to each other. In such a case, displacements of the stations could not be determined. In the second set of calculations, the measurements taken by Rangemaster II, used for the first time during the measurement, were omitted in order to control the directions and the distances in the 1979 period. The value of the mean error was reduced, and it was determined that there was a movement of appr. 4 cm in the south-east direction at the stations A and B with a possibility of 95%. This displacement is not conform with the general movement of the Northern Anatolian Fault Zone and may be caused by the complex geological situation.

Next season, the direction and distance measurements will be repeated so as to obtain more reliable results, subsequently the results obtained in 1976 will be compared with those of 1980.

LITERATURE:

AYTUN, A. 1972: İsmetpaşa İstasyonu Civarinda Krip Ölçümleri, Kuzey Anadolu Fay ve Deprem Kuşaği Simpozyumu.

NIEMEIER, W. 1976: Programmpaket DEFNA zur Analyse von Deformationsmessungen, Hannover, unveröffentlicht.

PELZER, H. 1971: Zur Analyse geodätischer Deformationsmessungen, DGK, Reihe C, Heft 164, München.

UĞUR, E. 1974: Kuzey Anadolu Fay Kusağinin Gerede-Cerkeş Bölümünde Yerkabugu Hareketlerinin Jeodezik Yöntemlerle İncelenmesi, İTÜ Doktora Tezi.

Detection of Crustal Motion by Repeated Geometric Observations

S. M. Nakiboglu and E. Clerici

Department of Surveying, University of Queensland, St. Lucia Q, 4067 Australia

A. M. Isikara

Department of Geophysics, Faculty of Earth Sciences, Istanbul University, Istanbul, Turkey

A method of statistical analysis of repeated geometrical observations is proposed for monitoring the crustal motions in the vicinity of active faults. The method takes into account the correlation between the change in position of points at a given epoch as a function of distance to the fault line, as well as the correlation in positions at successive epochs.

The covariances pertaining to the position and time dependant distortions in the geometric net are determined using the dislocation theory for a layered viscoelastic medium. The method is general in character however it depends on the fine details of the physical model.

A simulation study on a strike-slip fault is presented to illustrate the method.

1. Statement of problem and objectives

Detection of crustal motion can be made by comparing the coordinates of a geodetic net measured at time intervals Δt_{ij}. The net should cover an appropriate area of the fault and should have a geometric configuration to optimize the chance of crustal motion detection. The analysis of the vector of coordinate differences $\underline{d} = \underline{X}_{t_j} - X_{t_i}$ should indicate whether or not crustal motion has occurred as well as its magnitude in a specific time interval Δt_{ij}. In the analysis of the vector $\underline{d}$ the problem is to determine which of the elements d_i are of systematic nature, thus indicating crustal motion, and which are purely random due to noise in the geodetic observations. This problem can be solved by using statistical methods (Baarda, 1968), however the rigorous application of these methods requires the knowledge of the correlation between the coordinates $\underline{X}_{t_j}$ and $\underline{X}_{t_i}$ introduced by the physical phenomenon causing crustal motion. Correlation between the elements of these two vectors for any time interval Δt can be modelled as function of the position of the points in the net relative to the fault.

The objectives of this paper are:

. to derive a covariance function for crustal motion associated with dynamical processes in transform faults;

. to indicate a suitable statistical procedure which enables one to resolve the signal (crustal motion) from the noise in the vector of coordinate differences $\underline{d}$ as function of net configuration.

This approach makes it possible to establish a minimum time interval between geodetic observations to allow a lower bound of crustal motion to be detected with a given probability.

2. Covariance function

The crustal motion in the vicinity of a transform fault due to the motions of tectonic plates can faithfully be modelled as displacements associated with screw dislocations in an elastic layer overlying a visco-elastic half-space (Nakiboglu *et al.*, 1980). The fault is apparently locked down to a depth d and the dislocation (relative plate motion), which is confined into $d < X_3 < h$, loads the brittle part of the elastic lithosphere (figure 1). The failure occurs when the shear stresses around the locked section reach to the frictional resistance of the fault. The simple mechanical model in figure 1 implies that the surface displacements associated with plate tectonic movements are in X_1 direction only, and they are functions of X_2-coordinate. A more detailed treatment of the mechanical aspects is given in Nakiboglu, *et al.*, (1980). The crustal motion at a point $P(X_1,X_2,0)$ at time t resulting from a dislocation $\Delta u = \delta H(t)$ is

$$u_1 = \frac{\delta}{\pi} (f + \sum_{m=1}^{\infty} \tilde{\gamma}^m g_m), \qquad .....(1)$$

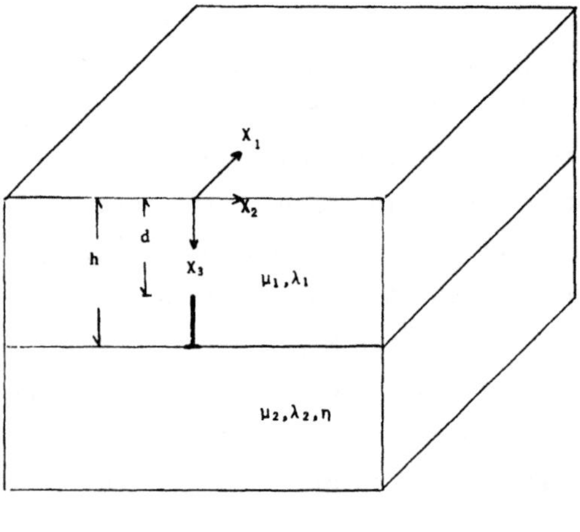

Figure 1:

Mechanical model of crustal motion along transform faults.

where $f = \tan^{-1}\frac{h}{X_2} - \tan^{-1}\frac{d}{X_2}$,

$$g_m = \tan^{-1}\frac{(2m+1)h}{X_2} - \tan^{-1}\frac{(2m-1)h}{X_2} - \tfrac{1}{2}\tan^{-1}\frac{d-2mh}{X_2} - \tfrac{1}{2}\tan^{-}\frac{d+2mh}{X_2},$$

$$\tilde{\gamma}^m = \gamma^m\left\{1 + \sum_{r=1}^{m} \frac{m!}{r!(m-r)!}(\gamma^{-1}-1)^r\left[1 - e^{-\frac{1+\gamma}{2t}t}\sum_{j=0}^{r-1}\left(\frac{1+\gamma}{2}\right)^j \frac{1}{j!}\left(\frac{t}{T}\right)^j\right]\right\},$$

$\gamma = \frac{\mu_1-\mu_2}{\mu_1-\mu_2}$ with μ_1 and μ_2 denoting the rigidities of the lithosphere and asthenosphere respectively, and T is the relaxation time which relates to the viscosity of the asthenosphere as $T = \eta/\mu_2$.

The mean value of γ can be taken to be ~- 0.27 on the basis of the parametric earth of Dziewonski *et al.*, (1975). Hence the series in equation 1 can be truncated at m=1 for the present purpose.

The viscoelastic relaxation of the asthenosphere can also be neglected if the time interval between two successive observations is small compared to the relaxation time. Hence, assuming a uniform plate motion and neglecting the contribution of viscoelastic relaxation into the strain accumulation, we can simplify equation 1 to obtain

$$u(X_2,t) = \frac{\dot{\delta} t}{\pi}\left(\tan^{-1}\frac{h}{X_2} - \tan^{-1}\frac{d}{X_2}\right). \qquad .....(2)$$

Here $\dot{\delta}$ is the relative velocity of the tectonic plates, and

t is the time elapsed since the last major earthquake on the fault.

Figure 2 shows the relative displacements of surface points as a function of distance to the fault line for a mechanical model applicable to the North Anatolian Fault Zone.

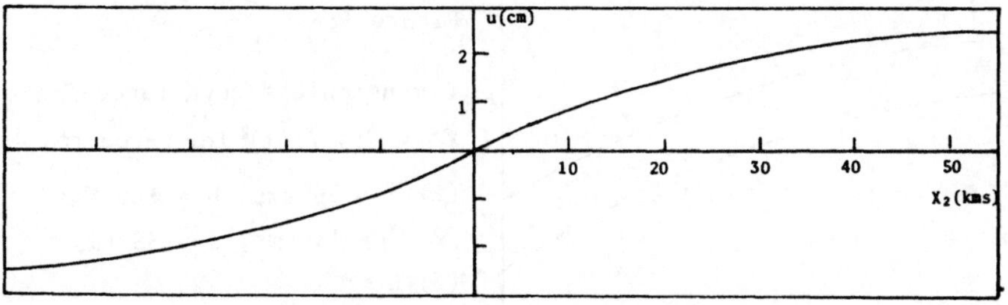

Figure 2: Relative displacements for t = 5 yr, based on a model with d = 50 kms, h = 100 kms and $\dot{\delta}$ = 5 cmyr^{-1}.

The covariance function pertaining to the relative displacements of control points at a given epoch can be found from

$$C(s) = \frac{1}{2l} \int_{-l}^{l} u_1(X_2,t)\, u_1(X_2 + s,t)\, dX_2, \qquad \text{.....(3)}$$

where s is the separation, in X_2-coordinate, between the points, and

l is the half-size of the fault area to be monitored.

Combining equations 2 and 3 and assuming that d<<h we obtain

$$C(u) = \frac{1}{2}\left(\frac{\dot{\delta t}}{\pi}\right)^2 \Bigg\{ \frac{1}{\varepsilon^3(4+\varepsilon^2u^2)} \left[\frac{2+\varepsilon^2u^2}{\varepsilon u}\, ln\, \frac{1+\varepsilon^2(u-1)^2}{1+\varepsilon^2} + 2\tan^{-1}\varepsilon - 2\tan^{-1}\varepsilon(u-1)\right]$$

$$+ \frac{\pi^2}{8} + \frac{\pi}{4}\, ln\, \frac{1+(1-u)^2}{2} - \frac{\pi}{2}(1-u)\tan^{-1}(1-u) - \frac{1}{4\varepsilon^2u^2+\left[1-\varepsilon^2(\varepsilon^2+u^2)\right]^2} \cdot$$

$$\Bigg[u\frac{1+\varepsilon^2(\varepsilon^2+u^2)}{2}\, ln\, \frac{(1+\varepsilon^2)^2}{\left[1+\varepsilon^2(u-1)^2\right]\left[\varepsilon^2+(u+1)^2\right]} - \varepsilon\left[1-\varepsilon^2(\varepsilon^2+u^2)\right]\left(\tan^{-1}\frac{u+1}{\varepsilon} -\right.$$

$$\left. -\tan^{-1}\frac{1}{\varepsilon}\right) + \frac{1-\varepsilon^2(\varepsilon^2-u^2)}{\varepsilon}\left[\tan^{-1}\varepsilon + \tan^{-1}\varepsilon(u-1)\right]\Bigg]\Bigg\}, \qquad \text{.....(4)}$$

where $\varepsilon^2 = 0.28$, and $u = s/d$.

The area size in the preceding equation is taken to be d which seems reasonable considering the fact that d ≃ 5 ~ 15 kms and most of the monitoring activities around transform faults are confined into the fault zone. For the case where d/h is not negligible the covariance function can be obtained by numerical integration.

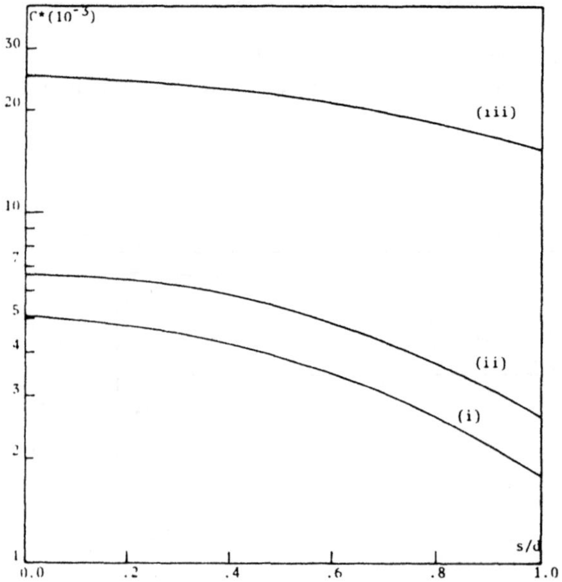

Figure 3:

Dimensionless covariance function $C^* = C(s)/(\dot{\delta t})^2$ for three models.

(1) d = 50 kms, h = 100 kms;

(2) d = 15 kms, h = 35 kms;

(3) $h \to \infty$.

The behaviour of the dimensionless covariance function $C^* = C/(\dot{\delta t})^2$ is shown in figure 3 for two extreme models and a simplified model corresponding to an elastic half-space. The first two curves are obtained by numerical integration and the third one is drawn using equation (4) which is also verified by direct numerical quadrature of equation (3).

3. Statistical analysis

As indicated in section 2 the vector of coordinate differences analysed for a time interval Δt_{ij} will involve only the X_2 coordinates (see figure 1). Let this vector be defined as

$$\underline{d} = X_j - \underline{X}_i, \quad \text{or} \qquad .....(5)$$

$$\underline{d} = \underline{\varepsilon} + \underline{\Delta}\ ; \quad E\{\underline{\varepsilon}\} = \underline{0},\ E\{\underline{\Delta}\} = \underline{\Delta}$$

where $\underline{\varepsilon}$ is a random component due to noise in the geodetic measurements and $\underline{\Delta}$ is the component due to the crustal motion. E{.} is the symbol for statistical expectation. Similarly to the collocation problem of geodesy (Moritz, 1972), the covariance matrix of $\underline{d}$ can be defined as

$$C_{dd} = C_{\varepsilon\varepsilon} + C_{\Delta\Delta} \qquad .....(6)$$

where, $C_{\varepsilon\varepsilon} = V_i + V_j$

V are the matrices of covariances of the coordinates obtained from the net adjustment at times t_i and t_j respectively. It is reasonable to assume no correlation between the observations at the two epochs. The matrix $C_{\Delta\Delta}$ can be constructed using the covariance function given in equation (3).

Occurrence of crustal motion in a time interval Δt can be tested using classical statistical procedures by formulating the Null hypothesis that no crustal motion has occurred.

Formally this can be stated as

$$H_o : E\{\underline{d}\} = \underline{0}; \quad P\{\underline{d}^T C_{dd}^{-1}\underline{d} > F_{1-\alpha,b,\infty}\} = \alpha; \quad b = n - 1 \qquad .....(7)$$

with n : dimension of vector $\underline{d}$

where P indicates probability, α is the level of significance of the test and $F_{1-\alpha,b,\infty}$ is a critical value obtainable from statistical tables. If the Null hypothesis is true the quantity $\underline{d}^T C_{dd}^{-1}\underline{d}$ is an F - distributed random variable with b and ∞ degrees of freedom. If a systematic component $\underline{\Delta}$ is present in the vector $\underline{d}$ then (Baarda, 1968)

$$E\{\underline{d}\} = E\{\underline{\varepsilon}\} + E\{\underline{\Delta}\} = \underline{\Delta}, \text{ and } E\{d^T C_{dd}^{-1} d\} = b + \lambda, \qquad .....(8)$$

$$\text{where,} \quad \lambda = \underline{\Delta}^T C_{dd}^{-1}\Delta. \qquad .....(9)$$

The values of λ can be obtained from tables of non-central F distributions. λ is a function of the level of significance of the test α("producer's risk") and of the power of the test β(consumer's risk).

If we could solve equation (9) for the individual elements of vector $\underline{\Delta}$ these values would correspond to the smallest crustal motions detectable by statistical testing on vector $\underline{d}$ which, as indicated by equation (5), is made of signal and of noise. For a description of the actual method of testing on $\underline{d}$ as proposed by Baarda, reference is made to the literature (for example Baarda, 1968, 1976 and Albarda, 1976).

Equation (9) as such can obviously not be solved in terms of the individual elements of $\underline{\Delta}$ without some additional knowledge on their relative magnitudes (Alberda 1976). For the physical model considered here this information is provided by figure (2), with the help of this we can replace vector $\underline{\Delta}$ by a vector $\delta\underline{C}$ where the elements of $\underline{C}$ are the ratios of the magnitudes of the elements of $\underline{\Delta}$ and δ is an unknown constant to be determined. Substitution in equation (9) gives

$$\delta^2 \, \underline{C}^T C_{dd}^{-1} \underline{C} = \lambda, \text{ or } \delta = \pm\lambda/\sqrt{\underline{C}^T C_{dd}^{-1} \underline{C}} \qquad \text{.....(10)}$$

Equation (1) allows us to estimate the smallest detectable vector of crustal motion $\underline{\Delta}$ with a given probability β for a given observation time interval and a given goedetic net configuration. The influence of time and of net configuration is accounted for in vector $\underline{C}$ and in the covariance matrix C_{dd}.

4. Simulation studies

To illustrate the use of the technique of analysis discussed in the previous paragraph the coordinates of two simulated quadrilaterals were perturbed according to the model of figure (2). The definition of coordinate system used is the one of figure (1), the numerical values are given in table 1. The resulting covariance matrices and vectors are as follows

$$C_{dd_{(1)}} = \begin{pmatrix} 7.06 & 2.96 & 3.06 & 2.96 \\ 2.96 & 7.06 & 2.96 & 3.06 \\ 3.06 & 2.96 & 7.06 & 2.96 \\ 2.96 & 3.06 & 2.96 & 7.06 \end{pmatrix}, \text{where } C_{\varepsilon\varepsilon} = \sigma^2 \underline{I}, \quad \sigma^2 = 4 \text{ cm}^2$$

$$\underline{C}_1^T = (0.36 \; 1 \; 0.36 \; 1)$$

$$C_{dd(2)} = \begin{bmatrix} 7.06 & 3.00 & 3.05 & 2.61 \\ 3.00 & 7.06 & 3.05 & 2.89 \\ 3.05 & 3.05 & 7.06 & 2.73 \\ 2.61 & 2.89 & 2.73 & 7.06 \end{bmatrix}; \qquad C_{\varepsilon\varepsilon} = \sigma^2 \underline{I} \quad \sigma^2 = 4\ \text{cm}^2$$

$$\underline{C}_2^T = (0.3\ 1\ 0.7\ 1.8)$$

Testing both cases at $\alpha = 0.05$ (one-sided) and $1-\beta = 0.80$ and applying equation (10), resulted in the following two vectors of minimum displacements detectable by testing on $\underline{d}$

$$(\min)\ \underline{\Delta}_1^T = \pm(2.8\ 7.8\ 2.8\ 7.8)\ \text{cm}$$

$$(\min)\ \underline{\Delta}_2^T = \pm(1.0\ 3.4\ 2.4\ 6.2)\ \text{cm}$$

Comments

The conclusions reached in an analysis of the type presented here are affected by the choice of the mechanical model (figure 1). Such dislocation models, even in much simpler forms, have been known to represent the crustal motion around transform faults satisfactorily (Savage 1975, 1976). Therefore the details of the model rather than the model itself are of primary concern. As seen in equation (1) both the magnitude and the areal extent of the displacements increase with increasing h whereas d has just the opposite effect on deformations. Therefore the success of the analysis depends on the proper choice of the critical quantities d and h as well as on the right value for the plate velocity. Obviously geodetic data is not sufficient, and one has to resort to earthquake data (for d) and seismic profiles (for μ). Geodetic monitoring of plate motions in areas away from the deforming plate margins should yield accurate estimates of relative velocity of neighbouring plates.

The simulation study in the preceding section is not intended as a modelling of a specific transform fault rather, it is intended to illustrate the method, and to draw some conclusions pertaining to the net configuration and to the observation frequency.

We readily see that the maximum annual rate of surface displacement is about 30% of the plate velocity and occurs at a distance of ~2d from the fault line. The displacements accumulate linearly in time. It is only logical to assume that one has to set up an observation program on the basis of the plate velocity such that the accumulated displacements are above the noise level.

The comparison of the minimum detectable displacements for the two test models indicates the importance of the net configuration for successful monitoring of crustal motions. A trilateration net with sides aligned at ~45° to the fault trace will be able to detect the crustal motion in rebound phase more successfully than the one with lines oriented either parallel or perpendicular to the fault trace.

References

Alberda, J.E. 1976. Quality control in Surveying. *Chartered Surveyor, Land, Hydrographic and Minerals Quarterly*, No. 2.

Baarda, W. 1968. A testing procedure for use in geodetic networks. *Neth. Geod. Comm. Publ. on Geodesy*, No. 5, Delft.

Baarda, W. 1976. Reliability and precision of networks. *Monograph, Computing Centre of the Delft Geodetic Institute.*

Dziewonski, A.M., Hales, A.L., Lapwood, E.R., 1975. Parametrically simple earth models consistent with geophysical data, *Phys. Earth Planet. Int.*, 10, 12-48.

Nakiboglu, S.M., Isikara, A.M., 1980. Analysis of stress and strain around a transform fault, *Interdiscip. Conf. Earthquake Prediction Res. in North Anatolian Fault Zone*, 31.3.80, Istanbul, Turkey.

Savage, J.C. 1975. Comment on 'An Analysis of Strain accumulation on a Strike-slip fault' by D.L. Turcotte and D.A. Spence, *J. Geophys. Res.*, 80, 4111-4115.

Savage, J.C., 1976. Geodetic measurements of crustal motion. The changing World of Geodetic Science, *Proc. Intern. Symposium*, Ohio State University, 1976, 99-116.

Land Deformation Control Network in the Epicentral Region of the 1978 Earthquakes in Northern Greece

D. G. Vlachos

Institute of Topography, University of Thessaloniki, Greece

In summer 1978, Northern Greece and especially the city of Thessaloniki suffered from a series of earthquakes. The heaviest with a magnitude of M = 6.5 according to Richter's scale took place on June 20. The shocks of May 24 (M = 5.8), June 19 (M = 5.2) and July 5 (M = 5.1) should also be pointed out. These earthquakes have caused extensive destructions in the villages of the region and in the city of Thessaloniki where the collapse of one single building costed 40 human lives.

After these earthquakes, faults and cracks were found at the earth's surface of the epicentral region. The most important of these faults with displacements varying from 2 to 35 cm are presented in Fig. 1. Their appearance shows the extent of horizontal and vertical displacements which certainly have to be seen in relation to the earthquakes themselves. Two important facts, namely that 1) the epicentral region is situated in a well-known geological fault zone (Serbo-Macedonian Zone) which has given rise to considerable earthquakes in the past and which, unfortunately to say, most probably will produce others in the future, 2) the earthquakes of this region have a vital influence on the lives of the large city of Thessaloniki, have led us to the decision to install a triangulation network, the aim of which is to determine the land deformation by geodetic methods. We hope that the determination of displacements of control points by periodically repeated measurements will give useful hints to the researchers who are involved in earthquake prediction research in the Serbo-Macedonian Zone.

The installed network has got the name "Volvi Network". It is situated between the lakes "Lagada" and "Volvi" and includes 16 trigonometric points on pillars which have been settled deeply and incrusted with the ground (Fig. 1). The network which comprises an area of 150 km^2 was installed by the Department of Topography, University of Thessaloniki, together with a group of students. The scarcity of financial support until now did not allow us to cover a larger area.

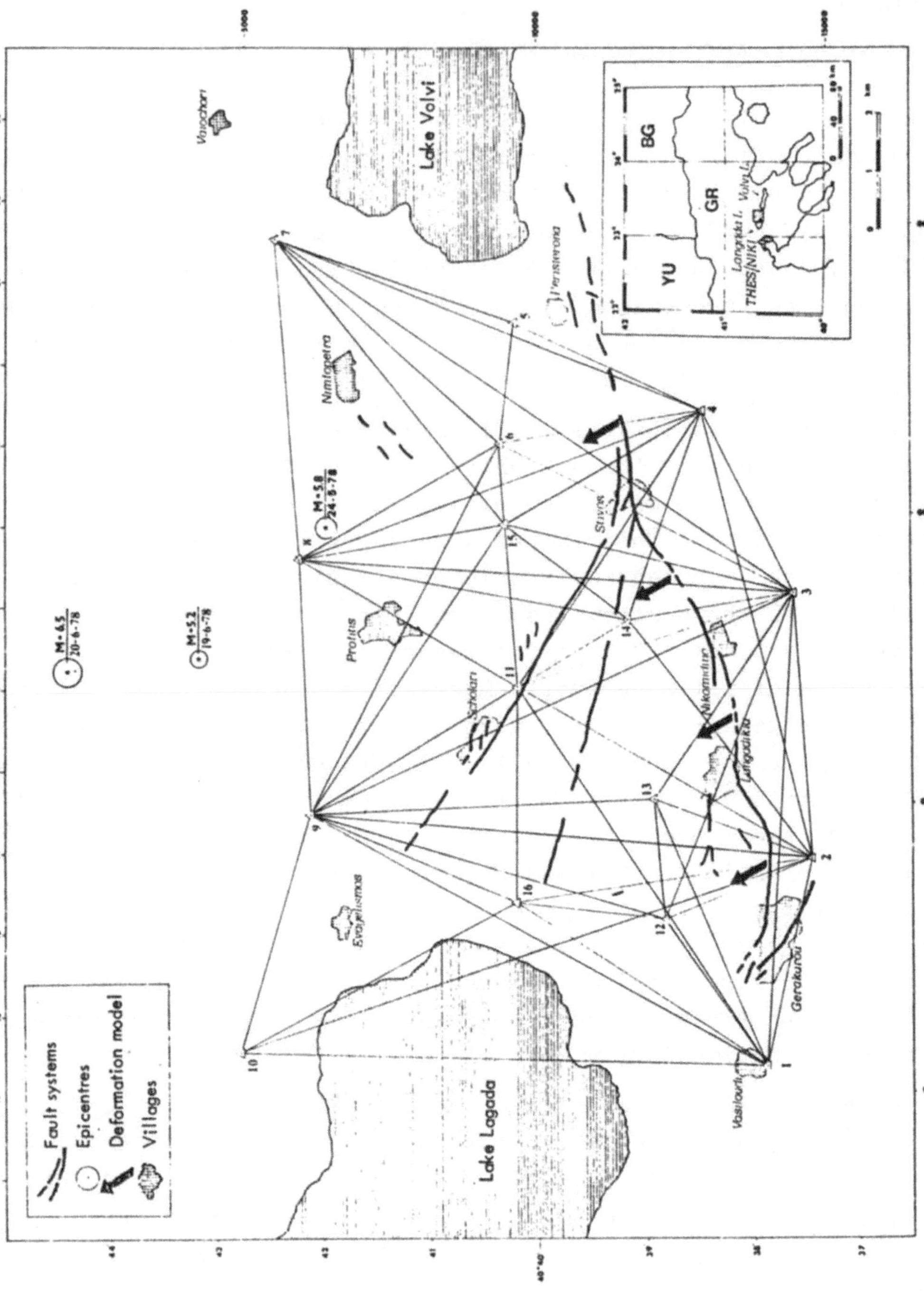
Fault systems
Epicentres
Deformation model
Villages
Lake Lagada
Lake Volvi
M=6.5 20-6-78
M=5.2 19-6-78
M=5.8 24-5-78
Protitis
Scholari
Stivos
Nimfopetra
Evangelismos
Gerakarou
Vasilouda
BG
GR
YU

A series of computer programs was used to select the observation points in order to optimize the network, taking into account

- the high precision required for the purpose,
- the topography of the terrain,
- the accessibility of the pillars.

In summer 1979, two groups of students have carried out the measurements of the geometric elements of the Volvi Network. One group measured the directions of the network by a theodolite Kern DKM 3 and the other a certain number of angles with a theodolite Wild T3. A third effort to measure the distances of the points of the network by a geodimeter AGA 8 failed because the instrument broke down. We expect that after the reparation of the geodimeter we shall be able to measure a certain number of sides of the network.

Altogether 106 directions and 97 isolated angles were measured. These points were taken as fixed ones for the following reasons:

a) Pillars 3 and 4 are situated in a rocky region.
b) The epicentres of the heaviest earthquakes are north of the southern fault (Fig. 1).

The correlation of this fault with the results of studies of focal parameters of all earthquakes registred in that region show that the deformation model is that of Fig. 1 (B. Papazahos et al., 1978).

After compensation of the observations (methods of direction) the weighted mean error was 1.2" and the corrections varied from 1.3" to 3.1". The mean errors of the final coordinates as calculated by aid of the elements Q_{xx}, Q_{yy} of the matrix of cofactors $Q_{28,28}$ of the unknowns are given in table I. In the same table the elements of the error ellipses are quoted for each point of the network. They are presented graphically in Fig. 2. This picture of the local precision for the network has been completed by calculation of the elements for ellipses of relative errors (D. Vlachos, 1970). A certain number of these ellipses is presented graphically in Fig. 3.

It has to be noted that the values of the long semiaxises of the error ellipses determined for the points (Fig. 2) vary from 1.7 cm to 6.6 cm, the latter values corresponding to point N° 10, the most remote point from the basis which is considered fixed.

TABLEAU I

Mean errors of coordinates and the elements of the ellipses of errors for the Volvi Network
(Direction method, 1979)

Σημειο	Q_{xx} / Q_{yx}	Q_{xy} / Q_{yy}	$\sigma_{\bar{x}} = \sigma_\delta \times \sqrt{Q_{xx}}$ / $\sigma_{\bar{y}} = \sigma_\delta \times \sqrt{Q_{yy}}$ (cm)	ϑ_0 / $\vartheta_0 + 100$ (g)	$\sigma_{y_0} = \sigma_\delta \times \sqrt{\lambda_1}$ / $\sigma_{x_0} = \sigma_\delta \times \sqrt{\lambda_2}$ (cm)
1	14.2193	0.0201	4.7	99.65	4.7
	0.0201	10.5839	4.0	199.65	4.0
2	4.8030	0.3301	2.7	76.62	2.8
	0.3301	4.0719	2.5	176.62	2.5
5	2.4401	0.2281	1.9	16.57	2.3
	0.2281	3.2356	2.2	116.57	1.9
6	2.2868	- 0.1725	1.9	142.00	1.9
	- 0.1725	2.1962	1.8	242.00	1.8
7	11.1916	3.1166	4.1	39.55	4.9
	3.1166	13.3152	4.5	139.55	3.7
8	6.3291	0.1308	3.1	10.34	3.3
	0.1308	7.1060	3.3	110.34	3.1
9	9.6188	- 0.6017	3.8	153.96	4.0
	- 0.6017	9.7693	3.9	253.96	3.7
10	25.6589	- 4.0089	6.3	134.36	6.6
	- 4.0089	21.3673	5.7	234.36	5.4
11	2.8143	0.2647	2.1	53.86	2.2
	0.2647	2.7499	2.1	153.86	2.0
12	6.6627	- 0.7378	3.2	137.11	3.3
	- 0.7378	6.0303	3.0	237.11	2.9
13	3.8614	- 0.4918	2.4	140.51	2.5
	- 0.4918	3.5591	2.3	240.51	2.2
14	1.2575	0.3341	1.4	27.30	1.7
	0.3341	1.8355	1.7	127.30	1.3
15	2.2332	0.0836	1.9	84.44	1.9
	0.0836	1.9190	1.7	184.44	1.7
16	7.7358	0.2179	3.4	84.63	3.5
	0.2179	6.9041	3.3	184.63	3.2

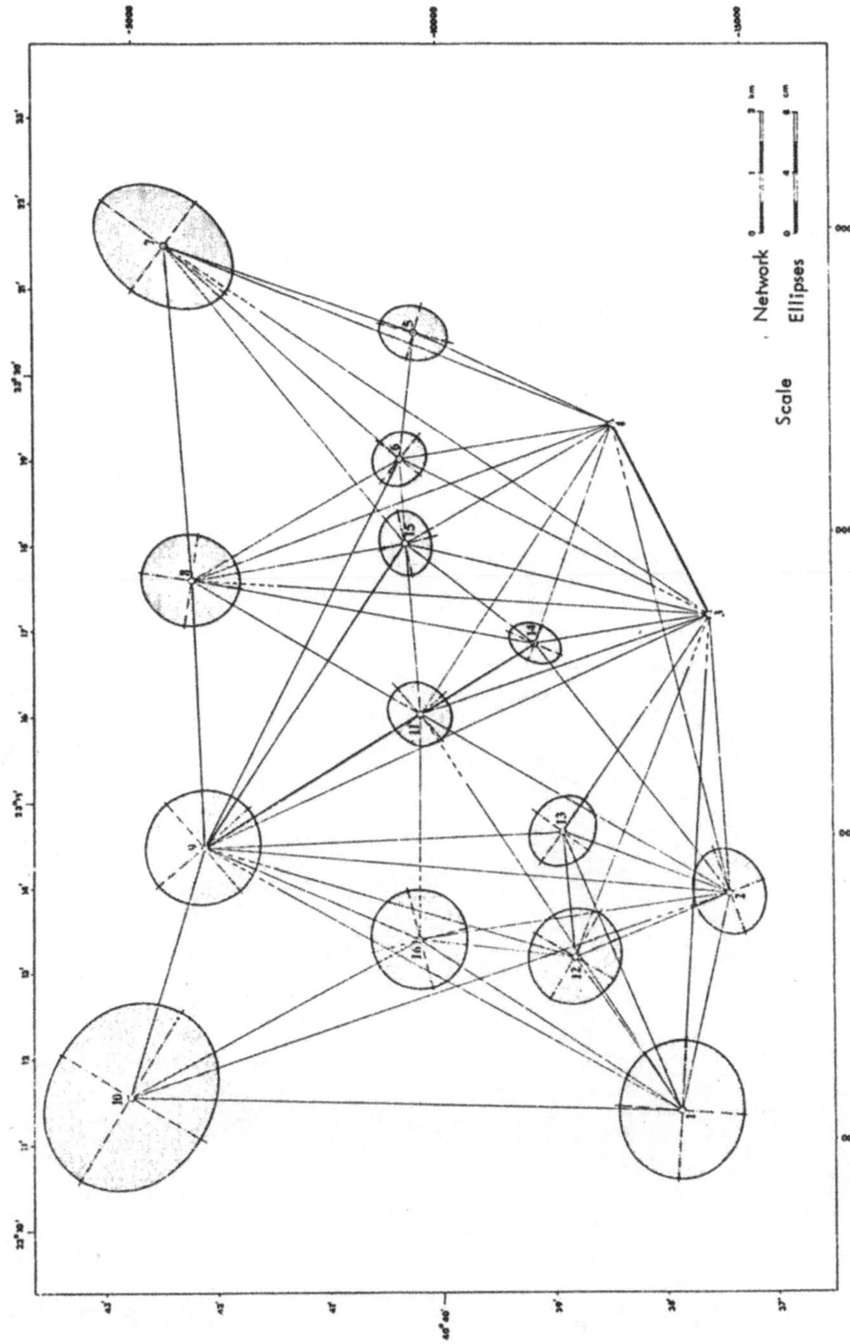

Figure 2

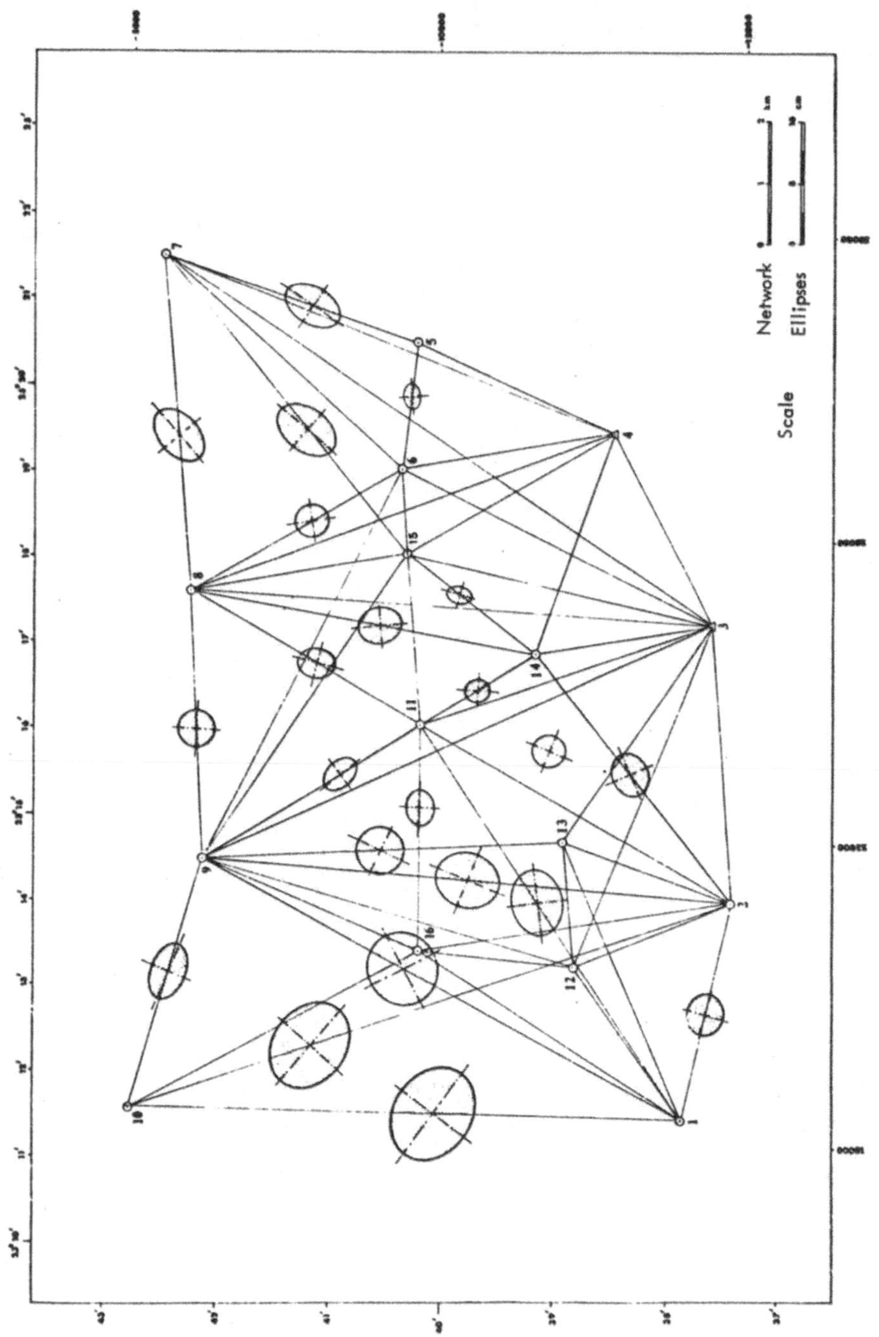

Figure 3

The fact that the difference between long and short semiaxises of the ellipses is fairly small, shows that the geometric structure of the network is satisfactory (criterion of homogeneity). The long semiaxises of the ellipses of relative error (Fig. 3) vary from 7.8 cm (pts 1-7) to 1.4 cm (pts 6-15) while the mean error of the calculated final distances are within the same limits. These results correspond to a relative precision (in terms of the longer semiaxises of ellipses of relative error or in terms of mean error of distances in the order of (0.5-1.0) x 10^{-5}.

The criteria of general network precision, expressed by the values of characteristic square roots of the matrix for the cofactors of unknowns and the principal eigenvectors, have also been calculated (H. Dupraz and W. Niemeir, 1979).

An optimization of Volvi Network, that may be achieved by application of a certain number of distance measurements (Fig. 4), when using the geodimeter AGA 8, has been studied (E. Grafarend, 1974). The results concerning the error ellipses are presented in table II and in Fig. 4. Furthermore the measurement of certain distances and isolated angles will give very useful information for interpretation and complete exploitation of the triangulation results (E. Livieratos, 1978).

As to the reduction of the sides which will be measured by a geodimeter, observations of the reciprocal zenithal distances between the summits of the network have been made. With the help of these observations the altitude of the trigonometric points has been calculated. The largest mean errors of the final altitude is 8 cm.

Besides the horizontal displacements, we intend to determine the variation of altitude of the control points within the Volvi Network. Therefore, we already have started to employ periodic high precision geometric levellings in spite of great difficulties as to the topography of the terrain.

The results of the 1979 measurements show that with the Volvi Network we got a practicable network giving us a chance to provide all information regarding the displacement of the control points with sufficient precision. Therefore, we hope that the series of triangulation and trilateration measurements which will

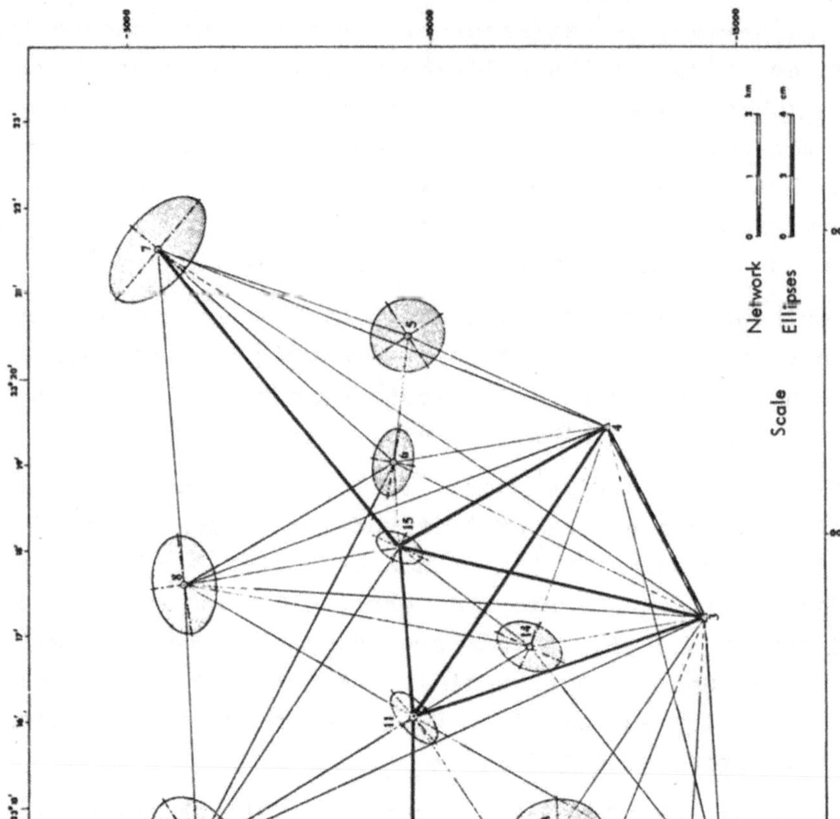

Figure 4

TABLEAU II

Mean errors of the coordinates and elements of the ellipses of error of the Volvi Network after the improvement proposed in Fig. 4

Σημείο	Q_{xx}	Q_{xy}	$\sigma_{\bar{x}} = \sigma_\delta \times \sqrt{Q_{xx}}$	ϑ_0	$\sigma_{y_0} = \sigma_\delta \times \sqrt{\lambda_1}$
	Q_{yx}	Q_{yy}	$\sigma_{\bar{y}} = \sigma_\delta \times \sqrt{Q_{yy}}$ (cm)	$\vartheta_0 + 100$ (g)	$\sigma_{x_0} = \sigma_\delta \times \sqrt{\lambda_2}$ (cm)
1	3.5094	- 1.9876	1.9	164.81	2.6
	- 1.9876	5.5043	2.3	264.81	1.5
2	1.8634	- 0.6656	1.4	154.19	1.6
	- 0.6656	2.0397	1.4	254.19	1.1
5	1.3593	- 0.1018	1.2	163.39	1.2
	- 0.1018	1.4504	1.2	263.39	1.1
6	1.2519	- 0.1436	1.1	111.37	1.1
	- 0.1436	0.4831	0.7	211.37	0.7
7	3.2111	- 1.7859	1.8	144.64	2.2
	- 1.7859	2.6041	1.6	244.64	1.0
8	2.7417	0.1943	1.7	91.16	1.7
	0.1943	1.3788	1.2	191.16	1.2
9	3.1857	1.0543	1.8	63.72	2.0
	1.0543	2.2161	1.5	163.72	1.2
10	6.5029	4.1358	2.6	53.94	3.2
	4.1358	5.4726	2.3	153.94	1.3
11	0.7136	0.3872	0.8	56.38	1.0
	0.3872	0.5564	0.7	156.38	0.5
12	1.2945	- 0.4225	1.1	189.04	1.9
	- 0.4225	3.6514	1.9	289.04	1.1
13	1.4181	- 0.0341	1.2	197.09	1.5
	- 0.0341	2.1620	1.5	297.09	1.2
14	0.6304	0.2961	0.8	27.28	1.1
	0.2961	1.1435	1.1	127.28	0.7
15	0.7104	0.0776	0.8	89.90	0.9
	0.0776	0.2378	0.5	189.90	0.5
16	1.0913	0.7176	1.0	22.33	1.7
	0.7176	2.7901	1.7	122.33	0.9

follow, exploited by the methods and techniques of free geodeti networks, will give us a chance to analyse the crustal deformations in the region under consideration.

In spite of the fact that the Volvi Network only covers a very limited area we hope that it will substantially contribute to earthquake prediction research in the Serbo-Macedonian Zone (T. Rikitake, 1976).

R e f e r e n c e s

H. DUPRAZ and W. NIEMEIR, 1979: "Un Critère pour l'Analyse des Réseaux de Contrôle". Mensuration, Photogrammétrie, Génie Rural, No. 4, pp 70-76.

E. GRAFAREND, 1974: "Optimisation of Geodetic Networks". Bolletino di Geodesia e Scienze, No. 4.

E. LIVIERATOS, 1978: "On the Error Analysis of Geodetical Derived Strains in Seismic Zones". Proc. ESA - Council of Europe Seismic on 'Earthquake Prediction', Strasbourg, ESA-SP-149, 27-32.

B. PAPAZAHOS, D. MOUDRAKIS, A. PSILOVIKOS and E. LEVENDAKIS, 1979: "Focal Properties of the 1978 Earthquakes in the Thessaloniki Area". Bulgarian Geophysical Journal (in press).

T. RIKITAKE, 1976: "Earthquake Prediction", Elsevier S.P.C., Amsterdam.

D. VLACHOS, 1970: "Détermination des déformations des grands barrages. Triangulations de haute précision" (in Greek). Technika Chronika, No. 8, pp 1-10.

Methods of Variance Analysis for the Evaluation of Geodetic Control Nets in Relation to Crustal Movements

Reinhard Baumer

FB Bau- und Vermessungswesen, Hochschule der Bundeswehr München, Werner-Heisenberg-Weg 39, D-8014 Neubiberg, Federal Republic of Germany

Summary

A review on actual methods of variance analysis for the detection of crustal movements using geodetic control nets is being given. These methods statistically render safe statements of deformations in geodetic nets. The analysis of movements of single points and blocks relative to each other is possible. The method is rigorous and gives a solid basis for further interpretations.

1. Introduction

Repeatedly observed geodetic nets are an accepted instrument for the detection of deformations of objects as well in engineering as in science.
An area to be investigated is represented by a number of discrete points on the surface. The position of these points is determined by an initial observation at the time t_0. When deforming forces are acting, the observation of the net at a time t_i is being repeated to draw conclusions from the changed position of the points with respect to one another.
During the last decade numerous methods for the statistical evaluation of deformations and their interpretations were developed. Some methods of variance analysis are here reported on.

2. Methods of deformation analysis

Basis of all deformation analysis is a direct comparison of geodetic measurements or the comparison of the derived point coordinates of the net. The comparison can be purely descriptive. This method was used until recently. The methods used nowadays represent a statistical test of congruence of the shape of the net (Niemeier, 1979) with the aid of the variance analysis: it is investigated whether two variances, computed independently from each other, differ significantly.

2.1 Analysis using differences of coordinates

The general idea of the variance analysis using differences of coordinates is the following (Pelzer, 1971):
Starting from the linear mathematical models of least squares adjustment of the observations in a net at different times

$$\underline{l}_i + \underline{v}_i = \underline{A}_i\underline{x}_i \qquad \text{functional model}$$

$$\underline{K}_{l_i} = s_o^2\underline{Q}_{l_i} = s_o^2\underline{P}_{l_i}^{-1} \qquad \text{stochastic model}$$

where

$\underline{l}$ vector of observations
$\underline{v}$ vector of residuals
$\underline{A}$ design matrix
$\underline{x}$ vector of the unknowns of coordinates
$\underline{K}_l$ covariance matrix of the observations
$\underline{Q}_l$ cofactor matrix of the observations
$\underline{P}_l$ weight matrix of observations
s_o^2 variance of the weight unit of the observations

one determines two independent estimation values for the variance of the observations.
One is computed by using the residuals of the single adjustments

$$s^2 = \frac{(\underline{v}^T\underline{P}_l\underline{v})_1 + (\underline{v}^T\underline{P}_l\underline{v})_2}{f},$$

where $f = f_1 + f_2$ is the sum of the degrees of freedom of the single adjustments.
The other is determined by the differences of coordinates of the adjusted points of comparison and is denoted by "mean gap":

$$\theta^2 = \frac{\underline{d}^T\underline{Q}_d^+\underline{d}}{h}$$

$h = u - \text{def}$ u number of differences of coordinates
def defect of datum (rankdefect of the design matrix)

$\underline{d} = \underline{x}_2 - \underline{x}_1$; $\underline{x}_2$, $\underline{x}_1$ unknowns of coordinates of the epochs of comparison

$\underline{Q}_d^+ = (\underline{Q}_{x_1} + \underline{Q}_{x_2})^+$ weight matrix of differences of coordinates

It is possible to show (e.g. Caspary, 1979) that both variances s^2 and θ^2 are independent.

From that follows the test statistics, namely the ratio

$$\overline{F} = \theta^2/s^2,$$

of the Fisher-distribution, which is central, if the socalled null hypothesis is valid. The null hypothesis

$$H_o : E\{\underline{x}_1\} = E\{\underline{x}_2\}$$

says, that the two computed unknown vectors of coordinates are identical, that means the two compared nets are congruent. If the test statistics, accepting a certain confidence level

$$P\{\overline{F} \geq F_{h,f,1-\alpha} | H_o\} = \alpha$$

exceeds the corresponding confidence limits of the central Fisher-distribution, the null hypothesis cannot be supported any more. In the other case the null hypothesis and hence the statement the differences in the variances are caused by random but unvoidable measuring errors is justified.

2.2 Analysis with a linear hypothesis

The linear hypothesis $H_o : E\{\underline{x}_1\} = E\{\underline{x}_2\}$ is being tested with the method of analysis in 2.1 using the differences of coordinates following the single adjustment.
Another possibility is to consider it as a condition between the unknowns (Koch, 1975; Heck et al., 1977).
To the adjustment of observation equations of the two epochs of observation to be executed in one model, one adds the corresponding condition equations. From a combined adjustment one obtains a variance differing from the variance s^2, got out from an adjustment without condition equations. The two variances are used again for the derivation of test statistics. This method does not assume two different adjustments. The variances of the weight unit, computed from the adjustment of the linear hypothesis, can be split into two from each other independent parts. This will be explained briefly:
The adjustment without linear hypothesis leads to the weighted square sum of the residuals according

$$\underline{v}^T \underline{P}_1 \underline{v} = \Omega_o = (\underline{Ax-l})^T \underline{P}_1 (\underline{Ax-l}) = s^2 f.$$

The linear hypothesis additionally added to the adjustment model results in a change of the unknown coordinates, so the weighted square sum of the residuals is now

$$\underline{\tilde{v}}^T \underline{P}_1 \underline{\tilde{v}} = \Omega_H = (\underline{A\tilde{x}-l})^T \underline{P}_1 (\underline{A\tilde{x}-l}) .$$

As can be shown (Koch, 1975; Caspary, 1979), Ω_H consists of two independent components

$$\Omega_H = \Omega_o + R,$$

in which case R can be calcated by the linear hypothesis alone. As well Ω_o as R are χ^2-distributed stochastic variables whose quotient

$$T = \frac{R/h}{\Omega_o/f} = \frac{R}{h \cdot s^2}$$

follows again the F-distribution and is used as test statistics.

A comparison of the test variables $\overline{F}$ and T shows their identity (Heck et al., 1977).
By this it is shown that the two methods of analysis only differ by their formal statement.

3. Application to parts of the net

The analysis using the test statistic $\overline{F}$ or T gives only the general statistical indication whether there were deformations in the net as a whole. Therefore one talks about a global test.
Some special cases can be derived from the global test. The procedure depends mainly on the structure of the object in question.
There are cases in which a clear sectioning of different groups of points are possible from the beginning.
A typical example from the field of geophysics would be a control net which connects two blocks over a visible fault. The purpose of the net is the controlling of relative movements of these blocks to each other (Welsch, 1979).
Such movements of blocks are being investigated, e.g. by Caspary (1979).

3.1 Fixed partial net

In a common adjustment of two epochs the coordinates of the points of one block are regarded as reference values and are put up only once as unknowns. The points of the other block describe the relative movements by their changes of coordinates; therefore they are put up twice as unknowns in the adjustment.
The result of the adjustment are the unknown vector with the corresponding covariance matrix and an estimation s_1^2 for the variance of the observations. A significance test for the relative movements can be carried out again as a variance test. According to the earlier explained theory of

the linear hypothesis tests the variance s_1^2 can be split into two parts which are independent from each other

$$s_1^2 = s^2 + s_2^2 .$$

The first part is identical with the one which is computed from two independent adjustments without linear hypothesis

$$s^2 = \frac{(\underline{v}^T \underline{P}_1 \underline{v})_1 + (\underline{v}^T \underline{P}_1 \underline{v})_2}{f} .$$

The second part follows from the movements of the points $\underline{d} = \underline{x}_2 - \underline{x}_1$ of the block in relative movement

$$s_2^2 = \frac{\underline{d}^T \underline{Q}_d^+ \underline{d}}{u},$$

u number of the components of the coordinates.

The test statistic, computed from the ratio of the two variances is compared with the corresponding confidence limits

$$\overline{F} = s_2^2 / s^2 .$$

3.2 Helmert transformed partial net

But also within the reference-net there could be small changes of points because of local disturbances of points. The model of the fixed partial net can be refined insofar as the coordinates of all points of the net for both epochs are put up as unknowns. The connection of the epochs is achieved in the sense of the theory of the generalized inverse by minimizing the partial norm of the unknown vector which refer to the coordinates of the reference group. The coordinates of the points of the reference block from both epochs are adjusted to the same values by a similarity transformation whereas there is no pressure forced upon the points of the moved block.
In addition to the adjustment the following hypothesis can be tested

$$H_o : E\{\underline{x}_1^R\} = E\{\underline{x}_2^R\},$$

by which is being investigated whether there are significant movements of points inside the reference block, and

$$H_o' : E\{\underline{x}_1^2\} = E\{\underline{x}_2^2\},$$

by which is being determined whether the coordinates of the points of the moved part differ significantly, i.e. whether there are relative movements of the block.

Both hypothesis are carried separately by variance tests.

3.3 Interpretation of the relative movements

The applications sofar described lead to statements whether there occured relative movements inside or between partical nets. Caspary (1979) investigates now whether these relative movements can be represented by appropriate parameters. These can either be estimated in the adjustment or computed following the adjustment under strict consideration of all correlations.

The choice of the parameters depends on the problem to be investigated and represents therefore a problem- and objectoriented interpretation of the process of movement for which the known methods of filtering and smoothing can be applied.

4. Application of movements of points

The statements of deformations of partial nets were more of global nature: the hypothesis was tested whether there were changes in the whole net or in parts of the net.

But the question of movement of a single point can also be answered.

4.1 Localizing deformations with "gap partitioning"

The method of localizing deformation with "gap partitioning" can be derived from the global method. It was found by Pelzer (1971) and thoroughly described by Niemeier (1976). The mean gap is split into two independent items. For this the gap vector $\underline{d}$ as well as the covariance matrix are split:

$$\underline{d} = \begin{vmatrix} \underline{d}_F \\ \underline{d}_v \end{vmatrix} , \quad \underline{Q}_d^+ = \begin{vmatrix} \underline{P}_{FF} & \underline{P}_{Fv} \\ \underline{P}_{vF} & \underline{P}_{vv} \end{vmatrix} .$$

The quantities with the v indices include all components of the point which is suspected to have changed significantly. The partitioning leads to a division of the quatric form $\underline{d}^T\underline{Q}_d^+\underline{d}$ with the following components

$$\underline{d}^T\underline{Q}_d^+\underline{d} = \underline{d}_F^T \underline{P}_{FF}\underline{d}_F + 2\underline{d}_F^T \underline{P}_{vF}\underline{d}_v + \underline{d}_v^T \underline{P}_{vv}\underline{d}_v .$$

By the transformation

$$\underline{\bar{d}}_v = \underline{d}_v + \underline{P}_{vv}^{-1} \underline{P}_{vF} \underline{d}_F$$

$$\underline{\bar{P}}_{FF} = \underline{P}_{FF} - \underline{P}_{Fv} \underline{P}_{vv}^{-1} \underline{P}_{vF}$$

one gets two independent components

$$\underline{d}^T \underline{Q}_d^+ \underline{d} = \underline{d}_F^T \underline{\bar{P}}_{FF} \underline{d}_F + \underline{\bar{d}}_v^T \underline{P}_{vv} \underline{\bar{d}}_v .$$

The portion of the gap refering to the presumably changed point is computed by

$$\theta_v^2 = \frac{\underline{\bar{d}}_v^T \underline{P}_{vv} \underline{\bar{d}}_v}{k},$$

whereas k is the number of the components of the coordinates of the point. In this way it is possible to find a corresponding portion of the gap for each point. The point with the maximum portion of the gap is considered to have changed significantly and is responsible that the previous global test showed deformations in the net. In order to check whether there are further deformations in the net a new global test without the changed point is executed. This method of localizing and elimination of points significantly changed is executed as long as the global test does not show any further deformations.

4.2 Localization with relative error ellipses

Another, geometrically very illustrative method for localizing movements is the analysis with relative error ellipses (Heck et al., 1977).
If one determines the test quantity

$$T_v = \frac{\underline{\bar{d}}_v^T \underline{P}_{vv} \underline{\bar{d}}_v}{2 \cdot s^2}$$

for each point of a net, the hypothesis H_o (no movements of points) can be tested by comparing T_v with the corresponding confidence limits of the F-distribution. For the null hypothesis the probability relation applies

$$P\,\{T_v \geq F_{2,f,1-\alpha} | H_o\} = \alpha .$$

If one looks only at the equal sign the equation of an ellipse turns out which defines as a confidence ellipse the portion of the same significance level α. The semi-axes of the ellipse can be computed by the eigenvalues λ_i of the $\underline{P}_{vv}$ matrix

$$\Lambda_j^2 = 2 \cdot s^2 \cdot \lambda_j \cdot F_{2,f,1-\alpha}, \qquad j = 1,2$$

whereas the direction of the axes are given by the corresponding eigenvectors.

If the relative deformation ellipse is constructed around the point P_{v_1} of the first epoch and the point P_{v_2} of the second epoch lies inside the ellipse, there is no reason to reject the null hypothesis. If the point lies outside, one has to conclude deformations.

5. Outlook

The described methods of deformation analysis give the possibility to recognize global movements of points statistically and to interprete them locally (Caspary, Welsch, 1979). They are known to geodesists and are applied frequently in the field of engineering. Specially in fields where an interpretation of movement is important, they represent a solid basis. It must be tried to obtain the systematics in the movements of points by simultaneous treatment of several epochs in order to interprete and describe crustal movements completely and correctly. The work exists in comprehending and processing the correlations of movements of simultaneous and successive measurements supported by model conceptions to come to a mathematical statistical interpretation of the dynamic process.

6. References

Caspary, W. (1979): Zum Problem der stufenweisen Ausgleichungen. Allgemeine Vermessungsnachrichten 86, 6 (1979), pp. 218-226

Caspary, W. und Welsch, W. (1979), (editors): Seminar über Deformationsanalysen; Schriftenreihe des Wissenschaftlichen Studienganges Vermessungswesen an der Hochschule der Bundeswehr München, Heft 4, München 1979

Heck, B., Kuntz, E., Meier-Hirmer, B. (1977): Deformationsanalyse mittels relativer Fehlerellipsen. Allgemeine Vermessungsnachrichten 84, 3 (1977), pp. 78-86

Koch, K.-R. (1975): Ein allgemeiner Hypothesentest für die Ausgleichungsergebnisse. Allgemeine Vermessungsnachrichten 82, 10 (1975), pp. 339-345

Niemeier, W. (1976): Grundprinzip und Rechenformeln einer strengen Analyse geodätischer Deformationsmessungen. VII. Internationaler Kurs für Ingenieurvermessungen hoher Präzision, Darmstadt 1976

Niemeier, W. (1979): Hypothesentests in geodätischen Netzen, in: Pelzer, H. (editor), Geodätische Netze in Landes- und Ingenieurvermessung, Stuttgart 1979

Pelzer, H. (1971): Zur Analyse geodätischer Deformationsmessungen. Deutsche Geodätische Kommission, Reihe C, Nr. 164, München 1971

Welsch, W. (1979): A Geodetic Micro-Network for Monitoring Tectonic Movements in a Peruvian Earthquake Fault Zone, in: Vogel, A. (editor), Terrestrial and Space Techniques in Earthquake Prediction Research, Braunschweig/Wiesbaden 1979, pp. 531-542

A Geometric-Gravimetric Estimation Procedure for Most Accurate Determination of Relative Point Displacements in Local Areas – Proposal and First Study of Sensitivity

Rainer Kelm

Deutsches Geodätisches Forschungsinstitut, Abt. I, Marstallplatz 8, D-8000 München 22,
Federal Republic of Germany

ABSTRACT

An estimation procedure for the most accurate determination of relative point displacements on the earth surface after repeated geodetic-gravimetric surveys in local areas is proposed. As input data for each epoch all available informations of gravity-sensitive and gravity-nonsensitive types of observations are to be considered. In a first step a local disturbance gravity field is estimated for each epoch using means of nonlinear regression, in a second step a threedimensional geometric-gravimetric estimation of the point displacements for each epoch is performed using least squares theory when some parameters are random variables. In a third step hypothesis tests help to clarify with statistical significance if expected relative motions of earth surface points between the epochs are true.

A first study of sensitivity for the first two steps leads to the conclusion that for the most accurate determination of relative earth point displacements, even in local areas, gravity disturbance quantities have to be considered in the estimation model, otherwise systematical gravity error influence cannot be separated from the purely geometrical adjustments results. Numerical examples are given.

1. INTRODUCTION

In active tectonic areas motions of earth surface parts in relation to one another - relative motions - are expected in mm- to cm-magnitude per year. Simultaneously local disturbances of the earth gravity field - e.g. produced by shallow mass movements beneath the earth surface - may occur in the magnitude of 0.1 mgal or even more according to G. Johnson et al.(198o) For the earthquake prediction research an important indicator of those phenomena is the episodic or continuous change of the motion rate of the earth surface parts. In order to determine such motions significantly geodetic surveys in relatively short time intervals (epochs) have to be performed by using most accurate and reliable measuring instruments. From those measurements threedimensional relative-relative to a fixed local reference system (datum) - coordinates are determined, with which motions and motion rates of earth surface parts can be derived and analysed by statistical hypothesis tests.

The paper restricts on considerations to most accurate repeated determination of threedimensional relative point coordinates presuming that the local area is covered by a geodetic network with point distances between 1 to 20 km depending on the point displacements expected and on the size of the area. To obtain relative point coordinates on mm-accuracy level, observation types of this accuracy range are needed which are sensitive to point displacements in horizontal and vertical directions - here called geometric-sensitive observations. The accuracy demands are today fulfilled, e.g. by distance observations measured with multi-wave-laser instruments - horizontal-sensitive - and levelling with precision levelling instruments - vertical-sensitive -, both having a precision on mm-level. High precision direction measurements are only available on cm-accuracy level on about 10 km, zenith distances even with dm-precision for these distances because of unmodelled refraction influences. The last observation types are nevertheless useful for control and reliability purposes.

With distance measurements alone relative coordinates can be determined only in the near-horizontal region being independent of the gravity field. For vertical displacement determination levelling results are the most accurate available today, but they are also sensitive to local disturbances of the earth gravity field - opposite to the distance measurements. In order to separate these gravimetric influences from geometric displacements a sufficient approximation of the local earth gravity field is needed. For the determination of gravity field parameters observation types should be used which are highly sensitive to gravity field disturbancies and weakly sensitive to geometric variations - so-called gravimetric-sensitive observations. The absolute or relative gravity - today measurable with 10 μgal accuracy - is such a gravimetric-sensitive observation type.

Consequently, for the determination of relative point coordinates highly accurate distance for horizontal and levelling observation for vertical displacement determination, gravity measurements for the gravity field disturbance determination and other kinds of observations for control and reliability purposes are demanded. How can they be combined in a plausible estimation procedure?

2. PROBLEM DEFINITION AND ANALYSIS

For the determination of relative coordinates and local gravity field parameters as decribed in the introduction it is presumed to have approximated coordinates X and approximated gravity field values G for the network points. With geodetic observations $l_{\underline{i}\,\underline{k}}$ delivering some geometric and/or gravimetric information from point i to point k and knowing that $l_{\underline{i}\,\underline{k}}$ are depending on X and G - $l_{\underline{i}\,\underline{k}}(X, G)$ - we search for estimated values $\delta\hat{X}$ and $\delta\hat{G}$ so that the adjusted values $\hat{X} = X + \delta\hat{X}$ and $\hat{G} = G + \delta\hat{G}$ obtain a posteriori accuracies in mm- and µgal-range. A sharp view on the observations $l_{\underline{i}\,\underline{k}}$ (X, G) reveal some problems which have to be analysed at first.

2.1 THE LOCAL DATUM PROBLEM

The earth gravity field G is because of its physical nature referred to a global geocentric reference system R, which can be described by a three-dimensional orthogonal coordinate system $X_R := \{X, Y, Z\}$ with Z-Axis as rotation axis of the earth - $G(X_R)$. So it is sensible to look for $\hat{X}$ in R - $\hat{X}_R$ But when using observation $l_{\underline{i}\,\underline{k}}$ (X, G) with informations only about the configuration of the points - represented by relative coordinates $X_{\underline{i}\,\underline{k}}$ - only adjusted relative coordinates $\hat{X}_{\underline{i}\,\underline{k}}$ independent on R can be obtained by any estimation procedure. Fortunately this fact is not a restriction on our objective, because with relative coordinates $\hat{X}_{\underline{i}\,\underline{k}}$ all kind of relative point motions can be determined by suitable statistical hypothesis tests - e.g. K. R. Koch (1980) - using the unbiased estimators $\hat{X}_{\underline{i}\,\underline{k}}$.
For computational purposes and for the connection to R, however, a local datum D in R is needed which is fixed in all epochs and defined by a minimum set of X_R, called ${}_DX_R$. Studies of such a local datum are in preparation.

2.2 HORIZONTAL-VERTICAL SENSITIVITY PROBLEM

The observations $l_{\underline{i}\,\underline{k}}$ (X, G) can be separated in geometric-sensitive $g_{\underline{i}\,\underline{k}}(X, G)$ and gravimetric-sensitive $w_{\underline{i}\,\underline{k}}(X, G)$. $g_{\underline{i}\,\underline{k}}(X, G)$ can be divided in two groups, $h_{\underline{i}\,\underline{k}}(X, G)$ being horizontal-sensitive and $v_{\underline{i}\,\underline{k}}(X, G)$ being vertical-sensitive. Classically $h_{\underline{i}\,\underline{k}}$ and $v_{\underline{i}\,\underline{k}}$ are adjusted in different models presuming that the loss of information of $h_{\underline{i}\,\underline{k}}$ for the vertical direction and of $v_{\underline{i}\,\underline{k}}$ for the horizontal direction is negligeable. But to obtain the demanded mm-accuracy the full information of all observations $l_{\underline{i}\,\underline{k}}$ is needed. For example, a distance measurement $s_{\underline{i}\,\underline{k}}$ of about 4 km between points i and k with height difference of about 400 m gives an information also for the vertical direction in the mm-range. This significant value should be taken into account! Therefore it is proposed to combine $h_{\underline{i}\,\underline{k}}$ in a common threedimensional estimation model - even for local areas!

2.3 GEOMETRIC-GRAVIMETRIC SENSITIVITY PROBLEM

The observation values $l_{\underline{i}\,\underline{k}}(X, G)$ are depending on geometric influences X and gravimetric influences G. The problem is to separate both influences sufficiently. An impression of the sensitivity of $l_{\underline{i}\,\underline{k}}$ to X and G is given in table 1 - computed for a three-point example decribed later. Variations in δX_{hor}, δX_{vert} lead to significant disturbances in $\underline{h}_{\underline{i}\,\underline{k}}$ and $\underline{v}_{\underline{i}\,\underline{k}}$, variations in δG significantly only in $w_{\underline{i}\,\underline{k}}$. From this it follows that a separation between X and G influences can be achieved when estimating $\delta\hat{G}$ by gravimetric-sensitive observations $w_{\underline{i}\,\underline{k}}$ and $\delta\hat{X}$ by geometric-sensitive observations $g_{\underline{i}\,\underline{k}}$ mainly.

Some observation types - like $\delta V_{\underline{1}\,\underline{2}}$ in the example - are geometric- and gravimetric dependent. Because of the geodetic objective of with first priority determining geometrical values $\hat{X}_{\underline{i}\,\underline{k}}$, all information should be considered for estimating $\hat{X}_{\underline{i}\,\underline{k}}$ by using all observations $l_{\underline{i}\,\underline{k}}$, but for the estimation of $\delta\hat{G}_{\underline{i}\,\underline{k}}$ the informations from the gravimetric-sensitive ob-

Tab. 1: Sensitivity Scheme

		coordinates horizontal δX_h	coordinates vertical δX_v	gravity δG
parameter variation		~ 1 cm	~ 1 cm	~ 0,7 mgal
disturbance in	unit			
distance $\delta s_{\underline{1}\,\underline{2}}$	cm	14	2	0
potential difference $\delta v_{\underline{1}\,\underline{2}}$	$\frac{m^2}{sec^2}$	0,004	0,14	0,012
rel. gravity $\delta w_{\underline{1}\,\underline{2}}$	µgal	2	5	160

servations $g_{\underline{i}\,k}$ should be sufficient. In the following a sketch for the survey and estimation procedure as a result of this problem analysis is proposed.

3. PROPOSED SURVEY AND ESTIMATION PROCEDURE

3.1 THEORETICAL VIEW

Let the gravity potential of the earth be

$$W(\underline{X}_i) = V(\underline{X}_i) + C(\underline{X}_i) \tag{1}$$

with

$\underline{X}_i$: radius vector to a point $\underline{i}$ on the earth surface (fig.1)

$V(\underline{X}_i)$: gravitational potential

$C(\underline{X}_i)$: centrifugal potential

When regarding only relative gravity values in local areas the relative contrifugal potential is not significantly depending on rotation irregularities of the earth. Therefore $C(X_i)$ can be neglected in the further considerations.

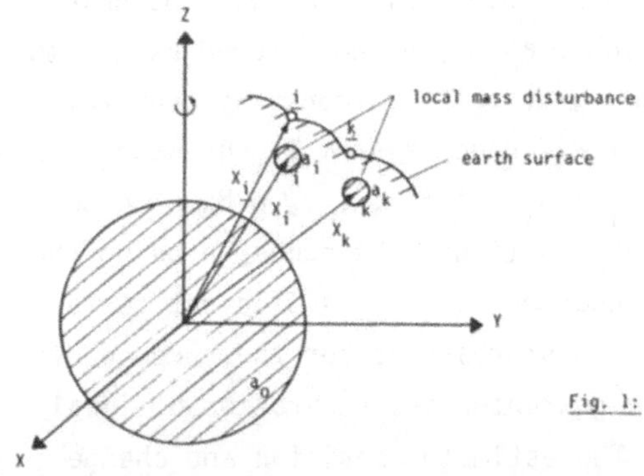

Fig. 1: Mass point model

The gravitational potential is given by

$$V(\underline{X}_i) = G \sum_{\text{Earth}} \frac{\bar{m}\,(X_k)}{|\underline{X}_i - X_k|}\, dX_k \tag{2}$$

with G : gravitational constant

$\bar{m}$: mass density of a mass point X_k

or for a mass point approximation

$$V(\underline{X_i}) = G \sum_{k=1}^{n} \frac{m_k\ (X_k)}{|\underline{X_i} - X_k|} \tag{3}$$

m_k : point mass

n : number of mass points depending on the degree of approximation.

For the determination of relative coordinates in local areas a splitting of $V(\underline{X_i})$ in a global and a local part is proposed:

$$V(\underline{X_i}) = \frac{a_o}{|\underline{X_i}|} + \sum_{k=1}^{n} \frac{a_k}{|\underline{X_i} - X_k|} \tag{4}$$

with

a_i : $G \cdot m_i$ $\quad i = 0, 1, \ldots, n$ (fig. 1)

n : number of network points .

The first term $\frac{a_o}{|\underline{X_i}|}$ represents the global potential which will be held fixed for all epochs. a_o may be considered as a homogeneous sphere with a constant mass and gravitational constant centered in the origin of the global terrestrial reference system R. For each epoch the local potential parameters a_i , X_i $\quad i = 1, \ldots, n$ have to be estimated with gravimetric-sensitive observations. The number n being the point number of the local network is regarded as to be sufficient for a significant estimation of local gravity disturbancies for the purpose of final coordinate estimation and should guarantee the separation of local gravimetric and geometric influences. The estimated position and change of position of the mass points $\frac{a_i}{|X_i|}$ could also be a helpful tool to the geophysists for the interpretation of possible mass movements underneath the earth surface.

3.2 SURVEY AND ESTIMATION PROCEDURE STEPS

For the realisation of the high accurate relative coordinate determination as proposed here the survey and estimation procedure can be divided into separated steps. According to the introducing character of this paper these steps will be presented in a useriptive way descriptive by a three-point example which is demonstrated in fig. 2 and table 2. Theoretical developments are not dealt with here. For the first epoch, following steps are proposed:

I Survey procedure

Step I/1

Measure gravimetric-sensitive observations $g_{\underline{i}\,\underline{k}}$ with highest accuracy, e.g.

absolute gravity $\gamma_{\underline{i}}$ $\pm$ 10 - 100 µgal

relative gravity $\gamma_{\underline{i}\,\underline{k}}$ $\pm$ 10 - 50 µgal

relative longitude $\gamma_{\underline{i}\,\underline{k}}$ $\pm$ 0,05"

relative latitude $\phi_{\underline{i}\,\underline{k}}$ $\pm$ 0,05"

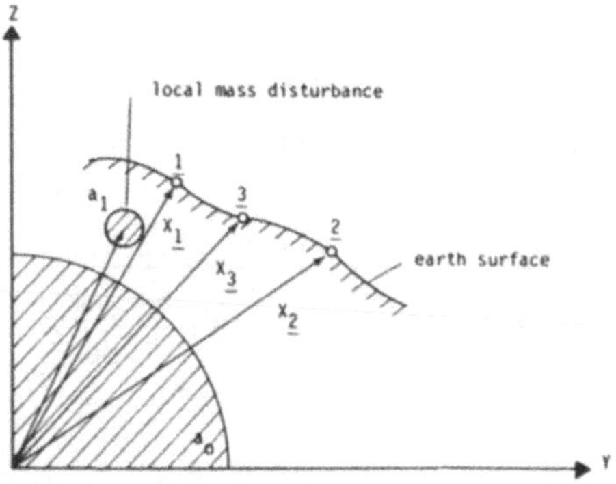

Fig. 2: Three-point example

Tab. 2: True values of the three-point example

epoch / parameters	unit	1 values	2a) value differences to epoch 1	2b)	2c)
$Y_{\underline{1}}$	m	4290307,000	0,010	0	0,010
$Z_{\underline{1}}$	m	4703236,000	0,020	0	0,020
$Y_{\underline{2}}$	m	4294003,000	0,020	0	0,020
$Z_{\underline{2}}$	m	4700490,000	0,040	0	0,040
$Y_{\underline{3}}$	m	4292150,000	0,030	0	0,030
$Z_{\underline{3}}$	m	4701508,000	0,020	0	0,020
a_0	m^3/sec^2	$3,9743\cdot10^{14}$	0	0	0
a_1	m^3/sec^2	2000	50	50	0
Y_1	m	$4,287\cdot10^6$	$-0,004\cdot10^6$	$-0,004\cdot10^6$	0
Z_1	m	$4,7\cdot10^6$	$0,001\cdot10^6$	$0,001\cdot10^6$	0

Note: Today relative gravity measurements of this accuracy can be performed without high expense. Absolute gravity measurements are more expedient; for the determination of relative longitude and latitude satellite space technics could be developed in connection with range and range rate measurements.

The $g_{\underline{i}\,\underline{k}}$ - observations could be not only partly or totally measured on or between the net points, but also on additional points within the local area. The only restriction is that approximated values of the points are known till an accuracy of 1 m within the global reference system relative to the local datum - a demand which is easily fulfilled.

Three point example

Note: Point $\underline{1}$ and $\underline{2}$ are network points
Point $\underline{3}$ is an additional point

true relative gravity values:

$$\gamma_{\underline{1}\,\underline{2}} = -\ 149{,}251_9 \cdot 10^{-5}\ \frac{m}{sec^2}$$

$$\gamma_{\underline{1}\,\underline{3}} = 7{,}1081 \cdot 10^{-5}\ \text{"}$$

$$\gamma_{\underline{2}\,\underline{3}} = 156{,}360_0 \cdot 10^{-5}\ \text{"}\ .$$

Step I/2

Measure all available observations $l_{\underline{i}\,\underline{k}}$ of highest accuracy between the net points, e.g.

distances $S_{\underline{i}\,\underline{k}} \pm 0.1$ ppm

geopotential differences $V_{\underline{i}\,\underline{k}} \pm 0{,}01\ \frac{m^2}{sec^2}$

(from geometric levelling and relative gravity measurements).

Example:

True observations between the network points $\underline{1}$ and $\underline{2}$

$$S_{\underline{1}\,\underline{2}} = 4604{,}447_0\ m$$

$$V_{\underline{1}\,\underline{2}} = -\ 4547{,}\ 746_6\ \frac{m^2}{sec^2}\ .$$

II Estimation Procedure

Step II/1

Choose a local datum within the global terrestrial reference system according to the observation types. When e.g. distances, azimuths, directions, potential differences are measured, a local datum with 3 translation parameters is needed. This datum would be the X, Y, Z-coordinates of an arbitrary net point.

Example

$S_{\underline{1}\,\underline{2}}$, $V_{\underline{1}\,\underline{2}}$ measured: 2 translation parameters needed

local datum: Point $\underline{1}$ with $Y_{\underline{1}} = 4290307{,}000$

$Z_{\underline{1}} = 4703236{,}000$ true values

Step II/2

Choose a suitable global constant a_o of (4).

Example

$$a_o = 3{,}9743 \cdot 10^{14} \frac{m^3}{sec^2}$$

Step II/3

Determine approximated values $X_{\underline{i}}$ for the network points by using geometric-sensitive observation types only, e.g. distances, directions, zenith distances etc. The geometric-sensitive observations leadto a most accurate a priori evaluation of the coordinates so that only a few iterations for the final estimation of the relative coordinates are expected. For the additional gravimetric points approximated coordinates on m-accuracy level are sufficient.

Example

Approximated values [m]	absolute differences to true values [m]
$Y_{\underline{2}} = 4294002{,}95$	0,05
$Z_{\underline{2}} = 4700489{,}95$	0,05
and	
$Y_{\underline{3}} = 4292150{,}50$	0,50
$Z_{\underline{3}} = 4701508{,}50$	0,50

Step II/4

Estimate the local potential parameters $\hat{a}_{\underline{i}}$, $\hat{X}_{\underline{i}}$ (i = 1, n) with gravimetric-sensitive observations only, e.g. with relative gravity values.

The parametric function of a gravimetric-sensitive observation can be regarded as an operator 0 operating on the potential $V(X_i)$ according to (4):

$$OV\,(X_i) = 0\,\frac{a_o}{|X_i|} + 0\,\sum \frac{a_k}{|X_i - X_k|} \qquad \begin{array}{l} i = 1, \underline{m} \\ k = 1, n \\ \underline{m} \geqq n \end{array} \tag{5}$$

The nonlinear regression model (5) can be solved by well-known regression models, especially when linearizing in a suitable way.

Example

With the relative gravity values $\gamma_{\underline{1}\,\underline{2}}$, $\gamma_{\underline{1}\,\underline{3}}$ and $\gamma_{\underline{2}\,\underline{3}}$ given in step I/1 a linear least square regression model is used to estimate $\hat{a}_i$ (i = 1, 2) for fixed coordinates X_i (i = 1, 2) the position of which is to be seen in fig. 3.

The estimated values are:

$\hat{a}_1$ = 2003,4

$\hat{a}_2$ = - 60,8 .

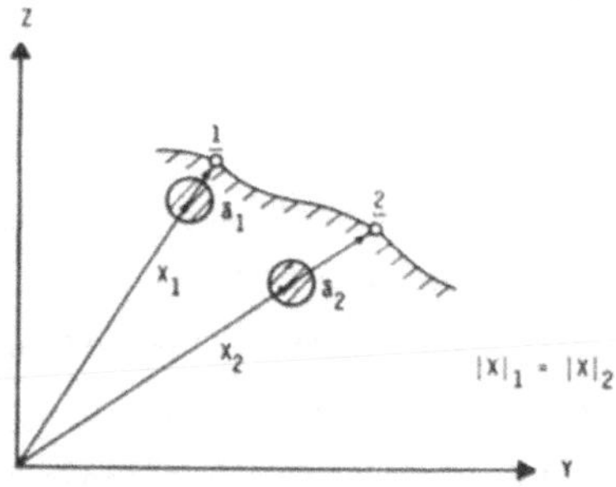

Fig. 3: Gravimetric estimation presumptions

The negative sign of $\hat{a}_2$ indicates that at this position the effect of a mass defect in regard to the global a_o-mass distribution seems to be acting. When analysing the true situation (fig. 2) this effect is explicable.

Step II/5

Estimate the relative coordinates $\hat{X}_{\underline{i}}$ of the network points with all available observations $l_{\underline{i}\,\underline{k}}$ between the network points and by considering the a priori knowledge of $\hat{a}_i$, $\hat{X}_i$ (i = 1, n) and their covariance matrices.

For this step a least square estimation method with random variables - e.g. C. Rao (1973) - is proposed which can be performed e.g. by collocation methods.

The advantage of this procedure is that using high accurate local potential parameters - estimated only by gravimetric-sensitive observations - for geometric estimator values the systematic influence of local gravity disturbancies can be significantly reduced, and the numerical system will be stabilized. An analysis of this estimation procedure is in preparation.

Example

The geometric and gravimetric estimation results for the three point example are - using the observation results of step I and the results of step II/1 to II/4:

$\hat{Y}_{\underline{2}} = 4294003{,}00_3$ m $\qquad \hat{Z}_{\underline{2}} = 4700490{,}000_4$ m

$\hat{a}_1 = 2003{,}4 \; \frac{m^3}{sec^2}$ $\qquad \hat{a}_2 = -\,60{,}8 \; \frac{m^3}{sec^2}$

$\hat{Y}_1 = 4{,}2872 \cdot 10^6$ m $\qquad \hat{Z}_1 = 4{,}6998 \cdot 10^6$ m

$\hat{Y}_2 = 4{,}2906 \cdot 10^6$ m $\qquad \hat{Z}_2 = 4{,}6968 \cdot 10^6$ m

For the next epochs the steps I/1 and I/2 of the survey procedure and steps II/4 and II/5 of the estimation procedure have to be repeated.

Example

2. Epoch

Step I/1

Observation values (true values) of the relative gravity

$\gamma_{\underline{1}\,\underline{2}} = -\,144{,}8115 \cdot 10^{-5} \; \frac{m}{sec^2}$

$\gamma_{\underline{1}\,\underline{3}} = 9{,}2177 \cdot 10^{-5}$ "

$\gamma_{\underline{2}\,\underline{3}} = 154{,}0292 \cdot 10^{-5}$ "

Step I/2

Observation values between the network points $\underline{1}$ and $\underline{2}$

$S_{\underline{1}\,\underline{2}} = 4604{,}443_1$ m

$V_{\underline{1}\,\underline{2}} = -\,4547{,}\,892_3$ m

Step II/4

Estimated potential parameters

$\hat{a}_1 = 554{,}8 \; \frac{m^3}{sec^2}$

$\hat{a}_2 = -60{,}8 \; \frac{m^3}{sec^2}$

Step II/5

Final estimation results

$\hat{\underline{Y}}_2 = 4294003{,}012_5$ m $\qquad \hat{\underline{Z}}_2 = 4700490{,}023_3$ m

$\hat{a}_1 = 554{,}8 \ \frac{m^3}{sec^2} \qquad \hat{a}_2 = -\ 60{,}8 \ \frac{m^3}{sec^2}$

$\hat{Y}_1 = 4{,}2872 \cdot 10^6$ m $\qquad \hat{Z}_1 = 4{,}6998 \cdot 10^6$ m

$\hat{Y}_2 = 4{,}2906 \cdot 10^6$ m $\qquad \hat{Z}_2 = 4{,}6968 \cdot 10^6$ m

When having measured the local network in several epochs, in step II/6 relative displacements of earth surface parts - as e.g. geophysically expected - have to be estimated by statistical hopothesis test methods. The relative coordinates of each epoch estimated in step II/5 are unbiased estimators, and so well-known statistical tests for unbiased estimation results can be used. In the following example the input of such a hypothesis test is presented for illustration.

Example

The differences ${}_{12}\Delta\hat{\underline{Y}}_2$, ${}_{12}\Delta\hat{\underline{Z}}_2$ of the relative coordinates between the 1. and 2. epoch are:

$${}_{12}\Delta\hat{\underline{Y}}_2 = 0{,}012_2 \text{ m} \qquad {}_{12}\Delta\hat{\underline{Z}}_2 = 0{,}022_9 \text{ m} \quad .$$

The a posteriori covariance matrix C $(\hat{\underline{X}}_2)$ is not computed in the example.

The null hypothesis H_o is: there is no relative displacement between point $\underline{1}$ and $\underline{2}$:

$$H_o : \{{}_{12}\Delta Y_{\underline{2}} = 0 \qquad {}_{12}\Delta Z_{\underline{2}} = 0\} \quad .$$

A qualitative anylysis reveals that with the both observation types, distance and potential difference, both having an a priori geometric-sensitive accuracy in mm-range, the H_o-hypothesis would fail, the differences as given above are significant.

The true values of this example are, indeed:

$${}_{12}\Delta Y_{\underline{2}} = 0{,}010_o \text{ m}$$

$${}_{12}\Delta Z_{\underline{2}} = 0{,}020_o \text{ m}$$

4. ANALYSIS OF SENSITIVITY

The question is still open if the results given in the proposed estimation procedure are approximating the reality in a better way than classical methods which do not consider local gravity field disturbances. We want to analyse this question only by a numerical analysis of the three-point example. Theoretical reasoning will be published later on.

Let us study four epochs, epoch 1 being the zero epoch, epoch 2a) presuming a geometric displacement of the two points in cm-level and a mass accumulation of 2,5 % with translation in km-level - imitating a magma flow and producing a mgal-level gravity anomaly -, epoch 2b) only presuming the local gravity disturbances of 2a) and epoch 2c) presuming only the geometric displacement of 2a). This "true" situation is illustrated by table 2. The true observation values are for the preestimation of the local potential parameters

	epoch 1	differences epoch 1 2a	2b	2c
$\gamma_{\underline{1}\,\underline{2}}$	$-149{,}2519\cdot 10^{-5}\ \frac{m}{sec^2}$	$+4{,}4404\cdot 10^{-5}$	$4{,}4408\cdot 10^{-5}$	$-0{,}0064\cdot 10^{-5}$
$\gamma_{\underline{1}\,\underline{3}}$	$7{,}1081\cdot 10^{-5}$	$2{,}1096\cdot 10^{-5}$	$2{,}1137\cdot 10^{-5}$	$-0{,}00041\cdot 10^{-5}$
$\gamma_{\underline{2}\,\underline{3}}$	$156{,}3600\cdot 10^{-5}$	$-2{,}3308\cdot 10^{-5}$	$-2{,}3271\cdot 10^{-5}$	$0{,}0023\cdot 10^{-5}$

and additionally for the final estimation

	epoch 1	2a differences to epoch 1	2b	2c
$V_{\underline{1}\,\underline{2}}$	$-4547{,}74661\ \frac{m^2}{sec^2}$	-0,14564	0,0652	-0,21084
$S_{\underline{1}\,\underline{2}}$	4604,446981 m	-0,003901	0	-0,003901

The estimation is performed first in the "classical" way: only a normal part of the earth potential constant for all epochs is considered, as it corresponds to the usual horizontal adjustment by the reduction of the observations and to the vertical adjustment to the height corrections. This procedure is simulated here by taking only the final estimation using only the global potential information of a_0 . Finally the proposed estimation procedure as described in the previous chapter is computed. Note that for this first numerical example only true - noiseless - observation values are taken into account for the computation, and so the results are affected by model errors only. As final results relative coordinates are estimated which are in this case the coordinate differences between point 1 and 2 : $\Delta\hat{Y}$, $\Delta\hat{Z}$. Additionally the estimation of the local mass parameters $\hat{a}_1$, $\hat{a}_2$ are given. Note that the distribution of the net points was not optimal for the evaluation of $\hat{a}_1$, $\hat{a}_2$. The results are given in table 3.

A sensitivity analysis of table 3 reveals: There are significant differences in the estimation of relative coordinates of one epoch. The classical method leads to 5 - 10 mm errors whereas the new one only to 0,3 - 3,3 mm errors (cp. row 3,4). The objective, however, is to estimate the relative point displacements between the two epochs. Here two cases can be seen (cp. row 11,12). When the local mass distribution is unaltered, both methods lead to correct results within reasonable tolerance intervals. In the case of mass disturbances the classical method delivers results with an 4 - 5 mm error, whereas the proposed method reduces the model error to 2 - 3 mm, even though the gravimetric estimation procedure is not optimal for this example. Nevertheless, the new method leads to a significant improvement of the estimation results.

5. CONCLUSIONS AND ASPECTS

A new survey and estimation procedure for the determination of relative three-dimensional point displacements in local areas is proposed. The new method combines horizontal-sensitive observations e.g. distances - and vertical-sensitive ones - e.g. potential differences as result of levelling and gravity measurements - and separates local gravity disturbances by using gravimetric-sensitive observations - e.g. relative gravity measurements - in a preestimation procedure. Only a first sketch of the estimation procedure with a numerical simulation study for a three-point-net is given. Some general conclusions may already be taken which have to be strengthened by theoretical and numerical examples of real geodetic networks:

1. For most accurate determination of relative three-dimensional point displacements in local areas horizontal-sensitive observations as distances, directions, azimuth-measurements and vertical-sensitive observations as potential differences derived from levelling and gravity measurements have to be combined in one estimation model.

Tab. 3: Estimation results of the three point example

row	parameters	unit	1 classical	1 proposed	2a classical	2a proposed	2b classical	2b proposed	2c classical	2c proposed
1	$\hat{a}_1$	$\frac{m^3}{sec^2}$	-	2003,4	-	554,8	-	554,8	-	2003,4
2	$\hat{a}_2$	$\frac{m^3}{sec^2}$	-	-60,8	-	-60,8	-	-60,8	-	-60,8
3	$\delta\Delta\hat{Y}$	m	$0{,}009_0$	$0{,}000_3$	$0{,}015_0$	$0{,}012_5$	$0{,}005_0$	$0{,}005_0$	$0{,}019_0$	$0{,}010_3$
4	$\delta\Delta\hat{Z}$	m	$0{,}012_2$	$0{,}000_4$	$0{,}026_7$	$0{,}023_3$	$0{,}006_8$	$0{,}003_4$	$0{,}032_1$	$0{,}020_4$
5	$\delta\Delta Y_{true}$	m	0,000		0,0100		0,0000		0,0100	
6	$\delta\Delta Z_{true}$	m	0,000		0,0200		0,0000		0,0200	
7	$12^{\delta\Delta Y}$	m	-	-	0,0060	$0{,}012_2$	-0,0040	$0{,}002_2$	0,0100	$0{,}010_0$
8	$12^{\delta\Delta Z}$	m	-	-	$0{,}014_5$	$0{,}022_9$	$-0{,}005_4$	$0{,}003_0$	$0{,}019_9$	$0{,}020_0$
9	$12^{\delta\Delta Y}{}_{true}$	m	-	-	0,0100		0,000		0,0100	
10	$12^{\delta\Delta Z}{}_{true}$	m	-	-	0,0200		0,000		0,0200	
11	Y-difference	m	-	-	$0{,}004_0$	$0{,}002_2$	$0{,}004_0$	$0{,}002_2$	0,000	0,000
12	Z-difference to true values	m	-	-	$0{,}005_5$	$0{,}002_q$	$0{,}005_4$	$0{,}003_0$	0,001	0,000

2. In order to separate local gravity disturbances from the geometric results gravimetric-sensitive observations like absolute andrelative gravity measurements have to be measured in addition. Especially in active tectonic regions mass displacements underneath the earth surface may occur which alter the gravity values on mgal-level without influencing significantly earth surface motions. These effects have to be modelled so that the geometrical results are not systematically falsified by gravimetric disturbance influences.

3. The proposed method allows a rigorous a posteriori error estimation which is needed for the hypothesis test phase. The classical method may feign too optimistic accuracy results, because the gravity field parameters are considered as constant for the epochs and errorless. The new method should lead to more realistic results for the parameters and their precision.

Further developments in theory and application of the proposed survey and estimation procedure may give hope for an essential step nearer to the best approximation of reality by using more heterogenous and accurate observation informations than before.

6. REFERENCES

1. Johnson, G.; A. Jörnsson; S. Sigurdsson (1980): Gravity and Elevation Changes caused by Magna Movement Beneath the Krafla Caldera, Northeast Iceland, J. Geophys. 47 (1980), 132 - 140

2. Koch, K.R. (1980): Parameterschätzung und Hypothesentests in linearen Modellen, Bonn 1980

3. Rao, C.R. (1973): Linear Statistical Interference and Its Applications, New York 1973

Suggestions for High Precision Gravity Measurements in Geodynamic and Earthquake Prediction Research

Bernd Richter

Institut für Angewandte Geodäsie, Richard-Strauß-Allee 11, D-6000 Frankfurt/Main 70, Federal Republic of Germany

Summary:

Based on experiences with precise gravity measurements along the Fennoscandian Gravity Line, vertical gradient determinations and measurements on the German Gravity Base Net (DSGN 76), instrumental improvements have been made to reduce the influences of temperature effects, levelling and periodic errors. Investigations have been carried out to study the drift behaviour and the linearity of the La Coste Romberg gravity meters to be used for the observation of high precision gravity nets.

1. Introduction

Measurements of temporal gravity variations can be used as an inexpensive and rapid means of detecting, monitoring and studying crustal deformation associated with many types of active geological processes. The following statement can be made concerning the study of variations of gravity at a fixed point on the surface of the earth in relationship to crustal deformations, if other effects as the earth- and oceanic tides, variations in atmospheric pressure, instrumental errors and changes in the watertable have been removed:

i. Gravity changes accompany most types of deformation and knowledge of such changes at distributed points can yield information on the spatial distribution of the deformation.

ii. Unambiguous estimates of elevation changes can be made on the basis of gravity data alone, because gravity can vary by displacement of the observation point along the free-air-gravity gradient or variation of the sub-surface density field.
But measurements of gravity and height variations (Δg and Δh) at a few points may permit the determination of the differential $\Delta g/\Delta h$ that then can be used to infer elevations changes from measured gravity changes.

iii. The wide range of possible values of the vertical gradient shows that the observed value can effectively constrain the interpretation of the causes of deformation (JACHENS 1977).

Studying all these questions, it is necessary to establish local gravity nets with a precision better than 10 µgal (BEETZ, BREIN 1973; TORGE 1978; GROTEN 1977). For these precise gravity measurements LaCoste-Romberg gravity meters Models D or G are normally used, so that the following instrumental investigations and experiences are restricted to these types.

2. Instrumental Investigations

2.1 Temperature and Voltage Stabilization

Influences of strong temperature changes on gravity measurements induce irregular drift-behaviour and effects on the scale of the meters (BREIN 1978; BOEDECKER 1979; GERSTENECKER 1973). To predict this behaviour using the adjustment of laboratory tests the result is not very satisfactory as the models do not correspond to the physical reality. Therefore, using an idea of Prof. Gerstenecker the gravity meters are put into an additional thermostated aluminium case in order to stabilize the temperature and so the influence on the gravity-meter to better than one degree.

The voltage of the batteries is regulated and stabilized due to the sensitivity of the temperature control to variations in the voltage.

2.2 Levels

An additional instrumental investigation was made, replacing the levels by more accurate ones (20 sec/2 mm) and, initially only for test purposes, adding electronic tilt-levels to replacing the glass-levels. In combination with new foot screws having a pitch of 0.5 mm the mis-levelling error should be reduced to a minimum.

3. Instrumental Effects

Dependent on the construction of the gravity-meter, errors may occur which can not be traced through instrumental investigations. The most important ones are the unclamping effect on the beam, the drift-behaviour of the spring, the scale-factor and the reading-errors.

3.1 Reading Error and Unclamping Effect

The reading error of normally ± 5 μgal can be reduced by a factor ten by using an external digital voltmeter (WENZEL 1977).

The unclamping effect is a more or less random process influenced by the attention of the observer and the delicacy with which the instrument is handled. The elastic feedback of the spring after unclamping is terminated after approximately fiveteen minutes. It is therefore very important for the observations to schedule a constant waiting time for the first reading after unclamping.

3.2 Drift Behaviour

Another problem is the determination of the drift. Due to the manner normally to connect two or more points it is possible to obtain an impression of the drift-rate by comparison of the two readings at the starting-point. The question arises whether a drift is a random process or, if not, how it can be calculated.

For this reason the daily drift for two measuring campaigns on the German Gravity Base Net (DSGN 1976) were determined, (BOEDECKER et al. 1979) normalized, and added together. In FIG.1 the result for one gravimeter throughout the two campaigns is shown. It is clearly seen that there is no random process and one can say that the drift can be described by a linear function for a single period of observations. However, this instrument displays a very strong driftrate. Turning to another example the situation is not so clear (FIG. 2). There a quite different drift behaviour can be seen, even during the two

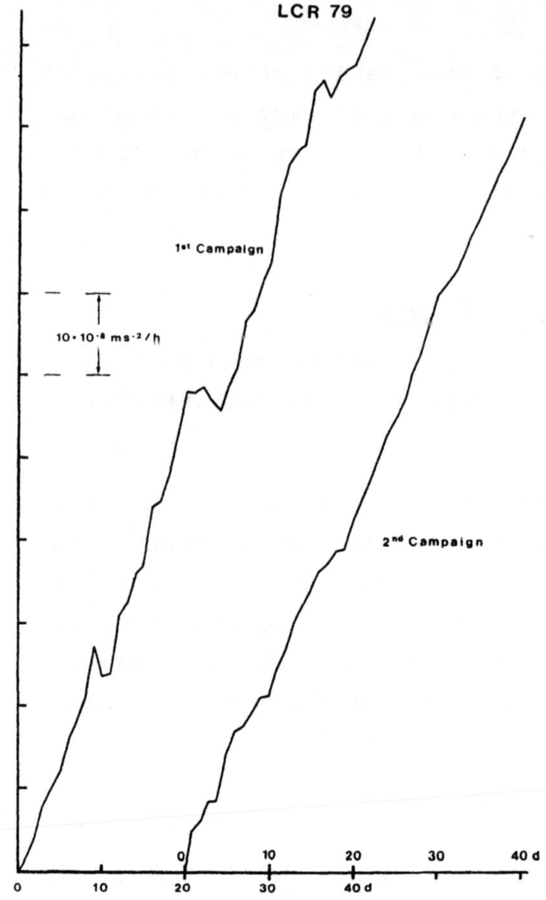

FIG. 1:

Drift behaviour during two campaigns

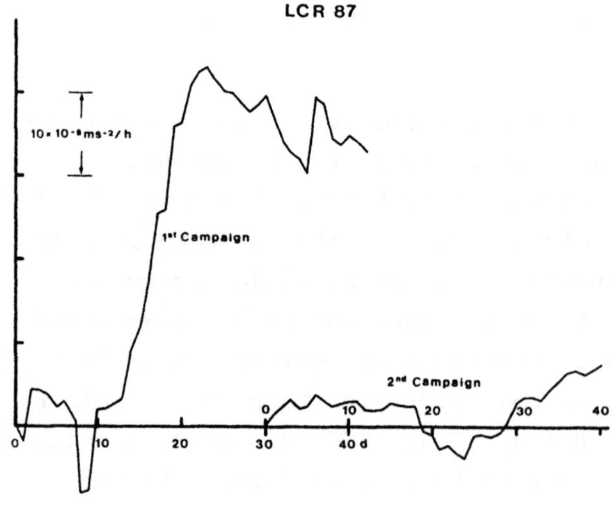

FIG. 2:

Drift behaviour during two campaigns (LCR 87).

campaigns. In this case it is not advisable to approximate the drift by a single linear function but perhaps by piece-wise linear of higher polynomial functions.

Nevertheless it can be seen that a significant drift behaviour is present. Taking these facts into account we conclude that drift may not be neglected and due to the complicated function a daily drift parameter has to be introduced. The disadvantage of this method lies in the risk of interpreting some real errors as drift.

3.3 Periodic Deviations in the Scale Factor

Turning now to a problem encountered during the adjustment procedure, namely the scale factor, both linear and periodic aspects have to be considered. In local test areas the linear scale factor is insignificant. This does not hold for the periodic one.

As an example take the experiences acquired the adjustment procedure with the German Gravity Base Net (DSGN 76). First point values were calculated for the total net in an adjustment using the observations for a single instrument. The weights for all observations were set to one and the results gave an m_0 between 6 and 11 μgal. Taking then all measurements together, and again setting the weights to one for each instrument, a m_0 of 18 μgal was obtained. This leads to the conclusion that there is an unmodelled systematic effect involved in the process. In the next diagram (FIG. 3a) the adjustment point values for two instruments during the two campaigns with respect to a mean value are shown. Following the lines the deflections from the median line are quite similar throughout the two measuring epochs. Corresponding curves are given for the other instrument.

Consider now the periodic errors due to the gear (KIVINIEMI 1974). The periods are 1, 3.94, 7.88, approximately 35, 70, 603 and 1206 mgal. Analysing these residuals with respect to the predicted periods one sees the amplitudes, the phases and the phase-shifts for all periods. Most significant are the

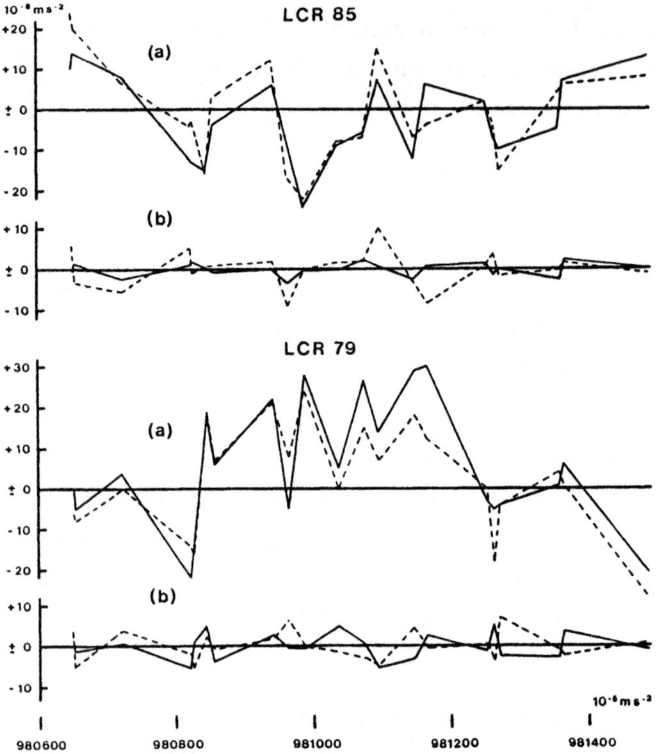

FIG. 3: Differences between the mean point values and the point values (Instruments: LCR 85 and LCR 79).

——— First campaign

------- Second campaign

(a) First adjustment

(b) Second adjustment taking the periodic errors into accou

thirty-five and seventy mgal periods. The amplitudes range up to 25 µgal. Taking the periodic errors into account, and repeating the procedure of single adjustment the following picture is obtained (FIG. 3b). The results of the point values are closer together and no systematic effect can be shown. The total adjustment using all instruments shows that, weighting all observations equal, the m_0 (11 µgal) is the same as that computed for the single adjustments.

These conclusions are based on the experience with four gravity meters observing in parallel during two campaigns only. Although the results are somewhat correlated. It is shown that periodic errors in the scale factor exist. This is very important and must be taken into account if results from several measuring campaigns with different instruments are to be compared.

4. Conclusions

If gravity measurements are used for the purpose of studying crustal deformations, it is necessary to establish a gravity base net which includes points with controlled heights such as those available in the geodetic network for the determination of the $\Delta g/\Delta h$ factor.

Beside the instrumental investigations it is necessary in designing the net and the observation scheme to take care of the drift-behaviour and the scale-factor. That means, that for the drift determination the minimum of two repeated readings at the starting point are necessary. To avoid problems with the scale factor a gravity net with small gravity differences is suggested. If this method is not applicable, one has to make considerable efforts to obtain gravity values precise enough for secular gravity variation studies.

Compared with levelling and extra-terrestrial techniques, gravity technique is relatively rapid and economical when applied to networks with dimensions of ten to a few hundreds of kilometers, for which motor transportation can be used. Modern relative gravity meters are capable of yielding measurements of gravity differences with standard deviations of 5 - 10 µgal, for observations with a single instrument.

5. References

BEETZ, H.; BREIN, R.: High Precision Gravity Measurements Along Special Lines. 2nd Int. Symp. Geodesy and Physics of the Earth, Potsdam 1973

BOEDECKER, G.: Instrumental Capabilities of La Coste-Romberg Gravity Meters for the Detection of Small Gravity Variations With Time. Paper pres. XVII I.U.G.G. Meeting, Canberra 1979

BOEDECKER, G.; REINHART, E.; RICHTER, B.: The New Gravity Base Net 1976 of the Federal Republic of Germany (DSGN 76). Paper pres. XVII I.U.G.G. Meeting, Canberra 1979

BREIN, R.: Some Experiments in High Precision Measurements of Small Gravity Differences to Determine Vertical Gradients of Gravity and to Investigate Recent Crustal Movements. Boll. d. Geo. Teo. e. App. Vol. XX, no. 80, Triest 1978

GERSTENECKER, C.: Die genaue Messung von kleinen Schwereunterschieden. Publ. Deutsche Geodätische Kommission, Reihe C, Heft 191, München 1973

GROTEN, E.: Results Obtained in High Precision Gravimetry 1975 - 1977. Pres. Int. Symp. on "Non-Tidal Gravity Variations and Methods of Their Study", Trieste 1977

JACHENS, R.: Temporal Gravity Changes as Applied to Studies of Crustal Deformation. Proceedings of Conference VI: Measurement of Stress and Strain Pertinent to Earthquake Prediction, Menlo Park 1977

KIVINIEMI, A.: High Precision Measurements for Studying the Secular Variations in Gravity in Finland. Publ. Finn. Geod. Inst. No. 78, Helsinki 1974

TORGE, W.: Zum Aufbau von Präzisionsschwerenetzen. Wiss. Arb. Lehrst. Geod., Phot. und Kart., T.U. Hannover, Nr. 67, Hannover 1978

WENZEL, H.-G.: Evaluation of a Digital Voltmeter Reading for La Coste-Romberg Model D and G Gravity Meter. Pres. Int. Symp. on "Non-Tidal Gravity Variations and Methods for Their Study", Trieste 1977

5 Physical state and processes of changing physical rock properties in the earthquake source region

Physical Parameters in the Earthquake Source Region and Their Temporal Changes

Tsuneji Rikitake

Department of Applied Physics, Tokyo Institute of Technology, Meguro-ku, Tokyo, Japan (152)

ABSTRACT

Short-term and imminent earthquake prediction depends on temporal changes of physical parameters in the earthquake source region. In this paper, characteristics of such changes or earthquake precursors of various disciplines such as land deformation, change in sea level, tilt, strain,crustal stress, foreshock, anomalous seismic activity, seismic gap, growth and decay of seismicity, b value, source mechanism, hypocentral migration, seismic wave velocity, earth tide, geomagnetic and geoelectric field, earth resistivity, gravity, underground water, geochemical factor such as radon concentration and the like are summarized on the basis of recent data taken approximately after 1975.

The precursor time versus earthquake magnitude relation is brought out. Some discussion about the probable cause of the temporal changes are made although nothing certain can be deduced.

1. INTRODUCTION

Geophysical investigation of physical parameters in the earthquake source region and of their time change is important for earthquake prediction. Especially, short-term and imminent precursors can only be detected by observing these parameters. In contrast to long-term prediction, that may be effected by analysing historical records of earthquake occurrence and by monitoring accumulation of crustal strain by means of geodetic surveys, short-term and imminent prediction should depend upon temporal changes in physical parameters in the focal region of coming earthquake, or, in other words, we are to observe the phenomena associated with preliminary rupture for the purpose of such prediction.

It is the aim of this paper to review existing data of temporal changes in such physical parameters on the basis of current studies. As a result of intensive work on earthquake prediction in China, Japan, the U.S.A. and the U.S.S.R. in recent years, precursory changes of amazingly large number have been reported by now. Only data taken in recent years will here be dealt with. Those who are interested in data approximately prior to 1975 are kindly requested to refer to Rikitake(1976).

2. EARTHQUAKE PRECURSORS

Rikitake(1979a, 1980) summarized the data relevant to earthquake precursor as taken in the later half of 1970's. The following is the brief outline of the data along with some additional data that reached the author quite recently.

2.1 Land deformation

There are two well-known ground uplifts of anomalous character at present. Castle et al.(1976) reported on an anomalous uplift amounting to 25 cm at maximum. Centering at Palmdale about 50 km north of Los Angeles, California, the uplift, that developed during 1960-1975, covers a 400 x 150 km area.

It turned out, however, that a bench mark about 10 km south of Palmdale subsided by 18 cm during 1964-1976 (Savage and Prescott, 1977). Meanwhile the uplifted area seems to extend to the California-Arizona border as can be seen in Fig. 1 (MacCabe, 1979). It is not at all clear whether the anomaly may be linked to an impending earthquake. Savage

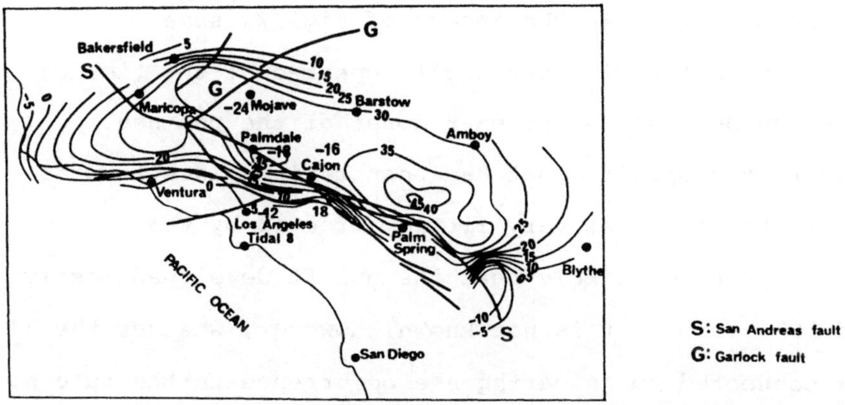

Fig. 1 Area of recent uplift in the Transverse Range and adjacent parts of the Mojave Desert in California, during the 1959-74 period. Uplift occurs along the San Andreas fault. Changes in elevation are shown with contours of 5 cm interval. Larger numerals indicate subsidence in cm since early 1974(McCabe, 1979).

and Prescott(1979) showed that no anomalous strain accumulation is taking place over the uplifted area. Castle(1978) argued that there is some evidence that a similar period of uplift and partial collapse may have taken place between 1902 and 1926. It appears to the author that there are a number of American seismologists and geophysicists who are now inclined not to believe that the uplift is connected with an occurrence of a major earthquake.

In contrast to such a large-scale uplift in California, a more localized uplift in the Izu Peninsula, about 100 km southwest of Tokyo, has been noticed since around 1975. Fig. 2 shows the uplift during a period from 1967-1969 to 1978(Geographical Survey Institute, 1978). The last survey being carried out immediately after the Jan. 14, 1978 Izu-Oshima Kinkai earthquake(M=7.0), of which the epicenter is indicated with a cross in the insert of Fig. 2, some subsidence associated with the fault appeared at the time of main shock can be seen on the east coast of the peninsula.

However, the uplift, that had been taking place towards the west of Ito City, was very little affected by the earthquake. It seems likely that the uplift developed mostly at around 1974-1975. It is not known, however, whether the uplift is connected to an earthquake occurrence in the future. In view of the fact that an anomalous land uplift is often observed prior to an earthquake, the Coordinating Committee for Earthquake Prediction, Japan is keeping a watchful eye on the uplift.

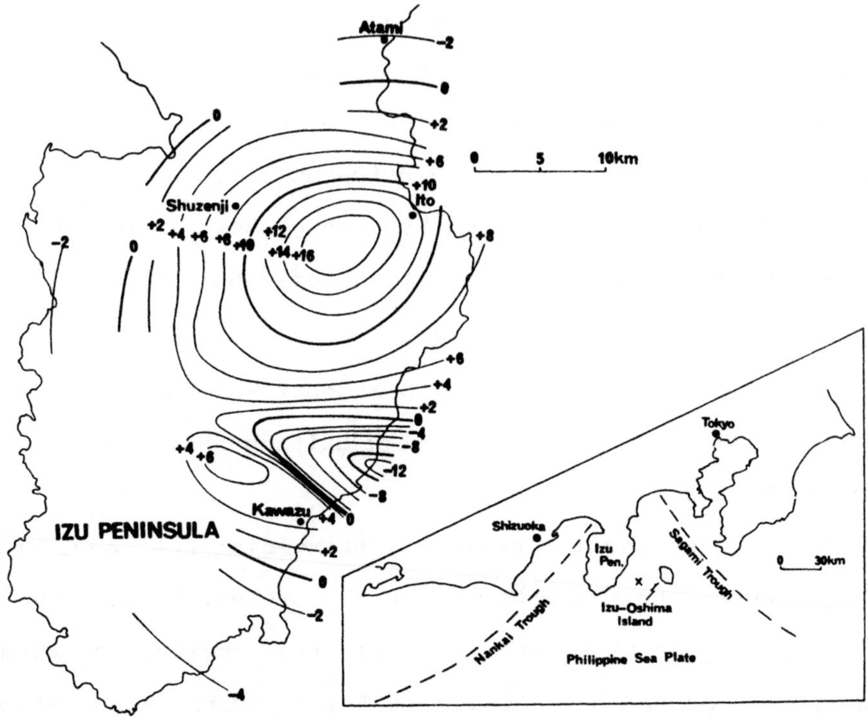

Fig. 2 Land uplift in units of centimeters in the Izu Peninsula during a period from 1967-1969 to 1978 (Geographical Survey Institute, 1978). The 1978 survey was carried out immediately after the Izu-Oshima Kinkai earthquake of magnitude 7.0 that occurred on Jan. 14, 1978. The geography around the peninsula is shown in the insert in which the epicenter of the above earthquake is shown with a cross.

2.2 Change in sea level

No outstanding changes in sea level forerunning an earthquake have been found in recent years although Sato (1977) and Wyss (1975, 1976a,b, 1977), who look back at the data in the past found considerable changes prior to some of the large-scale earthquakes.

2.3 Tilt, strain and crustal stress

In China, levelling surveys over a distance of several hundreds or tens of meters are popular. A remarkable change in ground tilting precursory to the Haicheng earthquake (magnitude 7.3, 1975) was detected by such a short levelling at Jin Xian about 185 km southwest of the epicenter (Raleigh et al., 1977).

Tiltmeters at Shenyang about 100 km north-northeast of the epicenter also recorded a reversal of tilting about two months prior to the earthquake.

An indoor base line for extremely short-distance levelling has been set up at the Dahuichang Crustal Movement Observatory about 25 km southwest of Peking. The base line crosses the Papa-oshan fault, one of the main active faults in North China althoug the length of the line amounts to only several tens of meters. An observer reads the scale of a mark. He then makes a 18o degree turn of the level and reads the scale of another mark. Making the difference in scale reading between the two marks, it is possible to observe the ground tilting. Measurements are usually made every 4 hours.

A water-tube tiltmeter and a strainmeter are set up in the basement of the observatory, too. Fig. 3 shows the changes in ground tilting as observed with the water-tube tiltmeter and short-distance levelling at the Dahuichang Observatory along with the short-distance levelling at Niuk'ouyu about 20 km further sothwest from Dahuichang. The changes in Fig. 3 may possibly be classified into the medium- and short-term precursors to the Tangshan earthquake (magnitude 7.8, 1976) respectively starting from the middle of the year 1975 and around June, 1976.

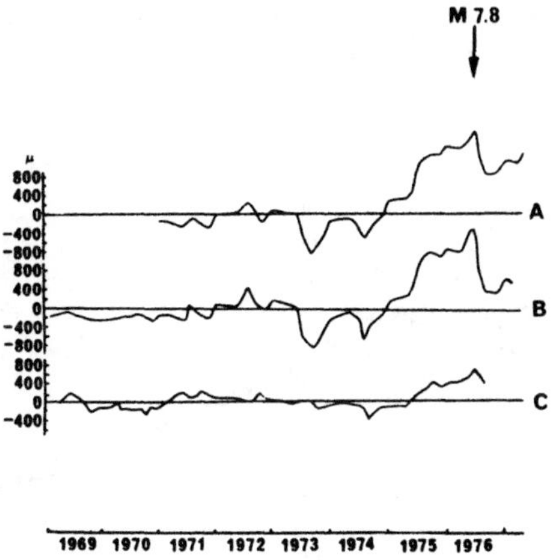

Fig. 3 Ground tilting preceded the Tangshan earthquake (M=7.8, 1976). A: Changes in ground tilt as observed at the Dahuichang Crustal Movement Observatory as measured by the water-tube tiltmeter. B: The same observed by the short levelling. C: The result of short levelling at Niuk' ouyu. (After the Dahuichang Crustal Movement Observatory).

Many tiltmeters are in operation in California. According to Johnston(1978), a remarkable change in tilt occurred in association with a magnitude 4.3 earthquake that occurred near the Briones Reservoir on Jan. 8, 1977. The epicentral distance was 5.5 km. The change, which is shown in Fig. 4, is characterized by a large deviation amounting to 2 μrad from the general change. Such a large change has not been observed during the last 3 years. It seems highly likely that the change is a kind of earthquake precursor.

Fig. 5 shows the precursor-like change recorded by a volume strainmeter buried at Irozaki, the southernmost tip of the Izu Peninsula. About 42 days prior to the Izu-

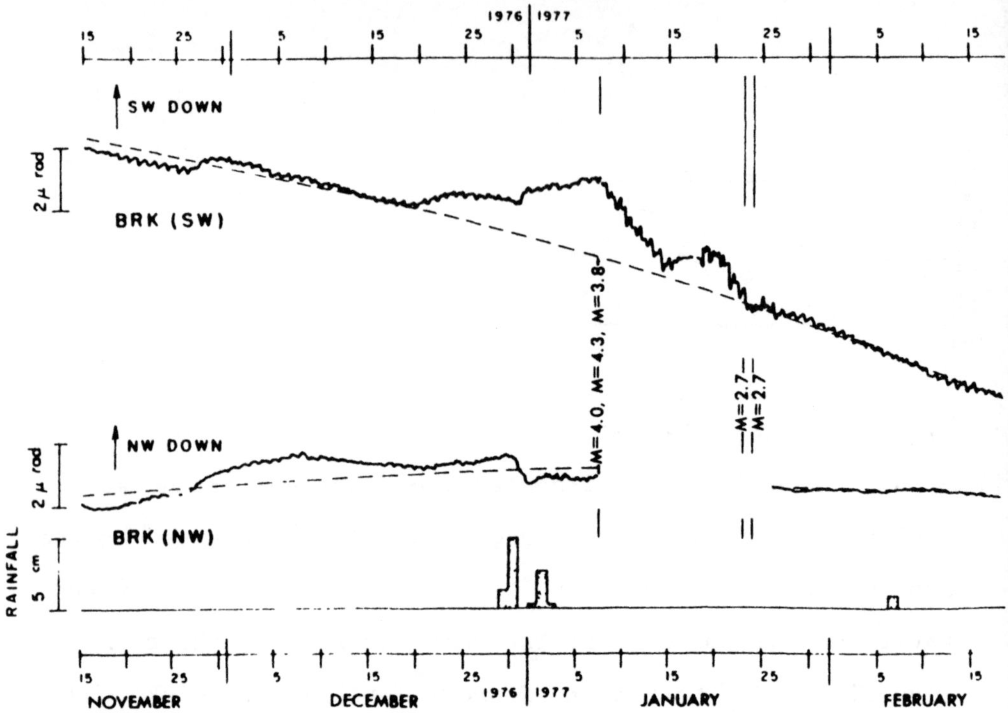

Fig. 4 The precursory change in ground tilting preceded the Briones, California, earthquake(magnitude 4.3, 1977) (Johnston, 1978).

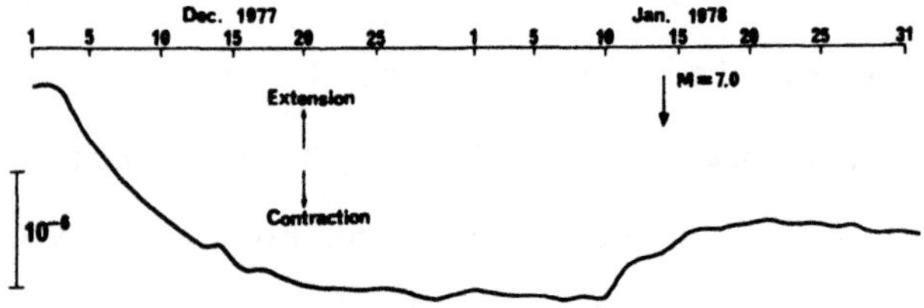

Fig. 5 Changes in the record of a volume strainmeter at Irozaki at the tip of the Izu Peninsula in association with the Izu-Oshima Kinkai earthquake(Japan Meteorological Agency, 1978).

Oshima Kinkai earthquake(magnitude 7.0, 1978), the strainmeter started to record a contraction as can be clearly seen in Fig. 5. Such a change tended to change its sign towards expansion about 4 days prior to the main shock. At about the same period, swarm activity of small earthquakes becomes conspicuous underneath the sea bottom between Izu-Oshima Island and Izu Peninsula. The distance between the epicenter of the main shock and Irozaki amounts to 32 km.

In situ measurements of crustal stress are being made in China(Tanaka, 1978 a,b). Those who are interested in details of measuring method, are asked to refer to Tanaka(1978, a,b). It has been reported that the crustal stress at a point 50 km distant from the epicenter of the Haicheng earthquake (magnitude 7.3, 1975) tended to increase 7 months prior to the earthquake reaching the maximum value amounting to 4.31 kg/cm^2. The direction of principal axis changed suddenly about 1 month preceding the main shock.

2.4 Seismological precursor

2.4.1 Foreshock

That a remarkable foreshock activity was observed on the occasion of the Haicheng earthquake and that such an activity provided the basis of the imminent earthquake prediction has been well known(Zhu, 1976).

In relation to the Izu-Oshima Kinkai earthquake that occurred on 12h 24m on Jan. 14, 1978, the following seismic activity has been reported. According to the observation by Japan Meteorological Agency, commencement of a swarm activity of small earthquakes around Izu-Oshima Island has been recognized since Jan. 12. The area inbetween Izu-Oshima

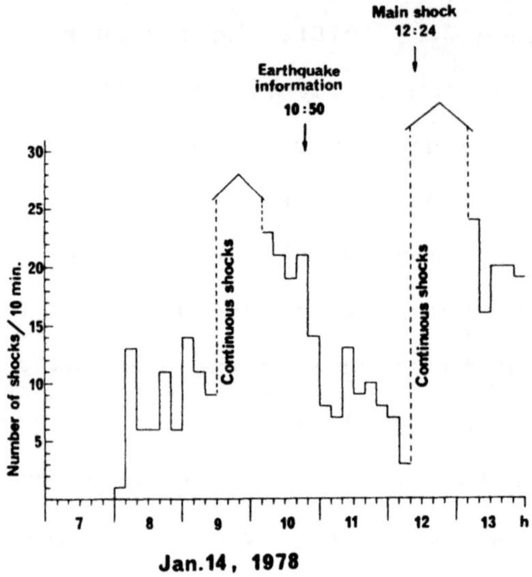

Fig. 6 Frequency per 10 minutes of foreshocks prior to the Izu-Oshima Kinkai earthquake as observed on Izu-Oshima Island(Japan Meteorological Agency, 1978).

Island and Izu Peninsula is well known for occasional occurrences of earthquake swarms although the activity appeared to be a little high this time. As can be seen in Fig. 6 (Japan Meteorological Agency, 1978), the number of small shocks began to increase so enormously that we observed almost continuous vibrations on the seismogram since 09h 30m in the morning of Jan. 14. However, a drastic decrease in the number of shocks began to be noticed at around 10 o'clock.

In light of such development of swarm activity, the Agency issued an earthquake information to the public at 10h 50m. The information reads "A moderately large earthquake, that would give rise to slight damage, may occur shortly". About 1.5 hours after the information issuance, the main shock actually occurred somewhere between Izu-Oshima Island and Izu Peninsula at 12h 24m.

Both the cases of Haicheng and Izu-Oshima Kinkai earthquakes provide typical examples of a characteristic pattern of foreshock sequences, namely an enormous increase in earthquake number followed by a sudden decrease culminating in the occurrence of main shock.

2.4.2 Anomalous seismic activity

Sekiya(1976, 1977) pointed out that a tendency that a swarm-like seismic activity sometimes occur in an area in which a large earthquake occurs later on.

Evison(1977a,b) also pointed out similar precursory seismicity for 4 and 9 earthquakes respectively in California and New Zealand. The logarithm of precursor time T of these anomalous activity seems likely to be proportional to the magnitude M of main shock. The relation between $\log_{10}T$ and M roughly agrees with that in Rikitake(1975, 1976, 1979a). This seems to have an important signigicance for earthquake prediction.

Ishida and Kanamori(1977) studied an anomalous seismic activity preceding the San Fernando earthquake(magnitude 6.4, 1971). Marza(1979) pointed out that even an intermediate-depth earthquake the Vrancea, Romania, earthquake(magnitude 7.2, 1977), was accompanied by an anomalous activity. The precursor time agrees well with Rikitake's relationship.

2.4.3 Seismic gap of the second kind

It is expected to have an area, where no large earthquake occurs over a fairly long period, preceding the occurrence of the next large earthquake because some period is required for strain accumulation. Such an aseismic area is called the seismic gap of the first kind(Mogi, 1978). The concept of seismic gap has been developed by Fedotov(1965), Mogi(1968), Sykes(1971), McCann et al.(1978) and others.

An area in which the activity of small earthquakes has decreased prior to occurrence of a large earthquake is named a seismic gap of the second kind(Mogi, 1978).

Ohtake(1976) and Sekiya(1976) noticed that a fairly wide area including Izu Peninsula was seismically calm during the period of several years before the Izu-Hanto Oki earthquake (magnitude 6.9, 1974).

Ohtake et al.(1977) pointed out that two Oaxaca, Mexico earthquakes of magnitude 7.3(1965) and 7.1(1968) were accompanied by a seismic gap of the second kind.

They further pointed out that an area between the focal areas of the 1965 and 1968 earthquakes had been in a state of conspicuous gap since June, 1973. In Fig. 7, they showed that a well-recognized gap can be observed during a period from June, 1973 to May, 1975.

On the basis of the above facts, Ohtake et al.(1977) pointed out a possibility of a large earthquake occurring in the Oaxaca area. The magnitude is supposed to take on a value of 7.5 with an allowance of ± 0.25, and the supposed location of epicenter is(16.5°N, 96.5°W) with a probable error of ± 0.5° for latitude and longitude. There was no guess about the time of earthquake occurrence although it was mentioned in the paper that the gap had been persisted during a period which is longer than that for the past earthquakes in the Oaxaca area.

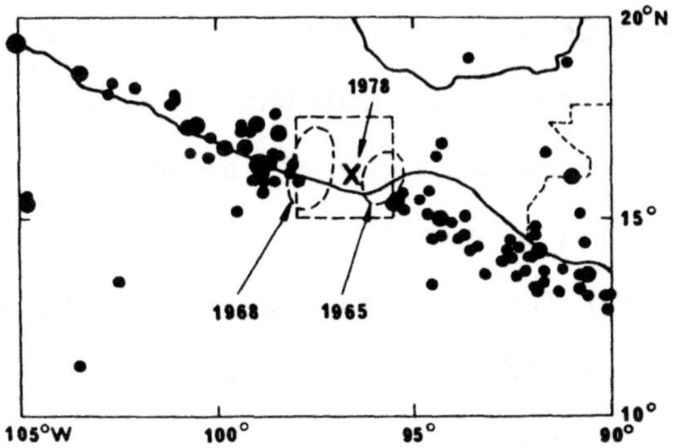

Fig. 7 The seismic gap near Oaxaca, Mexico during June, 1973 - May, 1975. Solid circles indicate earthquakes shallower than 60 km. The two ellipses are the aftershock areas of the 1965(magnitude 7.3) and 1968(magnitude 7.1) earthquakes. The cross shows the epicenter of the Nov. 29, 1978 earthquake as inserted by the present author(Ohtake et al., 1977).

An earthquake of magnitude 7.8 (probably surface wave magnitude) occurred at (16.07°N, 96.49°W) at 19h 52m 53.2s (UT), Nov. 29, .978 as predicted. The magnitude and location of epicenter agree very well with those guessed by Ohtake et al.(1977). The epicenter is shown with a cross in Fig. 7. The precursor time is estimated as 5.4 years in this case.

2.4.4 Other factors related to seismicity

Rikitake(1976) presented a number of premonitory increases in seismicity in a fairly wide area including the epicenter of main shock. The premonitory increase in the number of South Liaoning Province earthquakes that seems to begin about 1.5 years prior to the Haicheng earthquake is remarkable

(Zhu, 1976). In general, there is a tendency that the seismicity first increases in a wide area outside the epicentral area, and that a calm period similar to seismic gap of the second kind persists for a certain period of time.

The number of occurrence (N) for earthquakes of magnitude M in a region during a certain period is correlated to M as

$$\log_{10}N = a - bM \tag{1}$$

where a and b are constants. Equation(1) is called the Gutenberg-Richter formula(Gutenberg and Richter, 1944).

In Rikitake(1976) are quoted many an example of decrease in b value forerunning an earthquake. Further examples of decrease in b value are more recently put forward by Hasegawa et al.(1975), National Research Center for Disaster Prevention (1978), Tang(1978), Li et al.(1978) and Ma(1978).

In addition to the precursory changes in source mechanism, some more work related to precursory change in source mechanism have been reported in recent years. They are Lindh et al.(1978)on precursory change in source mechanism, and Ishida and Kanamori(1978) on change in wave form.

According to Engdahl and Kisslinger(1977), a several kilometer migration of microearthquake hypocenters towards the focus of a magnitude 5.0 earthquake was observed in Mid Aleutian Islands 35 days prior to the occurrence of main shock.

2.4.5 Change in seismic wave velocity

Many tests for checking the validity of premonitory change in V_P(P wave velocity) , V_S(S wave velocity) or V_P/V_S ratio have been made by many seismologists(Terashima, 1974; Utsu, 1975; Ohtake, 1976; Iizuka, 1976a,b,c; Wang et al., 1976;

Feng et al., 1974 1976a,b; Fen 1975; Duan et al., 1976). Rikitake(1979) summarized 17 example of changes in V_P/V_S ratio reported from China.

Ohtake and Katsumata(1977) and Yoshii(1978) pointed out that the accuracy of seismic observation required for detecting changes in seismic wave velocities should be unexpectedly so high that many of the results reported so far are not reliable. It turns out that it is no easy matter to detect changes in seismic wave velocities in spite of the optimistic view which was popular several years ago. There are many cases for which we have no changes in seismic wave velocities in the western United States as reported by Chou and Crosson(1978).

Wyss(1980) believes that the changes associated with the events in the Blue Mountain Lake area(Aggarwal et al., 1975) and Hawaii(Johnston, A.C., 1978) are reliable. He also pointed out that no precursory change in seismic wave velocity is observed for earthquakes caused by strike-slip and normal faults. Ohtake and Katsumata(1977) pointed out that the change forerunning the North Mino earthquake(magnitude 7.0, 1961) is reliable.

It has been known from rock breaking tests that dilatancy is produced immediately prior to rupture resulting in a decrease in P wave velocity(e.g. Matsushima, 1960). If so, it might rather be puzzling that no change in seismic wave velocity is associated with many of natural earthquakes.

2.4.6 Earth tide

Latynina and Rizaeva(1976) reported that the amplitude of M_2 constituent decreased by 6 % 30 days prior to a magnitude 4.5 earthquake that occurred nearby Dushambe in 1976. According to Mikuno et al.(1977) who analysed the data observed at the

Kamitakara Crustal Movement Observatory 60 km distant from the epicenter of the Gifu Chubu earthquake(M=6.6, 1969), the earth tidal admittance increased by 15 % 11 months preceding the earthquake.

2.4.7 Geomagnetic and geoelectric precursor

Smith and Johnston(1976) reported on a remarkable change in the geomagnetic total intensity amounting to 1 γ or a little larger in association with a magnitude 5.2 earthquake on Nov. 28, 1974 that occurred near Hollister, California. The anomalous change in the geomagnetic field was followed by a fault creep and a change in ground tilting. As the standard deviation of magnetic noise around this area is only 0.5 γ or so, the change observed must be significant and has something to do with the earthquake occurrence or fault creep.

Geomagnetic work aiming at monitoring possible seismomagneti effect and geomagnetic precursor, if any, has been conducted very intensively over the Izu Peninsula where the crustal activity had been very high as stated in Section 2.1. The Geomagnetic Survey Party(1977) and Sasai and Ishikawa(1978) reported on fluctuations of total intensity during 1976-1977.

The seismic activity in the Izu Peninsula culminated in the Izu-Oshima Kinkai earthquake in 1978. Honkura(1978) pointed out that a precursor-like change of several gammas took place at a station located in the middle of peninsula, the precursor time amounting to 64 days.

Summarizing geomagnetic surveys conducted over the peninsula at various periods, the Geomagnetic Group for Izu Peninsula(1980) found overall changes in the geomagnetic total intensity during the two six-month periods, i.e. July 1977

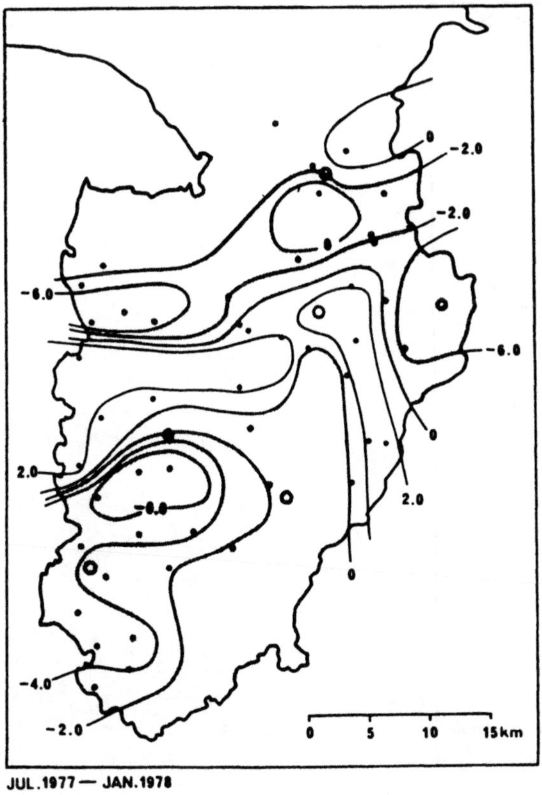

Fig. 8 Changes in the geomagnetic total intensity over the Izu Peninsula during July, 1977 - Jan., 1978. The last epoch is just after the Izu-Oshima Kinkai earthquake (Geomagnetic Group for Izu Peninsula, 1980).

to Jan. 1978 (just after the earthquake) and Jan. 1978 to July 1978 as shown in Figs. 8 and 9. It is interesting to note that the changes are as large as several gammas and that the pattern of change during the period after the quake is almost a reversal of that of the period including the earthquake. The geomagnetic changes thus brought to light seem to reflect the state of crustal stress probably caused by sporadic propelling of the Izu Peninsula due to the motion of the Philippine Sea plate.

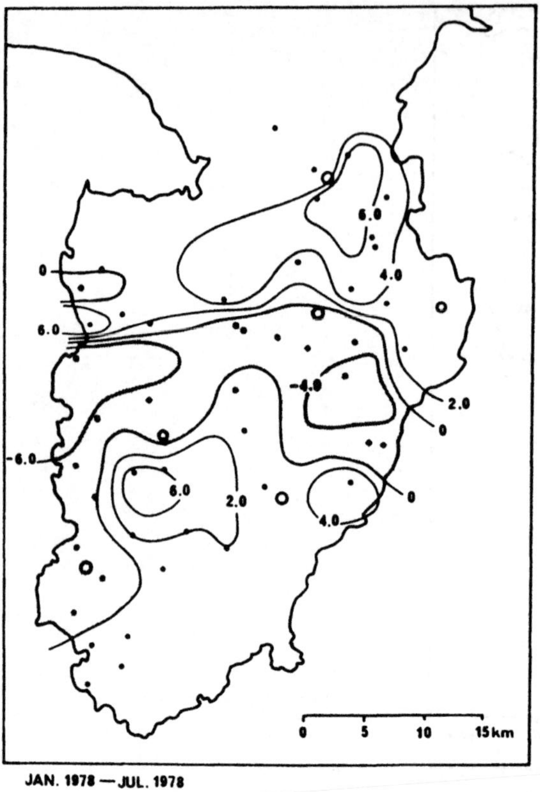

Fig. 9 Changes in the geomagnetic total intensity over the Izu Peninsula during Jan. 1978 - July, 1978. The first epoch is just after the Izu-Oshima Kinkai earthquake(Geomagnetic Group for Izu Peninsula, 1980).

The seismic activity in the Izu Peninsula did not cease in spite of the Izu-Oshima Kinkai earthquake. A precursor was clearly observed before the November, 1978, Kawazu earthquake of magnitude 5.0(Geomagnetic Survey Party, 1979). The observation has been conducted at a station located almost at the epicenter. Judging from the differences in the total intensity between respective stations as shown in Fig. 10, it is obvious that the magnetic field at Kawazu was subjected

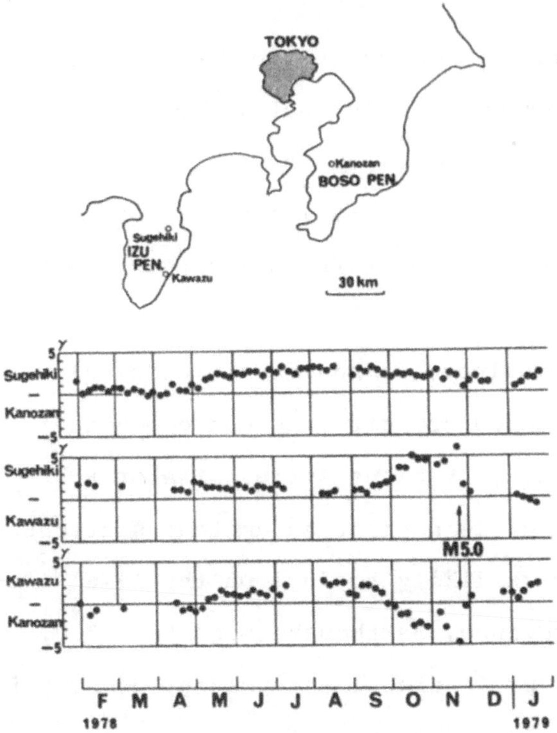

Fig. 10 Changes in the differences in the 5-day means of geomagnetic total intensity between the three stations; Kawazu, Sugehiki and Kanozan (Geomagnetic Survey Party, 1979).

to a premonitory decrease in the total intensity starting two months before the shock.

Geomagnetic observations aiming at detecting precursory changes are popular in China. It was reported that the geomagnetic field at Dalian, Liaoning Province, had appreciably changed relative to that at Peking. Comparing the vertical intensity at Dalian on May 22, 1974 to that on Oct. 27, 1973, an increase amounting to 21.5 γ was observed (Zhu, 1976; Raleigh et al., 1977).

As a result of comparing the geomagnetic field at 19 magnetic stations in September, 1974 to that in Feb. - Mar., 1975, it was known that a decrease amounting to 7 to 9 γ had been taking place at stations around Tangshan. This might be a precursor of the Tangshan earthquake(M=7.8, 1976) (Raleigh et al., 1977).

At Chengdu, Sichuan Province, Standard Seismological Observatory, Chinese colleagues are carrying out geomagnetic and geoelectric observations of various sorts. Use of a simple declination variometer is popular at amateur stations. Geomagnetic changes that were likely to forerun the 1976 Songpan-Pingwu, Sichuan Province, earthquake(magnitude 7.2, 6.7, 7.2) were observed by professional and amateur observations (Rikitake, 1979b).

In contrast to the above changes in the absolute value of geomagnetic field, precursory changes in the amplitude of geomagnetic variations of short period(Honkura and Koyama, 1978) and the transfer function(ratio of vertical component to horizontal one for geomagnetic variations of short period) (Rikitake, 1979c; Yanagihara and Nagano, 1976; Sano, 1978; Xu et al., 1978) have been intensively studied with some success.

Observation of changes in earth potential is very popular in China. One can find many reports on precursory changes in association with the Haicheng, Songpan-Pingwu and other earthquakes(Raleigh et al., 1977; Zhu, 1976; Noritomi, 1978a,b; Rikitake, 1979b). Recently, Corwin and Morrison(1977) and Koyama and Honkura(1978) reported on changes in earth potential precursory to Californian and the Izu-Oshima Kinkai earthquakes, respectively.

Much of precursory changes in ground resistivity as observed in the U.S.S.R., the U.S.A., China and Japan was reviewed by Rikitake(1976). It is intended here to add new findings of similar sort to the existing data.

In contrast to the precursory decrease in resistivity in the case of an earthquake(magnitude 3.9) and a swarm activity (magnitude 3.5 in total) on the San Andreas fault in 1973 (Mazzella and Morrison, 1974), no appreciable change in resistivity was observed for the Oct., 1977 earthquake (magnitude 3.8) and Dec., 1977 earthquake(magnitude 4.0), the epicenters of these quakes being located not very far from the last shocks that accompanied resistivity changes(Morrison et al., 1979).

Measurement of earth resistivity as a means of detecting earthquake precursor is very popular in China. Zhao and Qian (1978) presented a resistivity change that was brought to light a few years prior to the Tangshan earthquake.

Noritomi(1978a,b) summarized many Chinese examples of precursory resistivity change. Among them, the resistivity decrease observed at Changli, about 80 km distant from the epicenter of the Tangshan earthquake, is the most remarkable. The precursor time amounts to 5 days in this case. Changes in resistivity is observed at Chengdu and other places in Sichuan Province in association with the 1976 Songpan-Pingwu earthquake are summarized by Rikitake (1979b).

Sidorenko et al.(1979) reported on changes in earthquake resistivity preceding earthquake swarms and moderately large earthquakes observed at the Cherkey hydroelectric power plant on Sulak River in Dagestan.

It has often been reported that the highly-sensitive resistivity variometer at Aburatsubo, a near-shore station about 60 km south of Tokyo, registers precursory changes in resistivity having a precursor time of a few hours (Yamazaki, 1975; Rikitake and Yamazaki, 1976, 1977). When a strain of the order of 10^{-9} - 10^{-8} is observed in terms of resistivity change, it is amplified 10^4 times or so by a not-entirely-identified mechanism in the formation there.

Rikitake and Yamazaki(1979) further showed that a precursor having a longer precursor time was observed by the variometer in the case of the Izu-Hanto Oki earthquake (magnitude 6.9, 1974).

2.4.8 Gravity

Tanaka(1978a) and Hagiwara(1979) summarized the gravity changes associated with earthquakes in China. Decrease and increase of gravity amounting to a few hundred microgals have often been reported in China. If the changes are true, they cannot be accounted for either by vertical deformations of land or density changes due to compaction, so that it cannot be helped to assume some kind of mass transfer within the earth.

Hagiwara et al.(1976), Hagiwara(1977) and Geodetic Survey Party(1978) detected a gravity decrease of about 30 and 50 μgals at maximum respectively during Dec., 1974 - Mar., 1976 and Dec., 1974 - Feb., 1978 in association with the anomalous uplift as shown in Fig. 2 . The gravity change thus disclosed seems likely to be accounted for by the decrease of free-air rate due to the increase in height. There is no evidence that a mass of high density has come up close to the earth's surface from below.

According to the experiences in Japan and the U.S.A., no gravity change exceeding scores of microgals has been observed. The reason why they frequently observe extremely large changes in gravity reaching a few hundred microgals in China is not known.

2.4.9 Underground water

Very few systematic studies on anomaly of underground water have been conducted in relation to an earthquake in Japan. Wakita(1978c) presented 113 examples of anomaly of underground water in a summarized form.

Sato(1977) newly pointed out the anomalous change in the level of underground water preceded the Nankai earthquake (magnitude 8.1, 1946). The data have been overlooked for a long time. Seventeen examples of lowering of well-water level and/or muddy water were reported from the western Shikoku and the eastern Shikoku-Kii Peninsula respectively 5-7 and 1-2 days prior to the quake.

In the case of the Izu-Oshima Kinkai earthquake, Yamaguchi and Odaka(1978) reported on a precursor-like change in underground water level at a well about 35 km distant from the epicenter.

Wu(1975) reported that the pressure of gas-wells in Taiwan increased 15-16 % 10 days preceding an earthquake of magnitude 6.7, the epicentral distance being 20 km. Such a change appeared at 3 wells having a depth of 15 km. No such violent change has ever been observed in the past.

It is quite impossible to deal with all the underground anomalies reported in China because they are too numerous. There are many examples such as the well from which water spouted out before the Haicheng earthquake and the like.

Wakita(1978a,b) showed a number of underground water precursors observed in China.

It seems likely that we have to choose suitable wells for detecting precursory changes in underground water. Those wells for irrigation cannot be used for the purpose. When the underground water is running rapidly, it is also difficult to take precursory data. The best thing to do is to dig a deep well only for monitoring precursory changes in underground water.

2.4.10 Geochemical precursor

A remarkable change in radon content in the underground water was reported forerunning the Gasli, the U.S.S.R., earthquake(magnitude 7.3, May 17, 1976)(Sultanxodjaev et al., 1976). Fig. 11 shows the change in radon content observed at Ulegbek near Tashkent. The distance from the epicenter to the observation point amounts to 400 km.

In the U.S.A., a series of measurements of radon content have been conducted along the San Andreas fault(King, 1978; MacCabe, 1979). In general, annual change is usually predominating. However, changes that seem likely to have

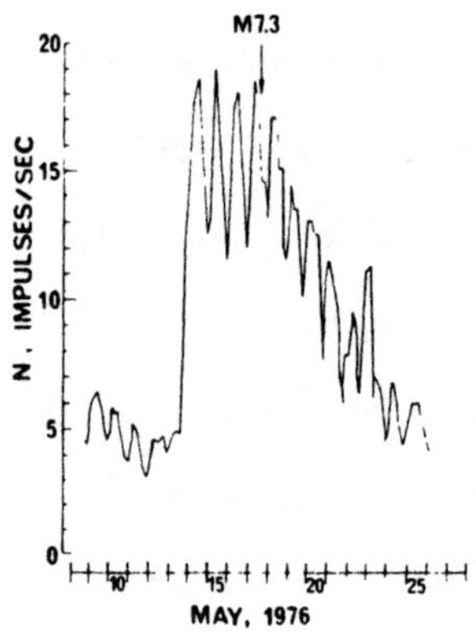

Fig. 11 Changes in radon concentration at Ulegbek preceded the Gasli earthquake (magnitude 7.3, 1976)(Sultanxodjaev et al., 1976).

something to do with occurrence of earthquakes of magnitude 4 or so have been noticed.

Measurement of radon content is now getting popular in Japan. Faculty of Science of the University of Tokyo, Geological Survey of Japan and Tokyo Metropolitan Office are doing such a measurement in the Tokai , South Kanto and Tokyo areas. It seems likely that the radon content observed at a station indicated a precursory change preceding the Izu-Oshima Kinkai earthquake(unpublished, Faculty of Science, University of Tokyo, 1978), the epicentral distance amounting to about 30 km.

In China, radon content observation is widely made for the purpose of earthquake prediction. According to Wakita (1978a,b), the observation is one of the routine work at 17 standard seismological observatories and about 60 seismological brigades. In addition to this there are countless amateur stations guided by experts.

There are so many reports of precursory changes in radon content in China that it is hard to cover all of them. Those who are particularly interested in them should refer to Wakita (1977, 1978a,b). According to the Chinese experience, it seems likely that there are two kinds of radon precursor; i.e. long-term precursor and short-term one having precursor time of 10 days or so.

In addition to radon measurement, measurement of helium concentration is also conducted in relation to earthquake prediction(Craig et al., 1975). Sugisaki and Shichi(1978), who measured He, Ar and N_2 contents in gases from a fault in a vault of a crustal movement observatory in Central Japan, found changes precursory to earthquakes.

3. LEAD TIME OF TEMPORAL CHANGES IN THE PHYSICAL PARAMETERS

Rikitake(1979a), who collected precursor data amounting to 391 in total number, examined the relation between precursor time, the lead time (T) of temporal change in the physical parameters preceding an earthquake occurrence, and earthquake magnitude (M) of the main shock. It is found out that there are precursors of the first kind for which $\log_{10}T$ (T is measured in units of days) is correlated to M as

$$\log_{10}T = 0.6.M - 1.01 \qquad (2)$$

on the average.

Meanwhile, there are precursors of the second kind of which the precursor time does not depend on M. The $\log_{10}T$ - M relation thus disclosed is shown in Fig. 12.

Rikitake(1979) also pointed out that precursor time of foreshock, tilt observed by a tiltmeter of horizontal pendulum type, earth-currents does not fit in the above empirical rule. It becomes apparent that the histogram of logarithmic precursor time has its peak at about T = 5 days although the spectrum is very broad.

It is highly likely that relation (2),which is obtained from all the available data, may be different from region to region.

4. CAUSE OF TEMPORAL CHANGES OF PHYSICAL PARAMETERS IN THE EARTHQUAKE SOURCE REGION

No definite reason why we sometimes observe a precursor before an earthquake has been worked out. It has been proposed, however, that a dilatancy model(e.g. Scholz et al., 1973)

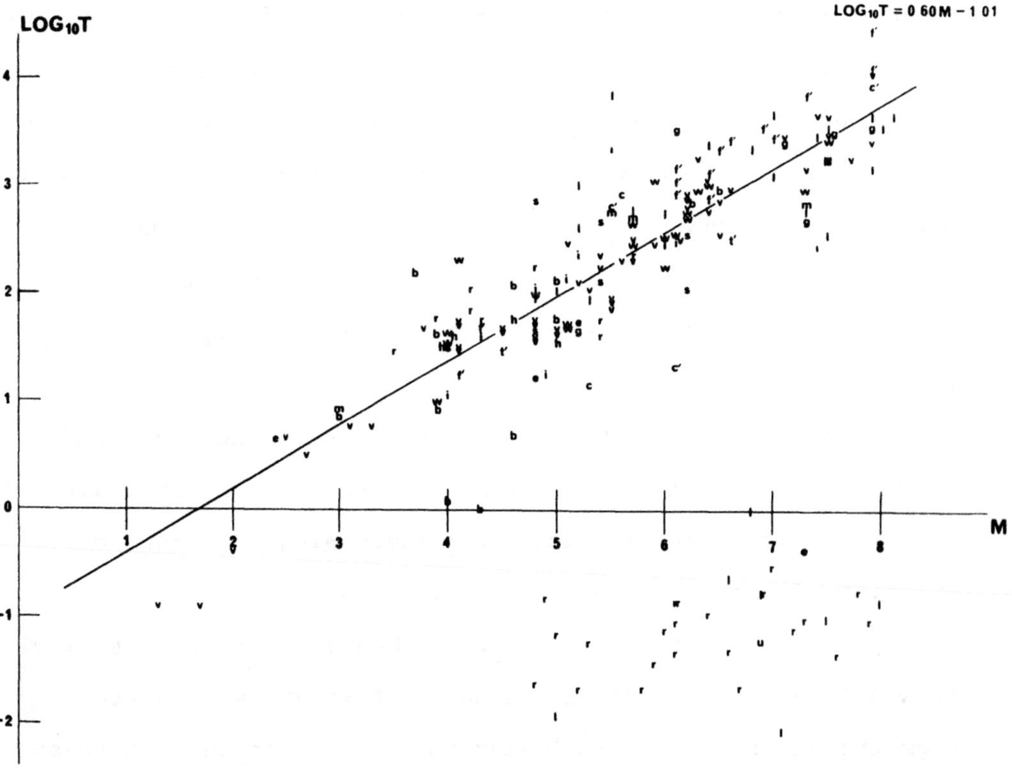

Fig. 12 The relationship between logarithmic precursor time T in units of days and earthquake magnitude M of main shock. The alphabets indicate various disciplines of geophysical and geochemical precursor different from one another (Rikitake, 1979).

could account for most effects precursory to an earthquake. According to the model, the process in the earth crust that leads to occurrence of an earthquake is as follows.

When the crustal stress increases to some extent, many microcracks are to be produced in the rocks composing the crust. This state is called the dilatancy which necessarily leads to an increase in volume. In that case, it is quite natural to expect an upheaval at the ground surface. Anomalous uplift often preceding an earthquake could therefore be accounted for as a result of dilatancy generation. It is known from laboratory experiments that the velocity of P wave being propagated through such a dilatant region becomes small although the velocity of S wave is affected very little. It is thus possible to understand the lowering of V_P and V_P/V_S ratio.

When underground water tends to fill the newly-opened cracks, the strength of the dilatant region recovers again. This state is called the dilatancy-hardening. In that case, quieting of seismic activity and lowering of b value could be explained. It is also expected that electric resistivity in a dilatant region lowers because of water being diffused from the surrounding. Such movements of water through fresh cracks may accompany an increase in radon concentration in underground water. The above series of events may well be expected to take place before the earthquake occurrence which is the result of a fault slip in the dilatant region. It may be said, therefore, that many of earthquake precursors can possibly be accounted for, at least qualitatively, provided a dilatancy process is assumed.

Mjachkin et al.(1975), Mogi(1974), Stuart(1974) and Brady(1974) proposed a dry model of dilatancy which is not necessarily related to water diffusion. In this case the appearance mode of precursors would be somewhat different from that for the wet model as described in the above.

There is no guarantee, however, that earthquakes are always associated with dilatancy generation. As stated in

Subsection 2.4.5, it has now become clear that there are many earthquakes that are not preceded by a change in V_P or V_P/V_S ratio. The idea that the dilatancy model could be applied to almost all earthquakes should be put aside.

It has been attempted to explain an anomalous uplift by assuming an aseismic pre-slip, especially in relation to subduction zone earthquakes. But it is not entirely known whether or not such a hypothesis is supported by the data of actual observation.

It should be emphasized, however, that the empirical relation for precursors of the first kind such as expressed by equations(2) can be approximately accounted for by taking the dilatancy model into account. On the contrary, nothing certain is known about the mechanism of precursors of the second and third kinds. It is only understood that some preliminary rupture begin to take place when precursors of the second kind break out. Precursors of the third kind may reflect the highly strained state of a focal region.

REFERENCES

Aggarwal, Y.P., Sykes, L.R., Simpson, D.W. and Richards, P.G., 1975. Spatial and temporal variations in t_S/t_P and in P-wave residuals at Blue Mountain Lake, New York; application to earthquake prediction. J. Geophys. Res., 80: 718-732.

Brady, B.T., 1974. Theory of earthquakes, 1. A scale independent theory of rock failure. Pure Appl. Geophys., 112: 701-725.

Castle, R.O., 1978. Leveling surveys and the southern California uplift. Earthquake Information Bull., 10: 88-92.

Castle, R.O., Church, J.P. and Elliot, M.R., 1976. Aseismic uplift in southern California. Science, 192: 251-253.

Chou, C.W. and Crosson, R.S., 1978. Search for time-dependent seismic travel times from mining explosions near Centralia, Washington. Geophys. Res. Lett., 5: 97-100.

Corwin, P.F. and Morrison, H.F., 1977. Self-potential variations preceding earthquakes in central California. Geophys. Res. Lett., 4: 171-174.

Craig, H., Lupton, J.E., Chung, Y. and Horowitz, R.M., 1975. Investigation of Radon and Helium as possible fluid-phase precursors to earthquakes. Technical Report, No. 1(SIO Ref. No. 75-15).

Duan Xing-bei, Zheng Jian-zhong, Zhou Zhi-qun, Yan Shou-min and Sun Ci-chang, 1976. Variations in teleseismic P wave residuals before the Haicheng earthquake. Acta Geophys. Sinica, 19: 286-294(in Chinese).

Evison, F.F., 1977a. Fluctuations of seismicity before major earthquakes. Nature, 266: 710-712.

Evison, F.F., 1977b. The precursory earthquake swarm. Phys. Earth Planet. Inter., 15: 19-23.

Fedotov, S.A., 1965. Regularities of the distribution of strong earthquakes in Kamchatka, the Kurile Islands, and northeastern Japan. Trudy Inst. Phys. Earth, Acad. Sci. U.S.S.R., No. 36(203): 66-93.

Feng De-yi, 1975. Anomalous variations of seismic velocity ratio before the Yongshan-Daguan earthquake(M=7.1) on May 1, 1974. Acta Geophys. Sinica, 18: 235-239(in Chinese).

Feng De-yi, Tan Ai-na and Wang Ke-fean, 1974. Velocity anomalies of seismic waves from near earthquakes and earthquake prediction. Acta Geophys. Sinica, 17: 84-98(in Chinese).

Feng De-yi, Zheng Si-hua, Sheng Guo-ying, Fu Zheng-xiang, Gao Shi-lei, Luo Rui-ming and Li Bing-can, 1976a. Preliminary study of the velocity anomalies of seismic waves before and after some strong and moderate earthquakes in Western China (1) —— The velocity ratio anomalies. Acta Geophys. Sinica, 19: 196-205(in Chinese).

Feng Rui, Pang Qing-yan, Fu Zheng-xiang, Zheng Jian-zhong, Sun Ci-chang and Li Bao-xiang, 1976b. Variations of V_P/V_S before and after the Haicheng earthquake of 1975. Acta Geophys. Sinica, 19: 295-305(in Chinese).

Geodetic Survey Party, 1978. Coseismic gravity change during Izu-Oshima Kinkai earthquake. Rep. Coord. Comm. Earthquake Prediction, 20: 56-60(in Japanese).

Geographical Survey Institute, 1978. Crustal deformation in Izu Peninsula. Rep. Coord. Comm. Earthquake Prediction, 20: 92-99(in Japanese).

Geomagnetic Group for Izu Peninsula, 1980. Changes in the geomagnetic total intensity in the Izu Peninsula with the Izu-Oshima Kinkai Earthquake of January 14, 1978. Rep. Coord. Comm. Earthquake Prediction, 23: 53-55(in Japanese).

Geomagnetic Survey Party, 1977. Repeated magnetic survey and observation of the total force intensity in the eastern part of the Izu Peninsula(2). Rep. Coord. Comm. Earthquake Prediction, 18: 47-51(in Japanese).

Geomagnetic Survey Party, 1979. Repeated magnetic survey and observation of total force intensity in the eastern part of the Izu Peninsula(4). Rep. Coord. Comm. Earthquake Prediction, 22: 75-78(in Japanese).

Gutenberg, B. and Richter, C.F., 1944. Frequency of earthquakes in California. Bull. Seismol. Soc. Am., 34: 185-188.

Hagiwara, Y., 1977. Gravity changes associated with seismic activities. J. Phys. Earth. 25, Suppl.: S137-S146.

Hagiwara, Y., 1979. Gravity and crustal deformation in China. Rev. Res. Disaster Prevention, 44: 66-69, 93-97, 121-124 (in Japanese).

Hagiwara, T., Tajima, H., Izutuya, S. and Hanada, H., 1976. Gravity changes in the eastern part of Izu Peninsula during the period of 1975-1976. J. Geod. Soc. Japan, 22: 201-209 (in Japanese).

Hasegawa, A., Hasegawa, T. and Hori, T., 1975. Premonitory variation in seismic velocity related to the Southeastern Akita earthquake of 1970. J. Phys. Earth, 23: 189-203.

Honkura, Y., 1978. On a relation between anomalies in the geomagnetic and telluric fields observed at Nakaizu and the Izu-Oshima Kinkai earthquake of 1978. Bull. Earthquake Res.Inst.,Univ. Tokyo, 53: 931-937(in Japanese).

Honkura, Y. and Koyama, S., 1978. Observations of short-period geomagnetic variations at Nakaizu(1). Bull. Earthquake Res. Inst., Univ. Tokyo, 53: 925-930(in Japanese).

Iizuka, S., 1976a. Temporal variations in V_P/V_S and some related phenomena before the 1969 Tokachi-oki earthquake, off Northeast Japan. Zisin(J. Seismol. Soc. Japan), Ser. 2, 29: 247-263(in Japanese).

Iizuka, S., 1976b. Temporal changes in V_P/V_S ratio associated with great submarine earthquakes, off eastern Hokkaido, Japan. Zisin(Seismol. Soc. Japan), Ser. 2, 29: 265-275(in Japanese).

Iizuka, S., 1976c. Temporal changes in seismic wave velocities related with the Matsushiro earthquake swarm. Zisin(J. Seismol. Soc. Japan), Ser. 2, 29: 365-374(in Japanese).

Ishida, M. and Kanamori, H., 1977. The spatio-temporal variation of seismicity before the 1971 San Fernando earthquake, California. Geophys. Res. Lett., 4: 345-346.

Ishida, M. and Kanamori, H., 1978. The foreshock activities of the 1971 San Fernando earthquake, California. Bull. Seismol. Soc. Am., 68: 1265-1279.

Japan Meteorological Agency, 1978. On the Izu-Oshima-Kinkai earthquake, 1978. Rep. Coord. Comm. Earthquake Prediction, 20: 45-50.

Johnston, A.C., 1978. Localized compressional velocity decrease precursory to the Kalapana, Hawaii, earthquake. Science, 199: 882-885.

Johnston, M.J.S., 1978. Tiltmeter studies in earthquake prediction. Earthquake Inf. Bull., 10: 182-186.

King, C.Y., 1978. Radon emanation on San Andreas fault. Nature, 271: 516-519.

Koyama, S. and Honkura, Y., 1978. Observations of electric self-potential at Nakaizu(1). Bull. Earthquake Res. Inst., Univ. Tokyo, 53: 939-942(in Japanese).

Latynina, L.A. and Rizaeva, S.D., 1976. On tidal-strain variations before earthquakes. Tectonophysics, 31: 121-127.

Li Quan-lin, Chen Jin-biao, Yü Lu and Hao Bai-lin, 1978. Time and space scanning of the b-value — A method for monitoring the development of catastrophic earthquakes. Acta Geophys. Sinica, 21: 101-125(in Chinese).

Lindh, A., Fuis, G. and Mantis, C., 1978. Seismic amplitude measurements suggest foreshocks have different focal mechanism than aftershocks. Science, 201: 56-59.

Ma Hong-ching, 1978. Variations of the b-values before several large earthquakes occurred in North China. Acta Geophys. Sinica, 21: 126-141.

MacCabe, M.P., 1979. Earthquake Hazards Reduction Program 1978. Open-File Rep. 79-387, U.S.Geol. Surv., 113pp.

Marza, V.I., 1979. The March 4, 1977 Vrancea earthquake seismic gap. Bull Seismol. Soc. Am., 69: 289-291.

Matsushima, S., 1960. Variation of the elastic wave velocities of rocks in the process of deformation and fracture under high pressure. Bull. Disas. Prev. Res. Inst., Kyoto Univ., 32: 2-8.

Mazzella, A. and Morrison, H.F., 1974. Electrical resistivity variations associated with earthquakes on the San Andreas fault. Science, 185: 855-857.

McCann, W.R., Nishenko, S.P., Sykes, L.R. and Krause, J., 1978. Seismic gaps and plate tectonics: Seismic potential for major plate boundaries. In: J.F. Evernden(Convener), B.L. Isacks and G. Plafker(Organizers), Proceedings of Conference VI Methodology for Identifying Seismic Gaps. U.S. Geological Survey, Open-file Rep. 78-943: 441-584.

Mjachkin, V.I., Brace, W.F., Sobolev, G.A. and Dieterich, J.H., 1975. Two models for earthquake forerunners. Pure Appl. Geophys., 113: 169-181.

Mikumo, T., Kato, M., Doi, H., Wada, Y. and Tanaka, T., 1977. Possibility of temporal variations in earth tidal strain amplitudes associated with major earthquakes. J. Phys. Earth, 25, Suppl.: S123-S136.

Mogi, K., 1968. Sequential occurrences of recent great earthquakes. J. Phys. Earth, 16: 30-36.

Mogi, K., 1974. Rock breaking and earthquake prediction. Zairyo(Materials), 23: 320-331(in Japanese).

Mogi, K., 1978. Rock mechanics and earthquake. In: H. Kanamori (Editor), Physics of Earthquake, Chikyu Kagaku, Iwanami Shoten, Tokyo, 8: 211-262(in Japanese).

Morrison, H.F., Fernandez, R. and Corwin, R.F., 1979. Earth resistivity, self potential variations, and earthquakes: a negative result for M=4.0. Geophys. Res. Lett., 6: 139-142.

National Research Center for Disaster Prevention, 1978. Anomalously small value of the Ishimoto-Iida's coefficient m for foreshocks of Izu-Oshima Kinkai earthquake of January 14, 1978. Rep. Coord. Comm. Earthquake Prediction. 20: 53-55 (in Japanese).

Noritomi, K., 1978a. Geoelectric and geomagnetic observations and phenomena associated with earthquake in China. Proceedings on the Chinese Earthquake Prediction by the 1977 Delegation of the Seismological Society of Japan, Seismol. Soc. Japan, Tokyo, 57-87(in Japanese).

Noritomi, K., 1978b. Application of precursory geoelectric and geomagnetic phenomena to earthquake prediction in China. Chinese Geophys., Am. Geophys. Union, 1, 377-391.

Ohtake, M., 1976. Search for precursors of the 1974 Izu-Hanto-Oki earthquake, Japan. Pure Appl. Geophys., 114: 1083-1093.

Ohtake, M. and Katsumata, M., 1977. Detection of premonitory change in seismic wave velocity. In: Z. Suzuki(Editor), Symposium on Earthquake Prediction Research, Sub-committee on Earthquake Prediction, National Committee for Geodesy and Geophysics, Science Council of Japan and Seismological Society, Japan, 106-115(in Japanese).

Ohtake, M., Matumoto, T. and Latham, G.V., 1977. Seismicity gap near Oaxaca, southern Mexico as a probable precursor to a large earthquake. Pure Appl. Geophys., 115: 375-385.

Raleigh, B., Bennet, G., Craig, H., Hanks, T., Molnar, P., Nur, A., Savage, J., Scholz, C., Turner, R. and Wu, F., 1977. Prediction of the Heicheng earthquake. EOS(Trans. Am. Geophys. Union), 58: 236-272,

Rikitake, T., 1975. Earthquake precursors. Bull. Seismol. Soc. Am., 65: 1133-1162.

Rikitake, T., 1976. Earthquake Prediction. Elsevier, Amsterdam, 357pp.

Rikitake, T., 1979a. Classification of earthquake precursors. Tectonophysics, 54: 293-309.

Rikitake, T., 1979b. Geomagnetic and geoelectric observations in Sichuan Province, China. Rev. Res. Disaster Prevention, 44: 125-127(in Japanese).

Rikitake, T., 1979c. Changes in the direction of magnetic vector of short-period geomagnetic variations before the 1972 Sitka, Alaska, earthquake. J. Geomag. Geoelectr., 31: 441-448.

Rikitake, T., 1980. Earthquake precursors —— A review. Proceedings of the UNESCO Symposium on Earthquake Prediction, in press.

Rikitake, T. and Yamazaki, Y., 1976. Resistivity changes as a precursor of earthquake. J. Geomag. Geoelectr., 28: 497-505.

Rikitake, T. and Yamazaki, Y., 1977. Precursory and coseismic changes in ground resistivity.J. Phys. Earth,25, Suppl.: S161-S173.

Rikitake, T. and Yamazaki, Y., 1979. A resistivity precursor of the 1974 Izu-Hanto-Oki earthquake. J. Phys. Earth, 27: 1-6.

Sano, Y., 1978. Changes in the CA transfer functions at Kakioka. Proceedings of the Symposium on Conductivity Anomaly, CA Group, Tokyo Inst. Technol., Tokyo, 143-149(in Japanese).

Sasai, Y. and Ishikawa, Y., 1978. Changes in the geomagnetic total force intensity associated with the anomalous crustal activity in the eastern part of the Izu Peninsula (2). —— The Izu-Oshima Kinkai earthquake of 1978 ——. Bull. Earthquake Res. Inst., Univ. Tokyo, 53: 893-923(in Japanese).

Sato, H., 1977. Some precursors prior to recent great earthquakes along the Nankai Trough. J. Phys. Earth, 25, Suppl.: S115-S121.

Savage, J.C. and Prescott, W.H., 1979. Geodimeter measurements of strain during the southern California uplift. J. Geophys. Res., 84: 171-177.

Scholz, C.H., Sykes, L.R. and Aggarwal, Y.P., 1973. Earthquake prediction: A physical basis. Science, 181: 803-809.

Sekiya, H., 1976. The seismicity preceding earthquakes and its significance to earthquake prediction. Zisin(J. Seismol. Soc. Japan), Ser. 2, 29: 299-311(in Japanese).

Sekiya, H., 1977. Anomalous seismic activity and earthquake prediction. J. Phys. Earth, 25, Suppl.: S85-S93.

Sidorenko, A.V., Sadovsky, M.A. and Nersesov, I.L., 1979. Soviet experience of prediction of earthquakes in the U.S.S.R. and the prospects for its development. Presented at the 1979 UNESCO Symposium on Earthquake Prediction (in Russian).

Sugisaki, R. and Shichi, R., 1978. Precursory changes in He/Ar and N_2/Ar ratios of fault gases prior to earthquakes. Zisin (J. Seismol. Soc. Japan), Ser. 2, 31: 195-206(in Japanese).

Sultanxodjaev, A.N., Chernov, I.G. and Zakirov, T., 1976. Hydrogeoseismic precursors to the Gasli earthquake. Rep. Acad. Sci. Uzbekistan, 7: 51-53.

Stuart, W.D., 1974. Diffusionless dilatancy model for earthquake precursors. Geophys. Res. Lett., 1: 261-264.

Sykes.,L.R., 1971. Aftershock zones of great earthquakes, seismicity gaps, and earthquake prediction for Alaska and the Aleutians. J. Geophys. Res., 76: 8021-8041.

Tanaka, Y., 1978a. Reports on observations of crustal stress, crustal deformation and gravity change, and their anomalous changes related to earthquakes in China. Proceedings on the Chinese Earthquake Prediction by the 1977 Delegation of the Seismological Society of Japan. Seismol. Soc. Japan, Tokyo, 89-111(in Japanese).

Tanaka, Y., 1978b. Reports on observations of crustal stress and crustal deformation, and their anomalous changes related to earthquakes in China. Chinese Geophys., Am. Geophys. Union, 1, 425-442.

Tang Chi-yang, 1978. Bases for the prediction of the Lungling earthquake and the temporal and spatial characteristics of precursors. Chinese Geophys., Am. Geophys. Union, 2: 400-424. Proceedings on the Chinese Earthquake Prediction by the 1977 Delegation of the Seismological Society of Japan. Seismol. Soc. Japan, Tokyo, 13-32(in Japanese).

Utsu, T., 1975. Detection of a domain of decreased P-velocity prior to an earthquake. Zisin(J. Seismol. Soc. Japan), Ser. 2, 28: 435-448(in Japanese).

Wakita, H., 1978a. Geochemistry related to earthquake prediction in China. Proceedings on the Chinese Earthquake Prediction by the 1977 Delegation of the Seismological Society of Japan. Seismol. Soc. Japan, Tokyo, 113-134(in Japanese).

Wakita, H., 1978b. Earthquake prediction and geochemical studies in China. Chinese Geophys., Am. Geophys. Union, 1, 443-457,

Wakita, H., 1978c. Changes in level and chemical composition of underground water. In: T. Asada(Editor), Method of Earthquake Prediction. Univ. Tokyo Press, Tokyo, 146-166 (in Japanese).

Wang, M.Y. Yang, M.Y., Hu, Y.L., Li, T.C., Chen, Y.T., Chin, Y. and Feng, J., 1976. Mechanism of the reservoir impounding earthquakes at Hsinfengkiang and a preliminary endeavour to discuss their cause. Eng. Geol., 10: 331-351.

Wu, F.T., 1975. Gas well pressure fluctuations and earthquakes. Nature, 257: 661-663.

Wyss, M., 1975. A search for precursors to the Sitka, 1972, earthquake: sea level, magnetic field, and P-residuals. Pure Appl. Geophys., 113: 297-309.

Wyss, M., 1976a. Local changes of sea level before large earthquakes in South America. Bull. Seismol. Soc. Am., 66: 903-914.

Wyss, M., 1976b. Local sea level changes before and after the Hyuganada, Japan, earthquakes of 1961 and 1968. J. Geophys. Res., 81: 5315-5321.

Wyss, M., 1977. The appearance rate of premonitory uplift. Bull. Seismol. Soc. Am., 67: 1091-1098.

Wyss, M., 1980. Recent Earthquake Prediction Research in the United States. In: T. Rikitake(Editor), Current Research in Earthquake Prediction I. Center for Academic Publications Japan / Japan Scientific Societies Press, Tokyo, in press.

Xu Wen-yao, Qi Kui and Wang Shi-ming, 1978. On the short period geomagnetic variation anomaly of the eastern Kansu Province. Acta Geophys. Sinica, 21: 218-224(in Chinese).

Yamaguchi, R. and Odaka, T., 1978. Precursory changes in water level at Funabara and Kakigi before the Izu-Oshima Kinkai earthquake of 1978. Rep. Coord. Comm. Earthquake Prediction, 20: 60-62. Bull. Earthquake Res. Inst., Univ. Tokyo, 53: 841-854(in Japanese).

Yamazaki, Y., 1975. Precursory and coseismic resistivity changes. Pure Appl. Geophys., 113: 219-227.

Yanagihara, K. and Nagano, T., 1976. The change of transfer function in the Central Japan Anomaly with special reference to earthquake occurrences. J. Geomag. Geoelectr., 28: 157-163.

Yoshii, T., 1978. Change in seismic wave velocities. In: T. Asada(Editor), Method of Earthquake Prediction. Univ. Tokyo Press, Tokyo, 78-92(in Japanese).

Zhu, F.M., 1976. Prediction, warning and disaster prevention related to the Haicheng earthquake of magnitude 7.3. Proceedings of the Lectures by the Seismological Delegation of the People's Republic of China. Seismol. soc. Japan, pp. 15-26(in Japanese).

Polarization Anomalies as Diagnostics of Dilatancy

Stuart Crampin

Institute of Geological Sciences, Murchison House, West Mains Road, Edinburgh EH9 3LA, UK

Abstract

Stresses before an earthquake may differentially open existing cracks in the Earth over an extensive area. This gives rise to dilatancy anisotropy, which causes shear-wave splitting, and may be recognised on seismograms as polarization anomalies in the shear wavetrains.

The crust and crystalline upper-mantle of the Earth are thoroughly permeated by cracks. Sedimentary rocks have joints and fractures, which are usually aligned, and in some circumstances have large enough crack densities to cause significant velocity-anisotropy (Bamford & Nunn 1979; Crampin, McGonigle & Bamford 1980). Igneous rocks have a variety of microcracks along grain boundaries and along cleavages and dislocations within grains (Simmons & Richter 1976), which may remain open for some distance into the upper mantle, if the pore pressures are large enough (Brace 1972). Under purely hydrostatic stress these various cracks will be closed and transparent to seismic waves. Under non-hydrostatic stress, such as occurs before an earthquake, these closed cracks may be opened in preferred directions, and such aligned cracks will produced anisotropic effects on seismic-wave propagation through such a dilatancy zone (Crampin 1978). The most diagnostic differences between seismic-wave propagation in isotropic and in anisotropic rock are the polarization anomalies caused by shear-wave splitting in anisotropic media. This splitting is also known as shear-wave birefringence and acoustic double-refraction. Crampin (1978) suggests that such polarization anomalies may be used to monitor the progress of dilatancy episodes.

The phenomena we are referring to by dilatancy, is not the dilatancy associated with high stress-concentrations near zones of weakness in strained regions, such as in observed before impending fractures in laboratory experiments (Brace, Paulding

& Scholz 1966; Gupta 1973a; Hadley 1975). Nur & Simmons (1969) demonstrated that open cracks close differentially, when an initially cracked specimen was loaded in the laboratory. Experiments showing that closed cracks open under non-hydrostatic stress are more difficult to arrange in the laboratory, and no experiments have yet been reported. Since the tectonic forces before a large earthquake act over a very large region, this type of dilatancy may occur at substantial distances from the immediate epicentral-area (up to 1000 km from a magnitude 7 earthquake, Dobrovolsky, Zubkov & Miachkin 1979). This type of dilatancy is the most likely cause of many of the precursory phenomena sometimes observed before earthquakes: V_P/V_S anomalies have been observed at 40 km from earthquakes in Garm, USSR (Semenov 1969) and 120 km from the San Fernando Earthquake in California (Whitcomb, Garmany & Anderson 1973); and Dobrovolsky et al. (1979) list many precursors occurring at larger distances, including geochemical changes in ground water up to 1000 km from epicentres of large earthquakes.

Garbin & Knopoff (1973; 1975a; 1975b) determined convenient theoretical velocity-variations for wave propagation through weak concentrations of thin aligned-cracks in dry and saturated conditions. Crampin (1978) obtained effective anisotropic elastic-constants for these velocity variations. Since the cracks in any dilatant rock have much smaller dimensions than the seismic wavelengths, and it appears that this dilatancy occurs over substantial regions, we can model the effects of dilatancy on seismic waves by means of these anisotropic elastic-constants, and Crampin (1978) has calculated synthetic seismograms through dilatancy zones. These synthetic seismograms demonstrate that, as shear waves pass through such a region of effective anisotropy, the shear waves split into two components. These components have orthogonal polarizations with respect to the direction of phase propagation, but not necessarily with respect to the ray or group velocity direction. These shear-waves have different velocities and consequently different arrival times. This causes polarization anomalies in the particle motion of the shear wavetrains, as one linearly polarized shear-wave is followed by another linearly polarized arrival with nearly orthogonal particle-motion. Generally, this causes abrupt changes in direction of the particle motion, and the delay between the two arrivals is proportional to the path length through the dilatancy and to the relative shear-wave velocity-anisotropy along that path. Shear waves in isotropic media have a unique velocity of propagation. Consequently, the delays and relative polarizations of anomalies caused by a path through a dilatancy zone will be preserved for any subsequent isotropic segment of the path.

Such polarization anomalies have been observed on unrotated horizontal seismograms in Armenia, USSR by Yegorkina, Rakitov, Garetovskaya & Yegorova (1977), and possibly in Nevada by Gupta (1973b) (but see Crampin, Evans, Doyle & Davis 1980). Polarization anomalies have also been recorded during the Turkish Dilatancy Project. A small centre of swarm activity was identified by Üçer, Crampin & Miller (1980, this volume), and a network of three-component seismometers was set up around the swarm during the summer 1979. Preliminary results of this experiment are discussed by Evans, Doyle, Üçer, Miller & Crampin (1980) in this volume.

This work has been supported by the Natural Environment Research Council and is published with the approval of the Director of the Institute of Geological Sciences.

References

Bamford,D. & Nunn,K.R., 1979. In situ measurements of crack anisotropy in Carboniferous Limestone of Northwest England, Geophys.Prosp.,27, 322-338.

Brace,W.F., 1972. Pore pressure in geophysics, in Flow and fracture in rocks, Geophys.Mon.,16, ed. Heard,H.C., Borg,I.Y., Carter,N.L., & Raleigh,C.B., AGU Washington.

Brace,W.F., Paulding,B.W. & Scholz,C., 1966. Dilatancy in the fracture of crystalline rocks, J.geophys.Res.,71, 3939-3953.

Crampin,S., 1978. Seismic-wave propagation through a cracked solid: polarization as a possible dilatancy diagnostic, Geophys.J.R. astr.Soc., 53, 467-496.

Crampin,S., McGonigle,R. & Bamford,D., 1980. Estimating crack parameters from observations of P-wave velocity-anisotropy, Geophys., 45, 345-360.

Crampin,S., Evans,R., Doyle,M. & Davis,J.P., 1980. Comments on papers about shear-wave splitting in dilatancy-induced anisotropy by I.N.Gupta and by A.Ryall and W.U.Savage, Bull.seism.Soc.Am.,(submitted).

Dobrovolsky, I.P., Zubkov, S.I. & Miachkin, V.I., 1979. Estimation of the size of earthquake preparation zones, Pure appl. Geophys. 117, 1025-1044.

Evans, R., Doyle, M., Üçer, S.B., Miller, A. & Crampin, S., 1980. An experiment to investigate polarization anomalies in North Anatolia, Proc. Interdis. Conf. on Earthq. Pred. Res. in the North Anatolian Fault Zone, Istanbul, Friedr. Vieweg & Sohn, Wiesbaden.

Garbin, H.D. & Knopoff, L., 1973. The compressional modulus of a material permeated by a random distribution of free circular cracks, Q. appl. Math., 30, 453-464.

Garbin, H.D. & Knopoff, L., 1975a. The shear modulus of a material permeated by a random distribution of free circular cracks, Q. appl. Math., 33, 296-300.

Garbin, H.D. & Knopoff, L., 1975b. Elastic moduli of a medium with liquid-filled cracks, Q. appl. Math., 33, 301-303.

Gupta, I.N., 1973a. Seismic velocities in rock subjected to axial leading up to shear fracture, J. geophys. Res., 78, 6936-6942.

Gupta, I.N., 1973b. Premonitory variations in S-wave velocity anisotropy before earthquakes in Nevada, Science, 182, 1129-1132.

Hadley, K., 1975. Azimuthal variation of dilatancy, J. geophys. Res., 80, 4845-4850.

Nur, A. & Simmons, G., 1969. Stress-induced velocity anisotropy in rock: an experimental study, J. geophys. Res., 74, 6667-6674.

Semenov, A.N., 1969. Variations in the travel times of transverse and longitudinal waves before violent earthquakes, Izv. Acad. Sci., USSR, Phys. Solid Earth (English transl.), 4, 245-248.

Simmons, G. & Richter, D., 1976. Microcracks in rocks, in Physics and chemistry of minerals and rocks, pp. 105-137, ed. Strens, R.G.J., John Wiley & Sons.

Üçer, S.B., Crampin, S. & Miller, A., 1980. Identification of swarm activity using the MARNET telemetered seismometer-network, Proc. Interdis. Conf. on Earthq. Pred. Res. in the North Anatolian Fault Zone, Istanbul, Friedr. Vieweg & Sohn, Weisbaden.

Whitcomb, J.H., Garmany, J.D. & Anderson, D.L. 1973. Earthquake prediction: variation of seismic velocities before the San Francisco Earthquake. Science, 180, 632-635.

Yegorkina, G.V., Rakitov, V.A., Garetovskaya, I.V. & Yegorova, L.M., 1977. Anisotropy of velocities of seismic waves and the stress state of the Earth's crust in the territory of Armenia, Izv. Acad. Sci. USSR, Earth Phys. (English transl.), 13, 554-562.

An Experiment to Investigate Polarization Anomalies in North Anatolia

Russ Evans and Stuart Crampin

Institute of Geological Sciences, Murchison House, West Mains Road, Edinburgh EH 9 3LA, UK

Mark Doyle

Department of Geophysics, University of Edinburg, James Clerk Maxwell Building, Edinburgh EH 9 3JZ, UK

S. Balamir Üçer

Kandilli Observatory, Çengelköy, Istanbul, Turkey

Alistair Miller

Overseas Development Ministry, London, UK

Abstract

A three-component seismometer network, 15 km in aperture, was set up over some swarm activity just East of the Sea of Marmara. In an eight week period several hundred small earthquakes were recorded beneath or immediately adjacent to the network. Almost all the shear wave trains of these earthquakes display polarization anomalies indicative of dilatancy anisotropy.

A three-component seismometer network, TDPNET, was set up during the summer of 1979 a few kilometres South of Izmit, near the Northern Anatolian Fault at the Eastern margin of the Marmara Sea. The network was designed to monitor the polarization anomalies suggested by theoretical and numerical investigations (Crampin 1978). A site was chosen on the basis of swarm activity recognised during routine epicentral locations at Kandilli Observatory (see Fig. 1 of Üçer, Crampin & Miller 1980, this volume). One of the MARNET stations, KLT, recorded a high level of local activity throughout the winter of 1978/79. This activity, centred a few kilometres to the South of the

Northern branch of the Northern Anatolian Fault, was monitored for eight weeks by a closely-spaced radio-linked network of six three-component stations recording on analogue magnetic-tape.

Several hundred earthquakes were recorded within and near TDPNET during this period (Fig. 1). They were located with HYPO71 (Lee & Lahr 1975) at depths between 8 and 15 km, by assuming an isotropic crustal-structure determined by using local quarry blasts. The magnitudes (M_L) range from 0.8 to 2.5, and an event of 3.5 M_L

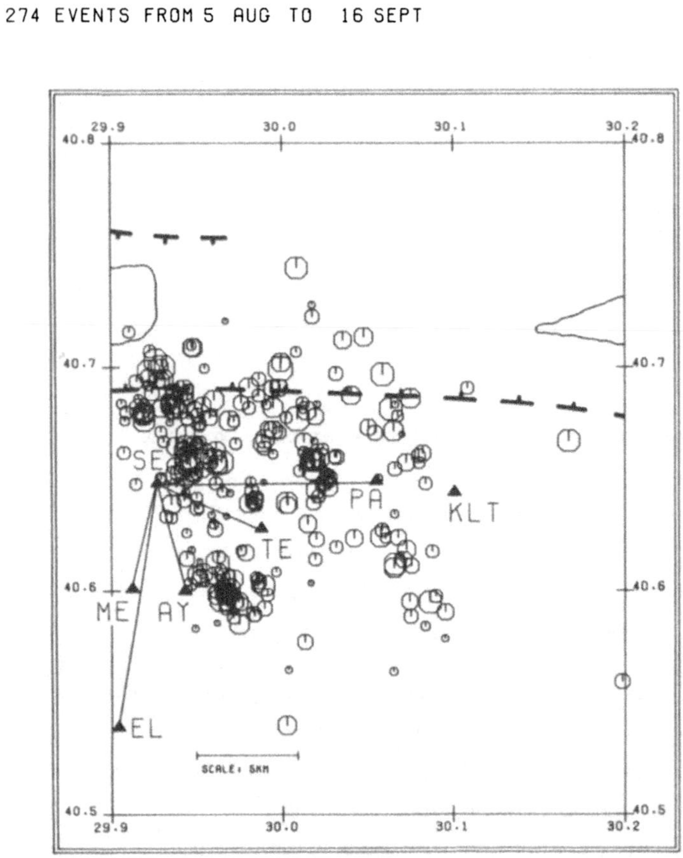

Fig. 1: Earthquakes located during the Turkish Dilatancy Project, with magnitudes ranging from 0.8 M_L to 2.0 M_L. AY, EL, ME, PA, SE, and TE are three-component TDPNET stations, and KLT is a vertical MARNET station. The dashed lines mark geological faults associated with the Northern Anatolian Fault.

occurred at the start of the recording interval. The majority of the events had impulsive P and S arrivals, and the records are very suitable for investigating polarization anomalies (Crampin, Evans, Üçer, Doyle, Davis, Yegorkina & Miller 1980).

The hypothesis of shear-wave splitting in seismic regions has been outlined by Crampin (1980, this volume): the crust of the Earth is permeated by a large variety of cracks, which will be opened and closed differentially by non-hydrostatic stresses, such as occur before earthquakes. On entry into such effectively anisotropic structures of aligned cracks, shear waves split into components with different velocities and nearly orthogonal polarizations (Crampin 1978). The delay between these arrivals is proportional to the differential shear-wave velocity-anisotropy and to the path length through the anisotropy. Such split arrivals can be recognised by abrupt changes in direction of the shear-wave particle-motion, when the wave forms are drawn in polarization-diagrams. These diagrams are projections of the particle motion on to the three orthogonal-planes defined by the rotated seismograms. On the evidence of the high level of seismic activity in this swarm, the whole region around TDPNET seems to be stressed and crack dilatancy is expected along all paths near the swarm activity.

The majority of shear wavetrains recorded by TDPNET display the abrupt changes in direction of particle motion indicative of dilatancy (see Crampin et al. 1980). In a small proportion of shear arrivals, not necessarily from Cardinal directions, the shear-wave splitting is clearly visible as separate arrivals on the unrotated horizontal seismograms. This phenomenon is similar to that observed by Yegorkina, Rakitov, Garetovskaya & Yegorova (1977) in seismic regions in Armenia, USSR. Thus it seems that dilatancy may be a comparatively widespread phenomenon in seismic regions.

Although the data processing is still at a preliminary stage, we suggest that the observed shear-wave splitting is due to propagation through dilatancy anisotropy. A demonstration that the splitting has appropriate three-dimensional variations, and that the delays between the two shear arrivals vary with time (that is with stress accumulation and release) would confirm our interpretation.

This work has been supported by a variety of organization, but the major support has come from the Natural Environment Research Council, U.K., and Kandilli Observatory, Turkey, and is published with the approval of the Director of Kandilli Observatory and the Director of the Institute of Geological Sciences.

References

Crampin,S., 1978. Seismic-wave propagation through a cracked solid: polarization as a possible dilatancy diagnostic, Geophys. J.R.astr.Soc., 53, 467-496

Crampin,S., 1980. Polarization anomalies as diagnostics of dilatancy, Proc.Interdis.Conf.Earthq.Pred.Res.in N.Anatolian Fault Zone, Friedr.Vieweg & Sohn, Wiesbaden.

Crampin,S., Evans,R., Üçer,S.B., Doyle,M., Davis,J.P., Yegorkina,G.V. & Miller,A., 1980. A preliminary report on observations of dilatancy-induced polarization-anomalies: the potential for earthquake prediction, Nature, submitted.

Lee,W.H.K. & Lahr,J.C., 1975. 'HYPO71 (revised): A computer program for determining hypocenter, magnitude and first motion pattern of local earthquakes, US Geol.Survey, Open File Report 75-311.

Üçer,S.B., Crampin,S. & Miller,A., 1980. Identification of swarm activity using the MARNET telemetered seismometer-network, Proc.Interdis,Conf.Earthq,Pred.Res.in N.Anatolian Fault Zone, Friedr. Vieweg & Sohn, Wiesbaden.

Yegorkina,G.V., Rakitov,V.A., Garetovskaya,I.V. & Yegorova,L.M., 1977. Anisotropy of velocities of seismic waves and the stress state of the Earth's crust in the territory of Armenia, Izv.Acad.Sci.USSR.Earth Phys.(English tranl.), 13, 554-562.

Possible Information on Local Features of the Focal Region Based on Geomagnetic Field Data Recorded After the Main Shock

Mirjana M. Pantović

Geomagnetic Institute, 11306 Grocka, Yugoslavia

SUMMARY

Some results obtained on the basis of analysis of continuous records of the geomagnetic field in the epicentral area after the earthquake of April 15, 1979 (M=6.6) at Montenegro coast are discussed in this work. Simple and weighted differences in H and Z component between temporary station at Montenegro and Observatory at Grocka as reference station, as well as the results of Principal Component Analysis (PCA), are interpreted with respect to the number of aftershocks with $M \geq 3$ recorded at the same period.
Obtained results seem to indicate that the geomagnetic field was locally disturbed in the period considered, which is very likely to be connected with numerous aftershocks i.e. the result of seismic activity after the main shock. Therefore, a possibility is suggested that geomagnetic field data, although recorded after the main shock, could be used for studying local features of the focal region. Here, as the geomagnetic field data are concerned, we primarily have in mind the possibility of answering the question whether a particular region is favorable for identification of tectonomagnetic effect.

1. INTRODUCTION

In order to study tectonomagnetic effect, as one of precursory phenomena, local behaviour of the geomagnetic field in particular area should be observed before an earthquake, i.e. in the stage of its "preparation". As it is known, magnetic anomalies tend to disappear after the occurrance of earthquake (TAZIMA, 1968). However, the interval of time in which anomalies disappear is not defined and an example is given about anomalous secular change of H component increasing before two earthquakes (M=6.1 and 6.4), and then showing decreasing tendency towards the normal level, as revealed by measurements after the earthquakes. Another example refers to the geomagnetic investigations in the seismoactive regions of Middle Asia (SHAPIRO at al., 1978). In January 1977 the earthquake happened in the vicinity of one of the profiles in Fergana polygon where measurements were carried out in 1973 and 1975 but, unfortunately, not before the earthquake. Measurements carried out 10 days after the earthquake revealed total field intensity change of about 30 nT in comparison previous survey, at the point of profile nearest to the epicenter. Therefore, it does not seem unreasonable to expect to identify certain local effects even if the geomagnetic field is obserevd after the earthquake.

Considering the above mentioned, the idea was to see whether any local changes of the geomagnetic field in the epicentral area of April 15, 1979 earthquake can be identified, on the basis of observations carried out after the main shock, particularly because a sequence of aftershocks continued in this period, some of them being of magnitudes almost 5. So, nine days after the main shock geomagnetic field survey (D, H and F) began, at all together 25 points in the area affected by the earthquake. At the same time, geomagnetic field variations (D, H and Z) were continuously recorded from April 24th to May 10th. Results presented in this work are based on the analysis of continuous records of the geomagnetic field at temporary station at the Montenegro Coast.

2. DISCUSSION OF THE RESULTS

Considering that the whole length of geomagnetic field records is only 16 days, avaliable data are not sufficient to be analyzed most adequately in every way. Yet, few exanples will be given in order to illustrate local behaviour of the geomagnetic field and an attempt will be made to relate it with aftershock activity in this area.

Simple differences of hourly mean values, both in H and Z, between temporary station at Montenegro (φ=42°12′, λ=18°59′) and Observatory at Grocka (φ=44°38′, λ=20°46′) show typical daily type of variation, especially in H component. It was to be expected according to relatively great distance between two stations which is about 300 km, i.e. about 2.5° in geographical latitude. This daily type of variation is more regularly expressed on magnetically quiet days - measured by the daily sum of K indices, especially when seismic activity was also lower on those days - measured by the number of aftershocks with M≥3. Fig. 1 shows the differences of hourly mean values of H component on 4th of May (ΣK=13, no aftershocks with M≥3) that can be considered as typical example of daily type of variation. In contrast to this example, Fig. 2 shows daily variation of ΔH differences for three consecutive days, 26th, 27th and 28th of April, with denoted number of aftershocks and ΣK. As many aftershocks followed each other it is difficult to separate their individual effects. But, it can be seen that the daily course of ΔH differences is somewhat disturbed in comparison with the one which is considered to be typical, on such a days when more aftershocks were recorded.

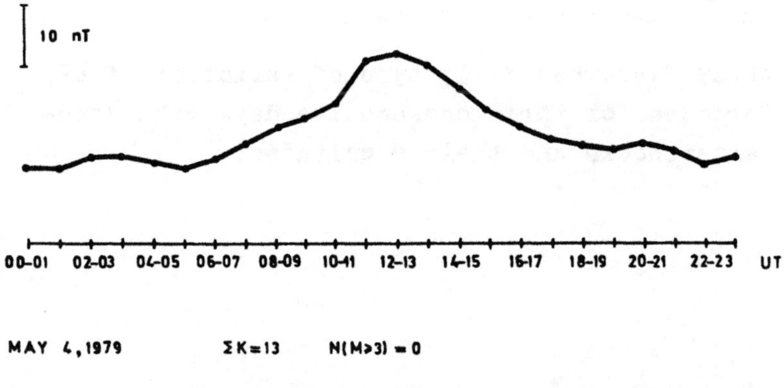

Fig. 1 Differences of hourly mean values of H component between temporary and permanent station, as an example of undisturbed daily type of variation.

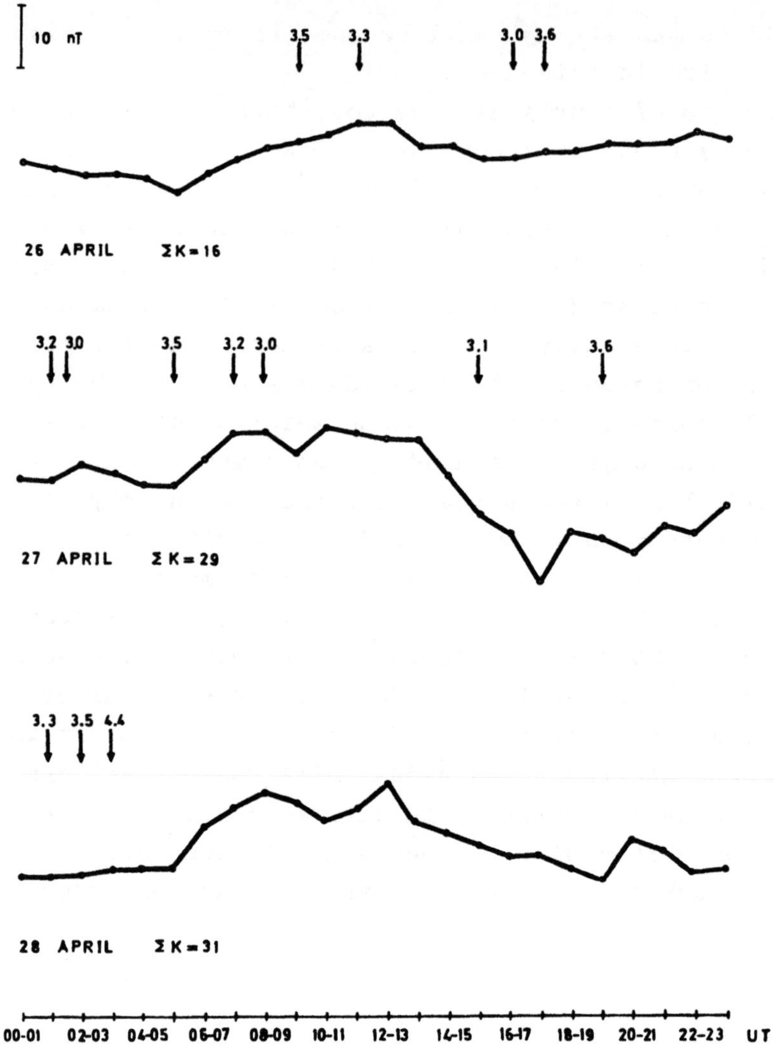

Fig. 2 Slightly disturbed daily type of variation of ΔH differences for three consecutive days with denoted aftershocks and their magnitudes.

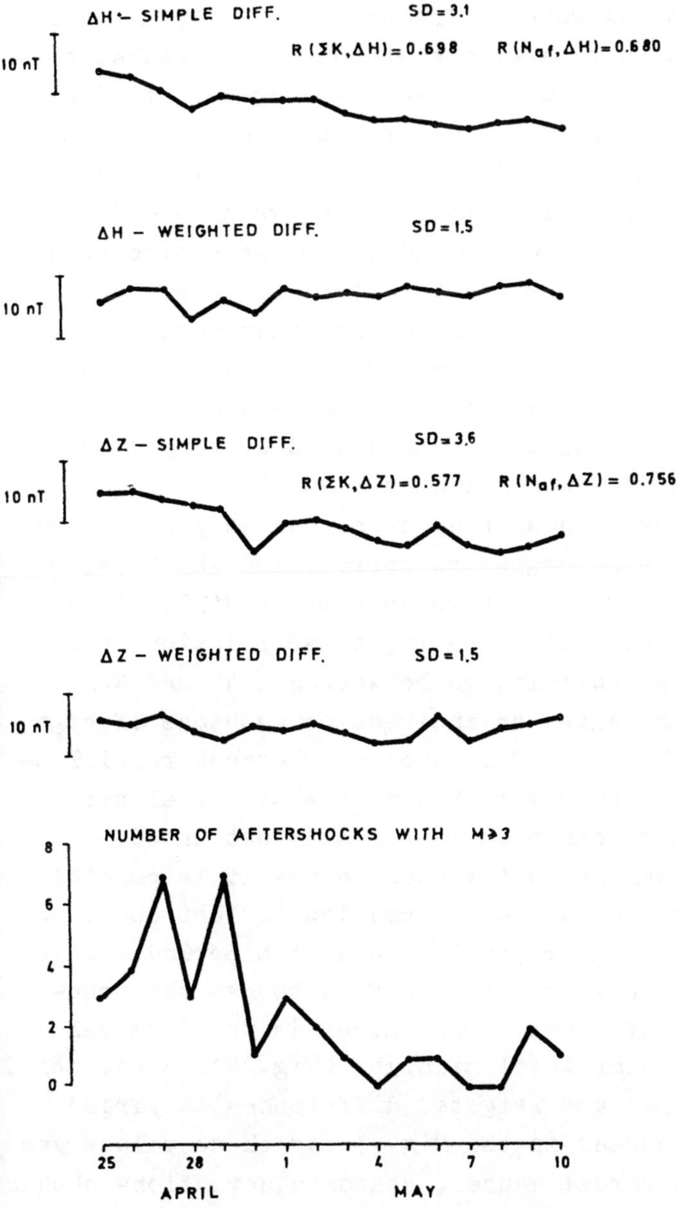

Fig. 3 Simple and weighted differences of daily mean values of H and Z component with corresponding number of aftershocks in the same period.

In Fig. 3, simple and weighted differences of daily mean values of H and Z component between Montenegro and Grocka, together with denoted standard deviations and number of aftershocks, are shown. By appearance, there is no obvious connection between ΔH, ΔZ and N(M) curves. However, it turns out that there is high degree of correlation between ΔH - N(M) and ΔZ - N(M), which equals 0.680 and 0.756 respectively. On the other hand, changes of ΔH and ΔZ are highly correlated with corresponding values of ΣK and correlation coefficients are R(ΣK, ΔH)=0.698 and R(ΣK, ΔZ)=0.577. It is obvious that external influences are still present in ΔH and ΔZ differences to a great extent. Therefore, one should be very careful while interpreting high correlation between ΔH, ΔZ and N(M). Besides this, it is well known that even obviously unrelated phenomena may exibit high degree of correlation which has no physical meaning, that turns out to be here as R(ΣK, N(M))= 0.404. For that reason, while treating these relations purely statistically, high correlation between ΔH, ΔZ and N(M) could be to a certain degree an indirect consequence of high correlation between ΣK and N(M). Later on a procedure will be suggested by which, in some way, the degree of correlation between ΔH, ΔZ and N(M) can be more reliably estimated.
In the case of dealing with daily mean values it is possible that, by averaging the data, an information is lost about some local changes of the geomagnetic field with period shorter than 24 hours. It is interesting therefore to see the behaviour of differences of hourly mean values in the interval 21.00-22.00 UT, i.e. near local midnight (Fig. 4). Here, standard deviation of simple and weighted differences is larger than in the case discussed in the Fig. 3. As these values are less influenced by external causes, standard deviations should be smaller unless we assume that ΔH and Δ Z reflect changes of the geomagnetic field of local character in connection with aftershock activity. It can be noticed that dependence of ΔH and ΔZ on ΣK is less expressed, especially for Z, but at the same time ΔZ is poorly correlated with N(M). Yet, it can be said that changes of ΔH and ΔZ are most significant at the beginning of the period when there was series of aftershocks with $M \geq 3$.

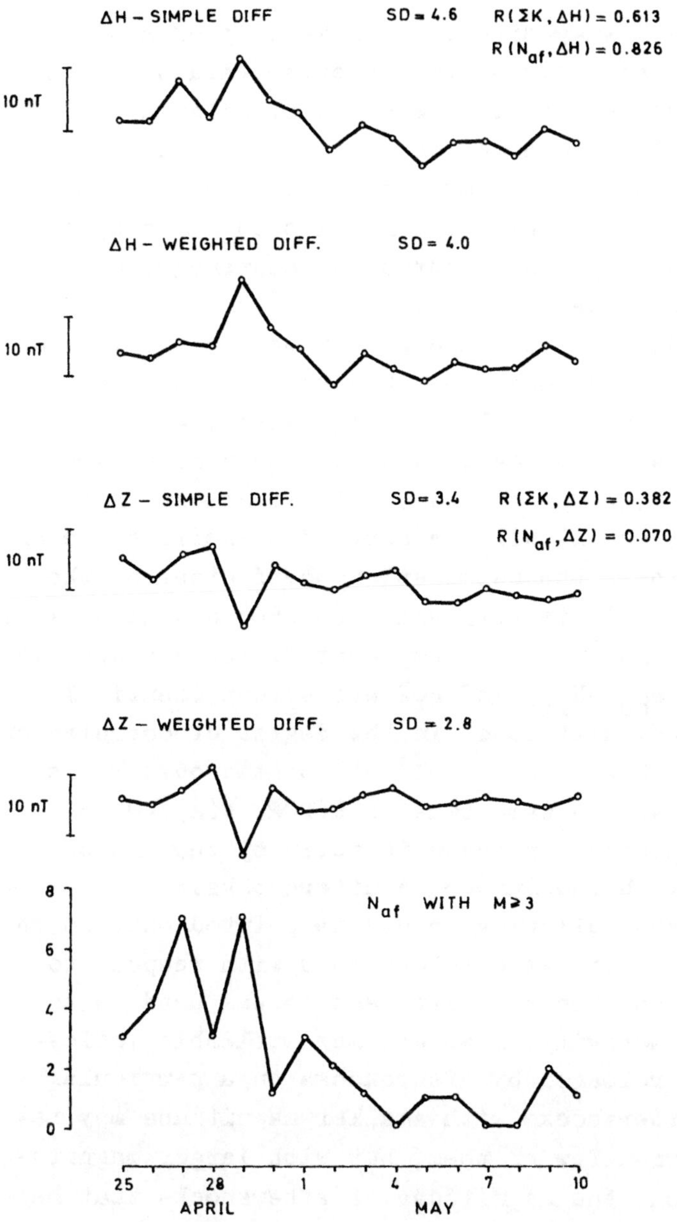

Fig. 4 Differences of hourly mean values of H and Z component in the interval 21.00-22.00 UT.

ISIKARA at al. (1979) suggested the possibility of applying Principal Component Analysis on the observed values of the geomagnetic field and one of the measures of solar activity (R or A_p) as variables, in order to obtain quantities that would reflect the behaviour of external and internal components of the geomagnetic field. The results obtained applying PCA method on H component recorded at Montenegro are shown in Fig. 5. As there was a very pronounced trend in the data in this period, observed values of H were approximated by the first degree polynomial and then values $\Delta H = H_{obs} - H_{fit}$, together with corresponding values of A_p, were used in the analysis. As it is seen in the figure, two components are obtained - PC1 and PC2. One of them, PC1, is almost in 1:1 correlation with A_p and it will be assumed to reflect the external component, while the correlation coefficient of the second one, PC2, with A_p is very small (0.056) and it will be assumed to reflect the internal component of the geomagnetic field. Although $\Delta H = H_{obs} - H_{fit}$ and PC2 are almost identical curves by appearance, difference in the degree of correlation of ΔH with A_p and PC2 with A_p should be pointed out. While $R(\Delta H, N(M)) = 0.104$, at the same time is $R(PC2, N(M)) = 0.335$. So, PC2 is less dependent on magnetic activity than ΔH but also more related with the number of aftershocks.

At this instance, the following should be pointed out. So far whenever obtained results were interpreted with respect to seismic activity, the number of aftershocks was used as an indication of this activity. However, more reliable indication is the energy released by aftershocks on a particular day besause more aftershocks with smaller magnitudes may release less energy than few of them, but with larger magnitudes. For that reason, the magnitudes of aftershocks that happened on a particular day were converted into energy, taking that zero level refers to such a days when there were no aftershocks with $M \geq 3$. Energy curve is shown in Fig. 6 together with curves PC1(Gr)/PC1(Mn) and PC2(Gr)/PC2(Mn), whose meaning and relation will be discussed. When PCA is also applied on H component for Grocka and ratio of corresponding components PC1 and PC2 for Grocka (Gr) and Montenegro (Mn) found,

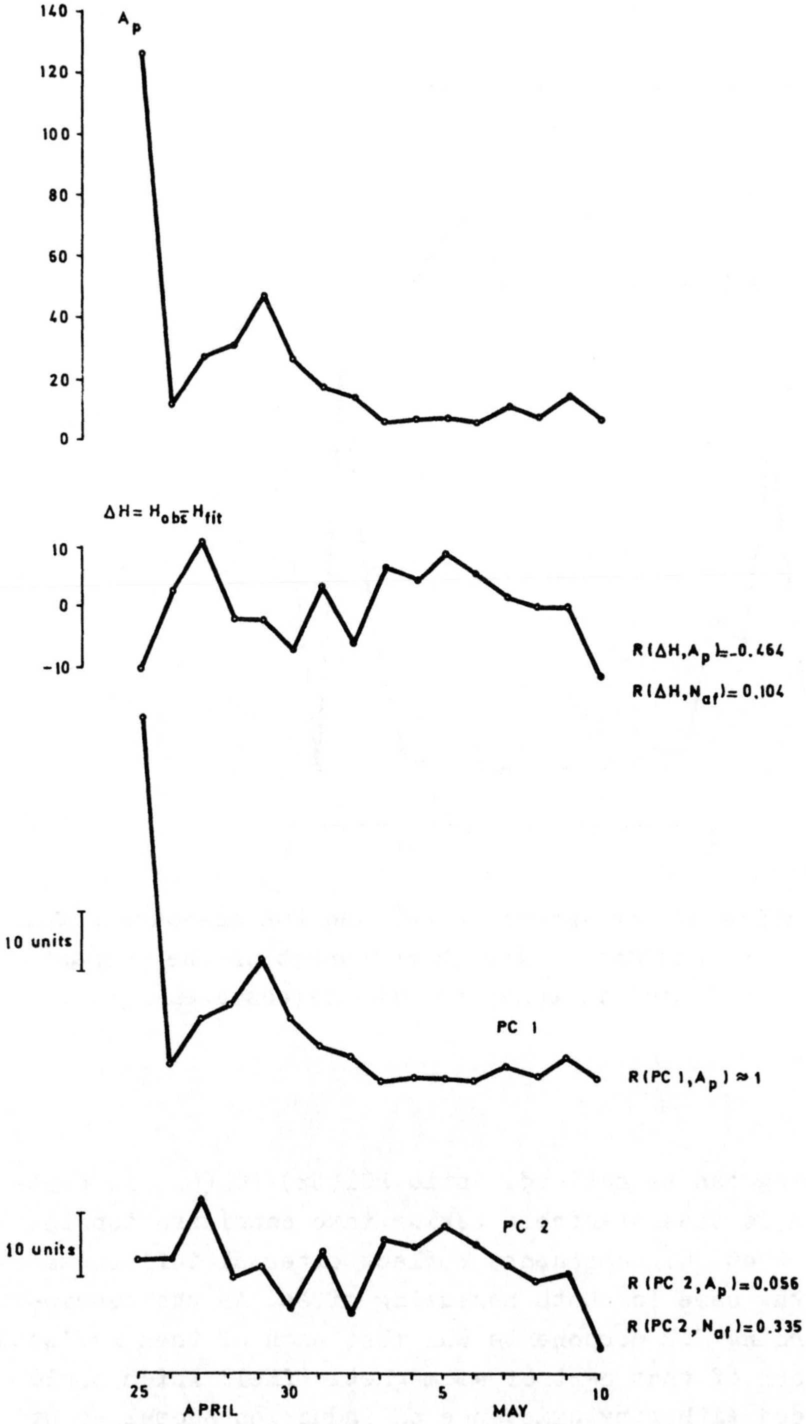

Fig. 5 Result of application of PCA method on H component recorded at Montenegro temporary station.

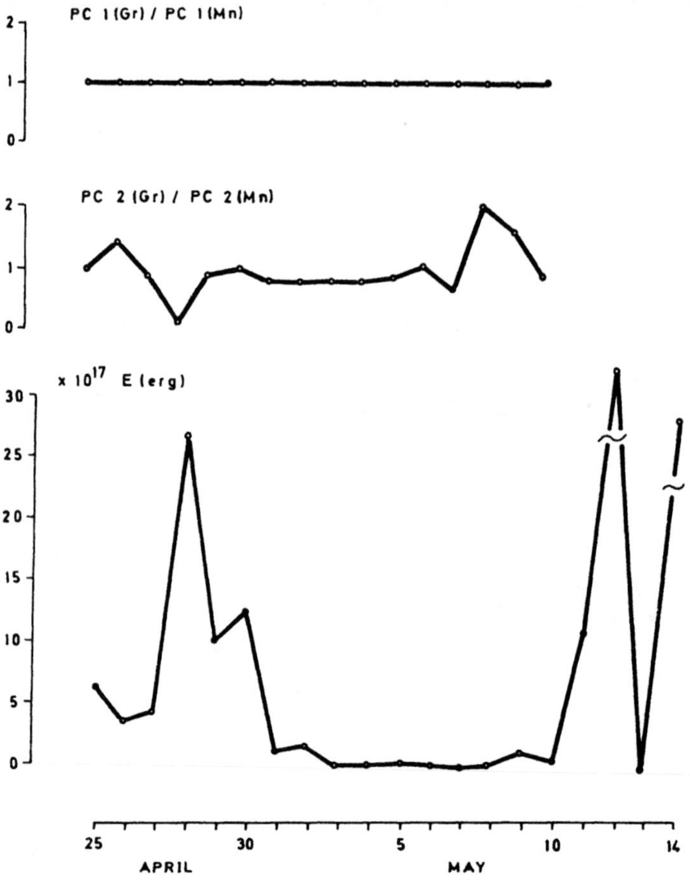

Fig. 6 Ratios of corresponding PC1 and PC2 components for Grocka and Montenegro (H component of the geomagnetic field) in relation with released energy.

the following can be noticed. Ratio PC1(Gr)/PC1(Mn) is constant, which is understandable taking into consideration the assumption that PC1 components reflect external influences which are the same for both measuring sites. As the assumption concerning PC2 components was that each of them reflects The behaviour of that part of geomagnetic field which could be influenced either by existance of induction anomalies or some tectonic processes beneath the measuring site, if there are no such influences, the ratio of PC2(Gr) and PC2(Mn) should also be constant. And indeed, it can be seen that the period of relatively constant ratio PC2(Gr)/PC2(Mn) coincides with the period in which the level of released energy is also constant. On 28th of April this ratio drops considerably below one and this minimum corresponds to the maximum of released energy in this period. Also, it should be noticed that this minimum was preceded by an increase of PC2(Gr)/PC2(Mn) ratio over one. At the end of the period considered an increase of the ratio can be noticed again. Energy curve has been deliberately extended beyond the date when the recording of the geomagnetic field at Montenegro was completed because on 11th and 12th of May two stronger aftershocks happened, with M=4.2 and 5.1 respectively. Assumption is that the second maximum of PC2(Gr)/PC2(Mn) curve would be again followed by a minimum which would, possibly, coincide with large peak on energy released curve on 11th and 12th of May.

And finally, let us come back to the comment concerning correlation between certain quantities considered here. It is suggested that instead of dealing with original values of given quantities, differences between successive values, i.e. successive differences from day to day, should be taken. Fig. 7 illustrates the result of application of such a procedure on PC2(Mn) and N(M). In the case of correlating successive differences correlation coefficient is considerably higher. Some more examples are given in Table 1. which permit us to make following conclusion. When taking successive differences we are dealing with values which are in some way less

Table 1

Illustration of the differences in correlation co-fficients between certain quantities when original values and successive differences are used.

	correlat with ΣK		correlat with N_{af}	
	original values	success differen	original values	success differen
ΔH daily mean	0698	-0368	0680	-0728
ΔZ daily mean	0577	-0130	0756	0564
ΔH 2100-2200 UT	0613	-0185	0826	-0723
ΔZ 2100-2200 UT	0382	0026	0070	-0698
$H_{obs} - H_{fit}$ (Mn)	-0438	-0165	0104	0528
PC 2 (Mn)	0022	0031	0335	0707

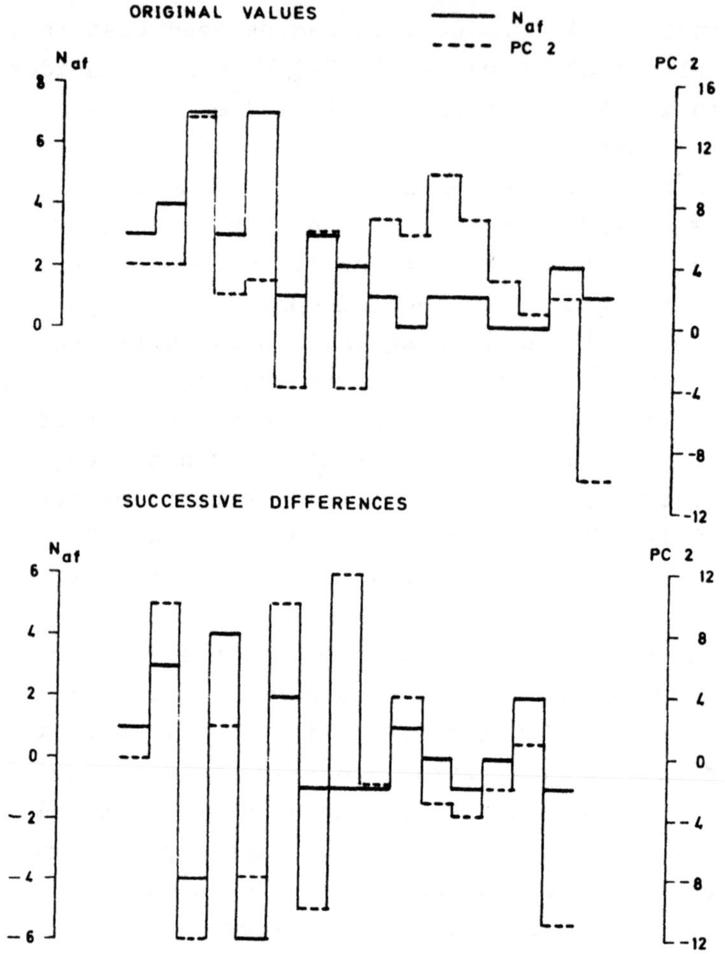

Fig. 7 Correlation between PC2(Mn) and N(M) when original values and successive differences are used.

dependent on external influences while, at the same time, their relation with N(M) is more pronounced. In other words, successive differences of ΔH for instance, are less correlated with successive differences of ΣK and therefore, high correlation coefficient between successive differences of ΔH and N(M) might have more real meaning. Somehow, as if what is imporatant are not the absolute values of certain quantities on particular day but the magnitude of their relative change from day to day.

3. CONCLUSION

Although there are cases when no local changes of the geomagnetic field in connection with some earthquakes and rather strong aftershocks were observed (SHAPIRO and ABDULLABEKOV, 1978), we dare say that results presented here indicate certain local features of the geomagnetic field behaviour at the Montenegro coast after the main shock of April 15, 1979, probably caused by continued seismic activity in this region.
In addition to already mentioned fact concerning the impossibility of adequate data analysis because of limited length of geomagnetic field records, one more thing should be noted as well. While analyzing these results, neither we had the data on the coordinates of epicenters of aftershocks, i.e. we did not know their location with respect to the point of observation of the geomagnetic field, nor we took into consideration aftershocks with $M<3$. On account of this, it is well known that the amplitude of local field changes depends not only on the focal depth and magnitude of the earthquake but also on the relative position of observation site and direction of magnetic field with respect to stress axes. Thus, a small earthquake having an epicenter favorably situated with regard to the observation site may produce well defined change of the geomagnetic field, while a larger one having differently situated hypocenter may not.
Fully respecting all possible remarks, we want to point out that some results presented here indicate very close connection between certain quantities which describe geomagnetic field behaviour and aftershock activity. Even if we accept the possibility that it might be a pure coincidence, it cannot be denied that it is a very striking coincidence, which draws our attention and points out the necessity of more rigorous analysis of the data, taking into consideration geological and all other relevant geophysical data concerning this region and the earthquake itself. If additional analysis happens to confirm some of the results obtained so far, we would be able to say that this particular region is favorable for observation of tectonomagnetic effect.

ACKNOWLEDGEMENT

The author wishes to express her thanks to Assoc. Prof. A. Mete Isikara for his valuable discussion as well as to Geomagnetic Institute's team for carrying out field work.

REFERENCES

1. ISIKARA, A.M., N. ORBAY and M. PANTOVIĆ, An attempt for elimination of non-local changes of the geomagnetic field for earthquake prediction, Proc. ESA-Council of Europe Seminar on Earthquake Prediction, Strasbourg 5-7 March 1979 - ESA SP-149.
2. SHAPIRO, V.A., A.N. PUSHKOV, K.N. ABDULLABEKOV, E.B. BERDALIEV and M. Yu. MUMINOV, Geomagnetic investigations in the seismoactive regions of Middle Asia, J. Geomag. Geoelectr., 30, 503-509, 1978.
3. SHAPIRO, V.A. and K.N. ABDULLABEKOV, An attempt to observe a seismomagnetic effect during the Gazly 17th May 1976 earthquake, J. Geomag. Geoelectr., 30, 487-492, 1978.
4. TAZIMA, M., Accuracy of recent magnetic survey and a locally anomalous behaviour of the geomagnetic secular variation in Japan, Bull. Geogr. Surv. Inst., 13, 1-78, 1968.

Investigations of Geophysical Parameters in the Area of Carinthia and Friuli in Relation to Seismicity

K. Aric, G. Duma and R. Gutdeutsch

Institut für Meteorologie und Geophysik, Währingerstr. 17, 1090 Wien, Austria

H. Friedmann

Institut für Radiumforschung der ÖAdW, Boltzmanngasse 9, 1090 Wien, Austria

Abstract

In consequence of the Friuli earthquakes since 1976 geophysical investigations in the area of N-Italy/S-Austria (Carinthia) have been carried out. It was intended to analyse the relation between the local stress field of the earth's crust and several physical parameters. There is an evidence of a correlation between seismicity and rockbursts in the mine of the BBU. The radon content of the observed spring depends on the precipitation and the groundwater level. If the investigations are restricted to a period where these influences are neglectible the time series of the radon content exhibits certain similarities with the earthquake activity.

Introduction

In view of possible applications to the Anatolian fault system measurements of the radon content of thermal spring water and the seismicity in Austria are described in the following paper. Austria in general has a low or moderate seismic activity. Nevertheless a great earthquake happened in May 1976 in Friuli approximately 30 km next to the Austrian border. This event provided the opportunity to observe the after shock series.

The Austrian lead-mine Bleiberg in Carinthia is situated about 50 km north of the Friuli earthquake region (N-Italy). After the large destructive earthquake in May, 1976, a seismic station was installed in the mine to record the numerous aftershocks in Friuli as well as the rockbursts in the mine. In addition to that, a continuous measurement of radon concentration of a thermal spring in Villach, near Bleiberg, was started (Friedmann, Hernegger, 1978). The influence of rain, groundwater level and atmospheric pressure on radon concentration was studied. The results are presented in detail by Aric et al. (1980).

Relation between Friuli earthquakes and rockbursts in the mine Bleiberg

Different types of seismic events were recorded in Bleiberg, namely

A: Shocks within a distance of about 5 km from the station which occur exclusively at working hours in the mine. These events were not taken into account.

B: Rockbursts in the mine, not restricted in time of day

C: Earthquakes in a distance of about 20 km which are not localized precisely so far

D: Earthquakes in Friuli

The minimum detectable ground velocity of the seismic station is 1μm/s within a range from 0.2 Hz to 20 Hz. The dynamic range amounts to about 40 dB.

Fig. 1 shows the frequency of the events B, C and D (weekly frequency) and the squares of measured amplitudes (below). Obviously there exists a relation between the seismicity in Friuli (type D), the earthquakes of type C and the rockbursts B. Essentially there are three epochs of increased

Fig. 1 Distribution of seismic activity near Bleiberg and rock bursts in the mine of Bleiberg

Above: events per week
B) rock bursts
C) near earthquakes ($\Delta \sim$ 20 km)
D) after shock activity in Friuli

Below: sum of squares of recorded amplitudes per week in mm^2 (S13-Geotech-vertical-seismometer)

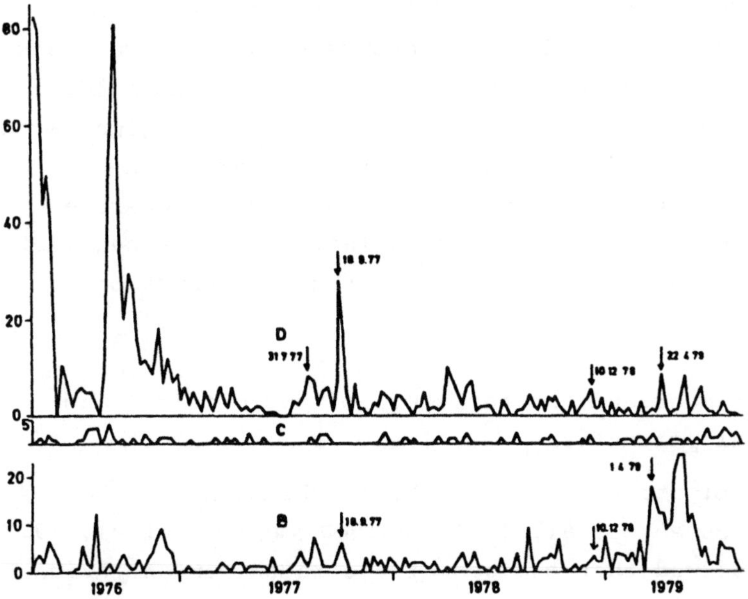
80
60
40
20
0
D
31.7.77
18.9.77
10.12.78
22.4.79
5
C
20
10
0
B
18.9.77
1.4.79
10.12.78
1976
1977
1978
1979

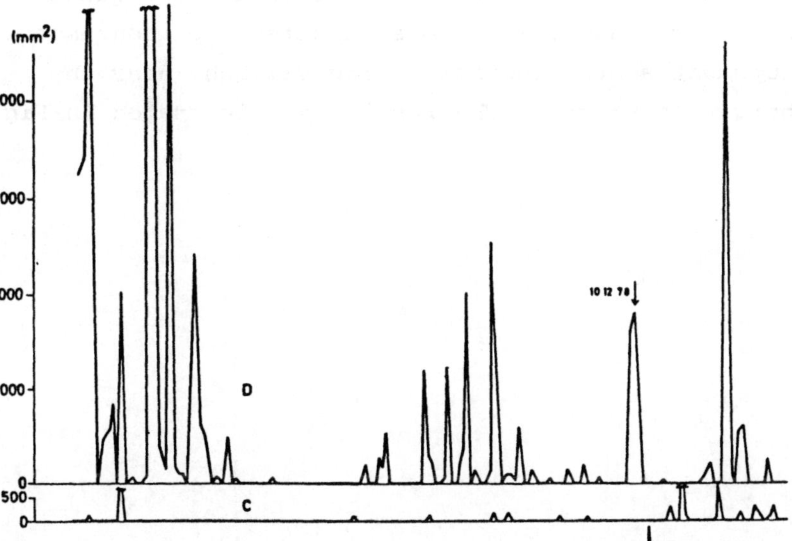
(mm²)
8000
6000
4000
2000
0
D
10.12.78
500
0
C
4000
2000
0
B
10.12.78
1976
1977
1978
1979

seismicity. The earthquakes in Friuli, single shocks or swarms, can be correlated with rockbursts in Bleiberg. This includes that both areas belong to the same stressfield. Another interesting point is that obviously the regions Friuli and South Carinthia,where also destructive earthquakes occured in the past time may not be regarded as seperate seismic regions. The question if an increasing frequency of rockbursts indicates an impending earthquake cannot be answered from this set of data.

Radon measurement and the influence of hydrological and meteorological parameters

Several investigations in the USSR, in China and in the USA (for example ULOMOV et al, 1971; SULTANKHODJAEV et al, 1977; KING, 1978) showed that changes of the radon concentration of springs or groundwater might be regarded as earthquake precursors in some cases. Therefore an attempt of continuous radon measurement at the thermal spring Villach, near the mine Bleiberg, was started. The result is illustrated in Fig.2

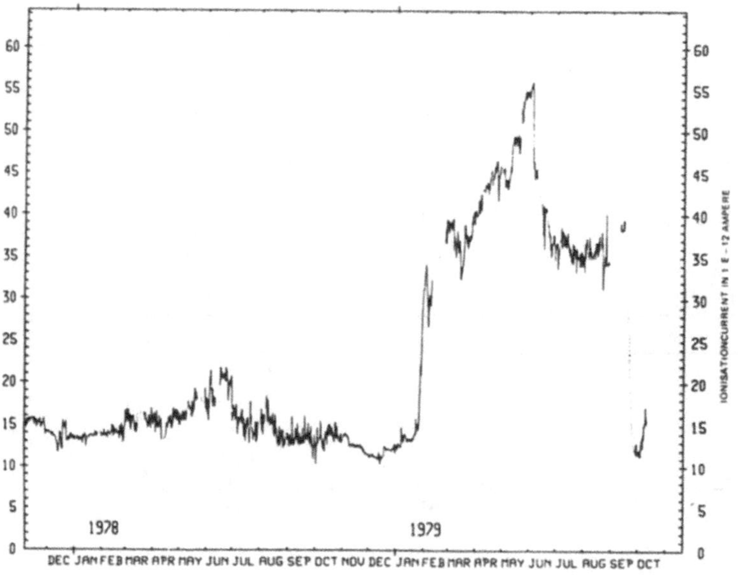

Fig. 2 Variation of the radon content of the thermal spring in Villach; 50 km north of Friuli. The radon content is proportional to the ionisation current of the measuring chamber

Results

It seemed convenient to examine if influences of rainfall, groundwater level or atmospheric pressure on the measurement exist. Hereby no effect of pressure could be found, but there were indications of an relation between rainfall and fluctuations of the radon concentration. To avoid that the correlation between concentration changes and earthquakes is falsified by this influence we restricted the investigation to the winter months where snow cannot penetrate into the soil. Fig.3 shows that an effect of seismic activity on radon concentration must not be excluded. Furthermore the coincidence between the radon anomaly in 1979, starting in February, and the increased rockburst activity (see Fig.2) seems to indicate a dependance of radon concentration on tectonic stress. It is also of interest to note that the large earthquake in Montenegro (distance 600 km) occured in April 1979.

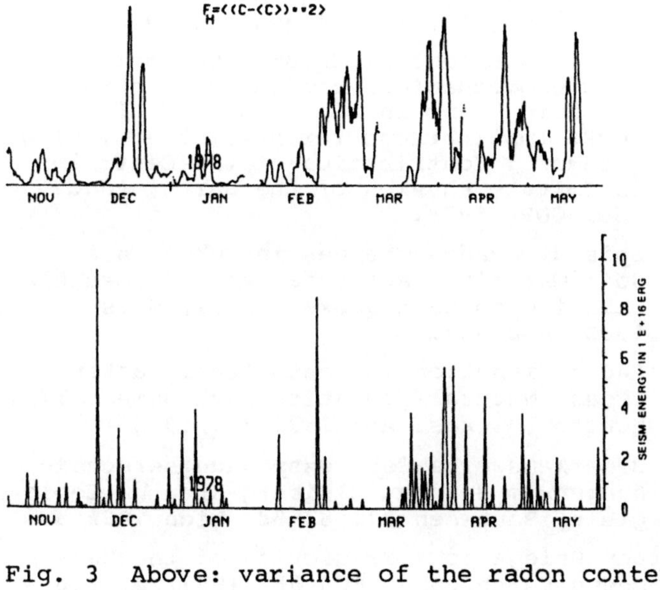

Fig. 3 Above: variance of the radon content 1978 - 1979

Below: seismic energy of events with 25 km $\leq \Delta \leq$ 100 km

Literature (extended list)

ARIC, K., G. DUMA, H. FRIEDMANN, R. GUTDEUTSCH and F. HERNEGGER: Untersuchungen geophysikalischer Parameter im Zusammenhang mit Erdbeben in Kärnten und Friaul, Festschrift, Zentralanstalt für Meteorologie und Geodynamik, Publikation Nr. 243, Wien 1980

ANDERLE, N.: Zur Schichtfolge und Tektonik des Dobratsch und seine Beziehungen zur Alpindinarischen Grenzzone. Jb. geol. B.-A., 94, Wien, 1950.

CASTEN, U. and A. CETE: Untersuchungen zur induzierten Seismizität im Bereich des Steinkohlenbergbaues des Ruhrgebietes. Glück Auf Forsschungshefte, 1980, Nr. 1.

DRIMMEL, J., E. FIEGWEIL and G. LUKESCHITZ: Die Auswirkung der Friauler Beben in Österreich. Arbeiten aus der Zentralanstalt für Met. u. Geodyn., Heft 38, Wien 1979.

FRIEDMANN, H. and F. HERNEGGER: A method for continuous measurements of radon in water of springs for earthquake prediction. Geoph. Res. Letters Vol. 5, No. 7 (July 1978)p. 565.

GUTDEUTSCH, R. and K. ARIC: Erdbeben im ostalpinen Raum. Arb. a. d. ZA.f. Met. u. Geodyn., Heft 19, Wien 1976.

HOLLER, H.: Gedanken zum Bau des Dobratsch in den östlichen Gailtaler Alpen. Carinthia, 166 bzw. (II) 86, Klagenfurt 1976.

HURTIG, E., P. BORMANN, P. KNOLL and F. TAUBER: Seismological and geomechanical studies of a strong seismic event in a potash mine of the GDR: Implications for predicting mining tremors. Contribution of the Central Institute Physics of the Earth, Adac. Sc. GDR, 1978.

KING, Chi-Yu: Episodic Radon Changes in Subsurface Soil Gas along activ faults and possible relation to earthquakes. Bull. Seis. 1980 in press.

--- Radon Emanation in tectonically active areas. Natural Radiation Environment III, Houston, Texas, April 23-28, 1978.

KOSTELKA, L.: Bemerkenswerte Gebirgsspannungserscheinungen im Bergbau Bleiberg-Kreuth, Geologie u. Bauwesen 27, 82-88, Wien 1962 a.

RALEIGH, B. and US Study Delegation: Prediction of the Haicheng Earthquake. EOS 58 (1977) p. 236.

SIMANE, J., J. SKLNAR, H. HRIBERNIGG, L. KOSTELKA and H. RAINER: Seismoakustische Untersuchungen zur Vorhersage von Gebirgsschlägen. Berg- und Hüttenmänn. Monatshefte 118, 375-384, Wien 1973.

SULTANKHODJAEV, A.N., I.G. TSCHERNOV and T. SAKIROV: Hydrogeoseismological Precursers of the Gasli Earthquake Proc. of the Uzb. SSR Academy of Sc. No. 7 (1977) p. 51.

TRAPP, E.: Die Erdbeben Österreichs 1961-1970. Mitt. d. Erdbebenkomm. 72, Österr. Akad. d. Wiss., Wien 1972.

TSCHERNIK, E.: Gebirgsschläge im Ostalpinen Blei-Zink Bergbau. Abhandl. dt. Akad. Wiss. Berlin Kl. VI, Bergbau etc. 1958, 2a, 138-142, Berlin 1958a.

ULOMOV, V.S. and B.Z. MAVASHEV: The Tashkent Earthquake of 26. April 1966. Akad. Nank. Uzbek. SSR FAN 188 (1971).

WIERZCHOWSKA, Z.: Untersuchungen von Gebirgsschlägen und Gebirgserschütterungen in Gornyslask unter Anwendung seismologischer Methoden. Freiberger Forschungshefte Geophysik im Bergbau, Teil 1, 1967, Seite 31-40.

Note on the Application of the PCA Method – A Possible Approach to the Problem of Elimination of Non-Local Geomagnetic Field Changes

A. Mete Işikara and N. Orbay

Department of Geophysics, Faculty of Earth Sciences, University of Istanbul, Turkey

M. Pantović

Geomagnetic Institute, 11306 Grocka, Yugoslavia

Abstract

The possible application of the PCA method for the separation of the seismo-magnetic effect from the secular and external variations of the geomagnetic field is discussed.

Extensive earthquake prediction research in the past decades has brought out very significant results regarding changes of some geophysical fields and parameters associated with earthquakes and tectonic activity in general. As a part of complex and multidisciplinary studies, geomagnetic methods certainly play an important role.

Since changes in the geomagnetic field associated with earthquakes have been demonstrated, different approaches are being applied in the analysis of the geomagnetic data in order to obtain the magnetic precursory events for particular earthquakes. But, the problem which still has not been solved completely is the elimination of the non-local changes of the geomagnetic field.

Regarding this question, we want to discuss briefly the justification of applying the Principal Component Analysis (PCA) to the geomagnetic data without giving any results concerning some particular examples, i.e. events. The statistical details of this method can be found elsewhere, e.g., Davies (1973).

If there is any magnetic precursor to an earthquake, it is likely to take the form of a relatively short-lived attention in the rate of change of the geomagnetic field due to the piezo-magnetic effect (in the present context this is called the seismo-magnetic effect). The main problem is to separate the seismo-magnetic effect from the secular variation which originates in the core and changes which are related to the solar radiation (both electromagnetic and particulate).

On the assumption that due to the conductivity of the mantle, the main field secular variation will change only slowly and smoothly with time, we may effectively remove this by fitting the data for an interval of registration with least squares low order polynomial, examining the residuals. This would not work during a secular variation impulse, such as occurred near 1970, but these events are not common.

The residuals contain the seismo-magnetic effect and a part that depends on solar radiation (from the diurnal variation to the 11 year cycle). We would expect the solar-dependent part to be highly correlated with some "appropriate parameter", whil the seismo-magnetic effect should be uncorrelated with this parameter. To separate the two parts, therefore, we may use PCA.

The first principal component describes the part of the variation that is most highly correlated with the chosen parameter and we interpret this as the sum of the external and induced parts of the magnetic field. The second principal component is orthogonal to the first, and we expect this to contain the seismo-magnetic effect.

In order to take full advantage of the method, it is important that the above mentioned parameter satisfies the following criteria:

(i) It is available as rapidly as the magnetic data with which it is to be correlated.
(ii) It correlates as closely as possible with the magnetic variations of non-tectonic origin.

So far, some results have been obtained by using sunspot number R or Ap indices as a possible "appropriate parameter" (Işikara et al., 1979, Pantovic 1981, Orbay et al., 1980). Presently we are working on selecting and texting the efficient "appropriate parameter" that can be used in the sense of the above discussion, i.e. to provide well observed magnetic precursory events.

Acknowledgement

The authors are gratefull to S.R.C. Malin for his advice and comments. Thanks are also due to A. Adam and all participants of the 5. IAGA Workshop on Electromagnetic Induction in the Earth and Moon.

References

Davies, J.C. (1973): Statistics and Data Analysis in Geology, John Wiley and Sons Inc., New York, London.

Işıkara, A.M., N. Orbay and M. Pantovic (1979): An Attempt for Elimination of Non-local Changes of the Geomagnetic Field for Earthquake Prediction. Proc ESA-Council of Europe Seminar on Earthquake Prediction, Strasbourg 5-7 March 1979, ESA SP-149.

Orbay, N., B. Dedehayır, Y. Andaç and A.M. Işıkara (1980): Interpretation of the Geomagnetic Data Recorded Around the Western Part of the North Anatolian Fault Zone in Terms of PCA Method (in preparation).

Pantovic, M. (1981): Possible Information on Local Features of the Focal Region Based on Geomagnetic Field Data; A.M. Işıkara, A. Vogel, H. Soysal (Eds.) Multidisciplinary Approach to Earthquake Prediction. Present volume, chapter V.

6 Space techniques in geodynamics and earthquake prediction research

Application of Space Technology in Earthquake Prediction Research – A Short Review

Andreas Vogel

Institut für Geophysikalische Wissenschaften, Freie Universität Berlin, Rheinbabenallee 49, D-1000 Berlin 33, Federal Republic of Germany

Abstract

Applications of space technology - the present status and future aspects - in support of earthquake prediction research is reviewed. In addition to space geodetic techniques, considerable contributions are to be expected from satellite remote sensing and satellite data transmission systems.

Space technology is expected to contribute to earthquake prediction research in three different ways:

1) Application of space geodetic techniques in support of ground-based geodetic surveys
2) Use of satellite remote sensing techniques to delineate neo-tectonic structures in seismotectonic active areas
3) Rapid data transmission for the transfer of data from observational networks in areas of potential earthquake risk to data collection and processing centers

In the following these various applications of space techniques are briefly discussed.

1. Space geodetic techniques

Deformation and displacements observed at the earth's surface provide information about changes of the strain field preceding earthquakes. The North-Anatolian fault zone is a tectonic plate boundary where mainly strike-slip displacements have occurred. Little is known about the present rates of motion. From the point of view of earthquake prediction research it is important to know whether these motions are continuous or episodic.

Geodetic ground surveys already have revealed considerable deformations and displacements near the plate edges in other parts of the world where earthquakes occur, and intensive efforts are being made to understand the puzzle of deformations, horizontal and vertical displacements which accompany earthquake activity. Accelerated crustal deformation and rapid or sudden displacements along faults have often been observed preceding, accompanying, and following large earthquakes. Observation of these phenomena is extremely important for a better understanding of earthquake generation and for providing prediction criteria.

A critical evaluation of the instrumental capabilities and natural limitation shows that the majority of ground geodetic measuring techniques have reached or will reach within a few years accuracies to meet basic requirements in geodynamic and earthquake prediction research.

Strain rates of 10^{-6} to 10^{-5} per year may be expected in seismotectonic areas. Multi-wavelength electronic distance measuring techniques reach an accuracy of 5×10^{-8} along baselines of tens of kilometers. The levelling accuracy is about 10^{-7} along distances of tens of kilometers.

However, we have to realize that ground geodetic surveys also are limited to a certain extent. Repeated ground geodetic surveys are not only expensive but also time-consuming and sometimes not fast enough to monitor pre-earthquake accelerated crustal deformation with required accuracy. A rough idea about the size of precursory deformations is given by the displacements associated with earthquakes. Magnitude eight earthquakes typically involve displacements of the order of ten meters over a fault length of 100 km. Moderate earthquakes of the order seven involve motion of perhaps tens of centimeters over fault lengths of tens of kilometers.

In order to "capture" strain precursors of such an earthquake of moderate size it would be necessary to extend geodetic surveys into areas of several hundred times hundred kilometers. Within such an area, a grid net of at least a twenty kilometer spacing, a repetition time of at least once per month, and an intersite measuring accuracy of about one centimeter would be needed to detect precursory strain accumulation. When strain accumulation is accelerated in the months and weeks prior to an earthquake, the repetition rate should be much faster and the grid spacing still smaller. There is another drawback of ground geodetic surveys when displacements have to be measured over large distances. The gross features of plate motion on a geological time scale are understood; however, their contemporary behaviour which is the basic mechanism of earthquake activity is poorly known. Detailed knowledge of interplate motion as well as intrinsic deformation of plates and transient strain events is important for a better understanding of the sources of earthquakes.

Plate motions are in the order of centimeters per year, and measuring techniques should be available to measure with an accuracy of a few centimeters relative positions and rates of position changes in three dimensions over hundreds and thousands of kilometers in time intervals of less than one year.

Conventional ground-based geodetic methods, such as levelling, triangulation and trilateration, are better suited to determining relative position and position changes over short distances. Over large distances the measurements must be made in small steps, and the measurement errors accumulate to limit the accuracy.

The space program has developed several new techniques which overcome certain limitations of the classical geodetic measurements and which seem to fulfil a great deal of the requirements in earthquake prediction research. Accurate laser and radiowave tracking with artificial satellites is one type and radio interferometric tracking of natural cosmic radio sources and radio signals from satellites is another type of such techniques. The atmospheric limit to accuracy for space systems appears to be about 1 cm.

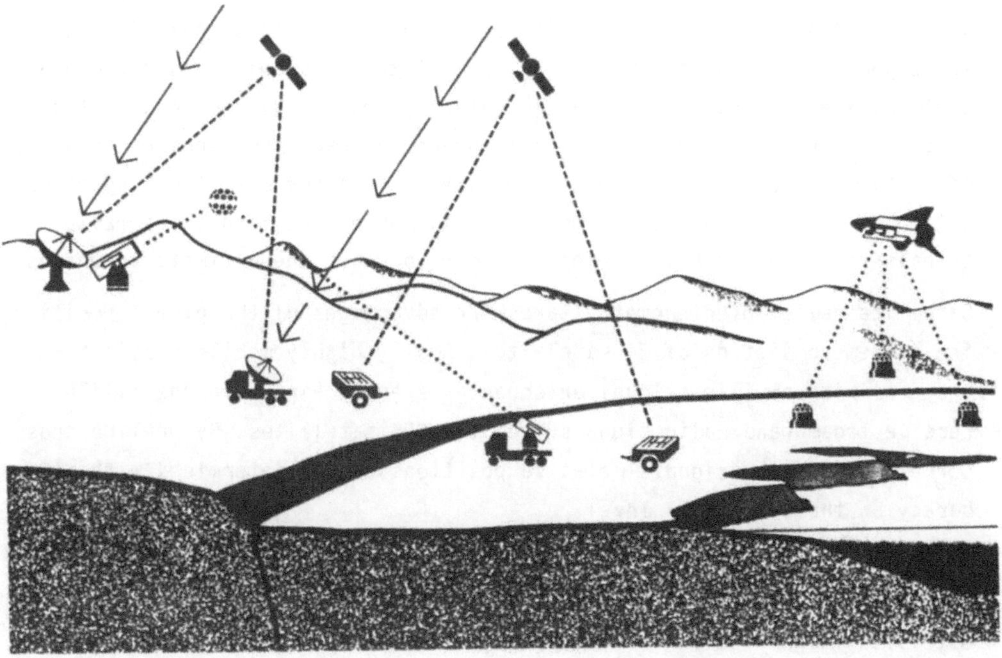

Fig 1 Schematical view of space geodetic techniques applicable in earthquake prediction research

The viewgraph of Fig. 1 shows in a simplified way various space techniques. Laser ranging measures the travel time of very short laser pulses to reflectors placed at a high-altitude satellite whose orbital position can be determined to the appropriate accuracy. Fig. 1 symbolizes to the left a fixed satellite laser ranging station operating in conjunction with a mobile station. The mobile station is moved from place to place to measure position changes on the level of centimeters when operated simultaneously with a high-precision fixed laser ranging system.

As an attractive alternative to mobile ground-based laser ranging the recently designed spaceborne laser ranging system has to be considered. A laser is placed into a space craft, to begin with into the space shuttle, as shown in Fig. 1 to the right. Ranging to a network of reflectors on the ground in areas of potential earthquake risk would be possible with only one laser system. In rapid succession intersite distances up to 25 km seem to be measurable at an error of only 1 cm. Test flights with the space shuttle have been planned for the beginning of the eighties. Unfortunately, because of the high costs involved the experiments have been postponed.

Very long baseline interferometry uses radio antennae for recording signals from deep-space radio sources. Fig. 1 shows to the left a fixed radio antenna and another one symbolizing a mobile one. Radio signals from deep space indicated by arrows are recorded in both sites and cross-correlated. Small mobile antennae of only a few meters diameter have to be operated together with large fixed antennae in order to reach the required accuracy of centimeters along base lines of hundreds of kilometers. Clock synchronization or still better real-time transmissions of the deep-space radio signals via satellites is needed in order to reach the ultimate accuracy.

Satellite radio interferometry takes the advantages of the global positioning system consisting of 24 satellites. Small, highly mobile stations equipped with omnidirectional antennae as seen in Fig. 1 in the middle receive broad band radio signals from the GPS satellites. By on-line cross correlation of the signals relative positions can be determined with accuracy on the centimeter level.

2. Seismotectonic application of satellite imagery

Satellite imagery helps to reveal neotectonic structures and to clarify seismotectonic conditions in zones of earthquake risk. Accurate mapping of geomorphological features adjoining lineaments reveal drag effects indicating active movement or recent tectonic activity along faults. Satellite imagery from the North-Anatolian fault zone already delineated fault systems of potential earthquake risk where destructive earthquake have occurred in historical time associated with fault displacements which, however, have not been clearly mapped by ground surveys. It is our hope that more detailed investigation by improved satellite remote sensing devices and digital processing systems for image enhancement and increased image resolution will further contribute to seismotectonic studies.

3. Data transmission via satellites

Rapid transmission of data from observation networks in earthquake zones to data collection and processing centres is required. There are two types of data from recording networks. Recording slow processes like changes of strain fields and displacements along faults is sufficient with only one information per minute whilst a sampling rate of 1/10-1/100 sec is needed for the recording of seismic events. Satellite data transmission systems like the METEOSAT have sufficient capacity for the transmission of data of the first type. Regarding the high frequency seismological data it seems necessary to analyse the seismogram records by a microprocessor in each observation site and to transfer a limited information extracted from the seismograms. There is no doubt that the satellite data transmission system will be essential for predicting and providing warning of impending earthquake hazards.

Further details and comprehensive references concerning the application of space technology in earthquake prediction research can be found in a special volume (Vogel (Editor) 1979)and a review paper (Vogel, 1980).

References

Vogel, A. (Editor), 1979: Terrestrial and space techniques in earthquake prediction research. Vieweg Braunschweig-Wiesbaden

Vogel, A., 1980: Contribution of space technology to earthquake prediction research. Adv. Earth Oriented Appl. Space Techn. Vol. 1, pp. 1-17

The Background to High Mobility Space Geodetic Techniques for Geodynamic and Earthquake Prediction Research

Peter Wilson

Institut für Angewandte Geodäsie, Frankfurt am Main, Sonderforschungsbereich 78 (Satellitengeodäsie) der TU München, Federal Republic of Germany

Summary: the paper summarises the requirements for space geodetic participation in geodynamic and earthquake prediction research and gives the conditions for their application.

The space geodetic contribution to geodynamic and earthquake prediction research, is based on the requirement for:

- a co-ordinate reference framework and determination of the parameters of earth rotation;
- an improved knowledge of contemporary relative plate motions;
- a better knowledge and understanding of regional deformations and strain accumulations at the plate boundaries;
- knowledge of the internal plate deformations.

For these purposes it is necessary to establish:

- global and continental networks of reference points;
- regional networks in areas of high seismic activity

and to

- improve the knowledge of the earth's gravity field to permit the identification of its short wave (approx. 100 - 150 km) characteristics.

The first of these tasks presumes the determination of 10 - 15 fixed reference points occupied by laser ranging or radio interferometry stations, a number of which should be co-located to ensure the uniqueness of the co-ordinate reference origin.

The second contribution comprises the determination of potential regional deformations and requires the establishment and periodic re-measurement of a large number of points, implying thereby the use of highly mobile measurement systems. These could be based on laser ranging, radio interferometry using natural or satellite sources and eventually on the direct use of the Global Positioning System (GPS) and space-borne laser techniques, should these be realised at some future date.

Finally, it is important to realise that there is a requirement for a precise knowledge of the gravity field, which in itself has historically been improved in a boot-strap fashion through the observation of large numbers of satellites. Direct improvements will come from satellite to satellite tracking; recent indirect improvements resulted from the use of satellite altimetry.

Apart from improvements to the gravity field, the techniques employed depend on:

1. base-line determinations for studies of plate motion (largely an absolute problem, due to the generally large separations involved) and local and regional deformations (which are determinable by relative measurements);
2. the monitoring of point positions in the absolute sense for the establishment of a global reference framework and for the determination of earth rotation parameters.

From this it can be seen that there is a strong requirement for the employment of a limited number of fixed tracking stations using different techniques (laser and radio interferometry for the foreseeable future), supported by a reasonably large number of mobile systems (laser, radio interferometer, G.P.S.....) for regional densification. The improvements in the knowledge of the gravity field will most probably be conducted by instrumentation in space. The use of different measurement techniques, employing independent means, serves to eliminate system dependent uncertainties and give added confidence that the results obtained are meaningful.

The Installation of a Spaceborne Ranging System for the Detection of Crustal Movements in the North Anatolian Fault Zone

C. Gerstenecker

Institut für Physikalische Geodäsie, Technische Hochschule Darmstadt, Federal Republic of Germany

Summary

In this paper the principles of a spaceborne ranging system are shortly explained and the model studies concerning the geodetic problems of such a system are summarized.

It is shown that under the commonly used assumptions, the monitoring of crustal movement in the North Anatolian Fault Zone with such a system is possible. Since some assumptions seem to be not realizable yet, the use of a laser ranging in an airplane and the combination of terrestrial and "airborne-measurements" is supposed. Such a solution should deliver a rather improved error situation

1. Introduction

In the early seventies the installation of a spaceborne ranging system was proposed (for example Mueller et al, 1975).

Such an "inverse" ranging system is built in a satellite whereby it is measuring the distances between the satellite and different ground stations. Depending on the measuring system the ground stations are equipped with inactive reflectors (if a laser system is used) or active receivers (for a microwave length system).

The spaceborne ranging system offers a lot of advantages above the common usual ground based ranging system especially in the economical point of view.

A great number of relative inexpensive ground stations can be observed by a single - of course very expensive - spaceborne ranging system, whereby the observations are made automatically and are repeatable at nearly any desired time.

The use of such a system is obviously senseful for geodetic and geodynamic purposes if cm-accuracy is achievable. On the other side the technical problems are serious. The ranging system must meet requirements, which hardly can be solved at the present state. However this paper deals only with the geodetic problems of such a system.

2. Geodetic model investigations

To clarify the geodetic problems and powers of the spaceborne ranging system several model investigations have run, which were concerned especially with the monitoring of crustal movements in the San Andreas Fault Zone (Mueller et al, 1973, Kumar, 1976, Kumar et al, 1978, SAO Report, 1977 , Pavlis, 1979). The results of these simulation studies can be summarized as follows:

- as ranging system at cm-accuracy level only a laser ranging system is suitable (SAO Report) because of refraction problems
- the resolution of the system must be better than 1 cm detecting a crustal motion of 1 cm
- as observation mode the short arc method is not senseful since the determination of the satellite orbit is too inaccurate (Kumar, 1976)
- the "geometric" observation mode delivers results at the cm-accuracy level (Kumar, 1976), whereby the accuracy depends on

  the numbers of ground stations observed during a single measuring event

  the topographic heights of the ground stations

  the selection of known fundamental stations, which are stable in time

  the combination of low and high altitude orbits

  the distance between the groundstations (Pavlis, 1979).

In every case the short arc and the geometric observation mode leads to not uniqueness or ill-conditioned adjustment problems, whereby the problems of the ill-conditioned matrices predominate the problem of not uniqueness.

On the other side the technical realisation of a laser ranging system working in the geometrical mode seems to be not possible (Pavlis, 1979), so only the short arc method can be used.

Further studies were carried out by (Cohen/Cook, 1979), to explore the possibilities to determine crustal strain rates from spaceborne ranging measurements. This investigation however assumes that the dislocations vectors of the crustal movement are known with cm-accuracy and does not solve the primary problem. Only the secondary power of a spaceborne ranging system is investigated.

3. The installation of a geodetic network in the North Anatolian Fault Zone; results of a model study

According to the simulation studies by (Kumar, 1976 and Kumar et al, 1978) in the San Andreas fault zone a similar investigation was carried out for a geodetic network in the North Anatolian fault zone, between the latitudes $\phi = 40^{\circ}$ N and 42° N vice versa the longitudes $\lambda = 32^{\circ}$ E and 36° E. Since a dense geodetic network in this region is missing 171 groundstations are arranged in a regular grid systems whereby the distance between the adjacent points is about 24 km. The topographic heights were estimated from a map. The lay out of the net is shown in Fig. 1.

For simulation study two different satellite orbits of low (100 km) and high altitude (1000 km) between were constructed passing the network in azimuths of 30°-45°, whereby the laser ranging system was working in the geometric mode. During one event the distances to 9 groundstations were measured simultaneously. For the least square adjustment the same approach given by (Kumar, 1976) was used.

Since the problems of such networks are described explicitly by (Muller et al, 1973, Kumar, 1976, Kumar/Cook, 1978) only the results of the calculations are given. The cm-accuracy level was achieved if

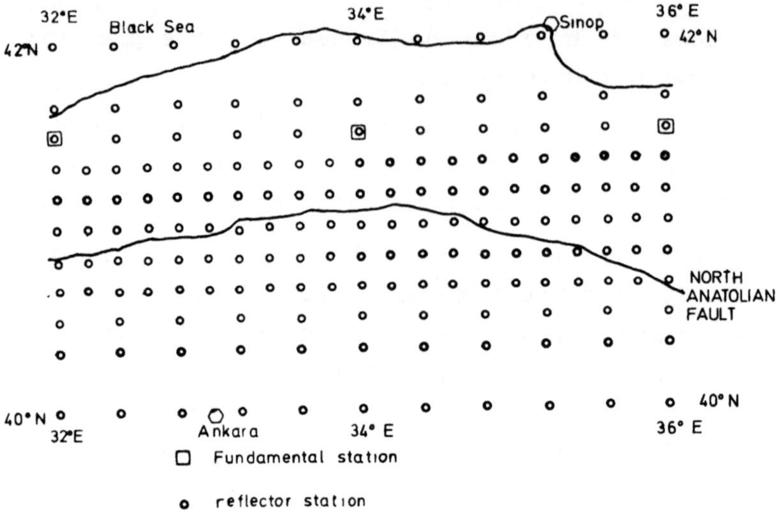

Fig. 1: Design of a groundstation net for the detection of crustal motion in the North Anatolian Fault Zone

- each groundstation was observed in the minimum more than 5000 times during one measuring period
- a standard deviation of ± 2.5 cm or better for the laser ranging system was assumed
- Gaussian noise was supposed
- three known or more fundamental stations (Kumar, 1976) were inserted in the adjustment.

Under these conditions the adjustment was repeated whereby the groundstations located on the southern part of the North Anatolian fault were drifted in the western directions with a continuous speed of 3 cm/year. Using the same statistical test-procedures as Kumar, the moving of the North Anatolian plate to west could be detected after 7 months significantly.

4. Discussion and conclusions

The study shows the validity of Kumars results for other regions as the San Andreas fault zone. A spaceborne ranging system performing the requirements of the model study is a global powerful measuring tool for crustal movements investigations.

From the practical point of view (geodetic and technological) however some assumptions are doubtful. Obviously the realisation of laser ranging system working in geometric mode in a satellite is not solvable (Pavlis, 1979). Because of the refraction problems a multiwave length laser is necessary (SAO-Report, 1977 , which is working nearly continously over a time period of several years. Even if the "Space Shuttle" is used to launch this system, the mechanical and optical performances of the device must be very stable to survive the accelerations during the start. On the other side the bad geometrical behaviour of the observed networks biases each deviations from the assumption of Gaussian normal distribution for the measurements, even if the number of observations to each groundstation must be very high. Especially it is hardly to show that the behaviour of the refraction residuals is gaussian and not greater than 1 or 2 cm. Any undetected systematic measuring error (instrumental or physical) can disturb the results seriously.

Therefore possibilities to improve the geodetic constraints of the whole system should be investigated. One of such a solution is the installation of the laser ranging system in an airplane. This proposal has some important aspects and the problems arising here seem to be solvable. A similar development to the short arc mode is started in the form of the "Aerial Profiling of Terrain System" (APTS) (Soltz, 1978) where a tracking laser is mounted on an inertial platform and build in an aircraft. An absolute demand then is that the laser ranging system is working in the geometric mode, since the position of the airplane cannot be determined by other means with sufficient (cm) accuracy. A further disadvantage is that the measurements cannot be carried out automatically at any desired time. The following advantages predominate these constraints. The range of the laser system can be diminished, the system can be easily repared and used at very different heights about the ground (as necessary to stabilize and improve the adjustment-results). Similary the azimuth of the flight can be changed. Additionaly the pointing mechanism to the different groundstations can be simplified because of the lower velocity of the airplane. The observations can be carried out, if all groundstations are visible. Multiple wavelength laser systems for terrestrial purposes are ready for use today and possess a sufficient range (Levine, 1978). The development of such a system is much cheaper as that of a spaceborne system. On the other side considerations to improve the geometrical structure of the network and the lay out of the measurements should be started. It seems to be possible that the significantly lower height of the airplane (10 km or less) above the ground improves the lay out of the measurements. But the most important improvements should come up, if airborne measurements are combined with terrestrial observations as distances and angles measurements. These aspects are looking very promising, so a model investigation should start, where the possibilities of an airborne ranging system should be analyzed in combination with terrestrial measurements.

5. References

Cohen, S. C., G. R. Cook — Determining Crustal Strain Rates with Spaceborne Geodynamics Ranging System Data, Manuscripta Geodaetica, Vol 4, No. 3, 1979

Kumar, M. — Monitoring of Crustal Movements in the San Andreas Fault Zone by a Satelliteborne Ranging System, Ohio State University Report of the Department of Geodetic Science, No 243, 1976

Kumar, M., I. Mueller — Detection of Crustal Motion Using Spaceborne Laser Ranging Systems, Bulletin Geodesique, Vol 52, p. 115-129, 1978

Levine, J. — Multiple Wavelength Geodesy, in Applications of Geodesy to Geodynamics, Ohio State University Report of the Department of Geodetic Science, No 280, 1978

Mueller, I., B. H. W. van Gelder, M. Kumar — Error Analysis for the Proposed Close Grid Geodynamic Satellite Measurement, Ohio State University Report of the Department of Geodetic Science, No 230, 1975

Pavlis, E. C. — Error Analysis for a Spaceborne Laser Ranging System, Ohio State University Report of the Department of Geodetic Science, No 290, 1979

Smith, D. E. — Spaceborne Ranging System, in Applications of Geodesy to Geodynamics, Ohio State University Report of the Department of Geodetic Science, No 280, 1978

Smithsonian Astrophysical Observatory, Staff (SAO) — Study of a close-grid geodynamic measurement system, Smithsonian Astrophysical Observatory, Reports in Geoastronomy, No 5, 1977

Soltz, J. A. — Applicability of APT Aided-Inertial System to Crustal Movement Monitoring, in Applications of Geodesy to Geodynamics, Ohio State University, Report of the Department of Geodetic Science, No 280, 1978

The Meteosat Data Collection System and its Potential as a Data Relay for Geophysical Measurements

A. Robson and H. Houet

European Space Operations Centre (ESOC), Darmstadt, Federal Republic of Germany

Abstract

The Meteosat Data Collection System provides a convenient method of relaying environmental data from any location within the coverage area of the Meteosat satellite (about 80° great circle from 0° lat, 0° long). The system consists of earth platforms, the Meteosat Satellite, a central receiving station and message distribution centre and conventional communications links to the platform owners.

The system's technical and operational performance is described with reference to actual applications and its potential as a relay for geophysical measurements discussed.

Introduction

Meteosat is a geosynchronous satellite which was launched in November 1977 and had three missions to fulfil:

1. The acquisition of images of the earth's disc every half hour in the visible and infrared parts of the spectrum. From these images a variety of meteorological parameters were to be determined e.g. wind vectors, sea surface temperatures, cloud top heights, etc.
2. The dissemination of corrected and annotated images to users' ground stations after retransmission by the satellite.
3. The collection and distribution of environmental data gathered by terrestrial platforms.

* This paper is based on an article published by the same authors in ESA Bulletin No. 21 (February 1980), pp. 44-48.

The Data Collection System (DCS) used to fulfil the third mission, could be used for the relay of geophysical data which in its turn could contribute towards earthquake prediction.

The present Meteosat in orbit suffered a failure in November 1979 which prevents Missions 1 and 2 from being continued. However the Data Collection System is still operational. A second Meteosat is scheduled for launch in September 1980 and plans are in preparation for a series of follow-on satellites to give mission capability until 1993.

The system

Because Meteosat is a geostationary spacecraft, it provides a link that is permanently available to all compatible data platforms within its field of view (about 80° of great circle from subsatellite point). This is an important advantage over the data-collection systems of near earth-orbiting satellites, to which the platforms are in view for only very limited periods.

The Meteosat Data-Collection System (DCS) has been designed to have the same characteristics as the data-collection systems carried by the other geostationary meteorological satellites presently in operation, namely the American NOAA/NESS Goes series and Japan's GMS spacecraft. This coordinated design approach permits the reports from Data-Collection Platforms (DCPs) carried on ships and aircraft moving from one satellite coverage area to another to be received and transmitted by any of the satellites in the network.

Meteosat is located over the equator and the prime meridian and therefore has line-of-sight contact with the geographical area shown in Figure 1. It is therefore able to communicate with mobile or static DCPs anywhere within this area.

There are three basic types of DCP:

- 'self-timed', which transmit their data at regular intervals based on an internal clock;
- 'interrogated', which transmit their data upon reception of a request signal from the satellite;

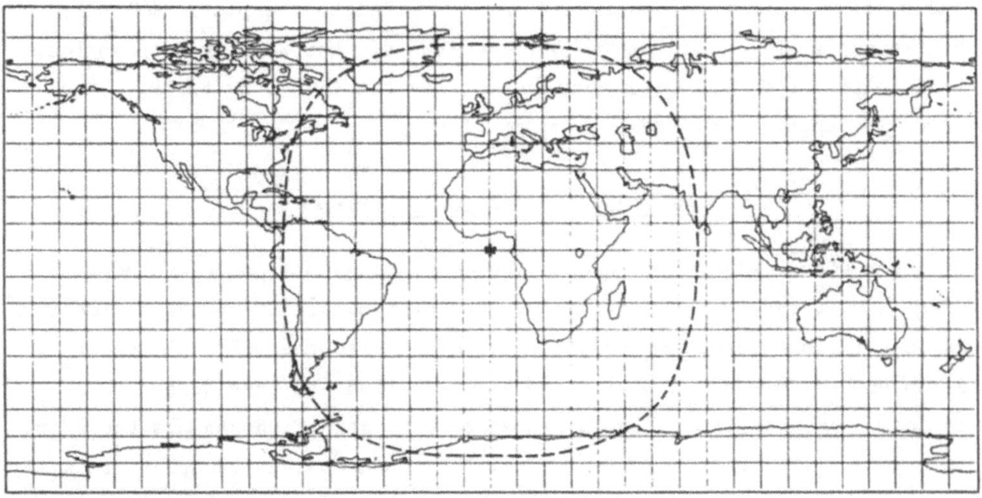

Figure 1 – Coverage area of the Meteosat Data-Collection System (DCS)

'alert', which transmit either a small amount of data or a request for interrogation when a particular parameter has been exceeded.

The DCPs transmit to Meteosat on any one of sixty-six 3 kHz reporting channels in the 402 MHz band. After conversion to the 1675 MHz band, the satellite retransmits the messages to its central ground station, in the Odenwald in Germany. There, the signals received are checked for quality before being transmitted to ESOC in Darmstadt for processing and distribution.

At ESOC, messages from self-timed and interrogated DCPs are collected, processed and disseminated every hour. Any message received from an alert DCP is processed immediately and made available for transmission to the platform's owners.

To interrogate a DCP, an interrogation message is generated in Darmstadt and sent to the Odenwald ground station. From there it is transmitted in the 2122 MHz band to Meteosat, which then retransmits the message on one of two possible frequencies in the 468 MHz band. The message includes the unique address of the DCP to be interrogated and only that particular platform will respond by transmitting the data it has collected.

The platform data received in ESOC are provided to the platform owners and others who wish to receive the data, either via the World Meteorological Organisation's Global Telecommunications System (GTS), by telex, or on magnetic tapes for non-real- time applications.

The data-collection platforms

Before a DCP is allowed to use the Meteosat Data-Collection System (Fig. 2), it has to be 'admitted' to the system. This involves allocating an identifying address and reporting frequency to the DCP, and making arrangements for data handling and dissemination. A prerequisite for admission is that the DCP's radio set be type-certified and has therefore undergone a stringent test programme. This ensures that the platform will

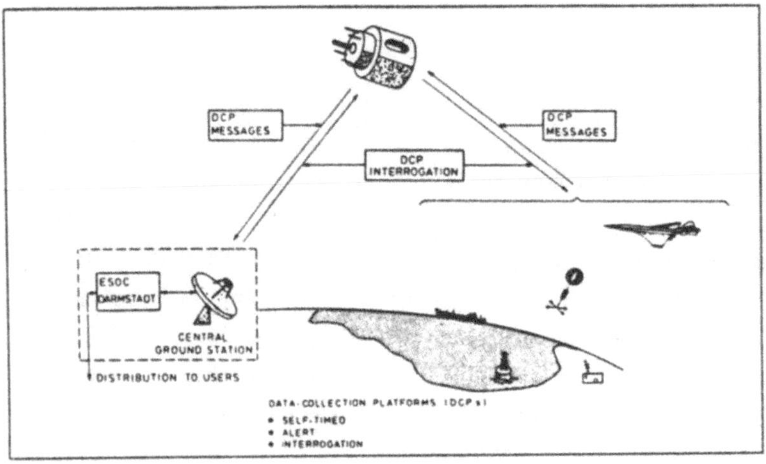

Figure 2 – The principle of the Meteosat Data-Collection System

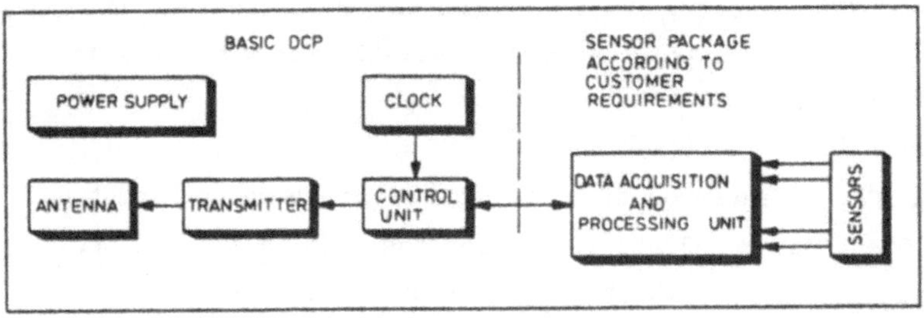

Figure 3 – Modular composition of a self-timed DCP with data-acquisition facility

function successfully within the DCS, and more importantly that it will not cause interference to other system users, even under extreme operating conditions. For platforms to be used solely within the Meteosat coverage agea, ESA conducts its own certification testing. For mobile DCPs, certification by any of the geostationary meteorological satellite operators is acceptable. Type-certified DCPs, are commercially available from several European manufacturers.

The most commonly used type of platform is the self-timed DCP, which consists of the following modules (Fig. 3):

Sensors Data Acquisition & Processing Unit	providing the measurements to be transmitted
Clock Control Unit Transmitter Antenna	providing the radio contact with Meteosat at pre-determined times
Power Supply	either batteries or mains power, depending on location

The platforms come in various shapes and sizes, but generally they are no more than 50x40x20 cm^3 and weigh less than 20 kg, excluding sensors and antenna. The size of the antenna depends upon the transmitter output power needed from the DCP; with a 5 W transmitter, a 1.5 m-long helical antenna suffices.

The Meteosat DCPs are designed to deliver 100 bits of data every second. Although there are 66 channels available on the satellite, only six are currently in use. Individual messages are allowed to last up to 1 min, which is long enough to transmit up to 100 measurements.

The frequency with which self-timed DCPs make their reports depends on their particular application, but hourly and three-hourly reporting is the most common.

Current applications of the Data-Collection System

The Meteosat DCS can be used to gather a wide range of environmental parameters, but it is especially suitable for recording slowly changing parameters. A number of experimental and pre-operational applications are currently enabling users to assess the system's capabilities and to gain operating experience. The following four examples of current operations might serve to demonstrate the broad range of the System's applications.

Meteorological data collection at remote land sites

The operational weather service of the Danish Meteorological Institute is presently operating a network of meteorological observing stations in Greenland. The information gathered, in addition to being important to the local population, provides the air traffic on the North-Atlantic and polar routes with essential weather data. Two types of station are in use: a complex manned type requiring a utilities infrastructure, 220 V power supply, etc., and a purely automatic type. The latter are intended to be visited only once per year for maintenance purposes, are completely self contained, and can be operated in locations far from human habitation.

A manned station located near Danmarkshavn (76 N 18 W) has been relaying its data via the Meteosat DCS since mid-1978. The unmanned automatic stations are a newer development and the first two were installed in the summer of 1979 at Aputiteq (67 N 32 W) and near Kap Cort Adelaer (62N 43 W). Both types of station provide three-hourly reports of air pressure, air temperature, air humidity, wind speed and wind direction. These reports are processed at ESOC and disseminated via the GTS to all meteorological centres that wish to receive them.

Meteorological and sea-state monitoring

The German Hydrographic Institute (DHI) is operating a measurement network in the North and Baltic Seas. As a part of this network, a buoy that uses a Meteosat DCP for data transmission is to be anchored at 55°05'N and 6°20'E, making it the most remote station in the DHI network. The buoy itself is rather large, with a draught of 10 m, a diameter of 10 m, and a weight of 60 t. It was originally conceived for use in a wide variety of oceanographic research projects, but its robustness makes it particularly suitable for permanent oceanmonitoring applications. The platform transmits air temperature and pressure, wind direction and speed, sea-surface temperature, water conductivity, depth and wave characteristics, as well as information on its own technical status (housekeeping). Except for the wave characteristics, the parameters undergo data reduction on board so that only averaged and maximum and minimum values are transmitted. The buoy transmits five minutes of data every three hours and, after some processing at ESOC, the measurements relayed to the DHI, the German Marine Weather Office (Seewetteramt), and to other users.

These data have several applications, in:

- forecasting sea conditions
- improving forecasting methods
- improving current-flow and wave-generation knowledge in the German Bight and continental-shelf areas
- supporting fishery research and marine traffic.

They can also provide a basis for decisions concerning the dumping of waste products at sea and other questions of environmental protection.

Hydrological monitoring

The effective management of freshwater resources continues to be one of mankind's most pressing needs. Even areas with abundant rainfall require considerable information on precipitation, flow rates of rivers, etc. to ensure that their population's needs are met.

A pilot study has been initiated by the UK Water Data Unit to examine the effectiveness of using the Meteosat DCS for the transmission of hydrological parameters (Fig. 4a,b). Five DCPs have been installed:

- at Bewdly on the River Severn (rural area)
- at Derby on the River Derwent (urban area)

  Both of these DCPs are connected to level meters, which make measurements every 15 min.
- at Blackwell Bridge on the River Rother

  This DCP is connected to ultrasonic discharge-gauging equipment, which measures water flow rate (m^3/s) at 30 min intervals.
- at Cirencester

Figure 4 a + b

- at Loch Quoich, Scotland
  Both of these DCPs are connected to rain gauges, which provide measurements every 15 min. The Loch Quoich installation supports the control of a hydroelectric power station.

Although the five platforms make frequent measurements, transmissions to Meteosat are normally made only at a fixed time every day, when all the measurements of the previous 24 h are transmitted. After processing at ESOC, the data are transmitted via the GTS to provide the user with a daily record. Under normal conditions, this daily reporting is sufficiently frequent. Exceptionally, however, conditions can develop which require a rapid reaction on the part of the user. For this reason, all the Water Data Unit DCPs have an alert mode, so that when a measurement exceeds a predetermined level a warning is transmitted on a separate alert frequency. These alert messages are processed immediately at ESOC and routed to the user.

Meteorological data collection from aircraft

The ASDAR (Aircraft to Satellite Data Relay) programme is an example of the use of DCPs in mobile applications, a use that would probably not have been realised without international coordination of system design specifications. The National Oceanographic and Atmospheric Administration (NOAA, USA) is conducting this pre-operational programme to collect weather data from aircraft, using specially developed, air-certified platforms installed on wide-bodied Boeing 747 transport aircraft (Fig. 5). These platforms store data from the standard aircraft avionics systems and transmit them hourly during the flight. A special stripline antenna has been developed because, for obvious reasons, the conventional DCP directional antenna cannot be used in this application. The lack of directionality of this strip antenna has to be compensated by increased transmitter power, namely 80 W compared with the 5 W necessary for a fixed DCP installation.

Every hour, each ASDAR platform transmits one message containing eight stored reports (Table 1) identifying

- the aircraft carrying the DCP
- aircraft position (latitude and longitude)

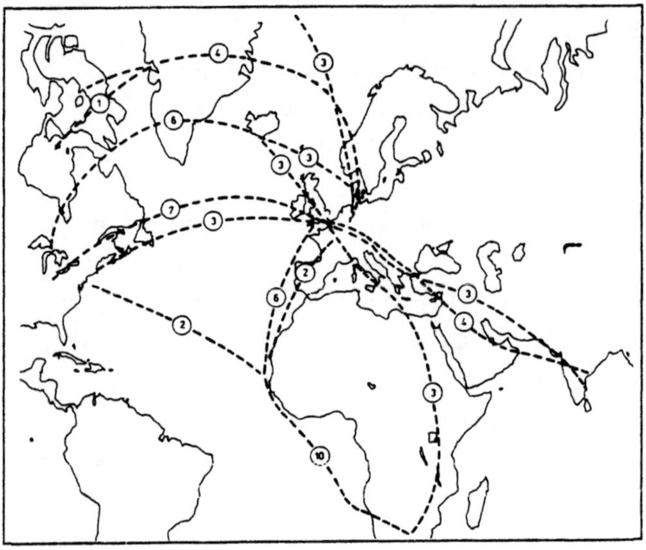

Figure 5 – ASDAR DCP messages du
the week 12-18 November 1979
..... Typical observation points
⑤ Number of times route flown during week

Table 1 – Example of ASDAR reporting

Aircraft identification PanAm/KLM	Aircraft position (degrees and minutes) latitude	longitude	Time of measurement (hour and minutes UT)	Flight level (hundreds of feet)	Outside temperature (minus deg. centigrade)	Wind direction (degrees of azimuth)	Wind speed (knots)
PA001Z	57 02 N	29 05 W	13 02	F 350	MS 37	259	081
PA001Z	57 04 N	27 32 W	12 54	F 350	MS 41	269	089
PA001Z	57 05 N	25 59 W	12 47	F 350	MS 37	279	086
PA001Z	57 05 N	24 27 W	12 39	F 350	MS 41	288	093
PA001Z	57 05 N	22 56 W	12 32	F 350	MS 41	295	096
PA001Z	57 03 N	21 26 W	12 24	F 350	MS 40	305	099
PA001Z	56 59 N	19 56 W	12 17	F 350	MS 43	310	110
PA001Z	56 54 N	18 26 W	12 09	F 350	MS 43	314	106
KL002Z	42 15 N	21 26 E	13 04	F 368	MS 48	308	011
KL002Z	42 52 N	20 19 E	12 56	F 368	MS 49	294	010
KL002Z	43 29 N	19 11 E	12 49	F 368	MS 50	293	014
KL002Z	44 09 N	18 04 E	12 41	F 368	MS 51	282	020
KL002Z	44 52 N	17 00 E	12 34	F 368	MS 51	276	025
KL002Z	45 32 N	15 51 E	12 26	F 368	MS 52	256	033
KL002Z	46 09 N	14 38 E	12 19	F 368	MS 53	246	037
KL002Z	46 53 N	13 36 E	12 11	F 368	MS 53	239	045

- time
- flight level (hundreds of feet)
- air temperature
- wind direction and speed

ASDAR messages can be received by all the geostationary meteorological satellites and after retransmission to the ground stations they undergo standard processing before distribution on the GTS. The data can be used by all meteorological agencies concerned with forecasting for air transport. Meteosat has been receiving ASDAR reports regularly since March 1978 and in the interim 20 aircraft have been equipped. Typically over 2000 messages per month are received and distributed by the Meteosat DCS.

Applications in Geophysics

Fundamental constraints of the DCS for all measurement applications is the information rate per channel and the maximum single message duration of 1 minute. This limits each individual message content to 649 eight-bit words. Arrangements can however be made to send several consecutive words.

Even with these constraints several types of instruments could be served by the DCS, e.g. tiltmeters, gravity meters, strain gauges, magnetometers, ranging systems. Here the technique would be to make measurements at suitable intervals, and store the data in the memory of the DCP until the allocated reporting slot when transmission to Meteosat would take place. For example a single 2-axis tiltmeter making measurements every 15 minutes would require only less than 1 minute report to be transmitted hourly to transmit all the raw data. Data Collection from seismometers would however be more problematical due to their inherently high data content. Before the DCS could be efficiently employed, it would be necessary to develop data compression algorithms for seismographs and the corresponding software for onsite data processing. Although little work seems to have been done in this field, it would appear that the difficulties are not very great.

Application to use the DCS

To use the Meteosat Data Collection System, a potential user would normally contact first the Meteorological Data Management Department (MDMD) of the European Space Operations Centre (ESOC), 6100 Darmstadt, F.R. Germany to learn about current channel and reporting time availability, suppliers of authorised DCPs, charges for the use of the system (these depend on the status of the user, e.g. whether a profit-making organisation, belonging to a State participating in the Meteosat Programme, etc.).

He would then be asked to apply for admission of his platform(s) to the System. The Agency, after satisfying itself that they could provide the requested support and that the DCPs to be used were of an authorised design, would then draw up a DCP Operations Support Plan for mutual agreement. This Support Plan would specify all technical details and the rights and obligations of both parties.

Conclusions

The Meteosat Data Collection System provides an effective means of transmission for environmental data. The system is especially useful where the source of the data is in inaccessible locations with poor or even no utilities infrastructure. This latter capability makes its survival potential in catastrophic conditions very high. The system allows regular reporting against a schedule, the interrogation of platforms on request and the immediate transmission of short messages whenever measurement thresholds are exceeded (Alert).

A consequence of the direct function of data relay is that regular monitoring of instrument performance takes place - this in turn can save the need to make regular inspection trips by personnel.

7 Theory of earthquake generation and prediction

Stress Accumulated Mechanisms on Strike-Slip Faults

D. L. Turcotte

Department of Geological Sciences, Cornell University, Ithaca, NY 14853, USA

Abstract

The tectonic framework causing seismicity on the San Andreas and North Anatolian faults can be understood in terms of plate tectonics. However, the mechanisms responsible for the distribution of seismicity in space and time on these faults are poorly understood. The upper part of the crust apparently behaves elastically in storing energy that is released during an earthquake. The relatively small distances from the fault in which stress is stored argue in favor of a plate with a thickness of 5-10 km. The interaction of this plate with a lower crust that is behaving as a fluid damps the seismic cycling in distances of the order of 10 km from the fault. Low measured heat flow also argues in favor of a thin plate with a low stress level on the fault. Future measurements of stress, strain, and heat flow should help to provide a better understanding of the basic mechanisms governing the behavior of strike-slip faults.

Introduction

The San Andreas and North Anatolian faults are two of the major strike-slip faults in the world that exhibit regular seismicity. Both lie in populated areas and result in high levels of seismic risk. In order to reduce this risk it is essential that an understanding of the seismicity on these faults be developed. Such an understanding should provide estimates of the probability of an earthquake occurring and may lead directly to the prediction of earthquakes.

Plate tectonics has provided a framework for understanding the seismicity on both faults. Ideally, a strike-slip fault accommodates lateral motion between two rigid surface plates. Stick-slip behavior on the fault leads to regular seismicity. The forces at accretional plate margins and at subduction zones drive the plate motion. The elastic surface plate acts as a stress guide. When a marginal fault locks stress accumulates locally until it reaches a level that causes the fault to slip. The oscillatory behavior at the plate boundary is damped out in a relatively short distance from the fault due to the interaction with an elasticoviscous region beneath the surface elastic plate. The small distance over which this damping occurs is strong evidence that the elasticoviscous region is in the lower crust.

Although this simple model for the behavior of strike-slip fault explains many features of the observed seismicity many questions remain unanswered. Why does stick-slip behavior occur? It has been explained qualitatively in terms of the difference between static and dynamic friction or, alternatively, in terms of a strain-softening rheology. A third mechanism for stick-slip behavior is a fault that "heals." An example of this mechanism is melting on a fault. The melting welds the two sides of the fault together until the accumulating stress is sufficient to break the weld. It is doubtful, however, that melt production is a usual occurrence during earthquakes (Cardwell et al., 1978). Other mechanisms for healing of a fault may be available.

Other important questions involve the stress accumulation on the fault. To what distance from a fault does cyclic stress accumulation occur and why? What is the structure of a major strike-slip fault at depth? Since continental lithosphere is estimated to have a thickness as large as 200 km the depth scale is large. Studies of eroded fossil faults (Weathers et al., 1979) may provide information on this question. A related question is the stress level as a function of time and depth on the fault zone. Is the stress drop in a great earthquake a small fraction of the initial stress or is the stress drop nearly complete?

In this paper some aspects of these questions will be addressed. Comparisons between the San Andreas and North Anatolian faults will be made. Unfortunately there will be relatively few answers but some suggestions for relevant observations will be made.

Tectonic Significance

San Andreas Fault

There is no question that the San Andreas fault represents a major boundary between the Pacific and North American plates. However, extensive seismicity occurs throughout the western United States. The entire region appears to accommodate the relative motion between these two plates as well as the Juan de Fuca plate (Turcotte and Oxburgh, 1978). An important question is what fraction of the relative motion between the two plates is accommodated in displacements on the San Andreas fault.

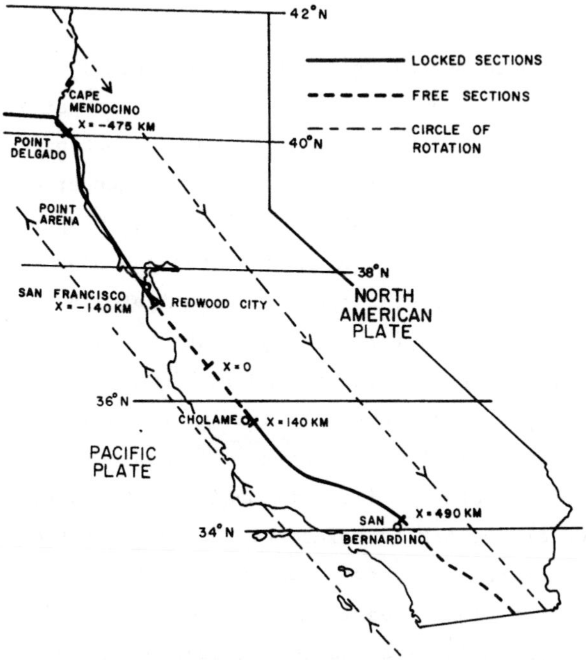

Figure 1. The locked and free sections of the San Andreas fault are illustrated. Also included are two circles of rotation for the motion of the Pacific plate relative to the North American plate.

It is shown in Figure 1 that the trend of the fault is in good agreement with small circles of rotation that define relative motion between the Pacific and North American plates. This agreement is strong evidence that a substantial fraction of the relative motion between these two plates is accommodated on the San Andreas fault. If this is the case the mean slip on the fault is predicted to be 5.5 cm/year.

However, the San Andreas fault has a number of bends. The large bend north of Los Angeles results in a thrusting component of motion. This motion is accommodated on a series of thrust faults which have created the transverse ranges. At the northern boundary of the transverse ranges the Garlock fault branches off the San Andreas fault. The region between the San Andreas and Garlock faults is being driven to the east in much the same way that the Anatolian plate is being driven to the west by displacements on the North and East Anatolian faults.

North Anatolian Fault

The tectonic significance of the North Anatolian fault is considerably less clear than the San Andreas. It is part of a long zone of continental collisions extending from southern Europe through Asia to India. McKenzie (1972) attempted to define the tectonics of Turkey in terms of a number of small plates and broad zones of regional deformation. More recent discussions of the tectonics of this region have been given by Dewey and Sengor (1979) and by Sengor (1979). In these formulations the North Anatolian fault is the boundary between the Anatolian (or Turkish) plate to the south and the Eurasian plate to the north as illustrated in Figure 2. The northward movement of the Arabian plate is apparently driving the Anatolian plate to the west resulting in right lateral strike-slip motion on the North Anatolian fault and left lateral strike-slip motion on the East Anatolian fault. The subduction of the African plate to the south of the Anatolian plate may also contribute to the westward movement of the plate. Subsidence of an oceanic plate can transmit tensional stresses to the adjacent continental plate. Marginal basins are one consequence of these tensional stresses. Subduction along the southern boundary of the Anatolian plate can therefore contribute to the westward movement of the plate.

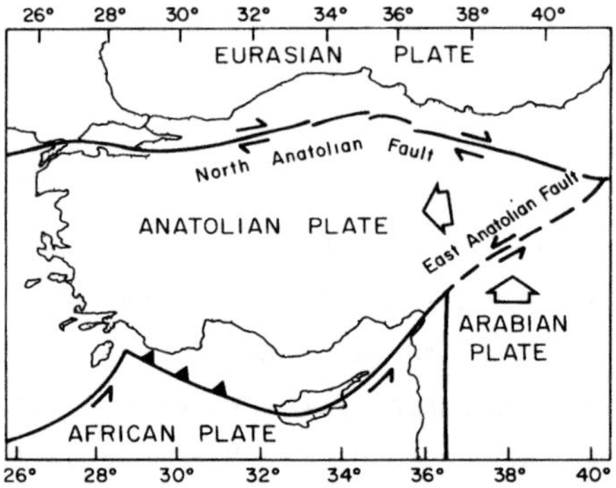

Figure 2.

Illustration of the tectonics of the Anatolian plate and its relationship to the North Anatolian fault.

Dependence of Seismicity on Space and Time

San Andreas Fault

The recorded history of seismicity on the San Andreas fault is relatively brief. However, it is clear that the fault can be divided up into sections where the fault is locked and great earthquakes occur and into sections where aseismic creep and small earthquakes relieve a significant fraction of the accumulating strain. These sections are illustrated in Figure 1.

The northern locked section extends about 330 km southward from the northern terminus of the fault at a triple junction between the Pacific, North American and Juan de Fuca plates near Cape Mendocino. A fault break was reported for the entire length of this section on the April 1906 San Francisco earthquake. The magnitude of the earthquake was probably about 8.3 and the mean displacement 4 $\pm$ 2 m. There have been no reports of fault creep or earthquakes on this section since 1906. A major earthquake occurred in northern California during late June 1838. Lauderback (1947) has carried out a study of this earthquake using available sources of information--diaries, books, etc. He concluded, "This study has convinced me that the earthquake of June, 1838, was the next preceding earthquake corresponding to the earthquake of April, 1906, involving a major fault-trace phenomenon along the central (and perhaps northern) Coast Range course of the San Andreas fault." If the 1906 earthquake relieved 68 years of accumulated strain at the rate of 5.5 cm/year the predicted displacement would be 3.74 m. This is in good agreement with observed slip in the 1906 earthquake. It should be noted, however, that there is considerable uncertainty whether the 1838 earthquake was a great earthquake. This is one of many examples where written accounts of historical earthquakes are of limited value.

The southern locked section of the fault extends some 350 km from Cholame to near San Bernardino. A fault break was reported for the entire length of this section in the Jan. 9, 1857 earthquake. Since then no fault creep or earthquakes have been reported on this fault although the magnitude 7.7 Kern County earthquake of 1952 occurred on an adjacent fault. It is estimated that the mean displacement during the 1857 earthquake was 7 $\pm$ 2 m (Sieh, 1978a). Recently Sieh (1978b) has dated displaced layers of sediments in river beds adjacent to the southern locked section. He concluded that great earthquakes had also occurred on this section in 1745^{+24}_{-55}, 1470 $\pm$ 40, 1245 $\pm$ 45, 1190 $\pm$ 45, 965 $\pm$ 50, 860 $\pm$ 35, 665 $\pm$ 80, and 545 $\pm$ 45. At a rate of 5.5 cm/year, 6.2 m of slip would have accumulated on the southern locked section between 1745 and 1857; this is in good agreement with the estimated slip in the 1857 earthquake. The mean interval between the ten documented earthquakes was 164 years; the longest interval was 275 years and the shortest 55 years. It has been 123 years since the last great earthquake occurred on the southern locked section.

Between the northern and southern locked sections there is a central free section. This section extends some 280 km from near Cholame to Redwood

City just south of San Francisco. Fault creep and small earthquakes occur on the section of the fault continuously. The mean slip appears to be 2.3 cm/year (Savage and Burford, 1971), about one-half the relative plate velocity. However in both the 1857 and 1906 earthquakes some slip occurred on this central section. Apparently a fraction of the accumulated strain is relieved continuously and the remainder in great earthquakes on the locked sections.

Great earthquakes on the two locked sections appear not to be related. The central free section acts as a buffer between the two sections. An important question is whether a great earthquake on one locked section transfers stress to the other locked section. This depends upon the distance from the fault over which strain accumulation occurs. If this distance is small then it is unlikely that an earthquake on one locked section will influence the other locked section.

The felt intensities of earthquakes associated with the San Andreas fault system are given in Figure 3 as a function of latitude for the time period 1880-1973. All earthquakes of intensity 5 or larger centered within 100 km of the fault are included (Turcotte, 1977). It is clear that there is some reporting bias in the data for the smaller events. This is expected since the great increase in the number of seismometers in California in the last 20 years has greatly improved the recording of earthquakes. However, despite the variation in the quality of the data some trends are clear. The most obvious is the reduction in the number of small and intermediate size earthquakes adjacent to the northern locked section following

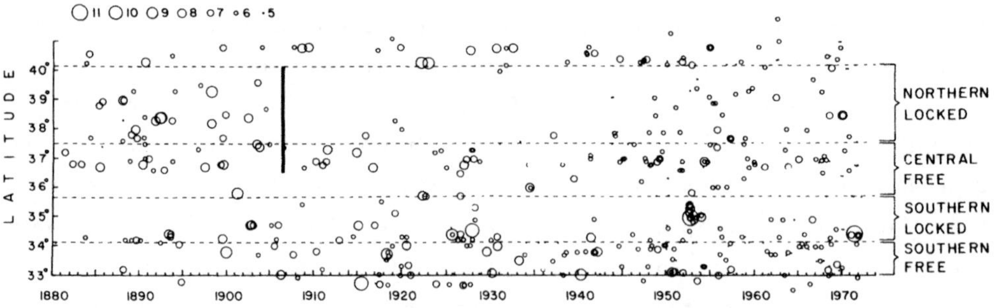

Figure 3. Felt intensities of earthquakes within 100 km of the San Andreas fault as a function of latitude from 1880 to 1972. The solid line shows the surface break of the 1906 earthquake.

the 1906 earthquake. This is strong evidence that the stress level in the lithosphere adjacent to the fault was substantially reduced by the earthquake.

North Anatolian Fault

The dependence of seismicity on space and time along the North Anatolian fault appears to be quite different than the San Andreas. During the period 1939 to 1967 seismicity apparently swept from east to west along the fault (Ambraseys, 1970). The dates and magnitudes of these earthquakes are given in Table 1; also included are the length of the surface break and the fault offset.

Table 1. Sequence of earthquakes with magnitudes greater than 7 along the North Anatolian fault.

Date	M	Length of Break, km	Offset m
Dec. 26, 1939	8	350	4.2
Dec. 20, 1942	7.3	50	2
Nov. 26, 1943	7.6	265	1.5
Feb. 1, 1944	7.6	190	3.6
Mar. 18, 1953	7.4	60	4.3
May 26, 1957	7.1	40	1.6
Jul. 22, 1967	7.2	80	2.3

The felt intensities of earthquakes associated with the North Anatolian fault system are given in Figure 4 as a function of latitude for the period 1880 to 1970 (Ergin et al., 1967, 1971). Also included are the surface breaks associated with the series of large earthquakes that have occurred since 1939.

The first and largest of these earthquakes was the Erzincan earthquake of Dec. 26, 1939. The surface break extended from near the junction of the North and East Anatolian fault some 350 km to the west. This was followed on Dec. 20, 1942 by an earthquake on a more northerly branch of the fault near the western end of the 1939 surface break. Another large earthquake occurred on this northerly branch on Nov. 26, 1943 extending the surface break 265 km to the west. This was followed on Feb. 1, 1944 by another large earthquake which extended the surface break another 190 km to the west. Two large earthquakes occurred in 1957 and 1967 near the western termination of the 1944 surface break. A large earthquake occurred farther to the west on March 18, 1953 but its relation to the other seismicity is not clear. Near the sites of the 1957 and 1967 earthquakes the fault appears to splay into several branches. An important question is whether the gap between the 1967 and 1953 earthquakes represents a seismic gap where a large earthquake can be expected.

The propagation of seismic activity along the fault can be attributed to the transfer of stress from a section of the fault where an earthquake has occurred to an adjacent section. This behavior is quite different than that of the San Andreas fault.

Assuming that the mean displacement in this recent series of earthquakes is 4 m and that the relative velocity across the fault is 1 to 2 cm/year the average interval between great earthquakes would be expected to be between 200 and 400 years. Therefore another series of great earthquakes on the North Anatolian fault would not be expected until approximately 2300 A.D.

Stress Accumulation

The problem of the cyclic accumulation of stress on a strike-slip fault has been considered by Spence and Turcotte (1979). These authors considered an elastic plate of thickness h overlying a semi-infinite elastico-viscous region. In order to simulate the coseismic slip on the fault an infinite series of displacements w_o were postulated to occur at time intervals T. The coseismic displacement was given by

$$\frac{w}{w_o} = 1 - \frac{2}{\pi} \tan^{-1} \left(\frac{x}{h}\right) \qquad (1)$$

A similar analysis has been given by Savage and Prescott (1978).

The coseismic displacements as a function of the distance from the fault associated with the 1906 San Francisco earthquake are given in Figure 5. Reasonably good agreement with equation (1) is obtained taking w_o = 1.5 m and h = 5 km. Also shown in Figure 4 are the strain accumulation between earthquakes assuming T = 60 years. The only parameter that appears in the analysis is the viscoelastic relaxation time η/G (η the viscosity of the elasticoviscous region beneath the plate and G the shear modulus). The results given are for η/G = 6 years (i.e., G = 7 x 10" dynes/cm^2, $\eta = 1.3 \times 10^{23}$ poise). The stress drop associated with the coseismic displacement given in Figure 5 is 130 bars.

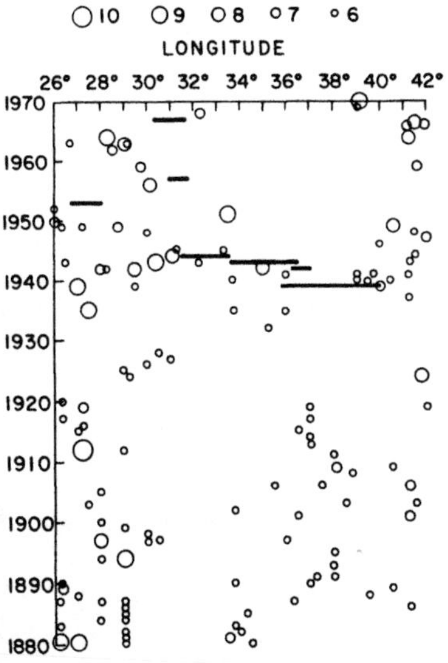

Figure 4.

Felt intensities of earthquakes on and adjacent to the North Anatolian fault as a function of longitude from 1880 to 1970. The solid lines are surface breaks of the series of large earthquakes that occurred on the North Anatolian fault between 1939 and 1967.

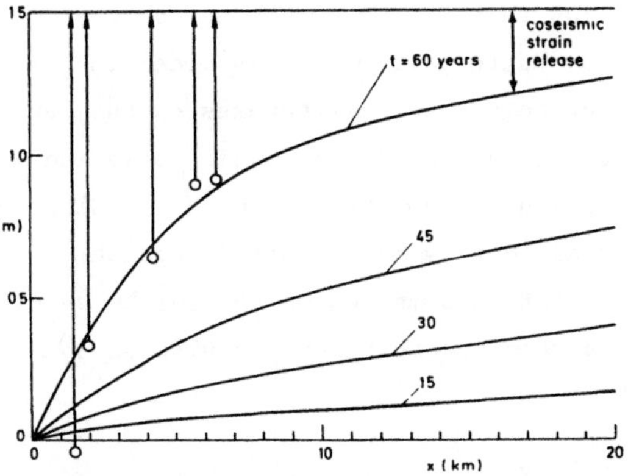

Figure 5. Illustration of the accumulation and release of strain during an earthquake cycle based on the elasticoviscous analysis by Spence and Turcotte (1979). Accumulated strain for intervals of 15 years is given. The coseismic strain release is compared with the observed values for the 1906 San Francisco earthquake. The period between earthquakes is 60 years, the thickness of the elastic plate 5 km, and the viscoelastic relaxation time is 6 years.

In order to have agreement with the coseismic strain it is necessary to consider a thin plate. This thin plate is also consistent with the relatively small displacements observed in major strike-slip earthquakes. A

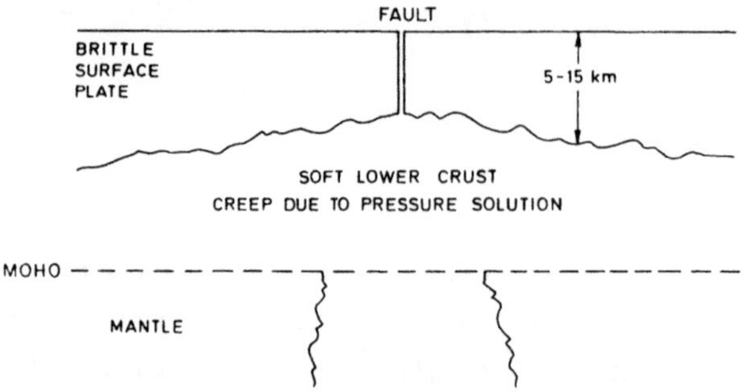

Figure 6. Idealized cross section of a strike-slip fault zone. A thin near-surface, brittle fault zone broadens into a broad zone of lower crustal deformation. A narrower zone of plastic deformation accommodates the displacement in the mantle.

thin rigid surface plate is consistent with crustal rheologies if the lower crust is assumed to deform by low temperature mechanisms such as pressure solution creep (Durney, 1976; Rutter, 1976). This model is illustrated in Figure 6. It should be noted that earthquakes on the San Andreas extend to depths of about 12 km (Barker, 1976). This would indicate a somewhat thicker plate than obtained from the strain data. Clearly further measurements of coseismic strain during major earthquakes on strike-slip faults will provide important information on the thickness of the elastic plate. Measurements of strain accumulation can also be compared with the theoretical analysis.

Heat Flow

The heat flow above a strike-slip fault places important constraints on the stress level on the fault. The heat generated per unit area and time on a fault q is related to the relative velocity across the fault u and the shear stress τ by

$$q = u\tau \qquad (2)$$

If the relative velocity is 5 cm/year and the shear stress 100 bars, then the heat generated is 0.38×10^{-6} cal/cm^{2}sec. Brune et al. (1969) have determined the distribution of the surface heat flow due to uniform heat generation on a fault with a prescribed depth. Comparing their results with the measurements of surface heat flow adjacent to the fault obtained by Henyey and Wasserburg (1971) they concluded that the stress on the fault is less than 100b.

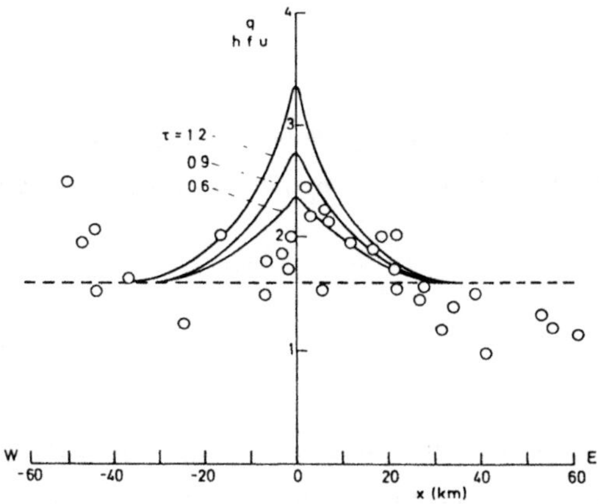

Figure 7. Heat flow measurements adjacent to the San Andreas fault (Lachenbruch and Sass, 1963) compared with a frictional heating model (Turcotte et al., 1980).

An extended series of heat flow measurements adjacent to the San Andreas fault have been given by Lachenbruch and Sass (1963). Their results are given in Figure 7. They find that the San Andreas fault is contained within a 100 km wide band of high heat flow (2 h.f.u.) but do not find a local anomaly at the main fault trace. Also included in Figure 5 is the predicted heat flow assuming a depth-dependent rheology on the fault (Tag et al., 1980). The upper part of the fault is assumed to undergo frictional sliding; a coefficient of friction is used to determine the stress level. At depth plastic deformation is assumed to occur and the rock rheology determines the temperature (Yuen et al., 1978).

The lack of an observed heat flow anomaly near the San Andreas fault also favors a thin surface plate. The broad region of elevated heat flow can be attributed to frictional heating on the base of the thin plate. Further measurements of heat flow adjacent to strike-slip faults would provide more constraints on fault behavior.

Sedimentary Basins

A characteristic feature of the San Andreas fault is a series of sedimentary basins. Examples are the Los Angeles, Ventura, and Santa Maria basins. It is likely that a similar basin is currently being formed in the Imperial Valley near the southern termination of the fault. These sedimentary basins are most easily explained as small spreading centers

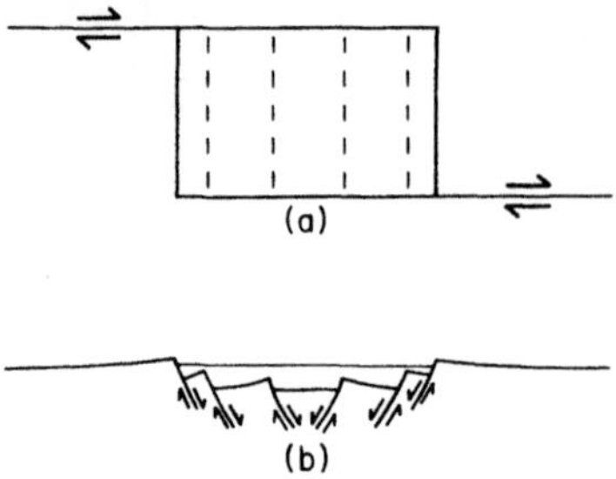

Figure 8.

Planform (a) and cross section (b) of an idealized pull-apart basin on a strike-slip fault.

(sections of oceanic ridge) between two transform faults. Actual oceanic crust need not be formed but the continental crust is thinned by the same process of normal faulting that is operative at rift valleys. This process is illustrated in Figure 8.

Volcanism is associated with the formation of these sedimentary basins. There is recent volcanism in the Imperial Valley and mid-Miocene volcanics have been found during drilling in the Los Angeles basin; these mid-Miocene volcanics coincided with the most rapid period of subsidence in the basin. Geothermal phenomena are associated with the volcanics. The Imperial Valley is an area of active geothermal development. Apparently the fault pattern responsible for the formation of these "pull-apart" basins is transitory and active basin formation ceases. In the cases of the Santa Maria, Ventura and Los Angeles basins the main branch of the San Andreas fault now lies to the east. After active basin formation ceases thermal subsidence occurs (Turcotte and McAdoo, 1979). These basins now contain large reservoirs of petroleum.

It is interesting to speculate whether the Sea of Marmara is a "pull-apart" basin on the North Anatolian fault system. The geometry appears appropriate in that the main branch of the fault approaches the sea on the south side. The prediction would be that the main branch would leave the sea on the west parallel to the north side of the sea. If the Sea of Marmara is a "pull-apart" basin active volcanism, geothermal systems, and high heat flow would be expected on its floor.

Conclusions

Major uncertainties remain concerning the physical processes associated with the cyclic accumulation of strain on major strike-slip faults such as the San Andreas and North Anatolian. Some measurements have provided important constraints on alternative fault models.

1. Coseismic strain

The displacements associated with large earthquakes make measurements relatively accurate. However, many measurements must be made to overcome local inhomogeneities which introduce noise into the data. It is also important to either measure absolute rotations or establish a base line at such a large distance from the fault that it is certain that it has not moved during the earthquake. Unfortunately these measurements require large earthquakes.

2. Heat flow

Measurements of heat flow place important constraints on the stress level on the fault. Groundwater circulation can influence the heat conduction calculations on which the interpretation of heat flow is based; measurements indicate that this is not a serious source of error.

There are other types of measurements that have been less successful.

1. Many direct measurements of strain accumulation have been carried out extensively adjacent to the San Andreas fault (i.e., Savage et al., 1979). It is clear that the present accuracy of the techniques is adequate to measure the strain. However, many measurements have been of questionable value because of the large noise level of the data. It is now becoming clear that this is due to slip on many small local faults. The near-surface rocks do not behave as an elastic half-space. A large fraction of the near-surface displacements are taken up by creep on faults. Useful information can be obtained by carrying out many measurements and averaging out the local strains, but great care must be exercised in interpreting the results.
2. Stress measurements suffer from the same problem as the strain accumulation problems. Measurements indicate that the measured maximum stress increases linearly with depth to depths of a kilometer or more. This can be explained in terms of the coefficient of friction on the many small faults that are present. As the lithostatic (overburden) pressure increases the stress that can be transmitted by a fault or joint increases linearly with depth. The crust behaves more like an elastic half-space as one goes deeper. The near-surface rocks constitute a brittle fabric.

As a better understanding of the mechanics of faults is developed many observations that now appear contradictory will begin to make sense. Such an understanding may provide the basis for the future prediction of earthquakes.

Acknowledgment

The research reported on in this paper has been supported by NASA under Cooperative Agreement ACC5-2.

References

Ambraseys, N. N. 1970. Some characteristic features of the North Anatolian fault zone, Tectonophy. 9, 143-165.

Barker, T. G. 1976. Quasi-static motions near the San Andreas fault zone, Geophys. J. Roy. Astron. Soc. 45, 689-705.

Brune, J. N., T. L. Henyey, and R. F. Roy, 1969. Heat flow, stress, and rate of slip along the San Andreas fault, California, J. Geophys. Res. 74, 3821-3827.

Cardwell, R. K., D. S. Chinn, G. F. Moore, and D. L. Turcotte, 1978. Frictional heating on a fault zone with finite thickness, Geophys. J. Roy. Astron. Soc. 52, 525-530.

Dewey, J. R., and A.M.C. Sengor, 1979. Aegean and surrounding regions: complex multi-plate and continuum tectonics in a convergent zone, Geol. Soc. Am. Bull. 90, 89-92.

Durney, D. W., 1976. Pressure solution and crystallization deformation, Phil. Trans. Roy. Soc. London A283, 229-240.

Ergin, K., U. Güclü, and Z. Uz, 1967. A Catalog of Earthquakes for Turkey and Surrounding Area, (11 A.D. to 1964 A.D.), Faculty of Mining Engineering, Technical University of Istanbul, Report No. 24.

Ergin, K., U. Güclü, and G. Aksay, 1971. A Catalog of Earthquakes of Turkey and Surrounding Area (1965-1970), Institute of Physics of the Earth, Technical University of Istanbul, Report No. 28.

Henyey, T. L., and G. J. Wasserburg, 1971. Heat flow near major strike-slip faults in California, J. Geophys. Res. 76, 7924-7946.

Knopoff, L., 1958. Energy release in earthquakes, Geophys. J. Roy. Astron. Soc. 1, 44-52.

Lachenbruch, A. H., and J. H. Sass, 1963. Thermo-mechanical aspects of the San Andreas Fault System, in Proceedings of the Conference on Tectonic Problems of the San Andreas Fault System, R. L. Kovach and A. Nur, eds., pp. 124-214, Stanford University Press, Stanford, California.

Lauderback, G. D., 1947. Central California earthquakes of the 1830's, Seis. Soc. Am. Bull. 37, 33-74.

McKenzie, D. P., 1972. Active tectonics of the Mediterranean region, Geophys. J. Roy. Astron. Soc. 30, 109-185.

Rutter, E. H., 1976. The kinetics of rock deformation by pressure-solution, Phil. Trans. Roy. Soc. London, A283, 203-219.

Savage, J. C., and R. O. Burford, 1971. Discussion of paper by C.H. Scholz and T. J. Fitch,'Strain accumulation along the San Andreas Fault,' J. Geophys. Res. 76, 6469-6479.

Savage, J. C., and W. A. Prescott, 1978. Asthenosphere readjustment and the earthquake cycle, J. Geophys. Res. 83, 3369-3376.

Savage, J. C., W. H. Prescott, M. Lisowski, and N. King, 1979. Geodolite measurements of deformation near Hollister California, 1971-1978, J. Geophys. Res. 84, 7599-7615.

Şengör, A.M.C., 1979. The North Anatolian transform fault; its age, offset and tectonic significance, J. Geol. Soc. London 136, 269-282.

Sieh, K. E., 1978a. Slip along the San Andreas fault associated with the great 1857 earthquake, Seis. Soc. Am. Bull. 68 1421-1448.

Sieh, K. E., 1978b. Prehistoric large earthquake produced by slip on the San Andreas fault at Pallett Creek, California, J. Geophys. Res. 83, 3907-3939.

Spence, D. A., and D. L. Turcotte, 1979. Viscoelastic relaxation of cyclic displacements on the San Andreas fault, Proc. Roy. Soc. London A365, 121-144.

Tag , P. H., D. L. Turcotte, and R. F. Cooper, 1980. A steady-state model for the distribution of stress and temperature on the San Andreas fault, submitted for publication.

Turcotte, D. L., 1977. Stress accumulation and release on the San Andreas Fault, Pure App. Geophys. 115, 413-427.

Turcotte, D. L., and D. C. McAdoo, 1979. Thermal subsidence and petroleum generation in the Los Angeles basin, California, J. Geophys. Res. 84, 3460-3464.

Turcotte, D. L., and E. R. Oxburgh, 1978. Intra-plate volcanism, Phil. Trans. Roy. Soc. London 288A, 561-579.

Weathers, M. S., J. M. Bird, R. F. Cooper, and D. L. Kohlstedt, 1979. Differential stress determined from deformation-induced microstructures on the Moine Thrust Zone, J. Geophys. Res. 84, 7495-7509.

Yuen, D. A., L. Fleitout, G. Schubert, and C. Froidevaux, 1978. Shear deformation zones along major transform faults and subducting slabs, Geophys. J. Roy. Astron. Soc. 54, 93-119.

Fracture Mechanics Modelling of Earthquake Generating Processes

Barry Kean Atkinson

Imperial College of Science and Technology, London SW7 2BP, Great Britain

Abstract

A review is given of some examples of Fracture Mechanics modelling of fracture and friction processes in earthquake source regions. Examples of possible theoretical approaches are preceded by an outline of Fracture Mechanics principles. Experimental determinations of Fracture Mechanics parameters are discussed and examples of some results are tabulated. Also dealt with are the interrelationships between micro-seismic emissions and Fracture Mechanics parameters. Some limitations of the Fracture Mechanics approach to earthquake modelling are noted and important future developments in experimental work are outlined.

Introduction

Despite rapid recent advances in other areas of rock deformation our understanding of fracture and friction processes has remained rather rudimentary. Until recently, the theoretical modelling of earthquake source processes was not developed to the point where useful guidelines could be given to the experimentalist in order to constrain rock fracture testing. The unifying approach of Fracture Mechanics (Irwin, 1958) has been extremely useful in overcoming these deficiencies and shows promise of being a powerful tool in the future modelling of earthquake generating processes. Physical models of earthquake source processes can be modelled using Fracture Mechanics principles and equations can be derived to describe them. Ideally, the constants in these equations can be determined by experiment and the combined predictions of theory and experiment can be tested by field-based studies.

This review is divided into five parts. In the first part some of the theoretical background to Fracture Mechanics is outlined. This is followed in the next section with some examples of how Fracture Mechanics notions can be used to model fracture and friction processes in earthquake zones. A third section deals with experimental methods for determining Fracture Mechanics parameters. A fourth section discusses the interrelationship between Fracture Mechanics concepts and microseismic emissions. The final section mentions briefly some limitations to the Fracture Mechanics approach to modelling earthquake source processes and outlines some important future developments in experimental studies.

Theoretical background to Fracture Mechanics

Before outlining some approaches to earthquake rupture based upon Fracture Mechanics it is important to start with a few basic preliminaries.

Fracture Mechanics is based upon Irwin's (1958) analysis of the stresses about a sharp crack tip in a linear elastic, isotropic and homogeneous brittle material (see also Paris and Sih, 1965). The stresses (σ_{ij}) near the crack tip in plane strain have the simple form:

$$\sigma_{ij} = K_n \, (2\pi r)^{-\frac{1}{2}} \, f^n_{ij} \, (\theta) \qquad (1)$$

where r is the distance from the crack tip, θ is the angle measured from the plane ahead of the crack and K_n are the stress intensity factors. The index n refers to the three fundamental modes of fracture: i.e. mode I, tensile; mode II in-plane shear; mode III, anti-plane shear. In this context, near the crack tip means within distances that are small with respect to the characteristic crack dimension. The stress intensity factors depend only on the applied loading and the crack geometry and therefore they determine the intensity of the local stress field. The remaining terms determine the distribution of the local stress field. The stress intensity factors for a given mode are additive. They have the dimensions of $[\text{stress}][\text{length}]^{\frac{1}{2}}$.

The stress intensity factors are related to another Fracture Mechanics parameter, G, the strain energy release rate for plane strain by

$$G = K_I^2 (1-\nu^2)/E + K_{II}^2(1-\nu^2)/E + K_{III}^2(1+\nu)/E \qquad (2)$$

where E is Young's modulus, ν is Poisson's ratio and K_I, K_{II} and K_{III} are the stress intensity factors for fracture in modes I, II and III, respectively. Thus the strain energy release rates for *different* modes are additive. G is defined as the energy released per unit area of crack advance during an infinitesimal increase in length of a crack in a linearly elastic body.

G (and K) can be related to the Griffith (1920) energy balance concept. If we assume that tensile fracture occurs at some critical value of G or K i.e. G_c, K_{Ic}, then in the absence of dissipative processes other than crack extension in plane strain

$$G_c = K_{Ic}^2 (1-\nu^2)/E = 2\gamma \qquad (3)$$

where γ is the Griffith thermodynamic surface energy.

Except for the case of an ideally brittle solid, however, there will be some dissipative processes occurring near propagating crack tips, e.g. acoustic emission, dislocation motion, and a more realistic fracture criterion is

$$G_c = K_{Ic}^2 (1-\nu^2)/E = 2\Gamma \qquad (4)$$

where Γ is the fracture surface energy (equal to γ for ideally brittle fracture).

Rice (1968) has extended the Irwin Fracture Mechanics approach to unify continuum descriptions of failure. He has been able to characterize the near-tip field of non-linear as well as linear elastic materials and finitely deformed elastic materials with a single parameter. This parameter is the path-independent J-integral for cracks in two-dimensional deformation fields. This integral is defined in the following manner. For a coordinate system in which the crack line coincides with the x-direction and the y-direction is perpendicular to the crack surface, the J-integral for contours (C) that begin and end on the traction free crack surfaces and which enclose the crack tip is given by

$$J = \int_C (Wdy - T.\partial u_j/\partial x)ds \qquad (5)$$

where σ_{ij} and u_j are the stress and displacement components, respectively; T is a traction vector on the contour C defined in relation to an outward normal unit vector; u is the displacement vector; ds is an element of arc on the curve C linking the lower and upper crack surfaces; and W is the strain energy density.

For small scale yielding J is equal to the strain energy release rate, G.

The Fracture Mechanics approach to fracture has the added advantage that it can be used to describe subcritical crack propagation at $K<K_c$ ($G<G_c$). Subcritical crack velocity (v) is uniquely determined for any given material and environment by the stress intensity factor (or G). Thus, subcritical crack extension can be characterized by a stress intensity factor - crack velocity diagram. A typical diagram for subcritical crack propagation by stress corrosion is shown in Fig. 1.

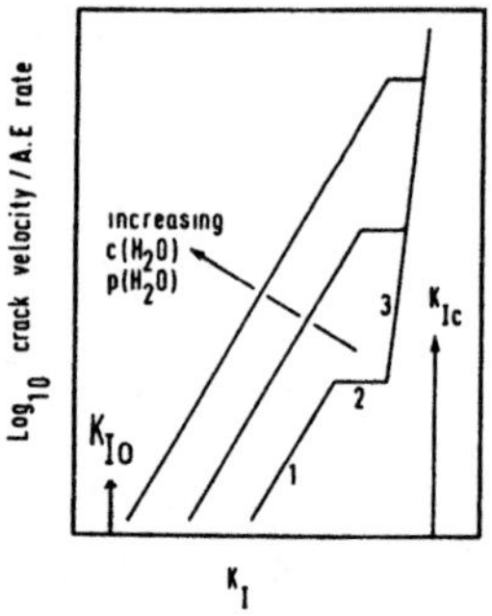

Fig. 1 Schematic crack velocity (v) or acoustic emission rate stress intensity factor (K_I) diagram for a material undergoing stress corrosion. The influence of crack tip "humidity" is shown. In region 1, reaction of a corrosive agent with crack tip material controls the rate of crack propagation. In region 2 the rate of transport of the reactive species to the crack tip controls the rate of crack propagation. In region 3 crack propagation is controlled by a mixture of corrosive and mechanical fracture. For a given set of environmental conditions, there may be a value of the stress intensity factor, K_{Io}, below which no appreciable crack growth occurs. This is known as the stress corrosion limit.

Of course stress corrosion is not the only means whereby subcritical crack extension might occur in rocks. Mobile species in mineral structures could migrate under high stresses to crack tips, there to react with the highly strained crack tip bonds. Dislocation motion could be blocked so that stress concentrations build up and cracks could nucleate and grow at sufficiently high homologous temperatures. For the likely environmental conditions of shallow earthquakes, however, stress corrosion is likely to be the most important mechanism of subcritical crack propagation.

Several attempts have been made to develop equations to describe K-v curves for stress corrosion. The one which is most commonly used to describe the results of experiments run on ceramics and rocks is the semi-empirical equation of Charles' (1958)

$$v = v_o' \exp(-\Delta H/RT)\, K_I^{\,n} \tag{6}$$

where v_o' and n are experimentally determined constants (n is sometimes known as the stress corrosion index), ΔH is the activation enthalpy, T is the absolute temperature and R is the gas constant.

The Charles equation can be used to describe all three regions of the schematic K-v diagram shown in Fig. 1 merely by making appropriate changes in the constants v_o', n and ΔH.

There is ample experimental evidence that mode I crack propagation can occur in rocks at subcritical velocities due to stress corrosion. I am currently working on the problem of subcritical crack propagation in mode II which also seems to occur by stress corrosion. It is not clear whether *any* subcritical crack propagation can occur by stress corrosion in mode III (see discussion section).

Theoretical approaches to fracture/friction processes

There have been a number of attempts in recent years to describe the fracture/friction processes of earthquake source regions in terms of Fracture Mechanics parameters. The most noteable of these are the works of Rice and co-workers, especially Rudnicki. See, for example, Rice (1979 a, b) and Rudnicki (1980). In addition, there have been many attempts made by materials scientists to describe in terms of Fracture Mechanics the time- and rate- dependent failure of structural engineering components, comminution, compressive splitting, and, of course, fast tensile failure. A comprehensive account of these is beyond the scope of this paper, but a few examples are presented below.

Fracture strength

To begin with, Fracture Mechanics can be used to replace the ambiguous quantity "fracture strength", which is not a fundamental material property, with a fracture equation based upon the stress intensity factor. For example, for a penny-shaped crack in an infinite body

$$\sigma_c = (\pi/2)\, K_{Ic}/\sqrt{c_o} \tag{7}$$

where σ_c is the tensile strength, c_o is half the crack length and K_{Ic} is the critical stress intensity factor. Thus tensile "strength" depends upon the size of the crack and a material property, K_{Ic}.

Comminution in fault gouge zones

Kendall (1978 a,b) has recently made a theoretical and experimental study of compression splitting which has important application to comminution. Kendall's (1978a) theory as modified by Karihaloo (1979) predicts that the force causing compression splitting (F_{split}) is given by

$$F_{split} = \left[b\,(2\Gamma dE/3)^{\frac{1}{2}}\right] / \;(1-w/d+8\alpha c/d) \tag{8}$$

where Γ is the fracture surface energy, E is Young's modulus, b and d are the thickness and width of the specimen, w is the width of the loading platten, c is the crack length, and α is a factor denoting the extent of lateral constraint. Compression splitting is thus controlled by the fracture surface energy and we can dispense with the ambiguous concept of compressive strength as a fundamental material property in this context. This theory has the further appeal, however, that not only does it predict that large bodies should be weaker than small ones and the stress at failure should vary as $d^{-\frac{1}{2}}$, a dependence often noted in the literature, but it predicts the impossibility of comminuting small particles by compression below a critical grain size, d_{crit}. This arises because as the specimen size is reduced towards

$$d_{crit} = 32E\Gamma/3\sigma^2_{yield} \left(1 + 8\ \alpha \left[{}^{c}/d\right] \right)^4 \qquad (9)$$

the splitting force rises towards infinity. At this critical particle size even the most brittle materials should deform plastically rather than fracture. In equation (9) σ_{yield} is the yield stress.

Application of this notion to the development of fault gouge zones suggests that there should be finite limit to the reduction in grain size that is possible due to comminution processes and that this critical size will be controlled by certain material properties, noteably the fracture surface energy and the yield stress. For example, if we assume for simplicity that $c/d \to o$ then equation (9) predicts a comminution limit for calcite at room conditions of approximately 1 micrometre (Kendall, 1978b), which compares well with the average particle size of 0.8 micrometres obtained after prolonged milling (see Kendall, 1978b).

Propagation of fault slip

An extension of Fracture Mechanics concepts to include cohesive forces at crack tips has been achieved by Dugdale (1960) and Barenblatt (1962) for the case of mode I crack propagation. The Dugdale-Barrenblatt approach allows a more microstructural approach to be made to crack propagation as it allows for non-linear processes in the vicinity of the advancing crack tip. Palmer and Rice (1973) have developed a cohesive force model for shear band propagation in clay slopes that can be applied to earthquake shear faulting to yield some important insights.

If τ_f is the residual frictional stress on a fault surface and the stress rises to τ_p near the end of the fault due to microstructural processes resisting slip initiation, an amount of sliding δ is required to reduce τ_p to τ_f (Fig. 2.). If the fault is semi-infinite and the end zone is small and

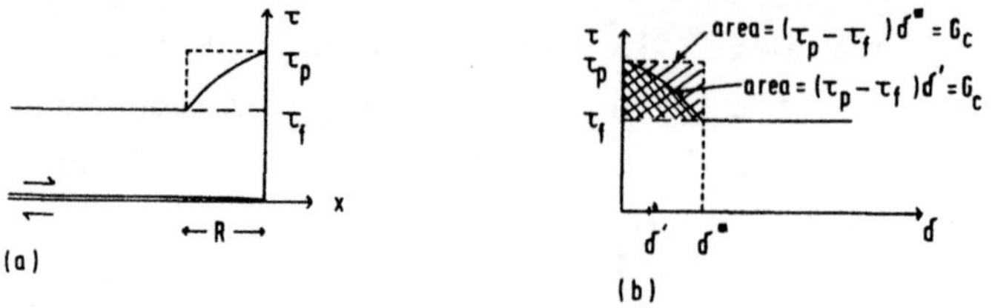

Fig. 2 (a) Fault surface stress (τ) as a function of the distance back from the fault tip (x). The breakdown zone, R, is shown. After Palmer and Rice (1973) and Rudnicki(1980). The two stress paths shown are for a constant end zone stress (dashed line and a linear distribution of stress in the end zone (solid line).

(b) Fault surface stress (τ) as a function of relative slip on the fault (δ). After Palmer and Rice (1973) and Rudnicki (1980). The two stress paths shown are the same as for Fig. 2 (a). S* = S' if end zone stress is constant.

contains a constant distribution of stress, the end zone size, R, is given by

$$R = \pi K_{II}^2 \left[8 (\tau_p - \tau_f)^2\right]^{-1} \tag{10}$$

where K_{II} is the stress intensity factor caused by the applied loads in the absence of an end zone. K for this geometry is given by (Rice, 1979a)

$$K_{II} = \tau_a (8\ell/\pi)^{\frac{1}{2}} \tag{11}$$

where τ_a is the excess of a remote stress over the fault friction ($\tau_\infty - \tau_f$), and ℓ is the total length over which the loads are applied and includes the end zone. Thus,

$$R = \ell\left[(\tau_\infty - \tau_f) / (\tau_p - \tau_f)\right]^2 \tag{12}$$

If the stress in the end zone decreases linearly, from τ_p to τ_f, instead of being constant, the expressions for R should be multiplied by 9/4.

The size of this end zone is of considerable interest if one wishes to use laboratory frictional sliding experiments as models of natural earthquake faults (Rice, 1979a). For laboratory friction experiments to be geometrically similar to slip zones in the field, the breakdown zone must be small by comparison with the length of the slipping zone. By means of standard Fracture Mechanics techniques and application of the J-integral the size of the breakdown zone can be expressed in terms of δ', a measure of a representative slip displacement or the breakdown process defined as

$$\delta' = G_c/(\tau_p - \tau_f) \qquad (13)$$

for a vanishingly small end zone (see Rice, 1979). For a small end zone and a linear decrease of stress in the end zone Rice (1979a) obtains

$$R = (9/16)\ \pi\mu\delta' \left[(1 - \nu) \quad (\tau_p - \tau_f) \right]^{-1} \qquad (14)$$

where μ is the shear modulus and ν is Poisson's ratio.

If $(\tau_p - \tau_f)$ is taken to be of the same order as a representative stress drop in an earthquake, then by inserting appropriate values for the material constants in equation (14), R can be expressed in terms of δ'. Rice (1979a) has inferred values of R = 12m when δ' = 2.5 micrometres for laboratory sliding experiments on flat-ground granite specimens (see Dieterich, 1979). However, much larger values of δ' may be more representative of natural conditions (see Rudnicki, 1979; 1980). Thus, Rudnicki (1980) suggests that if δ' = 2.5 cm, R may be of the order of 100 metres. Although δ' will depend strongly on fault surface conditions the implication of this is that for laboratory experiments to be geometrically representative of natural shear faulting either very large specimens will be needed or fault surfaces must be specially prepared to yield small values of δ' (Rudnicki, 1980).

Fault creep instabilities and their stabilization

Rice (1979) has commented on the widespread evidence that creep-like spreading of fault slip can occur on faults and that accelerating creep is widely regarded as a typical short-term precursor to earthquakes. He suggests that some rate-dependent aspect of material behaviour serves to stabilize the slipping region (i.e. an increasing stress intensity factor is required for an increasing, quasi-static, fault propagation speed) and allow creep-like slip events but that exhaustion of the stabilizing mechanism may allow the transition from precursory fault slip to an earthquake. Important mechanisms whereby this stabilization may occur include stress corrosion assisted microcrack growth at asperity contacts or from the time-dependence of gouge cementation (fault "healing"). The stable sliding of wet, pre-faulted sandstone specimens in laboratory tests may in certain cases be controlled by stress corrosion cracking of near-fault grains (see Rutter and Mainprice, 1978; Atkinson, 1980).

Fault stabilization arising from mechanical interactions with pore fluids in a fluid-infiltrated rock mass has also been proposed by Rice and Simons (1976). The time-dependence of deformation in a porous elastic solid arises from the diffusion of the infiltrating pore fluid. See also Rice (1979a) and Rudnicki (1980). The existence of pore-fluid diffusion effects introduces a factor D/V where D is the diffusivity and V is the steady fault propagation speed. Different regimes of behaviour are found depending upon the relative magnitudes of D/V, the end zone size and the fault length. For a semi-infinite fault loaded in plane-strain conditions over a fixed distance behind the fault tip by a shear stress $\tau_a = \tau_\infty - \tau_f$, Rice and Simons (1976) found that when propagating through a fluid-infiltrated elastic material the near-tip stress singularity field is of the same kind as could exist in a faulted solid under fully drained conditions (i.e. the pore pressure vanishes at the fault tip). The stress intensity factor, K of this singular field is, however, not the same as that for an identically loaded elastic solid without pore fluid effects, K_{nom}, but is given by

$$K = K_{nom}\ h\ (V\ell/D) \qquad (15)$$

The function h decreases monotonically from unity at $V\ell/D$ = o to $(1 - \nu_u) / (1 - \nu)$ as $V\ell/D \to \infty$. Here ν is the Poisson's ratio for an ordinary linear elastic solid without pore fluid effects (i.e. "drained") and ν_u is the Poisson's ratio for a linear elastic solid with pore fluid effects (i.e.

"undrained"). The stabilizing effect in fault propagation through a fluid infiltrated solid arise because $\nu \leqslant \nu_u \leqslant \frac{1}{2}$ (Rice and Cleary, 1976) and thus such a material is elastically stiffer for undrained than for drained conditions (see Rudnicki, 1980).

In a fluid-infiltrated solid K_{nom} only obtains at zero fault propagation speed. Rice and Simons (1976) explained this by noting that for any finite propagation speed the region near a fault-tip without an end zone responds in a drained fashion but that the surrounding material away from the fault responds in an undrained fashion and as a result, is elastically stiffer. Thus, for the case of a vanishingly small end zone the simplest criterion for fault propagation is to require that K equal some critical value K_{crit} given by

$$K_{crit} = K_{nom}\, h\,(V\ell/D) \qquad (16)$$

i.e. the K_{crit} that would pertain under fully drained, ordinary elastic conditions. Fault propagation is stabilised because an increase in driving force (K_{nom}) is necessary to increase the speed of propagation.

If the breakdown zone is *not* small compared to the drained region at the fault tip (i.e. $R \ll {}^{D}/V$ does not hold) then, for a finite value of R/ℓ and assuming a constant va--lue of the end zone stress, fault propagation is stable *only* for a limited range of values $V\ell/D$. Plots of K_{nom}/K_{crit} against $V\ell/D$ do not rise monotonically (Fig. 3) but instead they peak and then decrease to an asymptotic value at large $V\ell/D$ (Rice and Simons, 1976). These peaks mark instability points. In the absence of any other stabilizing mechanism they represent the largest possible K_{nom}-values and hence the highest possible speeds of fault creep.

Rice (1979a) has shown that from estimates of the values of parameters in equation (16) based on field observations of at least some natural fault creep events the values of $V\ell/D$ do fall on the stable, rising portion of a (K_{nom}/K_{crit}) vs ($V\ell/D$) curve. Thus he suggested that some stable fault creep events are made possible, at least in part, through quasi-static effects involving pore fluid stabilization. Of

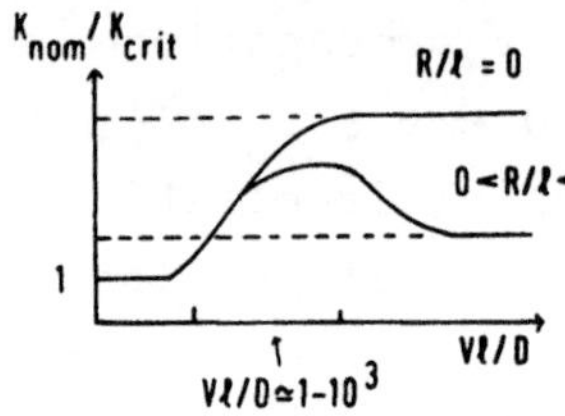

Fig. 3 Plots of K_{nom}/K_{crit} vs. $V\ell/D$ for pore fluid diffusion stabilization of fault slip. The region marked $V\ell/D \simeq 1 - 10^3$ corresponds to estimates of this parameter for natural fault creep events (see Rice, 1979). After Rice (1979), Rudnicki (1980).

course, this does not rule out other stabilizing mechanisms such as viscoelastic effects (Rudnicki, 1980), time-dependent friction (Dieterich, 1978) or fault healing (Rice, 1979a,b). There may well be a complex interplay of all these mechanisms in nature.

Critical energy release rate during earthquakes

Studies of the critical energy release rate during earthquake faulting may provide important information on the rupture process (Rudnicki, 1980). Following an analysis of Palmer and Rice (1973) the G_c for the initiation of the 1857 California earthquake has been estimated by Rudnicki (1980) through application of the J-integral.

For a long strike-slip fault locked beyond a given point P_L but undergoing steady relative displacement behind that point the relative displacement at some point Q in the region undergoing steady slip is δ_Q. The fault is loaded by displacements u at a distance h/2 from the fault imposed due to large scale inter-plate motions. Calculation of the J-integral about Q gives the following expression (Rudnicki, 1980)

$$J_Q - \tau_f \, \delta_Q = \mu \delta_Q^{\,2} / (2h) \qquad (17)$$

For a small end zone

$$J_Q - \tau_f \delta_Q = G_c \quad (18)$$

For the San Andreas fault in central California point P_L is near Cholame. To the northeast of Cholame the fault is creeping and to the southwest the fault is, generally speaking, locked. Taking the displacement rate of the creeping portion to be 3 cm/year and an average recurrence time for 1857 - type events in California of 160 years, δ_Q = 480cm. Substituting this value as well as $\mu = 2 \times 10^4$ MPa and h = 60 km into equation (17) gives an order of magnitude estimate of $G_c = 4 \times 10^6$ J.m^{-2} (Rudnicki, 1980). A smaller value of G may be all that is necessary to continue rupture.

G_c for frictional sliding on natural faults has been estimated as 1 - 10^4Jm^{-2} by Husseini and others (1975) and for natural fault creep as $\simeq 3 \times 10^2$Jm^{-2} by Rice and Simons (1976). As pointed out by Rudnicki (1980) these values for G_c are comparable in magnitude to those obtained for laboratory tests on tensile fracture (see Atkinson and others, 1979). For fresh fracture on natural faults G_c is much higher; values fall in the range 10^4 - 10^7J.m^{-2} (Husseini and others, 1975; Ida, 1973). However, even the largest values of G_c when multiplied by the fault area yield an energy which is much less than empirical estimates of the seismic energy released in earthquakes. For small earthquakes the fracture energy may be more comparable to the total seismic energy (Rudnicki, 1980). As Rudnicki points out this suggests that small earthquakes may in some way be different to large ones. Although estimates of G_c in earthquakes are presently rather crude its importance demands further attention.

Time - and rate - dependence of fracture stress

K_I-v diagrams have been used by Evans (Evans, 1972; 1974; Evans and Johnson, 1975) to predict the time - and rate - dependent behaviour of the tensile strength of materials undergoing slow crack growth. Essentially, this analysis involves integrating the area under the K_I-v curve. Because the times spent in the regions 2 and 3 of the schematic K_I-v curve (Fig.1) are often short, in general no great loss of accuracy is introduced if only region 1 is considered. This simplified analysis has been employed by Atkinson (1980) to predict the influence of strain rate on the tensile strength of a fine-grained quartz rock in water from the appropriate experimentally determined K_I-v curve. Henry and Paquet (1976) have made a similar study of calcite rocks.

Starting from the observation that each region of a K_I-v curve can be characterised by the relationship

$$v = AK_I^{\,n} \qquad (19)$$

where A and n are constant for each region of the curve. Evans and Johnson (1975) find that for constant stress rate conditions

$$(da/a^{n/2}) = (AY^n \, \sigma^n \quad d\sigma/\dot{\sigma}\,) \qquad (20)$$

Here a is the crack length and σ is the maximum stress supported by the material, $\dot{\sigma}$ is the stress rate and Y is a geometrical constant. For crack propagation in region 1 of the typical K_I-v curve the fracture stress, σ_f is given by

$$\sigma_f = \left[\, 2\dot{\sigma}\,(n_1 + 1)/A_1 Y^{n_1}(n_1 - 2)a_i^{(n_1-2)/2} \,\right]^{1/(n_1+1)} \qquad (21)$$

where a_i is the initial flaw size and the subscript 1 refers to values of the parameters A and n for region 1 behaviour.

The time-to-failure (T) at constant tensile stress (σ) is given by a similar integration of the K_I-v diagram (Evans, 1972; 1974). It is found that

$$T = 2/\sigma^2 Y^2 \int_{K_{Ii}}^{K_{Ic}} (K_I/v)\, dK_I \qquad (22)$$

where K_{Ii} is the stress intensity factor associated with the initial flaw size. The integration can be evaluated either numerically or analytically

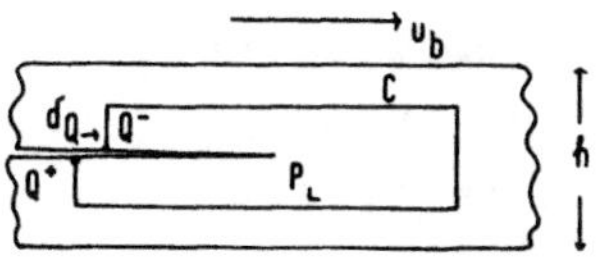

Fig. 4 Grossly simplified sketch of a long strike-slip fault loaded by displacements u_b. Relative slip at point Q is δ_Q. The contour marked C is that around which the J-integral is computed. The fault tip P_L corresponds to Cholame on the San Andreas Fault with the NE being to the left of this point.

In the previously mentioned study of the rate-dependent strength of a fine-grained quartz rock in water by Atkinson (1980) a close agreement was noted between the theoretical predictions of equation (21) and experimental observations of the tensile strength. It was found that at strain rates faster than approximately $6 \times 10^{-5} s^{-1}$ stress corrosion does not influence the tensile fracture stress, but at lower strain rates there is a monotonic reduction in strength from a maximum value of 72 $MN.m^{-2}$ in the absence of stress corrosion effects to approximately 40 $MN.m^{-2}$ at a strain rate of $10^{-10} s^{-1}$. On this evidence Atkinson (1980) concluded that this material was unlikely to show a fundamental tensile fracture stress at even the slowest tectonic strain rates (10^{-11} to $10^{-14} s^{-1}$).

The implications of K_I-v diagrams for the strength of rock in macroscopic compression can be appreciated by considering the growth of locally tensile microcracks at the tips of closed cracks oriented at an angle to the maximum compressive stress and sliding under frictional resistance. The interaction of many such tensile microcracks is widely held to give rise to compressive failure. If the frictional resistance on the sliding crack is S, and the resolved shear stress in the direction of sliding is τ, then when the tensile crack is short compared to the sliding portion of the crack the stress intensity at the microcrack tip, K_I, is approximately proportional to $(\tau - S)$(Rice, 1979a). Thus, for wet rocks susceptible to stress corrosion during deformation at very slow rates or under very long term creep loading the required value of (τ-S) to produce microcracking will be proportional to a value of K_I that is much less than K_{Ic} and may approach that of the stress corrosion limit (see Fig.1). In short-term loading microcrack growth will only begin at larger values of (τ-S) appropriate to K_I values close or equal to K_{Ic}. Using these concepts Rice (1979a) has noted that for wet rock compressed at stress levels high enough for microcrack growth to occur the level of the stress -strain curve in the regime of microcrack growth is determined by the crack growth velocity at prevailing crack-tip K_I-levels, and is thus determined by the appropriate K_I-v curve. Rice (1979a) has also used these insights to discuss the compressive stress-strain data of Wawersik and Brace (1971) on granite and diabase.

While the foregoing examples are not in any sense exhaustive at least they do illustrate the potentially wide application that Fracture Mechanics can have in studies of the physics of earthquake generating processes. For more extensive accounts the reader is referred to Rice (1979a, b) and Rudnicki (1980). A major barrier to the application of Fracture Mechanics concepts to earthquake faulting has been, until recently, the absence of experimental data on the values of Fracture Mechanics parameters for rocks and minerals. In the next section I discuss some of the recent experimental work that has been performed in an attempt to rectify this state of affairs.

Experimental determination of Fracture Mechanics Parameters

This section is divided into four parts; each part discusses experimental methods and some results for determinations of Γ, K, J and K-v data for geological materials.

Experimental determinations of Γ

Most early workers used the Obreimov-Gilman double cantilever beam specimen geometry (Gilman, 1960) or the three-point bend test (Friedman et al., 1972). For studies of minerals or to probe the variation in Γ over inhomogeneous surfaces these techniques are inadequate as they require quite large, well machined specimens. To overcome this problem Swain and Lawn (1976) developed a technique for the rapid determination of fracture surface energy based upon an analysis of well-developed localised fracture about indentations (see also Swain and Atkinson, 1978; Atkinson and Avdis 1980). For a pointed indenter the extent of cracking, 2c at a given load P is related to Γ by (Swain and Lawn, 1976)

$$\Gamma = K_P P^2/2Ec^3 \qquad (23)$$

where $K_P = (1 - \nu^2)/\pi^3 \tan^2 \psi$, ν is Poisson's ratio, ψ is the half-angle of the indenter and E is Young's Modulus. Despite uncertainties in the value of K_P and the fact that dynamic and kinetic effects are ignored this technique is likely to gain considerable appeal due to its speed and simplicity as well as its potential for probing inhomogeneous surfaces.

A partial compilation of experimental determinations of Γ is presented in Table 1. The test method is indicated and where the experiments were on single crystals an estimate of γ, the thermodynamic surface energy of Griffith (1920), is given if available. It will be seen that even for single crystals $\Gamma \neq \gamma$. For rocks the discrepancy is probably due to microcrack formation away from the fracture plane and, perhaps, local plasticity. For single crystals in which $\Gamma > \gamma$ the difference may be due to dissipative processes such as acoustic emission and plastic deformation.

Experimental determinations of K_c

A large number of techniques have been employed to measure K_c values for rocks and minerals. To date they have concentrated exclusively on determinations of K_{Ic}, i.e. the critical stress intensity factor for tensile crack propagation. Mostly, these are the same sort of techniques as used by structural ceramic engineers. Although no one technique is entirely satisfactory the most popular ones are the double torsion (DT) geometry (e.g. Atkinson, 1979a) and the single edge notched bend (SENB) specimen (e.g. Schmidt, 1976). Schematics of the double torsion specimen and the

Table 1: Compilation of experimental determinations of fracture surface energy, Γ, for geological materials. Values are compared with theoretical estimates of the thermodynamic surface energy, γ, taken from Brace and Walsh (1962) except for the datum on calcite which is taken from Gilman (1960). Test method codes: VI, Vickers indentation; OGC, Obreimov-Gilman double cantilever beam; 3PB, three point bending.

Material	γ $J.m^{-2}$	Γ $J.m^{-2}$	Test Method	Reference
Calcite $(10\bar{1}1)$				
20°C	0.38	0.27	VI	Atkinson/ Avdis (1980)
		0.35	OGC	Santhanam/ Gupta (1968)
150°C		2.50	VI	Atkinson/ Avdis (1980)
Quartz $(10\bar{1}0)$	0.48	1.34	VI	" "
		1.03	OGC	Brace/ Walsh (1962)
Quartz (0001)				
20°C		0.46	VI	Atkinson/ Avdis (1980)
200°C		1.77	VI	" "
Olivine $Fo_{88}Fa_{12}$				
{010}	0.37	0.98	VI	Swain/ Atkinson ('78)
{001}		1.26	VI	" "
Microcline (001)		3.20	VI	Atkinson/ Avdis (1980)
(010)		5.25	VI	" "
(110)		3.74	VI	" "
Orthoclase (001)	0.2	7.77	OGC	Brace/ Walsh (1962)
Tennessee Sandstone		38.0	3PB	Friedman and others (1972)
Lueders Limestone		9-12	3PB	" "
Westerley Granite		30	VI	Swain/ Lawn (1976)

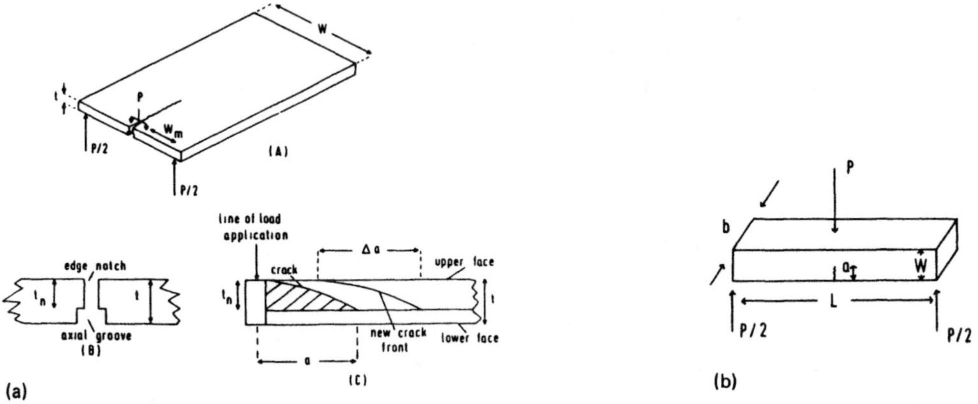

Fig. 5 (a) Schematic diagram of a double torsion specimen A. General view. B. Axial cross section C. Longitudinal cross section.

(b) Schematic diagram of a single edge notched bend specimen.

SENB specimen are shown in Fig. 5. Other specimen geometries which have a growing appeal and which may be especially suitable for experiments run on rocks under confining pressure are the radially symmetric short rod (SR) specimen (e.g. Barker, 1978) and the internally pressurised thick-walled cylinder (IPTWC) specimen (e.g. Clifton and others, 1976).

The stress intensity factor for the double torsion specimen is given by

$$K_I = PW_m \left[3(1 + \nu)/Wt^3 \, t_n \right]^{\frac{1}{2}} \qquad (24)$$

where P is the load on the specimen and the other variables are as shown in Figure 5. The value of P for catastrophic crack propagation is used to calculate K_{Ic}. Note that in this specimen the stress intensity factor is constant for a given value of P and is not a function of crack length. This property makes it ideally suited to the study of crack propagation in opaque rocks for which a determination of crack length might prove difficult. See also the section below on K_I-v determinations. Atkinson (1979a) has discussed the use of this specimen for tests on rocks.

The stress intensity factor for the SENB specimen is given by

$$K_I = Y\ (3PL/2bW^2)\ a^{\frac{1}{2}} \qquad (25)$$

where a is the initial crack length, Y is a geometrical constant and the other variables are as shown in Fig. 5. Y depends on the L/W ratio and is a function of $^a/W$. The biggest drawback of this specimen is the need to know the initial crack length. This is the crack length *immediately* prior to catastrophic crack propagation. Dalgleish and others (1980) have shown that in appropriate environments significant subcritical crack growth can occur and that this method may underestimate K_{Ic} by as much as 15%.

In Table 2 I present a partial compilation of experimental determinations of K_{Ic} for rocks and minerals. The testing method is indicated.

Experimental determinations of J_c

To date there have been very few experimental determinations of J_c for rocks. Wilkening (1978) has developed a method based on the compact tension (CT) specimen. Following the analysis of Rice and others (1973), J for this geometry is given by

$$J = 2/(Bb) \cdot \int_0^{\delta_c} F d\delta \qquad (26)$$

where B is the specimen thickness and b is the width of the unbroken ligament of the specimen, δ is the load-point displacement and δ_c the amount of displacement required before a linear load-displacement relation breaks down. For Barre granite Wilkening estimates J_{Ic} to be 50-90 $J \cdot m^{-2}$ and not 100-300 $J \cdot m^{-2}$ as estimated by other workers from Linear Elastic Fracture Mechanics - type analyses.

Weissinger and others (1980) have made simultaneous measurements of K_{Ic} and J_{Ic} on Nevada Tuff using three point bend specimens. They find that J_{Ic} varied from 29.5 $J \cdot m^{-2}$

Table 2: Compilation of experimental determination of K_{Ic}, the critical stress intensity factor for tensile crack propagation for geological materials. Test method codes: DT, double torsion; 3PB, three-point bending; CT, compact tension; IPTWC, internally pressurised thick-walled cylinder.

Material	K_{Ic} $MN.m^{-3/2}$	Test Method	Reference
Quartz			
a plane ⊥ *z*	0.852	DT	Atkinson(1979b)
a plane ⊥ *r*	1.002	DT	Atkinson(1979b)
Tennessee Sandstone	0.454	DT	Atkinson(1979a)
A porous Sandstone	0.57-1.46	IPTWC	Clifton and others (1980)
Arkansas Novaculite	1.335	DT	Atkinson(1980)
Carrara Marble	0.644	DT	Atkinson(1979a)
Falerans Micrite	1.01	3PB	Henry/Paquet (1976)
St. Pons Marble	0.70-1.39	3PB	Henry/Paquet (1976)
Westerley Granite	1.74	DT	Atkinson and Rawlings(1979b)
	2.50	CT	Schmidt/ Lutz (1978)
Chelmsford Granite	0.59-0.64	3PB	Peng/Johnson (1971)
Black Gabbro	2.884	DT	Atkinson/ Rawlings(1979a)

to 34.3 $J\cdot m^{-2}$ as specimen thickness was doubled. Schmidt and Lutz (1978) have determined J_{Ic} for Westerley Granite to be 114 $J\cdot m^{-2}$ using the compact edge-notched tension specimen.

Although Westerley Granite and Nevada Tuff both exhibit non-linear elastic behaviour K_{Ic} values calculated from J_{Ic} values using LEFM concepts show a good correlation with direct experimental determinations of K_{Ic}. This suggests that Linear Elastic Fracture Mechanics - type assumptions are justified for these materials.

Experimental determinations of K_I-v diagrams

Although static fatigue of quartz was first observed in the laboratory about ten years ago (Scholz, 1972; Martin, 1972) it is only recently that this process has been studied from the viewpoint of Fracture Mechanics (Atkinson, 1979b). Considering the wide range of earthquake and fault creep phenomena that static fatigue has been invoked to explain (Scholz, 1972; Martin, 1972; Anderson and Grew, 1977; Bonafede and others, 1976; Atkinson, 1980) it is surprising that so little work has been done in this area. Work in the writers' laboratory and elsewhere is currently underway to rectify this state of affairs and a growing body of knowledge is accumulating on K_I-v data for rocks.

Virtually all experiments run to date on rocks in order to establish K_I-v diagrams have used the double torsion specimen geometry (see Atkinson, 1979b). Wilkins (1980) has described a technique, however, that is based on time-to-failure data for stressed rock beams. Although this technique consumes large quantities of both test specimens and time it is easily capable of detecting very low crack velocities of ca. 10^{-12}m.s^{-1}. The double torsion technique, on the other hand, requires a very stable testing environment in order to get at velocities much below 10^{-10}m.s^{-1}.

The great advantage of the double torsion specimen is the speed and simplicity with which an *entire* K_I-v curve can be obtained from a *single* specimen. Provided that the initial crack length is relatively large, the crack velocity can be obtained from load relaxation data using the equation (Williams and Evans, 1973)

$$v = -\phi\, a_{i,f}\, P_{i,f}\, (1/P^2)(dP/dt) \qquad (27)$$

where a is the crack length and t is the time, and subscripts i and f denote measurements made at the beginning and at the end of a test, respectively. ϕ is a geometrical factor that takes into account the crack front shape.

For examples of studies using the double torsion specimen see Atkinson (1979b; 1980), Atkinson and Rawlings (1979a, b), and Henry and Paquet (1976). To illustrate typical

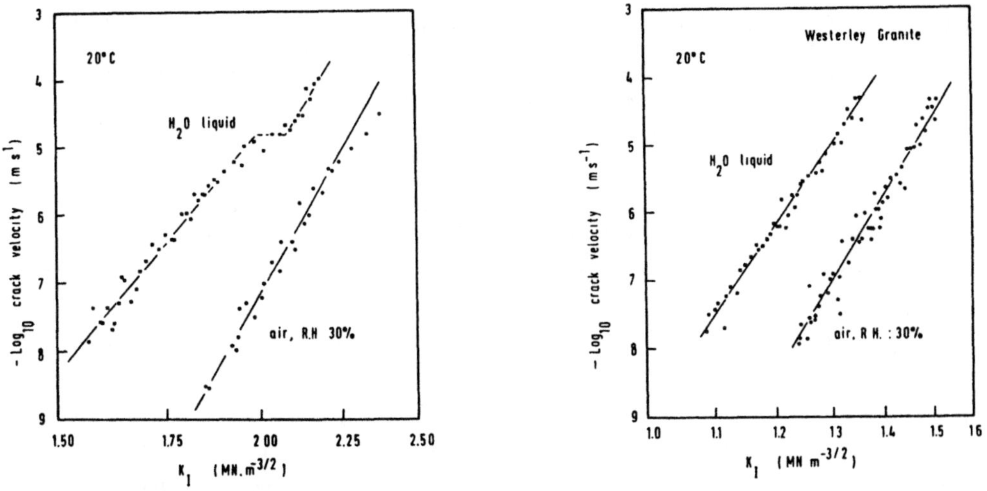

Fig. 6. Plots of $\log_{10}$ crack velocity (v) vs. $\log_{10}$ stress intensity factor (K_I) for (a) Black Gabbro and (b) Westerley Granite. Both sets of data determined with the double torsion specimen geometry.

behaviour, K_I-v curves for Westerley Granite and Black Gabbro are shown in Fig. 6. Note the trimodal behaviour of the gabbro and compare with Fig. 1. The fact that trimodal behaviour was not observed in Westerley granite does not mean that it does not occur; merely that it does not occur over the range of these experimental observations.

Most studies of K_I-v diagrams for rocks have fitted the results to equations of the same type as equation (6). A partial compilation of experimental determinations of the parameters in this equation is given in Table 3.

It is worth noting a few points about the data available so far. Firstly, even at the lowest crack velocities so far observed of ca. $10^{-12} m.s^{-1}$ (Wilkins, 1980) for a granite there is no unambiguous evidence for a stress corrosion limit in any earth material so far studied. Secondly, the activation enthalpies for crack propagation are rather low; suggesting that the rate of stress corrosion will be relatively insensitive to temperature changes. Thirdly, much more work is needed to tie down the K_I-v behaviour of rocks to the details of crack path and crack growth mechanisms.

Table 3: Compilation of K_I-v (opening mode stress intensity factor vs. crack velocity) data for geological materials. Data are described in terms of equation (6). ΔH is the activation enthalpy for crack propagation, and n is the stress corrosion index. Test method code: DT, double torsion.

Material	n	ΔH $kJ.mole^{-1}$	Conditions	Test Methods	Reference
Quartz					
a plane, ⊥ *z*	19.9		air, $20^{o}C$	DT	Atkinson (1979b)
a plane, ⊥ *r*	12	52.5	H_2O liq, 20°-$80^{o}C$	DT	Atkinson (1979b)
Arkansas Novaculite	25.1	69.5	H_2O liq, 20°-$80^{o}C$	DT	Atkinson (1980)
Falerans Micrite	26		H_2O liq, + $CaCO_3$, $20^{o}C$	DT	Henry/ Paquet(1976)
	130		air, $20^{o}C$	DT	Henry/ Paquet(1976)
St. Pons Marble	26-29		H_2O liq, + $CaCO_3$, $20^{o}C$	DT	Henry/ Paquet(1976)
Black Gabbro	32.1		air, $20^{o}C$	DT	Atkinson/ Rawlings (1979a)
	28.6		H_2O liq, $20^{o}C$	DT	Atkinson/ Rawlings (1979a)
Westerley Granite	39.1		air, $20^{o}C$	DT	Atkinson/ Rawlings (1979b)
	34.8		H_2Oliq, $20^{o}C$	DT	Atkinson/ Rawlings (1979b)

Acoustic emission and stress corrosion

Microseismic (acoustic) emission studies may prove to be an important bridge between the small-scale studies of the laboratory and the large-scale earthquake phenomena of nature. Insights gained in relating the acoustic response of rocks undergoing stress corrosion in the laboratory may prove of use in employing seismic techniques to study and monitor this process during earthquakes and fault creep.

Recently, Atkinson and Rawlings (1979a, b) have demonstrated that the rate of acoustic emission during stress corrosion cracking in Westerley granite and Black Gabbro is directly proportional to the crack growth rate. The acoustic emission event rate, dN_E/dt, and the ring-down count rate, dN_R/dt, follow equations of the form

$$dN_E/dt = \beta \, K_I^{n_E} \tag{28}$$

and

$$dN_R/dt = \Omega \, K_I^{n_R} \tag{29}$$

where β, Ω, n_E and n_R are constants.

Typical data for Westerley Granite are shown in Fig.7 which should be compared with Fig. 6.

To a good approximation

$$dN_E/da \propto K_I \tag{30}$$

and

$$dN_R/da \propto K_I \tag{31}$$

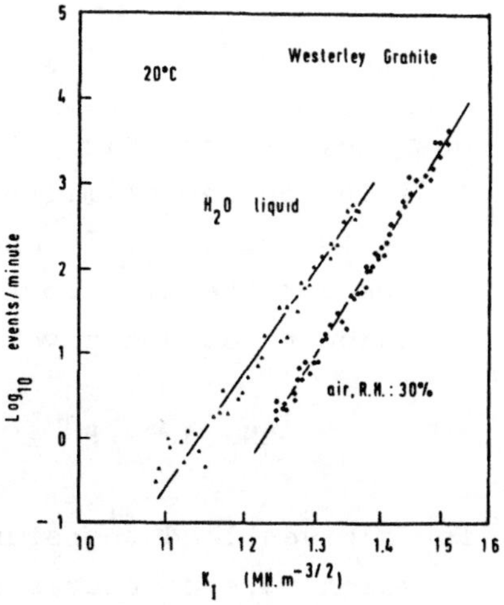

Fig. 7

Plots of log_{10} acoustic emission events per minute vs. log_{10} stress intensity factor for Westerley Granite.

This suggests that most of the acoustic emission is in response to the extension of the main crack and that the process zone around the crack tip is rather small. Otherwise dN_E/da and dN_R/da would have been more nearly proportional to K_I^2 (Evans and others, 1974). These observations have important implications for monitoring rates of stress corrosion in the earth. Significantly, acoustic emission was observed in the above experiments even at the slowest crack velocities studied, ca. $10^{-9} m.s^{-1}$.

Evans (1975) has analysed the acoustic emission to be expected from a proof-test cycle consisting of loading at a constant rate, holding for a time at the peak load and then unloading at a constant rate to zero load. The equations developed by Evans (1975) accurately predict the acoustic response given input data on the size distribution of microfracture-generating flaws and the dependence of crack velocity on the stress intensity factor. For a constant loading rate and provided that n >> m, where n is the stress corrosion index,

$$dN_E/dt \propto P^{(m-1)} \qquad (32)$$

P is the applied load and m is given by the Weibull distribution function

$$\eta = \chi \ (\sigma_i/S_o)^m \qquad (33)$$

where σ_i is the critical stress for microcrack formation, η is the number of microcracks, χ is the number of microcrack sources, and S_o and m are the shape and scale parameters for the distribution. During unloading the acoustic emission rates are given by

$$dN_E/dt \propto P^n \qquad (34)$$

For a given load (constant stress distribution) the acoustic emission rate decreases from its peak value with time according to

$$\log (dN_E/dt) \propto \left[(m - n + 2)/(n - 2)\right] \log t \qquad (35)$$

The above relations have been verified experimentally for ceramics (Evans, 1975) and basalt (Tittman, personal communication) in tension.

Earthquake foreshock rates often subside from their peak values immediately prior to the mainshock. McNally (personal communication) has noted that the decay rate of foreshocks is slower than the decay rate of aftershocks. McNally points to the close similarity between the decay rate of acoustic emissions in ceramics under the constant load portion of the proof-test cycle and the foreshock decay rates. These differ markedly from aftershock decay rates which, in turn, are similar to acoustic emission statistics of rocks under constant load. It seems reasonable, as McNally suggests, to hypothesize that the Evans (1975) equations can provide the basis of a model of acoustic emission statistics/earthquake statistics containing a flaw distribution that scales with heterogeneities in either laboratory specimens or the earth and a stress corrosion index that can be determined by experiment. Considerable scope may exist, therefore, for using laboratory studies of acoustic emission statistics as models of earthquake statistics and the subject should be given close attention.

Another area in which acoustic emission can yield important insights into earthquake phenomena is the study of amplitude distributions. Experimental studies have shown that amplitude distributions shift with increasing stress during both fracture and frictional sliding (e.g. Scholz, 1968; Weeks and others, 1978; Stesky, 1975). Recently, however, Atkinson and Rawlings (1979a, b) have shown that acoustic emission amplitude distributions for crack propagation in both granite and gabbro depend on crack propagation mechanism, and hence on stress intensity factor, crack velocity and "humidity" of the crack-tip environment. Fig. 8 shows typical results on the shifts in amplitude distribution characterised by the b-value, given by

$$n(a) = (a/a_o)^{-b} \qquad (36)$$

where n(a) is the fraction of the emission population whose peak amplitude exceeds amplitude a and a_o is the lowest detectable amplitude. Note that the b-value is higher in "wet" environments than in "dry" ones and that the rate of change of b with increase in stress intensity factor is different in

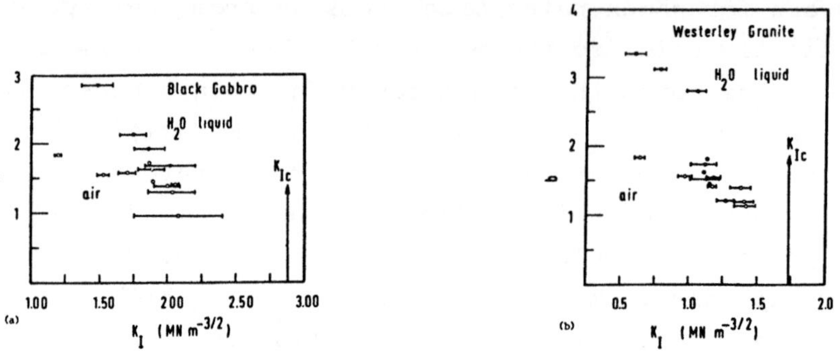

Fig. 8 Plots of the amplitude distribution exponent, b, vs. stress intensity factor (K_I) for (a) Black Gabbro, and (b) Westerley Granite.

the two environments. Note also that at high stress intensity factors close to K_{Ic} where stress corrosion is not important the two trends converge to common b-values.

Variations in magnitude-frequency relations of earthquakes are often observed in earthquake sequencies (e.g. Fiedler, 1974; Hasegawa and others, 1975; Morrison and others, 1976).

Since the log of the amplitude at the hypocenter is proportional to the earthquake magnitude, the possibility exists of using insights gained from laboratory acoustic emission studies to interpret changes in magnitude-frequency relations of earthquakes in terms of the physics of source processes.

Conclusions

Fracture Mechanics has been shown to be a powerful tool for modelling processes in earthquake source regions. This review has been limited for reasons of space to just a few areas in which this approach has been applied. Before ending, however, it is appropriate to mention a few limitations to the Fracture Mechanics approach and to highlight a few areas of important future research in experimental Fracture Mechanics.

The most notable limitations of Fracture Mechanics are in dealing with highly non-uniform stresses along fault surfaces or if stresses on an *entire* fault zone depend on relative slip. Modelling the fault zone as a series of dislocations may be a far sounder approach under these conditions (see, for example, Stuart, 1979). Furthermore, if non-linear behaviour in the vicinity of fault zones has a large volume and cannot be idealized as a crack, then some sort of inclusion model may be more appropriate (e.g. Rudnicki, 1977) in which an ellipsoidal zone of inelastically deforming material is embedded in an infinite, ideally elastic rock mass.

In the future considerable attention should be given to the determination by experiment of Fracture Mechanics parameters pertaining to mode II and III crack propagation. The absence of data for these modes clearly limits the application of Fracture Mechanics to earthquakes. Data for all three modes should be obtained for geologically realistic conditions of temperature, effective pressure and pore fluid chemistry. In view of the fact that precursory fault creep, aftershock behaviour and other time-dependent earthquake phenomena can be explained by proposing that stress corrosion is a fundamentally important mechanism in relatively shallow earthquake zones, it is vital that experimental evidence be obtained on this deformation process for crack propagation in modes II and III. It should be noted in conclusion that for a two-dimensional anti-plane (mode III) fault every element of the material deforms in pure shear. It is of considerable interest whether stress corrosion processes can be driven by shear stresses alone.

Acknowledgements

The preparation of this review was supported by the U.S. Geological Survey with contract No.: 14-08-0001-18325 under the National Earthquake Hazards Reduction Programme, and by the British Natural Environment Research Council with grant GR3/3716. Karen Cohen and Caroline Holloway skilfully helped prepare the manuscript.

References

Anderson, O.L. and Grew, P.C. (1977) Stress corrosion theory of crack propagation with application to geophysics. Rev. Geophysics. Space Physics 15, 77-104.

Atkinson, B.K. (1979a) Fracture toughness of Tennessee Sandstone and Carrara Marble using the double torsion testing method. Int. J. Rock Mech. Min. Sci. and Geomech. Abstr. 16, 49-53.

Atkinson, B.K. (1979b) A Fracture Mechanics Study of subcritical tensile cracking of quartz in wet environments. Pure and Applied Geophysics 117, 1011-1024.

Atkinson, B.K. (1980) Stress corrosion and the rate-dependent tensile failure of a fine-grained quartz rock Tectonophysics 64 (in press)

Atkinson, B.K. and Avdis, V. (1980) Fracture Mechanics parameters of some rock-forming minerals determined using an indentation technique. Int. J. Rock Mech. Min. Sci. and Geomech. Abstr. (in press).

Atkinson, B.K. and Rawlings, R.D. (1979a) Acoustic emission characteristics of subcritical and fast tensile cracking in rock. Proc. Institute of Acoustics (London) (in press)

Atkinson, B.K. and Rawlings R.D. (1979b) Acoustic emission during subcritical and fast tensile cracking of Westerley Granite and a Gabbro. EOS, Transactions American Geophysical Union, 60, 740.

Atkinson, B.K., Rutter, E.H., Sibson, R.H. and White, S.H. (1979) Some specific experimental and field studies pertaining directly to the mechanisms of seismic and aseismic faulting. Final technical report for FY1979 to US National Earthquake Hazards Reduction Program. Contract No: 14-08-0001-17662.

Barenblatt, G.I. (1962) Mathematical theory of equilibrium cracks in brittle fracture. Advances in Applied Mechanics 7, 55-129.

Barker, L.M. (1978) Short rod K_{Ic} measurements of Al_2O_3. *In* Bradt, R.C., Hasselman, D.P.H. and Lange, F.F. (eds.) Fracture Mechanics of Ceramics v3, 483-494. Plenum Press, New York.

Bonafede, M., Fazio, D., Mulargia, F. and Boschi, E. (1976) Stress corrosion theory of crack propagation applied to the earthquake mechanism. A possible basis to explain the occurrence of two or more large seismic shocks in a geologically short time interval. Bolletino di Geofisica, teorica ed applicata, XIX, 377-403.

Brace, W.F. and Walsh, J.B. (1962) Some direct measurements of the surface energy of quartz and orthoclase. Am. Mineralogist 47, 1111-1122.

Charles, R.J. (1958) Dynamic fatigue of glass. J. Applied Physics 29, 1657-1662.

Clifton, R.J., Simonson, E.R., Jones, A.H. and Green, S.J. (1976) Determination of the critical-stress-intensity factor K_{Ic} from internally pressurized thick-walled vessels. Experimental Mechanics 16, 233-238.

Dalgleish, B.J., Fakhr, A., Pratt, P.L. and Rawlings, R.D. (1980) Fracture toughness testing of ceramics and acoustic emission. J. Materials Science (in press).

Dieterich, J.H. (1978) Time-dependent friction and the mechanics of stick-slip. Pure and Applied Geophysics 116, 790-806.

Dieterich, J.H. (1979) Modelling of rock friction, part I: experimental results and constitutive equations. J. Geophysical Res. 84, 2161-2168.

Dugdale, D.S. (1960) Yielding of steel sheets containing slits. J. Mech. Phys. Solids 8, 100-104.

Evans, A.G. (1972) A method for evaluating the time-dependent failure characteristics of brittle materials - and its application to polycrystalline alumina - J. Materials Science 7, 1137-1146.

Evans, A.G. (1974) Slow crack growth in brittle materials under dynamic loading conditions. Int. J. Fracture 10, 251-259.

Evans, A.G. (1975) Residual stress measurement using acoustic emission. J. Am. Ceramic Soc. 58, 239-243.

Evans, A.G. and Johnson, H. (1975) The fracture stress and its dependence on slow crack growth. J. Materials Science 10, 214-222.

Evans, A.G., Linzer, M. and Russell, L.R. (1974) Acoustic emission and crack propagation in polycrystalline alumina. Mat. Sci. and Engineering 15, 253-261.

Fiedler, G.B. (1974) Local *b*-values related to seismicity. Tectonophysics 23, 277-282.

Friedman, M., Handin, J. and Alani, G. (1972) Fracture-surface energy of rocks. Int. J. Rock Mech. Min. Sci. 9, 757-766.

Gilman, J.J. (1960) Direct measurements of the surface energies of crystals. J. Appl. Physics 31, 2208-2218.

Griffith, A.A. (1920) The phenomena of rupture and flow in solids. Phil. Trans. Royal Soc. Lond. Ser. A. 221, 163-198.

Hasegawa, A., Hasegawa, T. and Hori, S. (1975) Premonitory variation in seismic velocity related to the southeastern Akita earthquake of 1970. J. Phys. Earth 23, 189-203.

Henry, J.P. and Paquet, J. (1976) Mécanique de la rupture de roches calcitiques. Bull. Soc. géol. France. XVIII, 1573-1582.

Husseini, M.E., Jovanovich, D.B., Randall, M.J. and Freund, L.B. (1975) The fracture energy of earthquakes. Geophys. J.R. Astr. Soc. 43, 367-385.

Ida, Y. (1973) The maximum acceleration of seismic ground motion. Bull. Seismol. Soc. Am. 63, 959-968.

Irwin, G.R. (1958) Fracture. *In* Handbuch der Physik (ed. S. Flügge) vol. 6, 55-1590, Springer-Verlag, Berlin.

Karihaloo, B.L. (1979) A note on the complexities of compression failure. Proc. Roy. Soc. Lond., Ser. A. 368, 483-493.

Kendall, K. (1978a) Complexities of compression failure. Proc. Roy. Soc. Lond. Ser. A., 361, 245-263.

Kendall, K. (1978b) The impossibility of comminuting small particles by compression. Nature, 272, 710-711.

Martin, R.J. (1972) Time-dependent crack growth in quartz and its application to creep in rocks. J. Geophys. Res. 77, 1406-1419.

Morrison, P.W., Stump, B.W. and Uhrhammer, R. (1976) The Oroville earthquake sequence of August 1975. Bull. Seism. Soc. America. 66, 1065-1084.

Palmer, A.C. and Rice, J.R. (1973) The growth of slip surfaces in the progressive failure of over-consolidated clay. Proc. Roy. Soc. Lond. Ser. A. 332, 527-548.

Paris, P.C. and Sih, G.C. (1965) Stress analysis of cracks. ASTM STP 381, American Society for Testing and Materials 30-76.

Peng, S. and Johnson, A.M. (1971) Crack growth and faulting in cylindrical specimens of Chelmsford Granite. Int. J. Rock Mech. Min. Sci. 9, 37-86.

Rice, J.R. (1968) A path independent integral and the approximate analysis of strain concentration by notches and cracks. J. Appl. Mech. 35, 379-386.

Rice, J.R. (1979a) Theory of precursory processes in the inception of earthquake rupture. Gerlands Beitr. Geophysik, Leipzig 88, 91-127.

Rice, J.R. (1979b) The mechanics of earthquake rupture. *In* Proc. Int. School of Physics "Enrico Fermi", Italian Physical Soc., Course LXXVIII (1979, Varenna on Lake Como, Italy) on Physics of the Earth's Interior edited by E. Boschi, North-Holland Publishing Co.

Rice, J.R. and Cleary, M.P. (1976) Some basic stress diffusion solutions for fluid-saturated elastic porous media with compressible constituents. Rev. Geophys. Space Phys. 14, 227-241.

Rice, J.R., and Simons, D.A. (1976) The stabilization of spreading shear faults by coupled deformation-diffusion effects in fluid-infiltrated porous materials. J. Geophys. Res. 81, 5322-5334.

Rice, J.R., Paris, P.C. and Merkle, J.G. (1973) Some further results of J-integral analysis and estimates. ASTM STP 536, American Society for Testing and Materials 231-245.

Rudnicki, J.W. (1977) The inception of faulting in a rock mass with a weakened zone. J. Geophys. Res. 82, 844-854.

Rudnicki, J.W. (1979) The stabilization of slip on a narrow weakening fault zone by coupled deformation - pore fluid diffusion. Bull. Seismol. Soc. Am. 69, 1011-1026.

Rudnicki, J.W. (1980) Fracture Mechanics applied to the earth's crust. Annual Rev. Earth and Planetary Sciences (editor F.A. Donath), Annual Reviews Inc. vol. 8 (in press)

Rutter, E.H. and Mainprice, D.H. (1978) The effect of water on stress relaxation of faulted and unfaulted sandstone. Pure and Applied Geophysics 116, 634-654.

Santhanam, A.T. and Gupta, Y.P. (1968) Cleavage surface energy of calcite. Int. J. Rock Mech. Min. Sci. 5, 253-259.

Schmidt, R.A. (1976) Fracture toughness testing of limestone. Experimental Mechanics 16, 161-167.

Schmidt, R.A. and Lutz, T.J. (1978) K_{Ic} and J_{Ic} of Westerley Granite - Effects of thickness and in-plane dimensions. 11th National Symp. Fracture Mechanics, Blacksburg, Virginia, 1978. To appear as future ASTM STP.

Scholz, C.H. (1968) Mechanism of creep in brittle rock. J. Geophys. Res. 73, 3295-3302.

Scholz, C.H. (1972) Static fatigue of quartz. J. Geophys. Res. 77, 2104-2114.

Stuart, W.D. (1979) Strain softening prior to two-dimensional strike-slip earthquakes. J. Geophys. Res. 84, 1063-1070.

Stesky, R.M. (1975) Acoustic emission during high-temperature frictional sliding. Pure and Applied Geophysics 113, 31-43.

Swain, M.V. and Atkinson, B.K. (1978) Fracture surface energy of olivine. Pure and Applied Geophysics 116, 866-872.

Swain, M.V. and Lawn, B.R. (1976) Indentation fracture in brittle rocks and glasses. Int. J. Rock Mech. Min. Sci. and Geomech. Abstr. 13, 311-319.

Wawersik, W.R. and Brace, W.F. (1971) Post-failure behaviour of granite and diabase. Rock Mech. 3, 61-85.

Weeks, J., Lockner, D. and Byerlee, J. (1978) Change in *b*-values during movement on cut surfaces in granite. Bull. Seismol. Soc. Am. 68, 333-341.

Weisinger, R., Costin, L.S. and Lutz, T.J. (1980) K_{Ic} and J-resistance-curve measurements on Nevada Tuff. Experimental Mechanics 20, 68-72.

Wilkening, W.W. (1978) J-integral measurement in geological materials. *In* Proc. 19th U.S. Symp. Rock Mechanics, Stateline, Nevada, 1978, Ed. Y.S. Kim, 254-258.

Wilkins, B.J.S. (1980) Slow crack growth and delayed failure of granite. Int. J. Rock Mech. Min. Sci. and Geomech. Abstr. (in press).

Williams, D.P. and Evans, A.G. (1973) A simple method for studying slow crack growth. J. Testing and Evaluation 1, 264-270.

Analysis of Stress and Strain Around a Transform Fault

S. M. Nakiboglu

Department of Surveying, University of Queensland, St. Lucia Q, 4067 Australia

A. M. Isikara

Department of Geophysics, Faculty of Earth Sciences, Istanbul University, Istanbul, Turkey

Abstract

Postseismic rebound and strain accumulation around a strike-slip fault is modelled as a screw dislocation within an elastic-viscoelastic medium. The stress and strain fields near the fault are obtained. The effects of various parameters such as depth of the locked section, thickness of the elastic lithosphere and the depth dependence of dislocations are investigated with conclusions that the elastically behaving lithosphere should be rather thin in order to confine displacements to the fault vicinity. The depth of the locked section together with asthenosphere viscosity are the governing factors of stress build-up and earthquake frequency.

1. Introduction

The North Anatolian fault is essentially a transform fault marking the boundaries of Anatolian and Black Sea tectonic plates. From the kinematical point of view the relative motion of the two lithospheric plates is a result of the push applied by the Arabian and African plates (McKenzie, 1972). However dynamical investigations of plate motions and slip phenomenon along transform faults are complicated by the yet unknown convective forces, and only qualitative studies of this nature seem to be possible at the present (for example see Richardson *et al.*, 1976 for a global investigation; Bird, 1978 for a treatment of intracontinental subduction; and Turcotte & Spence, 1974 for an analysis of a transform fault). In the present work we confine ourselves to the slip phenomenon and avoid the dynamical problem of plate motions. This way, we can investigate the slip, post-slip rebound and stress accumulation along a transform fault without going into the detailed modelling of convective forces acting underneath the lithospheric plates.

The transform fault zones are characterized by shallow earthquakes of mainly strike-slip type indicating that the elastic, brittle sections of the two adjoining lithospheric plates are locked down to a shallow depth whereas the plates are free to slide at greater depths in a ductile and non-catastrophic manner. The stresses gradually build up in the locked section due to the two complementing factors, namely the continuous slide in the ductile region and the transfer of stresses from lower sections to the brittle part as a result of the creep in the former one. This process continues until the failure of the brittle part marked by a major earthquake which then reloads the lower parts of the lithosphere and the cycle repeats itself. This scenario is obviously in agreement with the plate tectonics ideas as well as with the elastic rebound hypothesis, and it is by no means a new idea (see Turcotte & Spence, 1974; Budiansky & Amazigo, 1976, Spence & Turcotte, 1979; Turcotte *et al.*, 1979).

The slip processes in transform faults can be modelled in two phases using the theory of screw dislocations. In the post-slip period the dislocations are continuous in time and they are confined into the lower lithosphere. On the other hand, the coseismic events can be visualized as sudden individual dislocations in the brittle part. The postseismic deformation has two components in them; first is the ongoing gradual slip in the lower lithosphere and the second one is the creep response of the tightly coupled asthenosphere. Assuming that the pre and postseismic deformations involve small strains we can use the linear theory and the principle of superposition.

In the present work we critically examine the problem in order to assess the contributions of various factors such as the rheological modelling of the lithosphere and the coupling effects of the weak asthenosphere rather than interpreting the available observations on the basis of a comparison with a preferred theoretical model. We will assume the asthenosphere to be a linear viscoelastic (Maxwellian) half space which is most probably a first approximation for the power-law creep behaviour of the layer.

Turcotte & Spence (1974) treat the problem of strain accumulation on a transform fault by modelling the lithosphere as an elastic-perfectly plastic plate with free surfaces. Therefore they neglect the contribution of the asthenosphere into the strain accumulation as well as the contribution of the lower lithosphere which probably deviates from an idealized plastic body under intraplate stresses in short periods of time. In more recent works

(Turcotte, 1977; Spence & Turcotte, 1979; Turcotte *et al.*, 1979) they consider the contribution of the viscoelastic asthenosphere into the stress accumulation and indicate that the asthenosphere contributes significantly in damping out the time dependence at large distances.

Savage (1975) indicates that the observed postseismic deformation is much larger than either the preseismic or the coseismic one, and proposes a modification to the earlier Turcotte-Spence Model such that the strain accumulation is concentrated within a narrower zone along the fault. This modification is certainly compatible with the existence of a thin shear zone of higher temperature and lower viscosity along the transform faults as indicated by Yuen *et al.*, (1978). However a part of the postseismic deformation is due to the creep response of the asthenosphere, and it seems to be a worthwhile effort to investigate this effect.

Nur and Marko (1974) use a model of elastic layer overlying a viscoelastic half-space, and investigate postseismic rebound with the view of estimating the effective viscosity of the asthenosphere. They compare the observations with the viscoelastic creep response of the medium to a dislocation whose time-dependence is characterized by Heaviside's unit step function. However, as was pointed out earlier, the observations pertaining to postseismic response have other components besides the viscoelastic rebound. Budiansky & Amazigo (1976) treat the problem of cyclic stress accumulation along a transform fault in a homogeneous half space and estimate the effective lithospheric viscosity in California. The viscosity estimates on the basis of investigations of transform faults yield effective viscosity values whose physical interpretation is complicated by the existence of shear zones. Also, these shear zones differ from one transform fault to another depending on the rheology of the neighbouring plates, ambient temperatures, relative plate velocities and time (Yuen, *et al.*, 1978).

2. Theory

The problem of screw dislocations in an elastic layer overlying a viscoelastic half space have been treated in geophysical literature by Nur and Mavko (1974), Barker (1976), Rundle and Jackson (1977a,b), Rundle (1978). The development of Rundle and Jackson (1977a) is based on an approximate displacement Green function for a point source, and it is most useful in treating non-planar transform faults with finite dimensions. The convolution of the point-source Green's function with a given dislocation field yields the displacements associated with it. The transform faults are generally of large horizontal extend in comparison to the depth of the locked section, and their surface traces tend to have small curvature. Therefore we will suffice with the treatment of the problem of long rectilinear transform faults. However, detailed analysis of observations pertaining to individual transform faults requires the aforementioned refinement of the model.

The problem of screw dislocation in an elastic layer overlying a viscoelastic half space can be treated more simply by making use of the correspondence principle. Once the elastic problem is solved the corresponding problem for the viscoelastic case is obtained in a systematic manner by virtue of the principle (Bland, 1960). We will summarize the principle before proceeding with the mechanical formulation of the present problem.

The stress-strain relations for a homogeneous isotropic viscoelastic medium initially in a state free of deviatoric stresses are

$$\dot{\sigma}_{ij} + \frac{\mu}{\eta}(\sigma_{ij} - 1/3\, \sigma_{kk}\, \delta_{ij}) = 2\mu\dot{e}_{ij} + \lambda\, \dot{e}_{kk}\, \delta_{ij}, \qquad \text{.....(1)}$$

where λ,μ and η are the incompressibility, rigidity and the viscosity of the medium, σ_{ij} and e_{ij} are respectively the stress and strain tensors, δ_{ij} is Kronecker's delta, and ($\cdot$) indicates differentiation with respect to time.

Taking the Laplace transform of equation 1 and observing the initial conditions we obtain

$$\hat{\sigma}_{ij} = \hat{\lambda}\, \hat{e}_{kk}\, \delta_{ij} + 2\hat{\mu}\, \hat{e}_{ij}, \qquad \text{.....(2)}$$

where $\hat{\sigma}_{ij}$ and $\hat{e}_{ij}$ are the Laplace transforms of stress and strain tensors,

$$\hat{\lambda} = \frac{\lambda s + T^{-1}(\lambda + 2\mu/3)}{s + T^{-1}}, \qquad \text{.....}(3^a)$$

$$\hat{\mu} = \frac{\mu s}{s + T^{-1}}, \qquad \text{.....}(3^b)$$

s is the Laplace transform variable with dimension of frequency, and $T = \eta/\mu$ is the relaxation time.

Equations 2 are identical in form to the corresponding equations for an elastic body in which the Lame's constants are replaced by the differential operators given in equation $3^{a,b}$, and the stress and strain are replaced by their Laplace transforms. Since the equilibrium equations for quasi-static viscoelastic problem present the same analogy, the viscoelastic problem can be viewed as an elastic one in transform domain. Hence the solution of the elastic problem yields the Laplace transform of the solution for the viscoelastic case provided that they satisfy the Laplace transforms of the boundary conditions. First we shall treat the problem of screw dislocations in an elastic layer overlying an elastic half-space and then obtain the solution for the viscoelastic half-space using the principle outlined above.

The deformations associated with relative plate motions along a transform fault during the preseismic phase can be considered as being due to a continuous noncatastrophic slip in the lower part of the lithosphere which is locked at the upper part down to a depth marked by major seismic activity (figure 1).

Assuming a dislocation of $\delta(X_3)$ in X_2- plane along X_1-direction, the resulting displacement field at an arbitrary point $(X_1X_2X_3)$ can be determined as (Rybicki, 1971)

$$u_1(X_1X_2X_3) = \int_{\zeta_3=d}^{h} \delta(\zeta_3)\mu \left.\frac{\partial G(X_i, \zeta_i)}{\partial \zeta_2}\right|_{\zeta_2=0} \int d\zeta_3, \qquad .....(4)$$

$$u_2 = u_3 = 0.$$

Here u_i are the components of the displacement vector,

ζ_i are the coordinates of the moving point, and

$G(X_i,\zeta_i)$ is the Green's function.

Rybicki (1971) also gives the Green's function for the present case in which the dislocation is confined into a plane perpendicular to the free boundary. With the present notation, we write

$$G(X_i\ \zeta_i) = -\frac{1}{8\pi\mu_1}\sum_{m=0}^{\infty}\sum_{i=1}^{2}\sum_{j=1}^{2}(2-\delta om)\gamma^m \ln\{(\zeta_2-X_2)^2$$

$$+\left[\zeta_3 + (-1)^i\, 2mh + (-1)^{i+j}X_3\right]^2\} \text{for } X_3,\zeta_3 \lesssim h. \qquad .....(5)$$

Here $\gamma = (\mu_1-\mu_2)/(\mu_1+\mu_2)$, and

μ_1,μ_2 are the rigidities of the two layers (figure 1).

It is worthwhile to note that equation 4 is valid under the assumption of stresses being continuous across the dislocation area.

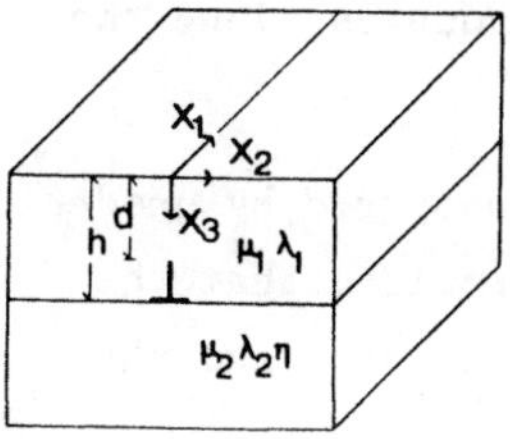

Figure 1:
A mechanical model of transform faults. The pre- and post-slip dislocation region is indicated as

The only two nonvanishing components of the stress tensor follow from the Hooke's law as

$$\sigma_{21} = \mu_1 \frac{\partial u_1}{\partial X_2} , \qquad .....(6)$$

$$\sigma_{31} = \mu_1 \frac{\partial u_1}{\partial X_3} \text{ for } X_3<h, \text{ with the strain components}$$

$$e_{21} = \frac{1}{2} \frac{\partial u_1}{\partial X_2} \text{ and } e_{31} = \frac{1}{2} \frac{\partial u_1}{\partial X_3} . \qquad .....(7)$$

3. Influence of mechanical model

The rheology of the lithospheric plate strongly depends on, among various other factors, the temperature and duration as well as magnitude of stresses. It is evident from the observations pertaining to the seamount loading of oceanic crust that a portion of the earth's crust behaves elastically under considerable shear stresses on geological time scales (Lambeck & Nakiboglu, 1980). Laboratory experiments indicate that the flow is a dominant mechanism at temperatures exceeding roughly one half the melting point temperature. If we assume the extrapolation to geological settings to be valid this would indicate approximate depths of 20 and 35 kms for oceanic and continental crusts respectively. Considering that the elastic lithosphere is marked by an isoterm the elastic thickness is variable, and it might be even smaller in the shear zones. Similar values for the elastic thickness of the lithosphere are proposed by Turcotte *et al.*, (1979). It is necessary to know the depth up to which the plate can be assumed to behave elastically at plate margins for the present problem. This depth together with the depth to which the fault is locked should be the determining factors on the magnitude as well as areal extend of the deformations on the surface for a given dislocation model. Since the frictional resistance changes with temperature and depth, the dislocations on the fault plane should be depth-dependant as well. Therefore a model with uniform dislocations along the depth may not be a good approximation for the real situation. Since the stress and strain fields depend on the thickness of elastic layer and the depth of the locked section, these parameters are also the governing factors, along with the relative plate velocity, of the frequency and the energy of major earthquakes along the fault.

The resulting deformation field on the surface as observed by geodetic methods should give an indication of these parameters provided that the geodetic data is supplemented by heat flow and earthquake data, and the

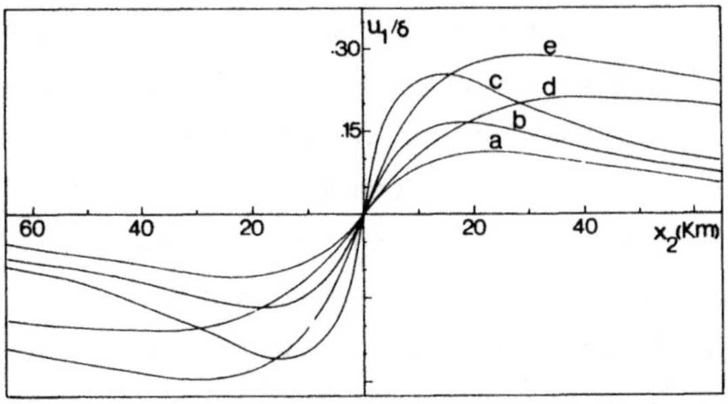

Figure 2: Dimensionless surface deformations for various values of d and h; (a) d = 15, h = 35, (b) d = 10, h = 35, (c) d = 5, h = 35, (d) d = 20, h = 100, (e) d = 10, h = 100. All values are in kms.

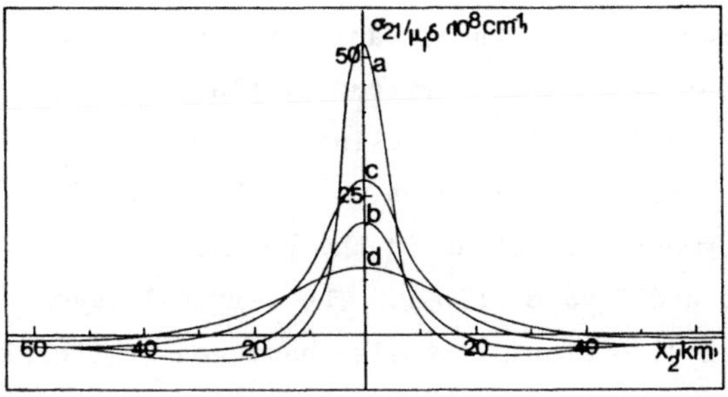

Figure 3: Variation of $\sigma_{21}|_{X_3=0}$ as a function of distance to the fault trace; (a) d = 5, h = 35, (b) d = 10, h = 35, (c) d = 10, h = 100, (d) d = 20, h = 100, all in kms.

observations pertaining to relative plate motions as monitored in the inactive mid-plate regions.

The effects of the locked section d and the elastic thickness h are illustrated in figures 2 and 3. The computations are done for the following simple dislocation model

$$\Delta u = \begin{cases} \delta\ H(t) & d \leq X_3 \leq h \\ 0 & 0 \leq X_3 < d \end{cases} \qquad .....(8)$$

Here

$$H(t) = \begin{cases} 0 & t<0 \\ 1 & t<0 \\ \tfrac{1}{2} & t=0 \end{cases} \quad \text{is the Heaviside's unit step function.}$$

It is sufficient, for the present comperative purposes, to investigate the influence of d and h at $t = 0^+$. Figure 2 shows the dimensionless surface deformation corresponding to various values of d and h. The thickness of the elastic layer determines the areal extend of deformations for a given d. As anticipated, the decrease in the depth of the locked section increases the surface deformations. The distribution of σ_{21} at $X_3=0$ reveals a similar behaviour for the stresses. Increase in the thickness of the elastic layer increases the wavelength whereas d seems to be the determining factor for stress magnitudes on the fault plane. A 100% increase in d causes a drop of about 100% in σ_{21}. Therefore estimates of viscosity and frequency of major earthquakes strongly depend on the value of d.

Obviously one single lithospheric model can not be applicable to every transform fault because of regional variations. The more readily noted differences between the general properties of the oceanic and continental crust should have a significant bearing on the behaviour of plates along the margins. The symmetry in figures 2 and 3 will be lacking if the two adjoining plates are of oceanic and continental types as in the case of San Andreas Fault. This lack of symmetry is clear in the geodetic observations presented in Prescott and Savage (1976). The vertical layering and rigidity contrast in horizontal direction might also have non-negligible effects in some cases (McHugh & Johnston, 1977).

Another important parameter in the mechanical modelling is the dislocation function u and its time dependency. A uniform dislocation of the type given in equation 8, and widely used in geophysical literature, is not a physically plausible one. This is partly due to the singularity at the tip of the dislocation. Also, a realistic dislocation model should allow for change with depth because the frictional stresses on plate boundary vary with temperature and pressure. The time dependence of the model is important if we try to account for the contributions of the creep behaviour in the asthenosphere.

A more realistic dislocation model could be the one with a parabolic variation i.e.

$$\delta^* = \begin{cases} \delta \left(\frac{X_3-d}{h-d}\right)^2 & d \leqslant X_3 \leqslant h \\ 0 & 0 \leqslant X_3 \leqslant d \end{cases} \qquad \text{.....(9)}$$

Here δ is the relative displacement at the bottom of the elastic part. The elastic displacements on the free surface and the stresses for the two

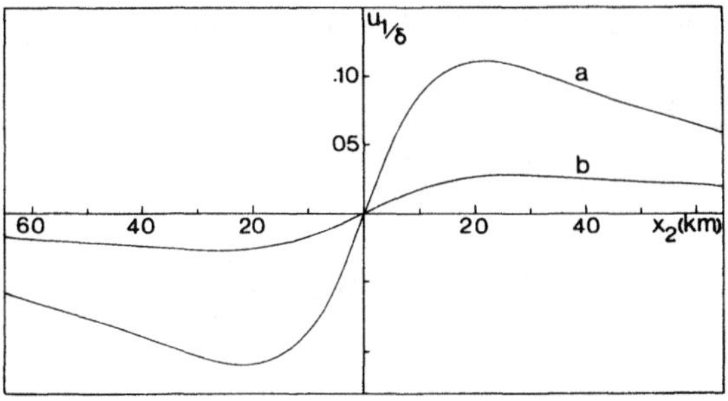

Figure 4: Elastic surface displacements for (a) uniform, and (b) parabolic dislocation models for d = 15 km, h = 35 km.

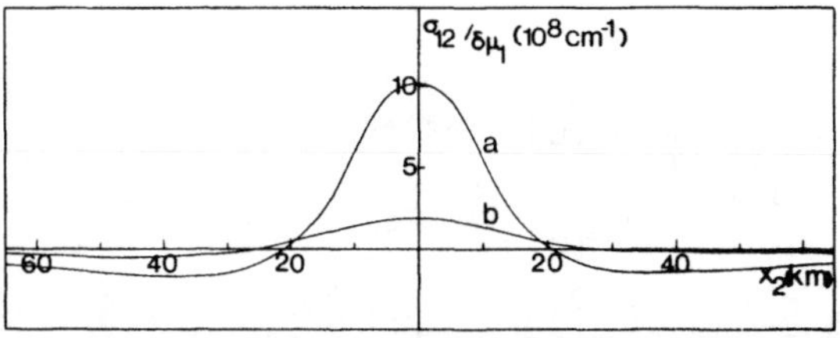

Figure 5: Elastic stresses $\sigma_{21}|_{X_3=0}$ for (a) uniform, and (b) parabolic dislocation models for d = 15 km, h = 35 km.

dislocation models (equations 8 and 9) are compared in figures 4 and 5. With the parabolic dislocation, the stresses on the fault plane are reduced by 4 folds and the maximum displacement becomes 5 times smaller. This comparison of the two extreme cases clearly indicates the importance of the dislocation model.

4. Preslip Phase: Strain Accumulation

After adding a note of caution in the previous section on the proper choice of the mechanical model we proceed with the investigation of deformations associated with the tectonic plate motions along transform faults. The creep behaviour and its effects on the surface displacements can best be illustrated by using the model

$$\Delta u = \delta^* H(t) \quad d \leq X_3 \leq h \qquad .....(10)$$

This model enables us to see the effects of time dependent deformation as we have isolated and left out the contribution of continuous plate motion.

Combining equations 4, 5 and 6 yields the solution of the elastic problem as

$$u_{1_e} = \frac{\delta}{4\pi(h-d)^2} \sum_{m=0}^{\infty} (2 - \delta om)\gamma^m \sum_{i=1}^{2} \sum_{j=1}^{2} f_{mij}\ (X_2, X_3), \qquad .....(11)$$

where

$$f_{mij} = (h-d)X_2 - a_{mij}\ X_2 \ln \frac{X_2{}^2 + b_{mij}^2}{X_2{}^2 + a_{mij}^2} + (a_{mij}^2 - X_2{}^2)(\tan^{-1}\frac{X_2}{a_{mij}} - \tan^{-1}\frac{X_2}{b_{mij}}),$$

$$a_{mij} = d + (-1)^i\ 2\ mh + (-1)^{i+j}\ X_3, \text{ and}$$

$$b_{mij} = h + (-1)^i\ 2\ mh + (-1)^{i+j}\ X_3.$$

The elastic stresses follow from equation 6 as

$$\begin{matrix}\sigma_{21} \\ \\ \sigma_{31}\end{matrix} = \frac{\delta}{4\pi(h-d)^2}\mu_1 \sum_{m=0}^{\infty} (2-\delta om)\gamma^m \sum_{i=1}^{2} \sum_{j=1}^{2} \begin{cases} g_{mij}^1\ (X_2, X_3) \\ g_{mij}^2\ (X_2, X_3). \end{cases} \qquad .....(12)$$

Here

$$g_{mij}^1 = b_{mij} - a_{mij} \ln \frac{X_2{}^2 + b_{mij}^2}{X_2{}^2 + a_{mij}} - 2X_2(\tan^{-1}\frac{X_2}{a_{mij}} - \tan^{-1}\frac{X_2}{b_{mij}}) - \frac{a_{mij}^2\ b_{mij} + X_2^2(d - h + a_{mij})}{X_2{}^2 + b_{mij}^2}, \text{ and}$$

$$g_{mij}^2 = (-1)^{i+j+1}\ X_2 \ln \frac{X_2{}^2 + b_{mij}^2}{X_2{}^2 + a_{mij}^2} + 2(-1)^{i+j}\ a_{mij}(\tan^{-1}\frac{X_2}{a_{mij}} - \tan^{-1}\frac{X_2}{b_{mij}}) + \frac{(-1)^{i+j}\ X_2\ a_{mij}\ (d - h - b_{mij})}{X_2{}^2 + b_{mij}^2} + (-1)^{i+j}\ X_2.$$

The solution of the corresponding problem in which the underlying half-space is viscoelastic can be obtained using the correspondence principle as outlined in the previous section.

The Laplace transforms of the relevant quantities are

$$\Delta\hat{u} = \delta/s,$$

$$\hat{\gamma}^m = \gamma^m \{1 + \frac{1-\gamma^2}{2T\gamma}\ (s + \frac{1+\gamma}{2T})^{-1}\}^m.$$

Substituting these into equations 11 and 12, and taking the inverse Laplace transforms results in

$$u_1 = \frac{\delta}{4\pi(h-d)^2} \sum_{m=0}^{\infty} (2-\delta om)\tilde{\gamma}^m \sum_{i=1}^{2} \sum_{j=1}^{2} f_{mij} (X_2 X_3), \qquad .....(13)$$

$$\begin{matrix} \sigma_{21} \\ \sigma_{31} \end{matrix} = \frac{\delta\mu_1}{4\pi(h-d)^2} \sum_{m=0}^{\infty} (2-\delta om)\tilde{\gamma}^m \sum_{i=1}^{2} \sum_{j=1}^{2} \left\{ \begin{matrix} g^1_{mij} (X_2, X_3) \\ g^2_{mij} (X_2, X_3), \end{matrix} \right.$$

where

$$\tilde{\gamma}^m = \gamma^m \left\{ 1 + \sum_{r=1}^{m} \frac{m!}{r!(m-r)!} (\gamma^{-1}-1)^r \left[1 - e^{-\frac{1+\gamma}{\gamma T}t} \sum_{j=0}^{r-1} \frac{(1+j)^j}{2^j j!} \left(\frac{t}{T}\right)^j \right] \right\}.$$

The figures $6^{a,b}$ show the displacements and stresses resulting from the present model at various epochs t/T. The crustal model is selected to be d = 15 kms, h = 35 kms, γ = -0.27. The viscoelastic creep contributes significantly into the deformation as well as into the stress accumulation as is observed in figures $6^{a,b}$. This indicates the necessity of including the stress transfer by creep from the lower layers into the locked section.

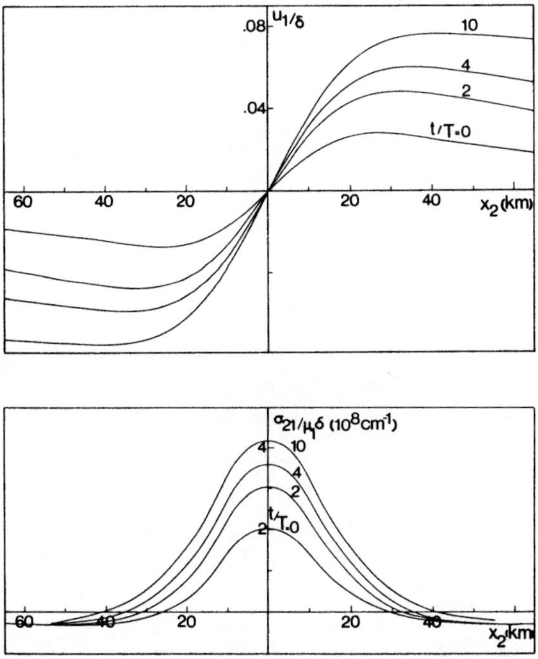

Figure 6: Accumulation of (a) surface deformation and (b) stress on the locked section of the fault due to the viscoelastic response of the substratum. The model is d = 15 km, h = 35 km, γ = -0.27.

The continual dislocation due to the plate motion will magnify this strain accumulation and the resultant displacements, the final ones being equal to the sum of the two superposed effects. Improved quality of geodetic data may provide enough resolution in the near future to enable one to isolate the viscoelastic response with exponential trend leading to a dependable estimate of viscosity.

A dislocation model probably more compatible with the actual plate motions is

$$\Delta u = \dot{\delta} t \left(\frac{X_3 - d}{h-d}\right) H(t), \quad d \leqslant X_3 \leqslant h, \qquad \text{.....(14)}$$

where $\dot{\delta}$ is the relative velocity of the neighbouring plates along the transform fault. The resulting pre-slip displacements and the stresses are as given in equation 13 in which $\tilde{\gamma}$ is replaced by the following expression.

$$\tilde{\gamma}^m = \gamma^m T \left\{ \frac{t}{T} + \sum_{r=1}^{m} \frac{m!}{r!(m-r)!} (\gamma^{-1}-1)^r \left[\frac{t}{T} - \frac{2}{1+\gamma} + \frac{2}{1+\gamma} e^{-\frac{1+\gamma}{2T} t} \left[r + \sum_{j=1}^{r-1} \sum_{k=1}^{j} \frac{(1+\gamma)^k}{2^k \, k!} \left(\frac{t}{T}\right)^k \right] \right] \right\}. \qquad \text{.....(15)}$$

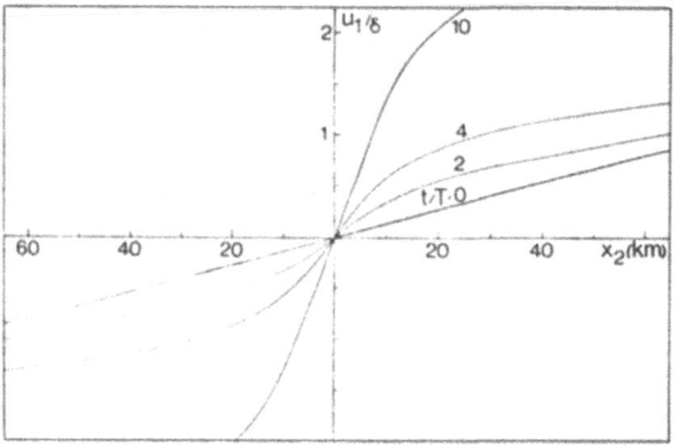

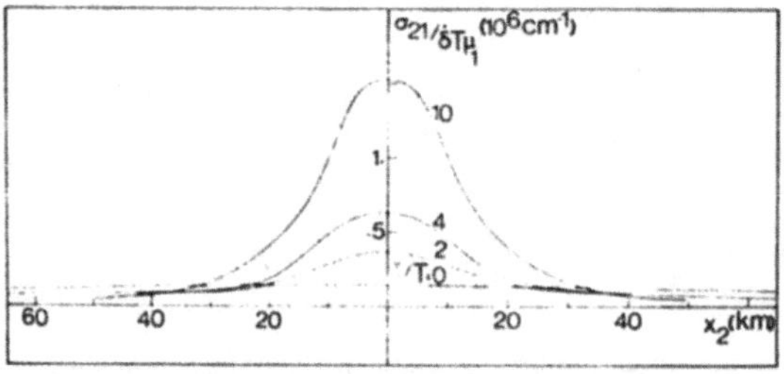

Figure 7: Accumulation of (a) deformation, and (b) stress due to creep and continuous plate motion. The model is same as in figure 6.

The accumulation of stress and strain for this case is shown in figures 7[a,b] for a model with d = 15 kms and h = 35 kms. This model is not intended as a representation of an actual transform fault, rather, it is for the purpose of illustrating the interaction between creep and the continuous plate motion. The strain accumulates at a much faster rate, and it seems possible to break the result into linear and exponential components yielding plate velocity and relaxation time. The displacements extend to larger distances for parabolic dislocation in comparison to the ones corresponding to the uniform dislocation.

5. Co-slip phase:

An appropriate model of co-slip displacements and associated stress relaxation can be obtained by introducing episodic screw dislocations in the locked section of the lithosphere, i.e. $\Delta u = \delta H(t-t_o)$ for $0 \leqslant X_3 \leqslant d$. The shear stresses reach to a critical value on the locked section at $t=t_o$ which result in failure and sudden stress relaxation.

The relevant expression for the stresses and the displacements can be obtained following the procedure used in the previous section. For example, assuming $t_o = 0$, the displacements become

$$u_1 = \frac{\delta}{4\pi} \sum_{m=0}^{\infty} (2-\delta om)\tilde{\gamma}^m \sum_{i=1}^{2} \sum_{j=1}^{2} \left(\tan^{-1} \frac{X_2}{a_{mij} - d} - \tan^{-1} \frac{X_2}{a_{mij}}\right) \quad .....(16)$$

The dimensionless co-slip displacements and subsequent viscoelastic deformations are given in Figure 8.

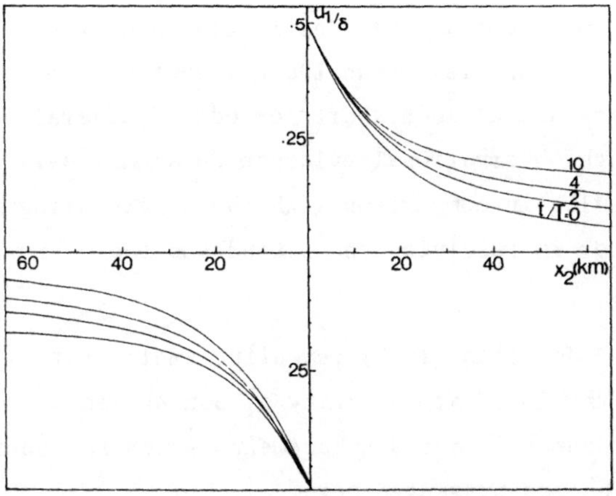

Figure 8: Co-slip displacements and subsequent viscoelastic deformations. The model is as in figure 6.

The post-slip behaviour of the fault can then be obtained by superimposing the creep and relaxation components as given in equations (13)&(16)

6. Comments

Geodetic observations supplemented with seismic data pertaining to focal depths of earthquakes and velocity profiles should yield valuable information about the rheology of the tectonic plate margins in the short time periods of $10\sim10^2$yr. The reliable estimate of the rheological parameters, in turn, can be used in a straightforward manner in determining the earthquake frequency along a transform fault. However the present state of investigations is far from being complete due to the neglected complexities of the problem. Specific investigations of a given transform fault should also consider the curvature of the fault trace and variable dislocations changing with depth as well as with the curvature. A possible dependency of the depth of the locked section on the stresses as indicated by Turcotte *et al.*, (1979) needs further investigations. One way of resolving this point is perhaps the analysis of seismic activity and focal depths around the fault zone during the strain accumulation phase. Another important point is the question of the level of reduction of stresses after a large earthquake. This point has a strong bearing on the postseismic behaviour. The ambient tectonic stresses are probably of the order of 10^2bars whereas the stress drops by major earthquakes along plate margins lie between 10 to 70 bars.

The observations along San Andreas Fault indicating that the strain is concentrated within a narrow fault zone, although it might be partly due to the formation of shear zones, indicate that the elastic response of the plate margins is confined into depths much less than 100 kms even at the prevailing stress levels of 70 bars and at such short periods of several years. At present there is not much observational evidence on anomalously high heat flow around transform faults in comparison with the neighbouring regions. This is an important factor in resolving the extend and the characteristics of the shear zones.

The observations of McHugh and Johnston (1978) revealing small tilt changes accompanying the creep in the immediate vicinity of San Andreas Fault indicates a small thrust component in the displacements which has been neglected in analyses similar to the one presented here.

Acknowledgement

Invaluable discussions with Kurt Lambeck are gratefully acknowledged.

References

Barker, T.G., 1976. Quasi-static motions near the San Andreas Fault Zone, *Geophys. J.R. astr. Soc.*, 45, 689-705.

Bird, P., 1978. Initiation of intracontinental subduction in the Himalaya, *J. Geophys. Res.*, 83, 4975-4987.

Bland, D.R., 1960. *The Theory of Linear Viscoelasticity*, Pergamon Press, N.Y.

Budiansky, B., Amazigo, J.C., 1976. Interaction of fault slip and lithospheric creep, *J. Geophys. Res.*, 81, 4897-4900.

Lambeck, K., Nakiboglu, S.M., 1980. Seamount loading and stresses in the ocean lithosphere, *J. Geophys. Res.*, in press.

McHugh, S., Johnston, M., 1977. Surface shear stress, strain and shear displacement for screw dislocations in a vertical slap with shear modulus contrast, *Geophys. J.R. astr. Soc.*, 49, 715-722.

McHugh, S., Johnston, M.J.S., 1978. Dislocation modelling of creep-related tilt changes, *Bull Seism. Soc. America*, 68, 155-168.

McKenzie, D., 1972. Active tectonics of the Mediterranean Region, *Geophys. J.R. astr. Soc.*, 30, 109-185.

Nur, A., Mavko, G., 1974. Postseismic viscoelastic rebound, *Science*, 183, 204-206.

Prescott, W.H., Savage, J.C., 1976. Strain accumulation on the San Andreas Fault near Palmdale, California, *J. Geophys. Res.*, 81, 4901-4908.

Richardson, R.M., Solomon, S.C. and Sleep, N.H., 1976. Intraplate stress as an indicator of plate tectonic driving forces, *J. Geophys. Res.*, 81. 1847-1856.

Rundle, J.B. and Jackson, D.D., 1977[a]. A three-dimensional viscoelastic model of a strike slip fault, *Geophys. J.R. astr. Soc.*, 49, 575-591.

Rundle, J.B., Jackson, D.D., 1977[b]. A viscoelastic relaxation model for post-seismic deformation from the San Francisco Earthquake of 1906, *Pure and Applied Geophys.*, 115, 401-411.

Rundle, J.B. 1978. Viscoelastic crustal deformation by finite quasi-static sources, *J. Geophys. Res.*, 83, 5937-5945.

Rybicki, K., 1971. The elastic residual field of a very long strike-slip fault in the presence of a discontinuity, *Bull. Seism. Soc. Am.*, 61, 79-92.

Savage, J.C., 1975. Comment on 'An Analysis of strain accumulation on a strike slip fault' by D.L. Turcotte and D.A. Spence, *J. Geophys. Res.*, 80, 4111-4115.

Spence, D.A. and Turcotte, D.L., 1979. Viscoelastic relaxation of cyclic displacements on the San Andreas Fault, *Proc. R. Soc. Land. series A*, 365, 121-144.

Turcotte, D.L., Spence, D.A., 1974. An analysis of strain accumulation on a strike slip fault, *J. Geophys. Res.*, 79, 4407-4412.

Turcotte, D.L., 1977. Stress accumulation and release on the San Andreas Fault, *Pure and Applied Geophys.*, 115, 413-427.

Turcotte, D.L., Clancy, R.T., Spence, D.A. and Kulhawy, F.H., 1979. Mechanisms for the accumulation and release of stress on the San Andreas Fault, *J. Geophys. Res.*, 84, 2273-2282.

Yuen, D.A., Fleitaut, L. Schubert, G., and Froidevaux, C., 1978. Shear deformation zones along major transform faults and subducting slabs, *Geophys. J.R. astr. Soc.*, 54, 93-119.

A Multiple Mode of Faulting Mechanism Proposed for the North Anatolian Fault and Prediction of Related Earthquakes

K. Ercin Kasapoglu

Yerbilimleri Enstitusu, Hacettepe Universitesi, Beytepe-Ankara, Turkey

SUMMARY

Investigation of the nature of strain, stress and displacement distributions in the North Anatolian Fault zone, has been attempted on an analytical model of the plates adjacent to the fault, utilizing the finite element techniques.

Concentration of tensile stresses occurs around the eastern end of the fault, as a consequence of the northward movement of the Arabian plate. Initial faulting occurs at this end, in the form of a tensile fracture, opens and propagates diagonally in NE-SW direction along the East Anatolian Fault. Movements along the East Anatolian Fault, releases the tensile stresses originally developed at the eastern end of the North Anatolian Fault; and thus causes a redistribution of strains and stresses in the North Anatolian Fault zone. This redistribution is responsible for the subsequent ruptures, predominantly in 'slip' mode and associated with a frictional strength criterion. These shear fractures, initiate at the eastern end of the North Anatolian Fault and propagate towards west, are described as 'progressive faulting'. Before these progressive ruptures could reach to the central portion of the fault zone, a third mode of faulting begins, again as shear fractures, at the western end of the North Anatolian Fault and propagates backward along the fault, towards east. This third mode of faulting is described as 'retrogressive'. So, the proposed mechanism for the North Anatolian Fault is a 'progressive and retrogressive strike-slip' mechanism.

The elastic strain energy, accumulated in the fault zone, is largely released by these progressive and retrogressive faultings in the eastern and western sections; and no important ruptures occur in the central section of the fault, to cause large earthquakes. Therefore, this central section of the North Anatolian Fault zone, specifically between the longitudes of 33^{0} E an 35^{0} E, may be considered as a 'seismic gap'. However, there are evidences of continuous creep events in the central section of the fault zone which indicates a continuous strain energy accumulation in this area. Therefore, the possibility of a large earthquake to occur in this seismic gap area in the future shoul be seriously considered.

INTRODUCTION

The North Anatolian Fault (NAF) zone, approximately more than a thousand kilometers in length and about forty kilometers in width, stretches almost the entire length of Turkey from east to west as shown in Figure 1. Large earthquakes are known to occur in the NAF zone, the last on August 19,1966 at Varto with magnitude 6.9 . In terms of plate tectonics, the NAF is believed to accomodate a right-lateral sliding between the plate of Black Sea in the north and the Turkish plate in the south (Figure 2). The new theory of global tectonics visualized the NAF zone as a region of intense deformation where large crustal plates are moving horizontally past one another. The NAF has been more active in the recent decades. From 1939 through 1979 seven earthquakes of magnitude 6.8 or greater have occured on the NAF be-

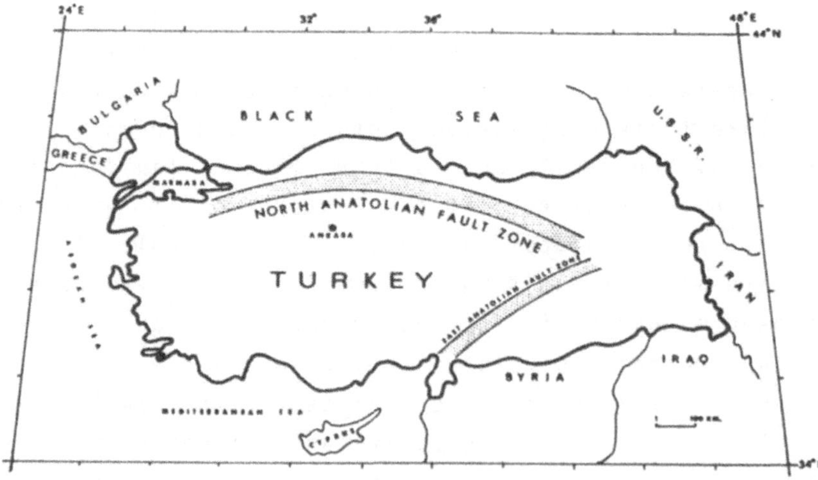

Figure 1: General outline of the NAF and the EAF zones

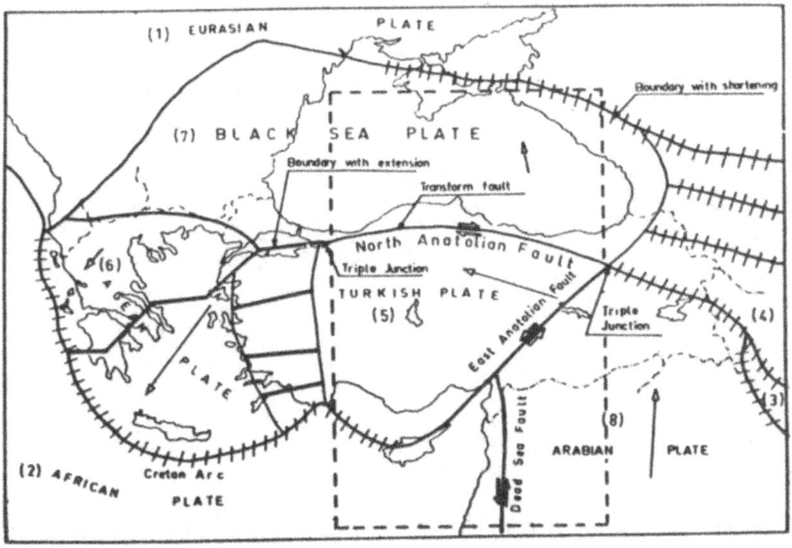

Figure 2: Plate tectonics model of McKenzie (1972) (Dashed lines indicate the analytical model boundaries)

tween 30° E and 40° E causing surface faulting almost continuously in that region. There have also been numereous earthquakes near the intersection of the NAF and the East Anatolian Fault (EAF).

The relocated earthquake data (Figure 3) indicate that different sections of the NAF have intrinsically different seismic characteristics. Earthquake epicenters tend to occur in clusters along the fault. For example, the section of the fault near 33°E was active before, during and after the occurence of major earthquakes there on November 26,1943 and February 1,1944. On the other hand, long sections of the fault that ruptured the ground surface during these same large earthquakes, did not produce any magnitude 5.0 or greater foreshocks or aftershocks. In three largest earthquakes, December 26,1939, November 26,1943 and February 1,1944, the epicenters or points of initial faulting slip, lay along section of the fault that experienced moderate earthquake activity before and after the larger earthquakes (Dewey, 1974).

An understanding of the strain, stress and displacement distributions along the NAF is essential to understanding the

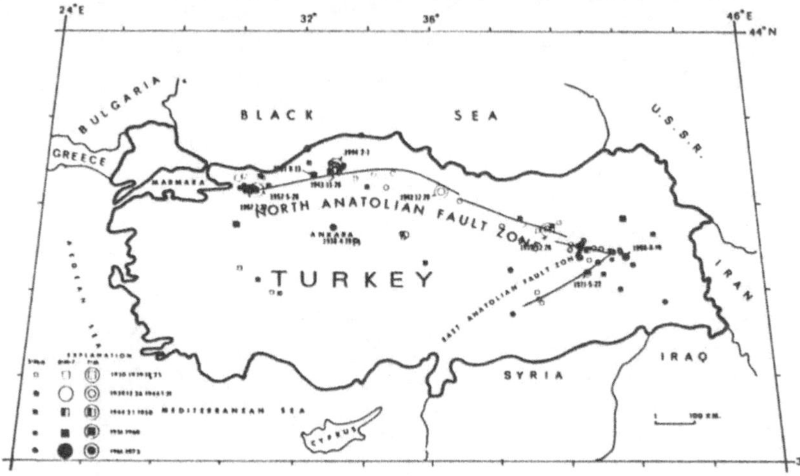

Figure 3: Seismicity of the NAF and vicinity (after Dewey,1974)

mechanical behavior of the fault. In the central section of the NAF zone there is little seismic activity, where the fault appears to be locked and strain is being accumulated. In the eastern and western sections of the fault zone however, small earthquakes and a seismic creep relieve atleast a fraction of the accumulating strain. In the locked section of the fault zone the accumulated strain appears to be relieved in large earthquakes in the past.

Investigation of the nature of strain, stress and displacement distributions in the NAF zone, has been attempted on an analitical model of the plates adjacent to the fault, utilizing the finite element techniques. The data presented herein, are based on very preliminary theoretical results of the investigation which is still in progress. The main objective of this presentation is to open a discussion on the proposed mechanism of faulting and prediction of related earthquakes in the NAF zon

MODEL FORMULATION

A basic computer model (Figure 4a,b) for the plates adjacent to the NAF was developed for the theoretical analysis of strain, stress and displacement distributions across the fault by finite element techniques and used during the course of this investigation.

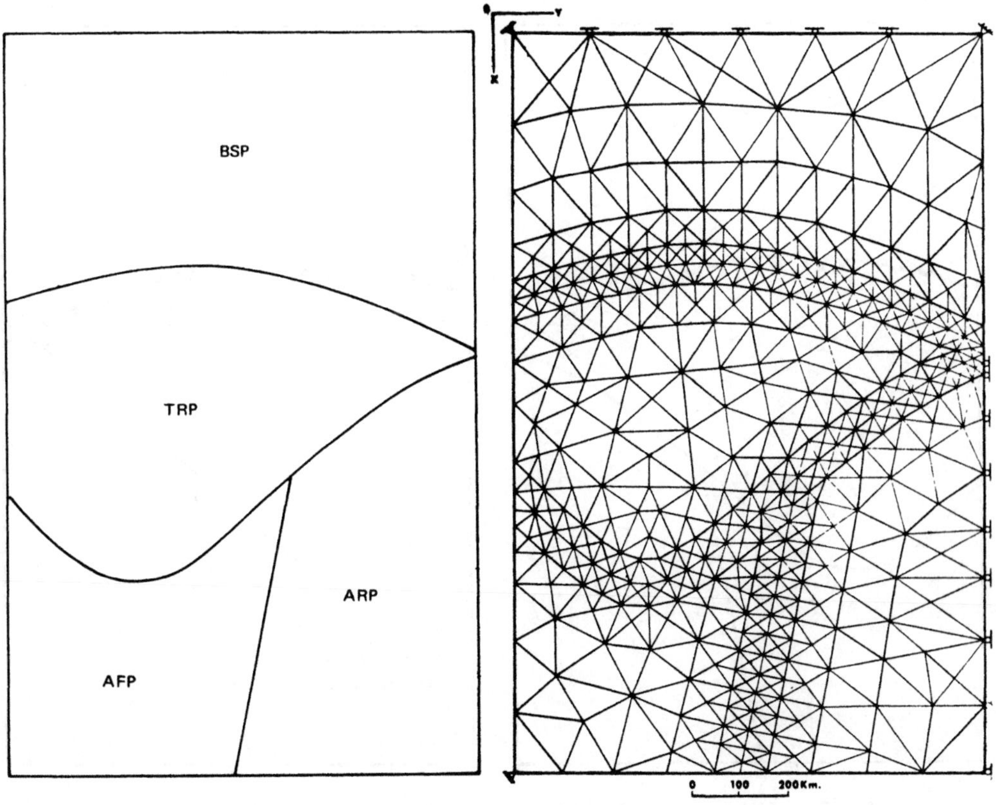

Figure 4: The analitical model of the plates adjacent to the NAF. a) Plate boundaries, b) Finite element idealization of the model.

The concept utilized here involves the ordinary finite element partitioning of a solid model into a descrete number of two-dimensional elements with 'dual nodal points' used along prescribed planes of discontinuities (i.e., the interplate boundaries or the fault lines). A Coulomb-Navier representation with a tension cut-off has been utilized (Kasapoglu,1976). The two-dimensional (plane strain) idealization was assumed to provide an adequate approximation to prototype conditions. The finite element mesh used was comprised of 955 elements and 502 nodes.

The material properties used for the model were estimated taking a plate thickness of 40 kilometers and using standard continental and oceanic sections. For the boundary conditions,

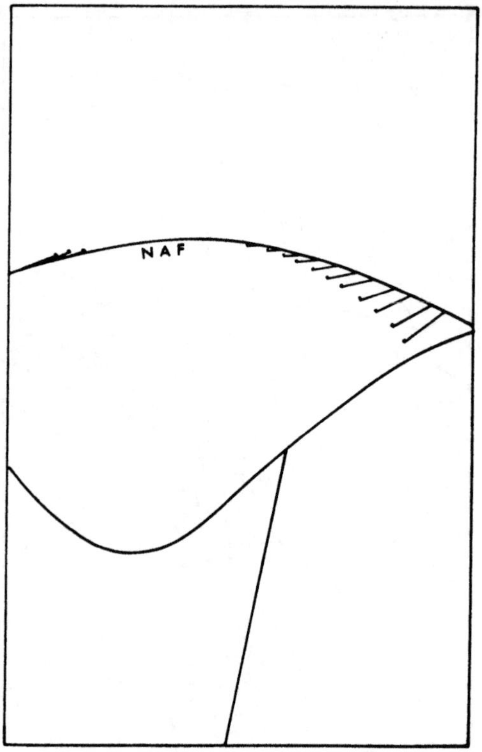

Figure 5: Relative displacements at some particular points along the NAF.

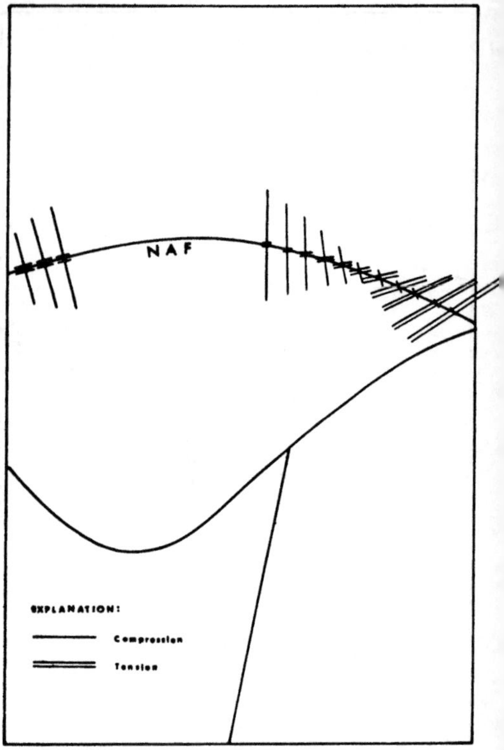

Figure 6: Major and minor principal stress at some particular points along the NAF.

the plate of Black Sea was assumed as almost stationary and the Arabian plate moves towards North, relative to the African plate; and in fact, the motion of the Arabian plate was used as a driving force for the model, in the form of displacement boundary conditions.

STRAIN-STRESS AND DISPLACEMENT DISTRIBUTIONS

Both strains and stresses along the NAF are distributed non-uniformly for the proposed boundary conditions. In general both normal and tangential strains, as well as the normal and tangential stresses have tensile values in the eastern end of the NAF and in the northern section of the EAF.

Distortion characteristics and the type of deformation of the model basicly depends upon the nature of applied boundary conditions. The displacement and the mode of deformation of the model are primarily reflecting the differences in stiffnes of the plate interiors and the fault zones.

For the proposed boundary conditions, concentration of the maximum shear stresses in the model occurs near the eastern and western ends of the fault zone, but not necessarily at the plane of the fault. Concentration of the maximum shear stress contours around the eastern end of the fault zone also points out the effect of rotational deformation of the plate on the mechanism of faulting which will be discussed later.

The direction and magnitudes of the major and minor principal stresses are inhomogeneously distributed throughout the model, with significant variations across the fault planes, and significant stress concentrations at the eastern and western ends of the NAF. The profound effect of boundary conditions is apparent. Direction of relative displacements, and major and minor principal stresses, at some particular points on the fault line are shown in Figure 5 and 6, respectively. Some of this data are more or less in agreement with the data derived from focal mechanism solutions by Canıtez (1973) and Alptekin (1973).

In general, the NAF zone is mostly under compression; only at the very eastern end of the fault, a small zone of tension exist due to the turning moment generated by the northward pushing of the Arabian plate.

MECHANISM OF FAULTING

Initial faulting occurs at the eastern end of the NAF, in the form of a tensile fracture, opens and propagates diagonally in NE-SW direction along the EAF. Movement along the EAF, releases the tensile stresses originally developed at the eastern end of the NAF; and thus causes a redistribution of strains and stresses in the NAF zone. This redistribution is responsible for the subsequent ruptures,predominantly in 'slip' mode and associated with a frictional strength criterion. These shear fractures, initiate at the eastern end of the NAF and propagate towards west, are described as 'progressive faulting'. Before these progressive ruptures could reach to the central portion

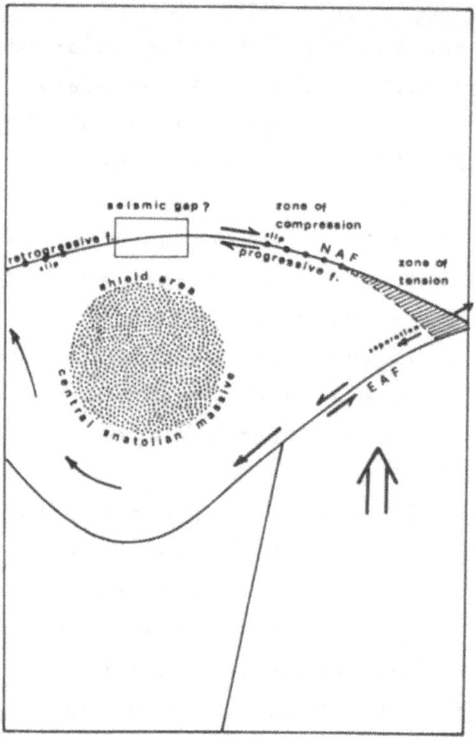

Figure 7:

The multiple mode of faulting mechanism for the NAF.

of the fault zone, a third mode of faulting begins, again as shear fractures, at the western end of the NAF; and propagates backward along the fault, towards east. This third mode of faulting is described as 'retrogressive'. So, the proposed mechanism for the NAF is a 'progressive and retrogressive strike-slip mechanism (Figure 7).

CONCLUSIONS

The elastic strain energy, being accumulated in the NAF zone, is largely released by the progressive and retrogressive strike-slip faulting in the eastern and western sections; and no important ruptures occur in the central section of the fault to cause large earthquakes. Therefore, this central section of the NAF, specifically between the longitudes of 33°E and 35°E, may be considered as a 'seismic gap'. However, there are evidences of continuous creep events in the central section of the fault which may indicate a continuous strain energy accumulation in this area. Therefore, possibility of a big earthquake to occur in this seismic gap area in the future should be seriously considered.

REFERENCES

Alptekin,O., 1973, Focal mechanism of earthquakes in Western Turkey and their tectonic implications: PhD. Thesis, New Mexico Mining and Technology Inst., U.S.A.

Canıtez, N., 1973, Recent tectonic studies and North Anatolian Fault problem: Symp. on the North Anatolian Fault and Earthquake Belt, M.T.A.-Ankara, 35-58 (in Turkish).

Dewey, J.W., 1974, Turkey's North Anatolian Fault... a comparison with the San Andreas Fault: U.S.G.S., Earthquake Information Bull., 6,3,12-16.

Kasapoglu, K.E., 1976, Use of analitical methods in determination of fault mechanism and prediction of probable earthquakes: Yerbilimleri, 2,1,93-96 (in Turkish).

McKenzie, D.P., 1972, Active tectonics of the Mediterranean region: Geophys. Jr. Astr. Soc., 30,109-185.

A Thermal Window in the Lithosphere Underneath North Anatolia and California

C. Froidevaux and L. Fleitout

Laboratoire de Géophysique et Géodynamique Interne, Université Paris Sud, 91405 Orsay, France

The seismicity along major strike-slip plate boundaries like the San-Andreas or the North Anatolian fault zone is relatively shallow. It is restricted to depths not exceeding 15 or 20 km so that the lower crust and the mantle below it are sheared throughout the lithospheric thickness, say down to about 100 km. This ductile deformation generates heat by an amount equal to the product of the velocity across the plate boundary, say 5 cm/yr or 1 cm/yr, times the shear stress. This last quantity is not measurable and has to be estimated on the basis of laboratory creep data. This requires the knowledge of the exact temperature within the shear zone. Indeed creep laws are highly temperature sensitive. However, this temperature is likely to be strongly modified by the heat dissipated by the deformation itself. This brief argument illustrates the important feedback between thermal and mechanical quantities within the lithosphere. These can only be computed by solving doubled thermal and mechanical equations (Yuen et al., 1978; Fleitout and Froidevaux, 1980).

The numerical models show the rapid growth of a thermal peak after the onset of shearing. The stress required is high in the initial stage, but rapidly decreases as the generated heat makes the material softer. Finally the temperature levels off in the central part of the shear zone and the heat diffuses out so that the thermal anomaly gets wider. The value of the temperature maximum depends upon rock type and upon differential velocity u. For u = 10 cm/yr it amounts to 920°C for olivine, 700°C for diabase and 560°C for wet quartzite. In all three cases it corresponds to a viscosity around 5.10^{20} poises. The sheared lithosphere acquires a mechanical strength similar to that of the asthenosphere.

In order to gain a more detailed view of the thermal structure we present some new solutions. They refer to the physical conditions prevailing in the cases of San-Andreas and of North Anatolia. The starting equations and computational procedures are specified elsewhere (Fleitout and Froidevaux, 1980). Here the creep law is taken from laboratory data on dry olivine.

Figure 1 shows the evolution of the thermal profile at a depth of 60 km where the initial background temperature is assumed to amount to 580°C. After 1 Ma a pronounced narrow temperature peak has formed. The anomaly is seen to have a smaller amplitude for the smaller velocity difference. Its width broadens with time and is quite larger than the width of the sheared zone. The latter islimited to the region where the material temperature is within a few tens of degrees of

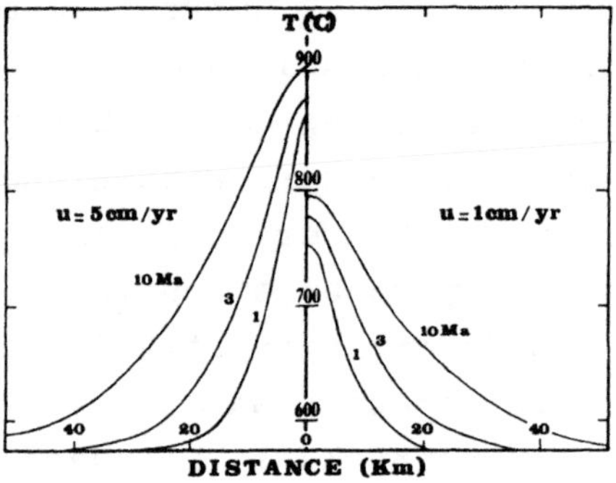

Figure 1

Temperature profiles at 1, 3 and 10 Ma after the onset of shearing in a dry olivine medium. The temperature at time zero is 580°C, except for a 20°C anomaly around the origin which localizes the initial position of the zone of weakness. The two sets of curves correspond to 2 differential velocities u. The rate of deformation is

$$= \frac{1.3 \times 10^{12}}{T} \; {}^{3}\exp - \frac{125\ (Kcal) + pV^*}{RT}$$

(Kohlstedt and Goetze, 1974) with an activation volume $V^* = II\ cm^3/mole$ and a pressure p corresponding to 60 km depth. The stresses found for the 10 Ma solutions are 545, resp. 1870 bars, for u = 5 cm/yr, resp. 1 cm/yr.

the maximum value. Similar one - dimensional time dependent solutions were computed for 5 other depths.

Figure 2 gives the amplitude of the temperature anomaly within the lithospheric mantle for a 10 Ma old shear zone. A pronounced thermal window extends down to about 90 km, over a half width of the order of 30 km.

What geophysical observations could detect the predicted anomalous thermal structure? Heat flow data (Lachenbruch and Sass, 1980) are certainly sensitive. However, a deep seated heat source only yields a very diffuse surface effect. A sharp heat flow anomaly could only derive from shear heating within the crust where the physical processes are not yet clearly understood. Two approaches are potentially more promising: the determination of the corresponding electrical conductivity or seismic velocity anomalies. Up to now mapping of 3-dimensional structures has been more accessible to seismology

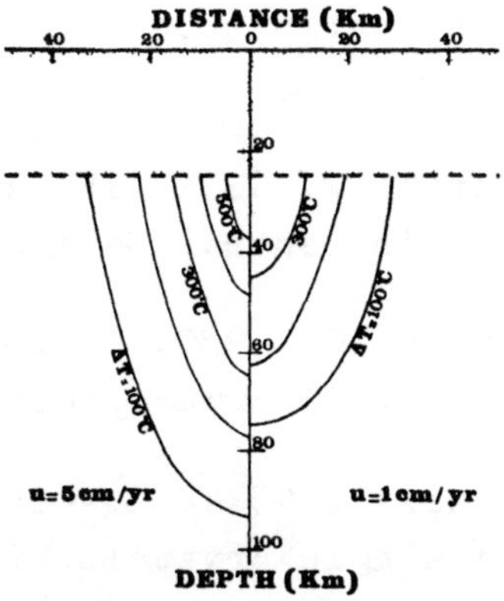

Figure 2

Amplitude of the temperature anomaly vs. distance from the center of the shear zone and vs. depth in the lithospheric mantle at time t = 10 Ma. The shaded region represents a 25 km thick crust. Again 2 solutions are pictured for 2 different u values. These are not true two-dimensional solutions but derive from one-dimensional computations at several depth. The assumed initial mantle geotherm is $T(^{o}C) = 60 + 8.7 \times z$ (km). Other specifications are as in Figure 1.

(Aki et al., 1977) than to electro-magnetic methods. The dense network of seismometers in California has revealed a very conspicuous low velocity body under the San-Andreas fault (Cockerham and Ellsworth, 1979). It extends from depths of 25 km to 75 km and has an Eastward dip. The temperature window shown in the left of Figure 2 yields a seismic velocity anomaly of about 2% and encompasses the appropriate depth range. This compares well with the 3% effect observed in California. Under Northern Anatolia the anomalous structure ought to be somewhat less pronounced if the differential plate velocity is closer to 1 cm/yr. Its exploration would greatly enrich our physical insight into the dynamical behaviour of the lithosphere.

References

Aki, K., Christofferson, A. and Husebye, E.S. 1977. Determinatic of the three dimensional structure of the lithosphere, J. Geophys. Res., 82, 277-296.

Cockerham, R. and Ellsworth, W. 1979. Three-dimensional large-scale mantle structure in central California, EOS, 60, 875.

Fleitout, L. and Froidevaux, C. 1980. Thermal and mechanical evolution of shear zones, J. Struct. Geol., 2, in press.

Kohlstedt, D.L. and Goetze, C. 1974. Low-stress high-temperature creep in olivine single crystals, J. Geophys. Res., 79, 2045-2051.

Lachenbruch, A. and Sass, J.H. 1979. Heat Flow and Energetics of the San-Andreas Fault, Proceedings of Conference IX - Magnitude of Deviatoric stresses in the Earth's crust and Upper Mantle, U.S.G.S. edition, 47-146.

Yuen, C.A., Fleitout, L., Schubert, G. and Froidevaux, C. 1978. Shear deformation along major transform faults and subducting slabs, Geophys. J.R. astr. Soc., 54, 93-120.

8 Interdisciplinary approaches to earthquake prediction

The Concept of a Multidisciplinary Approach to Earthquake Prediction

Andreas Vogel

Institut für Geophysikalische Wissenschaften, Freie Universität Berlin, Rheinbabenallee 49, D-1000 Berlin 33, Federal Republic of Germany

Abstract

Earthquake prediction research requires a broad interdisciplinary approach. Recognition of the various factors which are controlling the generation and occurrence of earthquakes is of basic concern. From the methodological point of view the three main lines to be followed are studies of precursory strain, investigation of chemo-physical parameters, both the spatial distribution and temporal changes in the earthquake source region as well as analysis of the seismotectonic pattern. Experimental approaches and field measurements required are discussed in view of the North-Anatolian fault zone. Modelling of earthquake generating processes constrained by the complexity of observations such as outlined in the present article is considered as a necessary basis for earthquake risk assessment and finally prediction of earthquakes.

Earthquake prediction research - whether it concerns the intermediate step of seismic risk assessment or the final goal of prediction of place, time of occurrence and focal mechanism - requires a broad interdisciplinary approach. Recognition of the various factors controlling the generation and occurrence of earthquakes is basically necessary. The interdisciplinary approach necessitates a broad cooperation between earth scientists in various fields including geophysicists, geologists and geodesists. Experimental field studies, analysis of the observational data, modelling of earthquake generating processes, and numerical solutions constrained by the complexity of observational data should be the basic concept of research towards predicting earthquakes.

According to our present state of knowledge three main lines should be followed from the methodological point of view:

1) Studies of precursory strain events
2) Investigations of geophysical parameters and temporal changes of rock properties in the earthquake source region
3) Recognition of the seismotectonic pattern

Fig. 1 shows schematically the three lines to be followed in experimental research.

The first line is concerned with the observation of surface deformation and displacements as indicators of stress-induced strain. It is subdivided into three domains of instrumental methodology.

Continuous monitoring of strain in the earthquake source region requires high-precision recording instruments of sufficient, long-term stability. Recording tiltmeters, strainmeters, gravity meters, sea-level recorders in coastal areas and groundwater level recorders in wells today are reaching the sensitivity and accuracy which is necessary to detect precursory strain events. In the North-Anatolian fault zone where horizontal movements are predominant, strainmeters seem to be most appropriate.

Long-term changes of strain preceding and associated with earthquake activity are investigated by repeated geodetic ground surveys such as triangulation, trilateration and levellings, which are combined or in the future may be replaced to a certain extent by precise gravity surveys. In order to attain a more detailed knowledge of the strike-slip motion and the associated strain in the North-Anatolian fault zone, repeated surveys of the Turkish first order triangulation network and local horizontal resurvey networks are recommended in selected areas.

Space geodetic techniques are now able to measure position changes over distances of hundreds and thousands of kilometers with an accuracy of a few centimeters. Earthquake activity is intimately related to the large-scale motion of tectonic plates. Space geodetic observation sites such as VLBI antennas and satellite laser ranging stations to the north and the south of the North-Anatolian fault system would be necessary to relate the displacements along this plate boundary to the global reference system of tectonic plate motion. Highly mobile satellite laser ranging, spaceborne laser ranging and satellite radio interferometric measurements would be necessary in support of the geodetic ground surveys.

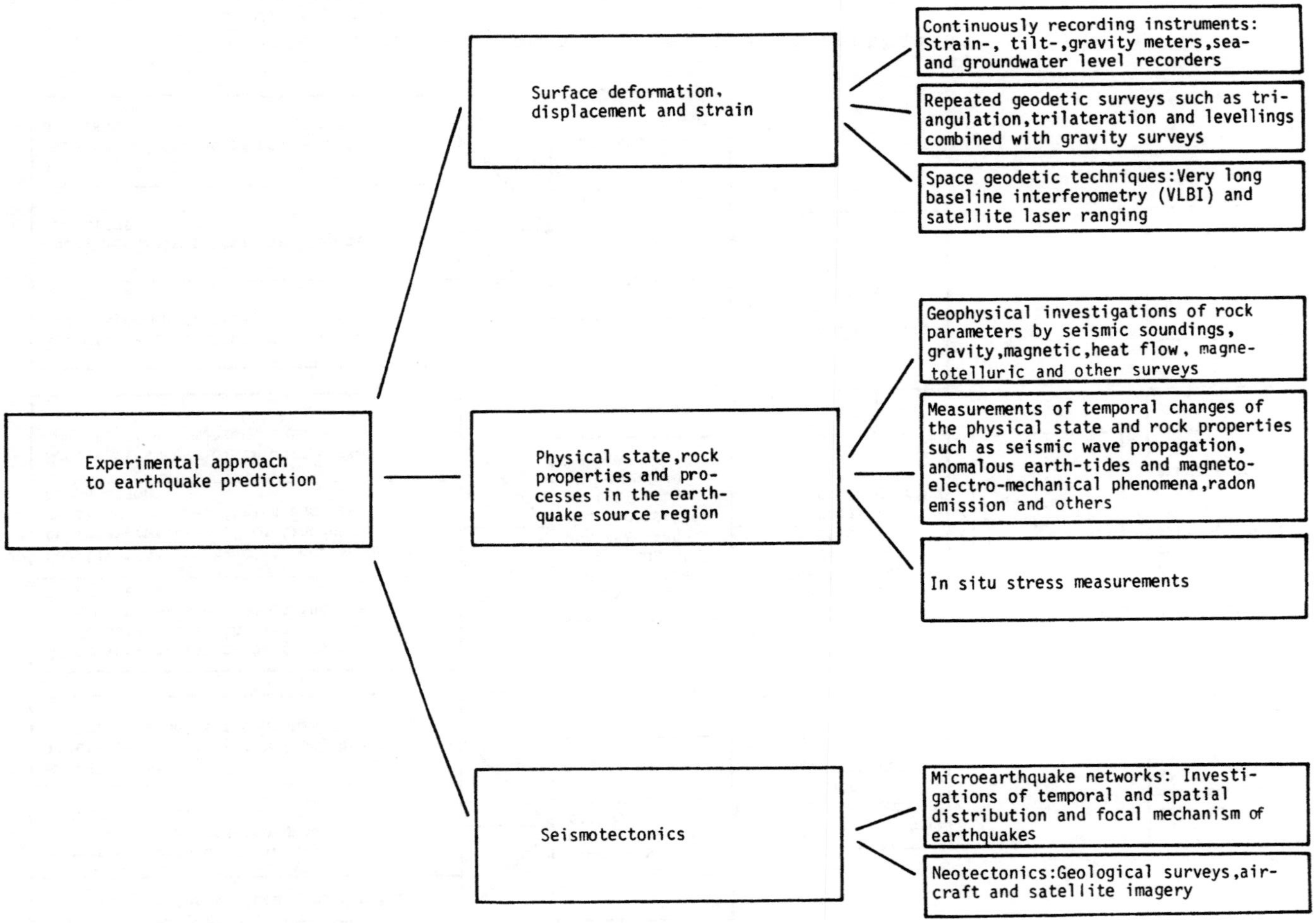

Fig 1 Experimental approach in earthquake prediction research

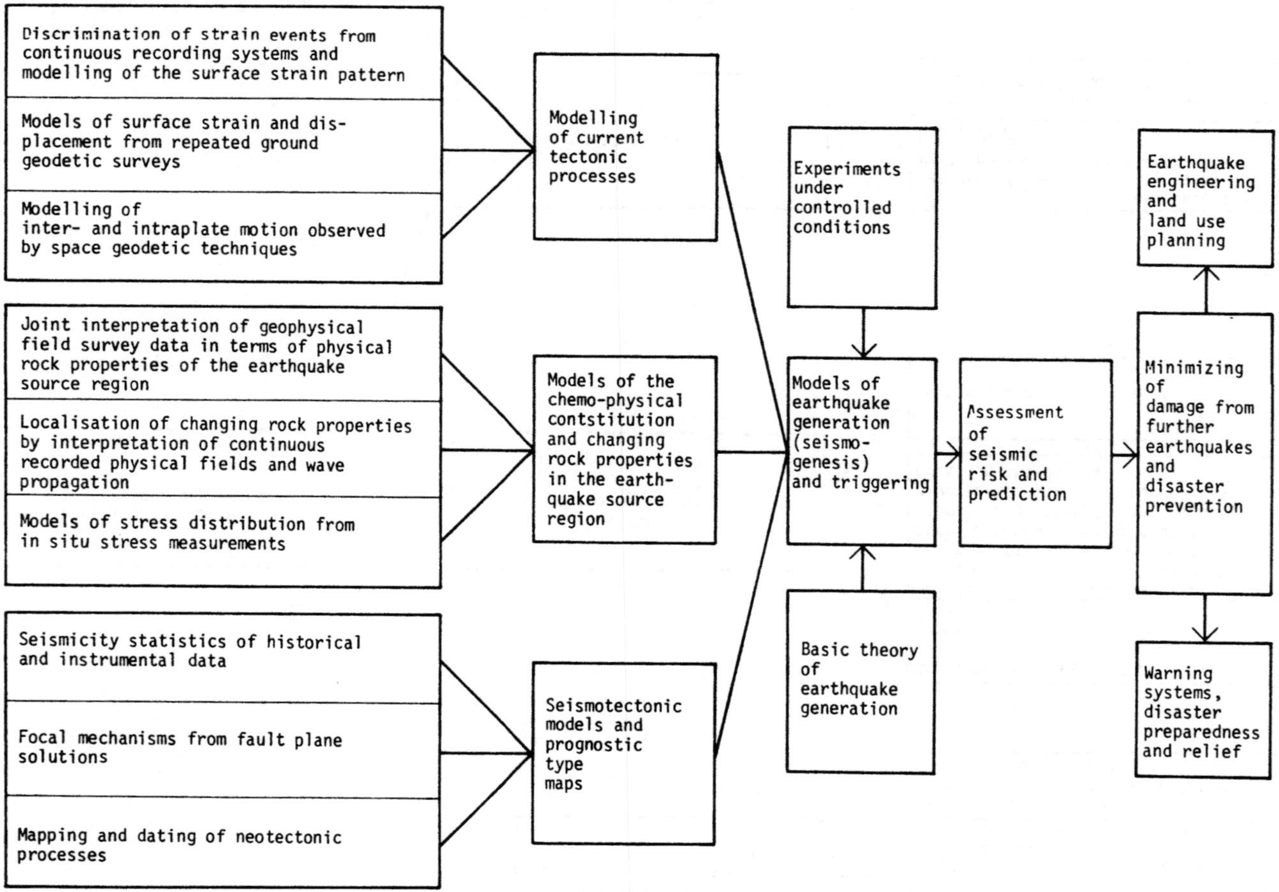

Fig 2 Complex data interpretation and modelling in earthquake prediction research

Experimental research along the second line shown in Fig. 2 is concerned with physical properties, state and processes in the earthquake source region. Three different fields of research can be distinguished.

One important requirement in earthquake prediction research is the knowledge of various physical parameters in the source region. Intensive geophysical surveys such as seismic soundings, gravity, magnetics, heat flow measurements and magnetotellurics should be carried out on the North-Anatolian fault zone in order to provide necessary field data.

Of special concern in earthquake prediction research are the phenomena of earthquake generation indicating changes of the physical state and rock properties in the source region. Regarding temporal changes of seismic wave propagation, polarization anomalies of earthquake waves have already been studied as diagnostics of dilatancy. Repeated measurements of seismic wave velocities under controlled conditions should be carried out in earthquake-prone areas of the fault zone. It would be of great value if mobile magnetographs and electrographs could be operated in order to look for time-dependent magnetotelluric transfer functions as well as magneto- and electromechanical phenomena.

Surveys of radon emanations in water wells would be furthermore required. A network of recording tilt-meters could provide information on anomalous earth-tides associated with dilatancy.

Stress is the key factor in earthquake prediction research. The overall stress distribution determined by in-situ measurements may point out earthquake prone areas of the fault zone. For measurements of temporal changes of the stress field as an indicator of impending earthquakes the capability of stress measuring techniques is still insufficient.

Experimental research along the third line in Fig. 1 is concerned with the different aspects of seismotectonics, comprising both neotectonics and the statistics of earthquake occurrence as well as focal mechanisms. Besides the presently operating network of seismographic stations highly sensitive microseismograph stations should be installed in earthquake prone areas of the North-Anatolian fault zone.

Additional neotectonic studies by aircraft and satellite imagery as well as ground surveys will certainly reveal still unknown evidence of neotectonic processes in the fault zone.

More details about the present capabilities of measuring techniques in earthquake prediction research can be found in the proceedings of a meeting on this subject (Vogel (Editor), 1979).

Earthquake prediction research will hardly be successful when the symptoms obtained from any of these various measuring techniques are considered separately. Models of earthquake generation constrained by the complexity of observational data are required for each specific area. Models of earthquake generating processes - even speculative ones - are a necessity for each area under investigation right from the beginning. A working hypothesis has to be followed when establishing survey networks and recording stations in areas of potential earthquake risk.

Fig. 2 shows the scheme of data analysis and modelling. The first steps of data analysis and interpretation corresponding to the various experimental and instrumental methods of Fig. 1 are lined up on the left in the scheme of Fig. 2.

Discrimination and evaluation of strain events from continuously recording instruments, modelling of surface strain and displacements based on ground geodetic surveys and modelling of inter- and intraplate motion, when sufficient space geodetic data will be available, are altogether contributing to attain a model of current geotectonic processes (upper line in the scheme of Fig. 2).

The distribution of physical rock properties obtained by interpretation of geophysical survey data, the localization of changing rock properties by analysis of continuously recorded physical fields and wave propagation as well as the stress pattern from in-situ measurements are the kind of information needed to establish a model of the chemo-physical constitution and processes in the earthquake source region (central line in the scheme of Fig. 2).

Seismicity statistics, fault plane solutions as well as mapping and timing of neotectonic processes finally lead to a seismotectonic model of the area in question (lower line in the scheme of Fig. 2).

The three lines of intermediate interpretation have to be brought together to one complex model of earthquake generation, triggering processes included, and supported by basic theoretical studies and experiments under controlled conditions in both the field and laboratory.

Models of earthquake generation constrained by the complexity of observational data finally provide the criteria for seismic risk assessment and prediction.

In the near future prospects are small of predicting time and place of occurrence, mechanism and energy release of earthquakes. To begin with, we have to deal with earthquake prediction in a wider sense, trying to acquire all kinds of information that may be helpful to minimize damage from further earthquakes. Necessary support of measures such as proper land-use planning and design of earthquake-resistant buildings is today realistic from the point of view of earthquake prediction research. Extensively increased activities in earthquake prediction research are needed to establish warning systems, to improve disaster preparedness and relief and to take various precautions for reducing casualties and damages from earthquakes.

References

Vogel, A. (Editor), 1979: Terrestrial and space techniques in earthquake prediction research. Vieweg Braunschweig-Wiesbaden

Deterministic Earthquake Prediction – Present Stage and Future Developments

R. Meissner

Institut für Geophysik, Universität Kiel, D-2300 Kiel, Federal Republic of Germany

Abstract

Deterministic prediction is understood as a rather short time decision based on a number of certain precursor phenomena. Virtually all precursors are related to the accumulation of stress. Critical stresses involving dilatancy seem to concentrate rather on some local asperities than on large parts of a future rupture zone. Modifications of today's measuring techniques should lead to a monitoring of stresses, even in the pre-dilatancy elastic range, and to a determination of the stress axes on the basis of anisotropic behaviour of certain parameters like seismic velocities, seismic absorption, strain, and electric conductivity. Fast data-collecting facilities, such as a transfer via a geostationary satellite, are considered prerequisites of a successful earthquake prediction research.

1. Introduction

A deterministic earthquake prediction seeks answers to the three questions WHEN, WHERE, and HOW LARGE. Obviously it is impossible to determine the time, the location, and the magnitude of an impending earthquake several years or even months before the event takes place, although certain precursor phenomena are observed such a long time span in advance. Successful

---

Publication No. 202, Institut für Geophysik, Kiel

predictions like that of the famous Haicheng quake in 1975 with M = 7.3 (Wu et al. 1978) are necessarily rather short-time decisions. They are based on a dramatical increase and intensification of certain precursor phenomena. Statistical methods of earthquake prediction and the determination of seismic gaps, which are important for long time assessment of earthquake risk and for selecting sites for deterministic studies are not subject of the present study.

There seems to be a common consensus that the physical parameter stress (or strain) plays a critial role in the earthquake mechanism. It is well known from experimental work that a critical stress must be exceeded in order to arrive at a "point of no return", a begin of enhanced microcrack- growth which will lead to the final rupture. It is also commonly agreed among workers in the field of stress-strain experiments that the critical stress depends heavily on parameters like anisotropic structure, microcracks and macrofaults of a rock sample (Nur and Simmons, 1969, Sobolev et al. 1978, Soga et al. 1978).

The famous "dilatancy" phenomenon, i.e. an inelastic increase in volume connected with an irreversible opening of microcracks starting between one or two thirds of the critical stress was postulated by Nerserow et al. 1969, Nur 1972, Wang et al. 1975, and has been verified under experimental conditions (Sobolev et al. 1978, Rummel and Frohn 1980) Rupture starts after a fast coalescence of microcracks with or without weakening by intruding water.

In the last 2 years severe doubts have arisen regarding the validity of the dilatancy concept in nature (Brune 1979). These doubts are based on two observations: First, many strike slip earthquakes along the San Andreas fault system, the best monitored area in the world, have shown no precursor phenomena connected in any way to dilatancy (such as e.g. velocity anomalies of seismic waves), second, many strike slip earthquakes seem to act under extremely low stresses, stresses much below the threshold where dilatancy could possibly start. The old stick-slip concept of Reid (1911) is coming to life again. The question whether dilatancy might be concentrated only on very local "zones of difficult slip" (Wesson et al. 1973) or "asperities" (Jones and Molnar, 1979) seems to be crucial. Also the difference between the "static critical stress", necessary for the begin of rupture, and the smaller "dynamical critical stress", necessary for the propagating of rupture seems to be important for determinations of magnitude and fault length (Brune 1979).

2. The precursor phenomena

The observation and monitoring of several or at least some precursor phenomena in an endangered area, e.g. a seismic gap, is the precondition of a successful prediction. According to the concept that a critical stress – be it the static stress σ_c at a local asperity or a more regional lower limit - must be exceeded, one might think of measuring the stress directly in an area which is a candidate for future quakes on the basis of statistical evaluation of the fore-shock and microseismic activity (see e.g. Wyss 1980). Such a determination of stress in a certain area, e.g. along a prominent fault zone, requires the drilling of many holes with the installation of strain or stress measuring devices. Although such measurements might be helpful for measuring stress and strain variations and for changing the rheology by water injection they certainly allow a measurement of stresses only at certain locations. On the other hand, there are many physical parameters closely related to stress and representing larger areas, that other measurements and parameters might be more adequate. Table 1 gives an overview of the physical parameters which are related to stress, some of them being well established precursor phenomena. As far as these phenomena are related to a deterministic prediction they will be discussed in the following in more detail.

Table 1: PRECURSOR PHENOMENA AND THEIR VARIATIONS

Physical parameter	Method of measurement
(1) stress	direct measurements in boreholes
(2) dislocation strain creep	geodetic meas. on the surface; geologic mapping; measurements in boreholes; future spaceborne techniques
(3) tilt	same as under (2) plus measurements by tiltmeters in boreholes
(4) seismic velocity seismic anisotropy	travel time measurements of foreshocks and artificial sources under various azimuths
(5) seismic absorption	measurement of amplitudes and velocities of artificial sources
(6) seismicity, foreshock activity	seismic array methods by net-works of seismometers
(7) gravity field	repeated gravity measurement; possible future space-borne techniques
(8) magnetic field	repeated or continuous magnetic measurements
(9) electric conductivity electric anisotropy	repeated or continuous measurement of conductivity under various azimuths; electromagn. measurement
(10) ground water level	continuous measurement of ground water table in wells and lakes
(11) gas content of ground-water and soil	physical and chemical measurements in wells and soils
(12) anomalous animal behaviour (non-physical precursor)	systematic physiolog. investigations and tests of certain animals; systematic questioning of people; relation to physical parameters

Many of the precursor phenomena listed in table 1 have already been measured or monitored, others, like seismic anisotropy and absorption or electric anisotropy have shown to be extremely important under laboratory conditions but have not yet been tested sufficiently in the field. It seems to be crucial that seismic velocities and electric conductivity decrease only in certain directions under stress. (Rummel and Frohn 1980, Sobolev et al. 1978, Fuji and Hamano 1977). This may explain a large part of the negative results of field experiments which so far have not yet been continuously observed under different directions.

3. Control of stress and stress variations by monitoring selected precursor phenomena

Not all precursor phenomena listed in table one are adequate for continuous observations. Levelling, gravity, ordinary magnetic and electric surveys (7), (8), (9) - however important for statistical hazard assessment - fail to be of use for a deterministic short time prediction because of the large time period necessary for measurements and re-measurements. Starting with point (2) of table 1 a geodetic electro-optical distance measurement device for distances up to 30 km seems to be most adequate to monitor distances, distance changes and angles. Such an experiment using a network of several markers on both sides of a suspicious fault zone could determine zones of accumulation of strain, i.e. zones of diffficult slip and zones of enhanced seismic or aseismic slip. Displacement vectors can be mapped, and asperities can be detected (Bonatz et al. 1980).

Point (3) of table 1 certainly deserves an increased attention. First it is possible to determine aperiodic changes of tilt resulting from aseismic creep events. In some cases these events mark the latest stage of critical strain accumulation. Second and more important is the different response of tilt towards the tidal forces if dilatancy modifies the elastic and inelastic constants of some part of the subsurface, as theoretically postulated by Zschau (1979). Although a careful correction with regard to rainfall, temperature and loading tides is necessary, the above mentioned tilt modification induced by the build-up of critical stresses in an area can ertainly be measured by sensitive instruments.

The points 4, 5 and 6 of table 1 are the seismic methods which can be and will be improved in the coming years. In order to measure velocity and velocity anisotropy certainly the classical method of using V_p - velocities or the ratio of V_p/V_s of natural events is not adequate and must be modified by using artificial sources. These can be Vibroseis or explosive sources on land or - better - non-explosive sources in lakes. With non-explosive sources the signal amplitude can be controlled. For all artificial sources, the time-break and the coordinates can be determined 100 to 1000 times more accurate than it is possible for natural seismic events. Consequently, the velocity can be determined 100 to 1000 times more accurate; different directions parallel and perpendicular to the suspected axes can be observed. With such a high accuracy even the elastic velocity - increase with stress, well below the beginning of dilatancy, should be observable. Also the seismic attenuation which is strongly related to stress under experimental conditions and which should theoretically be even stronger effected than velocity by a stress variation can be measured by using controlled sources. Shooting to a network of seismometers at a fixed time interval of a day or a week, a stress accumulation should certainly be observed by monitoring travel times and amplitudes. Certainly the foreshock activity, or generally the seismicity, must be carefully studied in an area. This is important for long-time risk assessment as suggested by Wyss (1980) and also for the sometimes dramatic increase in activity a day or a few days before the main shock (Jones and Molnar 1979). These "passive" seismic methods together with the "active" ones, using artificial sources should certainly help to clarify the problem where stress accumulates. Together with the geodetic and tilt observations the important question whether dilatancy and high stresses concentrate on rather local "asperities" or whether a whole region (of a future rupture) exhibits a weak - but certainly observable - increase of stresses. Fig. 1 shows the asperity concept with zones of "easy" and "difficult" slip (see also Wesson et al. 1973 and Molnar and Jones 1979). In the upper part the actual stress together with the "static" and the "dynamic" friction, necessary for the begin and the continuation of rupture are shown (see also Brune 1979). The asperity concept is supported by spectral analyses of large rupture zones which seem to show that the release of seismic energy - from asperities - often takes place at only some places of a rupture area (Wyss 1980). The low stresses of many strike slip quakes along known fault planes and the dramatic concentration of quakes along these faults support the idea that fault zones have special rheologic properties. This has

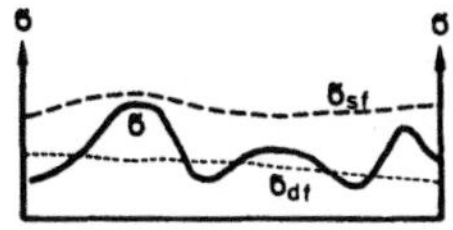

(1) = Asperities=Locations of difficult slip

(2) = Planes of past slippage

(3) = Slip by foreshock or creep

(4) = Tectonic stress $\sigma_t = 2\eta_z \dot{\bar{\varepsilon}}_z = 2\eta_z \frac{\bar{V}}{\Delta z}$

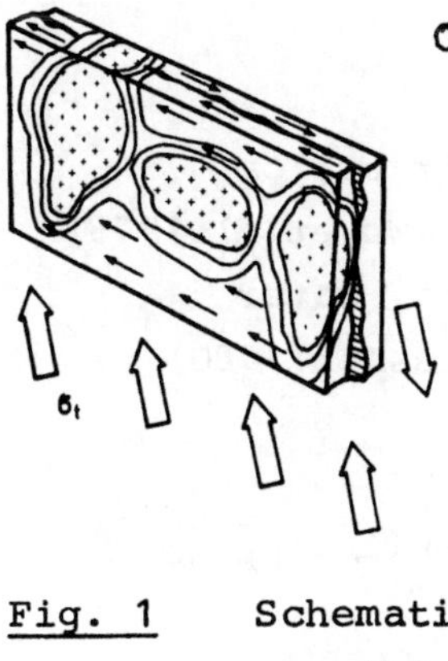

$\sigma_F + \sigma_{as} = 2\eta_{FZ} \dot{\bar{\varepsilon}}_{FZ} = 2\eta_{FZ} \frac{\bar{V}_{FZ}}{\Delta y} =$ stress released by foreshocks +aseismic creep

$\sigma = \sigma_t - \sigma_F - \sigma_{as}$ = local stress

σ_{sf} = static friction

σ_{df} = dynamic friction

Fig. 1 Schematic cross section along a fault zone and possible stresses

been shown by a number of recent papers (Vetter and Meissner 1977 , Meissner1980) where also the limited depths of most quakes is discussed. Depending on the temperature and the water content, maximum stresses can accumulate at depth of 3 - 10 km, fault zones showing roughly one order of magnitude lower values. The theoretical stress peaks in the upper crust are reflected in the peaks of frequency of occurrence of quakes with depth. Fig. 2 shows a comparison of the source-depth distribution of Haicheng shocks (Wu et al. 1978) with the model calculation of maximum stresses (Meissner, 1980). Water-saturated upper crusts of normal temperature and normal creep rates (10^{-14} - 10^{-17} sec^{-1}) all show sharp stress maxima at these depths. Here, apparently stresses can be preferably accumulated and released. These depths of about 10 km must be reached by active seismic (and possibly electric) experiments in order to control stress changes at places where they preferably take place. The special "weak" rheology of fault zones certainly make them the first can-

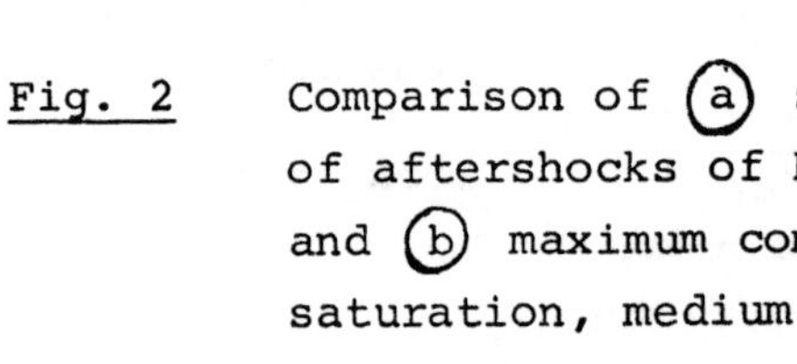

Fig. 2 Comparison of (a) source depth distribution of aftershocks of Haicheng - quake (Wu et al.1978) and (b) maximum compressive stresses, water saturation, medium temperature (Meissner 1980)

N = frequency of occurence

Z = depth

V + M = Vermiculith and Montmorillonit, supposed to characterise fault zones

τ = maximum compressive stresses

didates for a release of stresses and for detailed investigations. In general, theoretical relations between the inelastic parameters viscosity η and quality factor Q (Meissner and Vetter 1979) and between the elastic and inelastic parameters V_p, V_s and Q (Anderson and Minster, 1980, Zschau et al. 1980) may be used for a further control of the stress regime and for dilatancy effects.

Discussing briefly points (8) to (12) of table 1, a monitoring of magnetic field parameters and electric conductivity over a large area, the electrical methods with different azimuths, should certainly be applied. Present electrical investigations have only a limited depth of penetration, have only sporadically been carried out, and so far did not take azimuth effects into account. Groundwater levels and the gas content of groundwater and soil can be monitored by rather moderate efforts and should certainly be involved in the spectrum of precursor observations. Anomalous animal behaviour, finally, a non-physical and much debated subject, should not be expelled from the precursor methods, but specialists like animal physiologists and biologists should carry the burden of these investigations.

4. Future space-borne techniques

So far space-borne methods have not been used for measuring precursor phenomena. The necessary accuracy in the centimeter range is not yet reached by mobile laser nor by microwave techniques. The VLBI method, using cross-correlation of quasar signals, has obtained the necessary accuracy on a regional and global scale. Plate movements, important for the calculation of regional and long time stress build-up, can be infered from these measurements. For a short time prediction these methods will remain problematic.

Table 2 POSSIBILITIES OF SPACE-BORNE TECHNIQUES FOR EARTHQUAKE PREDICTION RESEARCH

Scale	Techniques	Goal
Continental - Regional	Network of fixed and mobile VLBI and laser stations observe quasars, moon or satellite	Absolute and relative positioning; Inter- and intraplate movements, deformation and stability of plates, establishment of reference points
	Satellite altimetry and satellite tracking	Long time change of gravity anomalies (?)
Local - Regional	Laser or microwave wave networks measure relative distances to satellite, Doppler (?)	Relative positioning; dislocations and their short time variations, creep, strain, displacement vectors
	Communication, data transfer of collected (and reduced?) field data via satellite (e.g. Meteosat) to data center	Fast accessibility of field data from remote areas. Concentration of all field data at an interpretation center.

A field where space methods could help earthquake prediction immediately is data communication. Whereas data from the instruments in remote areas to a central field station may be transmitted by telemetric methods, a data transfer from the central field station to a computer and interpretation center could effectively performed via a geostationary satellite. First tests seem promising. Table 2 summarizes possible advantages of space-borne techniques. Several steps towards the final goal of predicting earthquakes on a European scale have recently been suggested (Meissner and Zschau, 1978, Bonatz et al. 1980).

Conclusion

The technique for monitoring important precursor phenomena on the ground is certainly available, but concentrated efforts are essential. A modification of some well-established methods, e.g. the seismic techniques, is recommended and should certainly be carried out in the near future. Most important for all planned surveys is the concentration of many methods in only a few areas rather than observing only a few precursors in many areas. Even if the details of the start of a rupture - be it dilatancy at a local asperity or low-stress stick-slip behaviour at a fault zone - are not known, the observations of several precursor phenomena may help to predict an impending quake. Certainly, the stress regime is different for different stress patterns, i.e. for impending thrust-, strike-slip- and extensional rupture. The improved methods mentioned above should be able to determine all kinds of stress accumulations and even anistropies. A monitoring or a nearly continuous observation of precursors by fast data-collecting and interpretation facilities are needed for a successful earthquake prediction research.

References:

Anderson, D.L., Minster, J.B., 1980, The frequency dependence of Q in the earth and implications for mantle rheology and Chandler Wobble, Seismol Lab. CALTECH Calif. Inst. Technology Pasadena, Calif.

Bonatz, M., Meissner, R., Zschau, J., 1980, A first European joint effort in earthquake prediction research; Proc. 2nd workshop ESA Council of Europe, Strasbourg 31. Jan. - 1. Feb. 1980 (ESA SP-156), 107-122

Brune, S.M, 1979; Implications of earthquake triggering and rupture propagation for earthquake prediction based on premonitory phenomena; J. Geophys. Res. 84, 2195 - 3198

Fuji, N., Hamano, I., 1977; Anisotropic changes in resistivity and velocity during rock deformation; High Pressure Research, Applic. in Geophysics, 53 - 62

Jones, L.M, Molnar, P., 1975 Some characteristics of foreshocks and their possible relationship to earthquake prediction and premonitory slip on faults; J. Geophys. Res. 84, 3596 - 3608

Meissner, R., 1980, Bruch- und Kriechprozesse in kontinentaler Kruste, Geol. Rundschau, in press

Meissner, R., Zschau, J., 1978; Tectonics panel chairman's report, Proc. Europ. Workshop SONG, Schloß Elmau, 200 - 203

Meissner, R., Vetter, M. 1979; Relationship between the seismic quality factor Q and the effective viscosity η; J. Geophys. Res. 45, 147 - 158

Nerserow, I.L., Semyenow A.N., Simbireva, J.G; 1969, Physical basis of foreshocks. NAUKA, Moscow, 1969

Nur, A.,1972, Dilatancy, pore fluids, and premonitory variations of t_s/t_p travel times; Bull. Seism. SOC. Am 62, 1217

Nur, A. and Simmons, G., 1969; Stress induced velocity anisotropy in rock, an experimental study; J. Geophys. Res. 74, 6667 - 5576

Reid, H.F., 1911, The elastic rebound theory of earthquakes Bull Dept. Geology Univ. California G 413

Rummel, F., Frohn, C., 1980, Ultrasonic velocity variations during fracture of rock; Proc. ESC Symp. Phys. of Earthq. Sources and Earthq. Pred. Problems, Polish Acad. Science, in press

Sobolew, G., Spetzler, H., Salov, B., 1978; Prescursors to failure in rocks while undergoing anelastic deformations; J. Geophys. Res. 83, 1775-1784

Soga, N., Mizutani, H., Spetzler, H., Martin III, R.J.1978 The effect of dilatancy on velocity anisotropy in Westerly Granit; J. Geophys. Res. 83, 4451-4458

Vetter, U., Meissner, R., 1977, Creep in geodynamic processes; Tectonophysics 42, 37-54

Wesson, R.L., Burford, R.O., Ellsworth, W.L., 1973; Relationship between seismicity, fault creep and crustal loading along the central San Andreas fault; Proc. Conf. Tectonic Problems San Andreas Fault System, Standford Univ. Publ. Geol. Sci., 13 303 - 321

Wu, K.I., Yue, M, Wu, H., Chao, S., Chen, H., Huang, W., Tien, K., Lu, S., 1978: Certain characteristics of the Haichang earthquake (M = 7.3) sequence, in: Chinese Geophysics, Teng, T.L. Lee, W.H.K. editors, transl. from Acta Geophys. Simica, AGU publ. 289 - 308

Wyss, M., 1980 Recent earthquake prediction research in the United States; Developments in earth and planetary sciences, 2, 81.

Zschau, J. 1979; Earthquake prediction using tiltmeters. Proc. ESA-Conusil of Europe, Sem. Earthq. Pred. Strasbourg 5-7 March 1979, ESA-SP-149, 41-45

Zschau, J., Vetter, U., Meissner, R., 1980 Frequency dependence of Q from seismic, tidal and Chandler Wobble data, in prep. for J. Geophys. Res.

Earthquake Research on the San Andreas Fault Zone, California

Wayne Thatcher

U. S. Geological Survey, Menlo Park, CA 94025, USA

ABSTRACT

The United States Geological Survey (USGS) carries out an extensive program of seismic, geodetic, and geologic studies in California that may be relevant to designing new experiments on the North Anatolian Fault Zone. A telemetered microearthquake network of ∿350 stations monitors seismicity down to a threshold level of M=2 or better, accurately locates these events, and obtains fault-plane solutions for the larger earthquakes. Several types of repeated geodetic measurements determine the spatial and temporal pattern of crustal movements over a ∿100 km wide zone across the major faults. Historic triangulation measurements determine the horizontal deformation in California since ∿1900. Annually- repeated laser ranging (geodolite) meausrements of ∿600 line lengths have precisely monitored the strain field since 1971. Lines vary in length from 1 to 35 km and are measured to a precision of 3 to 8 mm. Repeated alinement array measurements (∿100 m aperture) and continuously recording wire strainmeters determine the slip rates on aseismically slipping (creeping) portions of the San Andreas fault system. Repeated leveling surveys monitor the vertical movement related to the compressional tectonics of the "big bend" region of the San Andreas in southern California. Both repeated and continuously recording magnetometer observations monitor magnetic field variations along major faults. Detailed geologic strip maps at 1:24,000 scale locate the most-recently active traces of major faults and are essential aids in planning geophysical measurement programs.

<u>General References on San Andreas Fault Zone Research</u>

Wesson, R. L. and others, Search for seismic forerunners to earthquakes in central California, <u>Tectonophys</u>, <u>42</u>, 111-126, 1977.

Eaton, J. E. and others, Aftershocks of the 1966 Parkfield-Cholame earthquakes: A detailed study, <u>Bull</u>. <u>Seismol</u>. <u>Soc</u>. <u>Am</u>., <u>60</u>, 1151-1197, 1970.

Savage, J. S. and Prescott, W. H., Precision of geodolite distance measurements for determining fault movements, <u>J</u>. <u>Geophys</u>. <u>Res</u>., 78, 6001-6008, 1973.

Thatcher, W., Strain accumulation on the northern San Andreas fault zone since 1906, <u>J</u>. <u>Geophys</u>. <u>Res</u>., <u>80</u>, 4873-4880, 1975.

Castle R. O. and others, Aseismic uplift in southern California, <u>Science</u>, <u>192</u>, 251-253, 1976.

Brown, R. D., and Wolfe, E. W., Map showing recently active breaks along the San Andreas fault between Point Delgada and Boinas Bay, California, <u>Miscellaneous Geologic Investigations</u> Map I-692, U. S. Geological Survey, Washington, D. C. 20242, 1972.

GPSR Compliance

The European Union's (EU) General Product Safety Regulation (GPSR) is a set of rules that requires consumer products to be safe and our obligations to ensure this.

If you have any concerns about our products, you can contact us on ProductSafety@springernature.com

In case Publisher is established outside the EU, the EU authorized representative is:

Springer Nature Customer Service Center GmbH
Europaplatz 3
69115 Heidelberg, Germany

Batch number: 10815974

Printed by Printforce, the Netherlands